The Handbook of Polyhydroxyalkanoates

The Handbook of Polyhydroxyalkanoates

Postsynthetic Treatment, Processing and Application

Edited by
Martin Koller

CRC Press
Taylor & Francis Group
Boca Raton London New York

CRC Press is an imprint of the
Taylor & Francis Group, an **informa** business

First edition published 2020
by CRC Press
6000 Broken Sound Parkway NW, Suite 300, Boca Raton, FL 33487-2742

and by CRC Press
4 Park Square, Milton Park, Abingdon, Oxon OX14 4RN

First issued in paperback 2023

First edition published by CRC Press 2021

ISBN 13: 978-0-367-54115-6 (pbk)
ISBN 13: 978-0-367-54107-1 (hbk)
ISBN 13: 978-1-00-308766-3 (ebk)

DOI: 10.1201/9781003087663

Publisher's Note
The publisher has gone to great lengths to ensure the quality of this reprint but points out that some imperfections in the original copies may be apparent.

Typeset in Times
by Deanta Global Publishing Services, Chennai, India

Dedicated to the fond memory of our father Josef Koller (1949–2019), who passed away during the creation of this book.

Contents

PART IV Degradation of PHA and Fate of Spent PHA Items

Foreword

Accumulation of plastic waste has caused several threats, including exploitation of fossil fuel resources, impact on global warming (due to incineration of crude oil-derived materials), and the dramatic pollution of aquatic environments. On the other hand, the enormous rise in the use of polymers in areas of great demand, such as packaging, has tremendously increased in the recent decade both in developed and developing countries.

For instance, the production of plastics in Europe reached 57 million tons in 2012, mostly divided between poly(ethylene), poly(propylene), poly(vinyl chloride), poly(styrene), and poly(ethylene terephthalate) production. These fossil-based plastics were consumed and discarded into the environment, generating 10.4 million tons of plastic waste, most of which ended up in landfills.

Issues related to sustainability, petroleum shortage, and fluctuating oil prices have intensified the development of new cost-effective, greener alternatives. So, the development of "environmentally friendly" materials will result in huge benefits to the environment and will also contribute to reduced dependence on fossil fuels. Polymers produced from alternative resources are a crucial issue, especially for short-life range applications, as they can be easily degraded by abiotic media or microorganisms. Nevertheless, the properties of these biomaterials often lag behind those of common thermoplastics, and some improvements are needed in order to make them fully operative for their industrial use.

The most optimistic forecasts have indicated that the potential for substitution of biopolymers with conventional synthetic ones can reach 15.4 million tons in the European Union (EU) (33% of the production of plastic at present). The development of biopolymer matrices and their use in common applications is the subject of increasing interest by numerous research groups, since these materials are considered capable of substituting certain synthetic thermoplastics. The reasons for this increase in the number of studies on these materials reside in the growing concern of society with regards to environmental aspects and sustainability of consumer goods, as well as in strict governmental regulations on the use of non-degradable thermoplastics.

There are many kinds of bio-based and biodegradable polymers, among which one of the most promising are microbial polyhydroxyalkanoates (PHA).

PHA can be synthesized by microorganisms and possess several applications in packaging, even in the food industry. However, their use is currently limited owing to their high production cost. The production cost of PHA depends on several key factors, including initial raw materials and substrates, as well as the selection of appropriate microorganisms, adaptation of biotechnological cultivation strategy, and downstream processing. The utilization of cheap substrates, modeling of the biological system, proper bioreactor and experimental design, as well as development of a new recovery method are pivotal to make PHA production efficient and competitive. It becomes clear that one has to consider the holistic aspects of ethics and sustainability, especially when choosing the appropriate raw material for biopolymer production. The opportunities for commercial competition of PHA in the global market have been addressed by several researchers in this second volume of *The Handbook of Polyhydroxyalkanoates.*

We are very confident that all these issues, covered by researchers with high expertise in PHA science and technology, will raise your interest, and all comments and suggestions for further research work will be highly appreciated.

We should acknowledge and thank all the authors for their hard work resulting in high-quality chapters, as well as the publishers (CRC Press, Taylor & Francis) for their confidence in the preparation of this volume.

Kianoush Khosravi-Darani
Shahid Beheshti University of Medical Sciences
Tehran, Iran

About the Editor

Martin Koller was awarded his Ph.D. degree by Graz University of Technology, Austria, for his thesis on polyhydroxyalkanoate (PHA) production from dairy surplus streams, which was embedded into the EU-FP5 project WHEYPOL (Dairy industry waste as source for sustainable polymeric material production), supervised by Gerhart Braunegg, one of the most eminent PHA pioneers. As senior researcher, he worked on bio-mediated PHA production, encompassing the development of continuous and discontinuous fermentation processes, and novel downstream processing techniques for sustainable PHA recovery. His research focused on cost-efficient PHA production from surplus materials by eubacteria and haloarchaea and, to a minor extent, to the development of PHA for biomedical use.

He currently holds about 80 Web of Science listed articles often in high ranked scientific journals, authored twelve chapters in scientific books, edited three scientific books and five special issues on the PHA topic for diverse scientific journals, gave plenty of invited and plenary lectures at scientific conferences, and supports the editorial teams of several distinguished journals.

Moreover, Martin Koller coordinated the EU-FP7 project ANIMPOL ("Biotechnological conversion of carbon-containing wastes for eco-efficient production of high added value products"), which, in close cooperation between academia and industry, investigated the conversion of the animal processing industry's waste streams toward structurally diversified PHA and follow-up products. In addition to PHA exploration, he was also active in microalgal research and in biotechnological production of various marketable compounds from renewables by yeasts, chlorophyte, bacteria, archaea, fungi, and lactobacilli.

At the moment, Martin Koller is active as a research manager, lecturer, and external supervisor for PHA-related projects.

Contributors

Grazyna Adamus
Centre of Polymer and Carbon Materials
Polish Academy of Sciences
Zabrze, Poland

Dan Åkesson
Swedish Centre for Resource Recovery
University of Borås
Borås, Sweden

Vera A. Alavarez
INTEMA
Composite Materials Group (CoMP)
Engineering Faculty
National University of Mar del Plata
Mar del Plata, Argentina

Emmanuel Asare
Department of Material Science and Engineering
Faculty of Engineering
University of Sheffield
Sheffield, UK

David Barsi
LMPE srl – Parco Scientifico di Capannori
Segromigno in Monte
Capannori
Lucca, Italy

Paweł Chaber
Centre of Polymer and Carbon Materials
Polish Academy of Sciences
Zabrze, Poland

Ilaria Chicca
LMPE srl – Parco Scientifico di Capannori
Segromigno in Monte
Capannori
Lucca, Italy

Emo Chiellini
LMPE srl – Parco Scientifico di Capannori
Segromigno in Monte
Capannori
Lucca, Italy

Federica Chiellini
Department of Chemistry and Industrial Chemistry
University of Pisa
Pisa, Italy

Patrizia Cinelli
Department of Civil and Industrial Engineering
University of Pisa
Pisa, Italy

Arianna Domenichelli
LMPE srl – Parco Scientifico di Capannori
Segromigno in Monte
Capannori
Lucca, Italy

Anabel Itohowo Ekere
Wolverhampton School of Sciences
Faculty of Science and Engineering
University of Wolverhampton
Wolverhampton, UK

Luiziana Ferreira da Silva
Laboratory of Bioproducts
Department of Microbiology
Institute of Biomedical Sciences
University of São Paulo
São Paulo, Brazil

Jorge A. Ferreira
Swedish Centre for Resource Recovery
University of Borås
Borås, Sweden

Filomena Freitas
UCIBIO-REQUIMTE
Chemistry Department
Faculty of Sciences and Technology
Universidade NOVA de Lisboa
Campus da Caparica
Caparica, Portugal

Annabelle Fricker
Department of Material Science and Engineering
Faculty of Engineering
University of Sheffield, UK

Vito Gigante
Department of Civil and Industrial Engineering
University of Pisa
Pisa, Italy

David Alexander Gregory
Department of Material Science and Engineering
Faculty of Engineering
University of Sheffield
Sheffield, UK

Tommaso Guazzini
LMPE srl – Parco Scientifico di Capannori, Segromigno in Monte
Capannori
Lucca, Italy

John W. Haycock
Department of Material Science and Engineering
Faculty of Engineering
University of Sheffield
Sheffield, UK

Baki Hazer
Kapadokya University
Department of Aircraft Airframe Engine Maintenance
Ürgüp, Nevşehir, Turkey

and

Zonguldak Buülent Ecevit University
Department of Chemistry
Zonguldak, Turkey

Vassilka Ivanova Ilieva
LMPE srl – Parco Scientifico di Capannori
Segromigno in Monte
Capannori
Lucca, Italy

Verena Jost
Fraunhofer Institute for Process Engineering and Packaging IVV
Freising, Germany

and

Technical University of Munich
TUM School of Life Sciences Weihenstephan
Munich, Germany

Vasiliki Kachrimanidou
Department of Food Science and Technology
Ionian University
Argostoli
Kefalonia, Greece

Kianoush Khosravi-Darani
Research Department of Food Technology
National Nutrition and Food Technology Research Institute
Faculty of Nutrition Sciences and Food Technology
Shahid Beheshti University of Medical Sciences
Tehran, Iran

Evgeniy G. Kiselev
Siberian Federal University
Krasnoyarsk, Russia

and

Institute of Biophysics SB RAS
Krasnoyarsk, Russia

Martin Koller
University of Graz
NAWI Graz
Office of Research Management and Service
c/o Institute of Chemistry
Graz, Austria

and

ARENA – Association for Resource Efficient and Sustainable Technologies
Graz, Austria

Tomasz Konieczny
University of Silesia
Katowice, Poland

Konstantina Kourmentza
Department of Chemical and Environmental Engineering
Faculty of Engineering
University of Nottingham
Nottingham, UK

and

Green Chemicals Beacon of Excellence
University of Nottingham
Nottingham, UK

Apostolos Koutinas
Department of Food Science and Human Nutrition
Agricultural University of Athens
Athens, Greece

Marek Kowalczuk
Centre of Polymer and Carbon Materials
Polish Academy of Sciences
Zabrze, Poland

Piotr Kurcok
Centre of Polymer and Carbon Materials
Polish Academy of Sciences
Zabrze, Poland

Dimitrios Ladakis
Department of Food Science and Human Nutrition
Agricultural University of Athens
Athens, Greece

Valérie Langlois
L'Université Paris-Est Créteil
CNRS
ICMPE
Thiais, France

Paul A. Lant
School of Chemical Engineering
The University of Queensland
Brisbane, Australia

Bronwyn Laycock
School of Chemical Engineering
The University of Queensland
Brisbane, Australia

Andrea Lazzeri
Department of Civil and Industrial Engineering
University of Pisa
Pisa, Italy

Elena Marcello
School of Life Sciences
College of Liberal Arts and Sciences
University of Westminster
London, UK

Christopher T. Nomura
Chemistry Department
SUNY College of Environmental Science and Forestry
Syracuse, NY, USA

Chrysanthi Pateraki
Department of Food Science and Human Nutrition
Agricultural University of Athens
Athens, Greece

Alexandra Paxinou
School of Life Sciences
College of Liberal Arts and Sciences
University of Westminster
London, UK

João R. Pereira
UCIBIO-REQUIMTE
Chemistry Department
Faculty of Sciences and Technology
Universidade NOVA de Lisboa
Campus da Caparica
Caparica, Portugal

Lucía Pérez Amaro
LMPE srl – Parco Scientifico di Capannori
Segromigno in Monte
Capannori
Lucca, Italy

Atahualpa Pinto
Department of Chemistry
Bard College
Annandale-on-Hudson
New York, NY, USA

Steven Pratt
School of Chemical Engineering
The University of Queensland
Brisbane, Australia

Olga Psaki
Department of Food Science and Human Nutrition
Agricultural University of Athens
Athens, Greece

Dario Puppi
Department of Chemistry and Industrial Chemistry
University of Pisa
Pisa, Italy

Iza Radecka
Wolverhampton School of Sciences
Faculty of Science and Engineering
University of Wolverhampton
Wolverhampton, UK

Ana T. Rebocho
UCIBIO-REQUIMTE
Chemistry Department
Faculty of Sciences and Technology
Universidade NOVA de Lisboa
Campus da Caparica
Caparica, Portugal

Maria A. M. Reis
UCIBIO-REQUIMTE
Chemistry Department
Faculty of Sciences and Technology
Universidade NOVA de Lisboa
Campus da Caparica
Caparica, Portugal

Estelle Renard
L'Université Paris-Est Créteil
CNRS
ICMPE
Thiais, France

Agustin Rios
L'Université Paris-Est Créteil
CNRS
ICMPE
Thiais, France

Ipsita Roy
Department of Material Science and Engineering
Faculty of Engineering
University of Sheffield
Sheffield, UK

and

National Heart and Lung Institute
Faculty of Medicine
Imperial College London
London, UK

Ryan A. Scheel
Chemistry Department
SUNY College of Environmental Science and Forestry
Syracuse, NY, USA

Maurizia Seggiani
Department of Civil and Industrial Engineering
University of Pisa
Pisa, Italy

Ekaterina I. Shishatskaya
Siberian Federal University
Krasnoyarsk, Russia

and

Institute of Biophysics SB RAS
Krasnoyarsk, Russia

Caroline S. Taylor
Department of Material Science and Engineering
Faculty of Engineering
University of Sheffield
Sheffield, UK

Fideline Tchuenbou-Magaia
School of Engineering
Faculty of Science and Engineering
University of Wolverhampton
Wolverhampton, UK

Cristiana A. V. Torres
UCIBIO-REQUIMTE
Chemistry Department
Faculty of Sciences and Technology
Universidade NOVA de Lisboa
Campus da Caparica
Caparica, Portugal

Davy-Louis Versace
Univ. Paris Est Creteil
CNRS
ICMPE
Thiais, France

Tatiana G. Volova
Siberian Federal University
Krasnoyarsk, Russia

and

Institute of Biophysics SB RAS
Krasnoyarsk, Russia

Alan Werker
Promiko AB
Lomma, Sweden

and

School of Chemical Engineering
The University of Queensland
Brisbane, Australia

Natalia O. Zhila
Siberian Federal University
Krasnoyarsk, Russia

and

Institute of Biophysics SB RAS
Krasnoyarsk, Russia

Magdalena Zięba
Centre of Polymer and Carbon Materials
Polish Academy of Sciences
Zabrze, Poland

The Handbook of Polyhydroxyalkanoates, Volume 3: Introduction by the Editor

This third volume of *The Handbook of Polyhydroxyalkanoates* is markedly application-oriented, involving the production of functionalized PHA biopolyesters, the post-synthetic modification of PHA, processing and additive manufacturing of PHA, development and properties of PHA-based (bio)composites and blends, the market potential of PHA and follow-up materials, different bulk and niche applications of PHA, and the fate and use of spent PHA items.

The 14 chapters of the third volume can be grouped into four different sections:

FUNCTIONALIZED PHA AND PHA MODIFICATION

The current state of the art in the development of chemically modifiable PHA, such as multistep modifications of isolated biopolyesters, short syntheses of monomer feedstocks, and metabolic engineering of monomer pathways is described in the first chapter of this section, written by Atahualpa Pinto, Ryan Scheel, and Christopher T. Nomura. Although modern studies have maintained a limited focus on the chemical properties of the feedstocks employed for PHA production, it is clear that in order to diversify their applications and value, concerted efforts must be applied to explore and augment the chemical tractability of these polymers, as shown in this chapter.

The second chapter in this section, provided by Estelle Renard, Agustin Rios, Davy-Louis Versace, and Valérie Langlois, focuses on the design of functionalized PHA-based polymeric materials by chemical modification as another emerging topic in the production of tailor-made biopolymeric materials.

The next contribution by Baki Hazer addresses the fact that PHAs' hydrophobic character is a significant disadvantage for its direct use. Specifically, the key to biocompatibility of biomedical implantable materials is to render their surface in a way that minimizes hydrophobic interaction with the surrounding tissue. As a way out, hydrophilic groups can be introduced into PHA biopolyesters in order to obtain amphiphilic polymers. This chapter is therefore focused on chemically modified PHA with enhanced hydrophilic character in order to use them as biomaterials for subsequent medical applications.

In this context, a highly specialized chapter by Anabel Itohowo Ekere and colleagues from Marek Kowalczuk's team describes the preparation of bioactive oligomers derived from microbial PHA and synthetic analogs of natural PHA oligomers, which are obtained by anionic ring-opening polymerization of β-substituted β-lactones. The promising properties of these bioactive oligomers for novel, high-value applications, especially in medicine, agro-chemistry, and cosmetology, are discussed in this chapter.

PROCESSING OF PHA

Global activities to convert microbial PHA to smart follow-up materials and the fate of these materials are described by different chapters of this volume, each specializing in a particular niche of these activities.

This encompasses a comprehensive overarching chapter by Vito Gigante, Patrizia Cinelli, Maurizia Seggiani, Vera A. Alavarez, and Andrea Lazzeri on processing and thermomechanical properties of PHA, including the preparation of composites by using compatible inexpensive fillers to achieve improved material properties.

Another contribution, provided by Dario Puppi and Federica Chiellini, deals with the additive manufacturing of PHA as one of the most future-fit directions of PHA processing, hence, methods of 3D-printing or electrospinning of PHA and mixtures thereof are comprehensively introduced and compared with established techniques for polymer processing.

Moreover, mechanical and permeation properties of films made of PHA-based composites and blends are described in detail in the next chapter by Verena Jost, who shows that the method of processing has a decisive impact on these properties, which can also be fine-tuned by diverse additives.

APPLICATION

Applications of PHA are not restricted by far to simple or smart packaging materials, e.g., in the food sector, but penetrates to an increasing extent into low volume–high value niche fields, where they outperform petrol-based plastic by far, and do not necessarily need to be as cheap as established plastics do. In the overarching chapter of this section, written by the team of Konstantina Kourmentza, the advantages of PHA against other bio-based and conventional polymers are discussed; moreover, current applications of PHA-based polymers are presented in this contribution, highlighting innovative products already available on the market, and the potential use of PHA monomers in diverse sectors is addressed.

The next chapter, dedicated to applications of PHA, was written by Ana T. Rebocho, João R. Pereira, Cristiana A. V. Torres, Filomena Freitas, and Maria A. M. Reis. It provides an overview of how PHA can be used in dependence on their material properties, and links those PHA properties to current and prospective applications. It is emphasized that public awareness of the environmental impact of plastics is expected to increase and will boost the interest in biodegradable materials like PHA and follow-up products.

The next chapter, provided by Tatiana G. Volova, Ekaterina I. Shishatskaya, Natalia O. Zhila, and Evgeniy G. Kiselev, presents the synthesis of various types of PHA by wild-type hydrogen-oxidizing bacteria. Recently developed technologies for synthesis of PHA with different chemical compositions (PHB homopolymer, co-, ter-, and quaterpolymers containing 3- and 4-hydroxybutyrate, 3-hydroxyvalerate, and 3-hydroxyhexanoate as well as copolymers containing 3-hydroxybutyrate and diethylene glycol or 3-hydroxy-4-methylvalerate) are described. Structure and physicochemical characteristics of these polymers are discussed, and the results for their processing and different target applications are presented. Moreover, technologies of PHA synthesis on various substrates (mixtures of electrolytic hydrogen with CO_2 and O_2, syngas obtained by conversion of natural gas and gasification of brown coal and hydrolyzed lignin, sugars, organic acids, lipids, and glycerol) have been implemented. It is shown how these PHA synthesis processes were already successfully scaled up to pilot production plants.

In the fourth chapter of the application-oriented section, special focus is dedicated by Emmanuel Asare and colleagues from the team of Ipsita Roy to the use of PHA and follow-up products in the biomedical and pharmaceutical field, where such materials are used as implants, surgical devices, wound-healing materials, drug delivery matrices, scaffolds for cell cultures, or, after (bio)chemical conversion, as active pharmaceutical compounds. Prospects of developing PHA-based medical prototypes for the repair and regeneration of damaged peripheral nerves, cardiac, bone, cartilage, skin, and pancreatic tissue, for medical device development such as coronary artery stents and nerve guidance conduits, as well as for drug delivery are comprehensively presented.

Finally, the last chapter in this application section is devoted by Lucía Pérez Amaro and other associated researchers of Emo Chiellini's team to the most obvious use of PHA, namely for production of biodegradable packaging materials; the chapter especially highlights the role of compostable polymeric materials in food packaging applications with a specific focus on items based on PHA as major components, and on biodegradability aspects of these materials.

DEGRADATION OF PHA AND FATE OF SPENT PHA

Apart from PHA biosynthesis, degradation of spent PHA-based products is a major task in the entire PHA field, especially when expecting an increasing amount of spent PHA-based items to be disposed of in the future. Therefore, a specialized chapter by Jorge A. Ferreira and Dan Åkesson deals with degradation of PHA by intracellular pathways inside the microbial cells, with aerobic PHA degradation pathways towards biomass, CO_2, and water, and by the microbiology underlying the anaerobic degradation towards methane and other valued products like biohydrogen or diverse organic acids in biogas plants. This chapter provides a holistic understanding of the fate of these biopolyesters during composting and other disposal strategies, and underlines the need to comprehend the relationship between PHA composition and its biodegradability.

Finally, the investigation of the fate and lifetime estimation of PHA and PHA-based composites under long-term exposure to different natural environmental conditions is described by Bronwyn Laycock, Steven Pratt, Alan Werker, and Paul A. Lant in the last chapter of this volume. This chapter analyses the complex changes associated with biodegradation in field or laboratory studies as guidance for estimating the practical lifetimes of PHA-based products. This is particularly the case for composite materials, where the controlling mechanisms can be very different to those of a homogeneous matrix.

At the end, I deeply hope that this final volume will meet the expectations of the scientific community, offer concrete answers to existing R&D questions, and, above all, will encourage further research activities in the intriguing realm of PHA biopolyesters!

Martin Koller

Part I

Functionalized PHA and PHA Modification

1 Recent Advances in Chemically Modifiable Polyhydroxyalkanoates

Atahualpa Pinto, Ryan A. Scheel, and Christopher T. Nomura

CONTENTS

1.1 INTRODUCTION

Poly-[(*R*)-3-hydroxyalkanoates] (PHA) are comprised of a set of natural polyesters originally found in bacteria. In native PHA-producing microorganisms, these materials are biosynthesized from intracellular fatty acid derivatives or central metabolites and accumulated as a response to increased levels of carbon-rich nutrients while experiencing a shortage of other essential elements in the environment (e.g., N, P, O, etc.) [1]. These biopolymers have been a boon to researchers worldwide as they have influenced and driven studies in a multitude of disciplines, such as microbiology, metabolic engineering, biochemistry, and materials chemistry. Much of the research to date has focused on the organisms, pathways, and enzymes used to produce PHA materials.

The predominant class of PHA materials is characterized by a polyester backbone product of the biosynthetic polycondensation of (*R*)-3-hydroxyacyl-CoA thioesters. These polymers and their bulk physical properties are differentiated and classified by the length of their hydrocarbon sidechains. Whereas PHA with sidechains containing 0–2 carbon atoms are referred to as short-chain-length (*scl*-) PHA, those with sidechains containing 3–13 carbon atoms are commonly identified as medium-chain-length (*mcl*-) PHA. The former display physical properties typically associated with crystalline thermoplastics, while the latter generally possess low crystallinity and elastomeric character [2].

Much effort has centered on expanding the capability of PHA-producing bacteria to incorporate, in a controlled fashion, MCL/SCL and unnatural monomers into copolymers by engineering biosynthetic pathways and expanding the substrate specificities of their enzymatic constituents. Numerous examples have given legitimacy to this approach, such as: chimeric PHA synthases with broad substrate specificity or the point mutations S325T and Q481K that enhance PHA synthase activity [3, 4], recombinant pathways capable of synthesizing unique copolymers [5, 6], and a multitude of PHA commercial enterprises such as Danimer Scientific, Mango Materials, Yield10 Bioscience, and others. These research efforts have yielded slow yet steady gains in the production of uncommon PHA copolymers from a variety of related and unrelated feedstocks.

An underdeveloped area of PHA research, however, has been that of finding solutions to the limited chemical tractability of PHA sidechains. Thus far this challenge has been circumvented by applying multistep synthetic routes post-polymerization to adapt them to specific applications and designs. Besides some excellent examples from a few laboratories, meaningful efforts to augment the chemical tractability of PHA via incorporation of unnatural or uncommon monomers have been sporadic and inadequate.

This chapter is not meant to be a comprehensive look at the chemical modifications of PHA, but will instead emphasize a few novel and unique strategies developed as a means to broaden the chemical properties and applications of PHA with hopes of encouraging the expansion of this budding field of research.

1.2 FLUORINATION OF THE PHA α-CARBON: A BIOSYNTHETIC APPROACH

Fluoropolymers are incredibly useful materials that have transformed the fields of polymer chemistry and materials science, with the classic example being that of poly(tetrafluoroethylene) (PTFE) which has extraordinary physical properties [7]. These properties arise from the C-F bond, which is the strongest single bond to carbon that can be made, and the unique electronic interactions that result from the high electronegativity of fluorine [8]. Organofluorine compounds have also been used extensively in the pharmaceutical industry, where substitutions with fluorine often dramatically alter enzymatic interactions [9, 10]. This utility is enhanced by the dearth of organofluorine compounds found in nature; very few metabolic pathways exist that can manipulate C-F bonds, and there are a limited number of biologically produced organofluorine structures [11–14].

Recently, a biosynthetic approach for producing fluorinated PHA was reported by Thuronyi *et al.* [15]. This novel recombinant pathway has been included here as an excellent example of the chemical modification of existing PHA through non-synthetic means (Figure 1.1), and because there is increasing interest in developing chemistries for modification of the notoriously unreactive C-F bond.

Exogenous fluoromalonate is imported into the cell by one of several potential transporter systems; Thuronyi *et al.* found that the most effective system was the two-component Na^+:malonate symporter MadLM isolated from *Pseudomonas fluorescens* [15, 16]. The intracellular fluoromalonate is then converted to fluoromalonyl coenzyme A (fluoromalonyl-CoA) in a two-step reaction by the malonate-CoA ligase MatB, isolated from *Rhodopseudomonas palustris* [17]. The unique acetoacetyl-CoA synthase NphT7 isolated from *Streptomyces sp.* CL190 catalyzes the

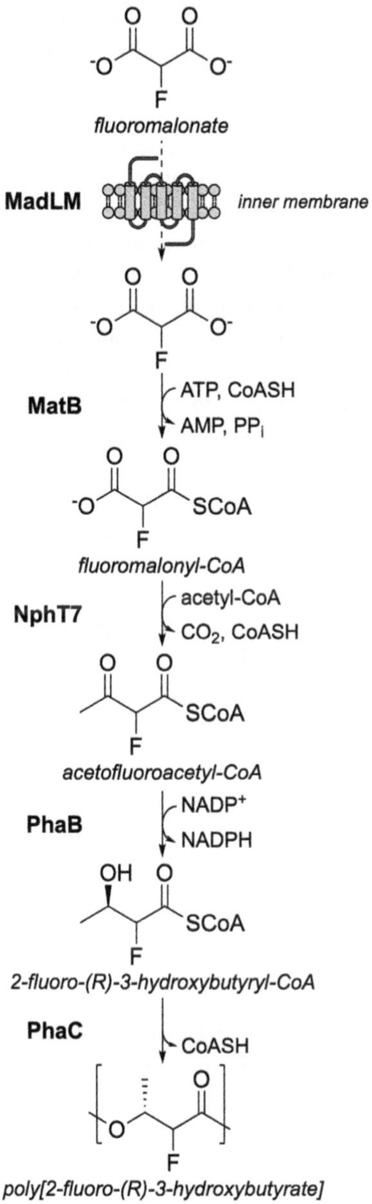

FIGURE 1.1 Recombinant pathway for 2-fluoro-PHA synthesis in *Escherichia coli*. Using biologically available fluoromalonate as a feedstock, poly[2-fluoro-*(R)*-3-hydroxybutyrate] can be synthesized as a copolymer with *(R)*-3-hydroxybutyrate comonomers [15].

condensation of one molecule of acetyl-CoA with one molecule of fluoromalonyl-CoA to yield one molecule of acetofluoroacetyl-CoA concomitant with the release of CO_2 and free coenzyme A [18, 19]. PhaB, an acetoacetyl-CoA reductase isolated from *Ralstonia eutropha* [20], catalyzes the chemoselective reduction of the beta keto group of acetofluoroacetyl-CoA to yield 2-fluoro-(*R*)-3-hydroxybutyryl-CoA.

This substrate, along with the non-fluorinated analog (*R*)-3-hydroxybutyrate which occurs due to the presence of the key fatty acid biosynthesis metabolite malonyl-CoA, is finally polymerized by the PHA synthase PhaC isolated from *Ralstonia eutropha* [21] to generate the copolymer poly[2-fluoro-(*R*)-3-hydroxybutyrate-*co*-(*R*)-3-hydroxybutyrate]. This is the first described biosynthesis of a fluoro-PHA, and represents a novel approach to diversifying the portfolio of chemically modified PHA by manipulation of the α-carbon.

With the increasing number of synthetic organofluorine compounds, driven in large part by the development of pharmaceuticals, there has been increasing interest in the activation and scission of C-F bonds for further modification. This is a challenge however, given the strength of the C-F bond and its unsuitability as a leaving group [8], which necessitates the use of harsh reaction conditions such as strong nucleophiles (e.g., organolithium reagents) and high temperatures [22–24]. More recently, mild reaction conditions have been found that exploit the fluorophilicity of silicon to cleave non-activated aliphatic C-F bonds [25].

Defluorosilylation of alkyl fluorides with relatively mild conditions has been recently achieved independently by several groups [26–28]. Two examples of these defluorosilylation reactions with alkyl fluorides are shown in Figure 1.2. It is possible that these reaction conditions could be applied to poly[2-fluoro-(*R*)-3-hydroxybutyrate] in which the fluorine is in an activated position on the α-carbon, allowing further modification of the organosilicon moiety [29]. However, as the ester linkage in PHA exhibits poor acid/base resistance, more research is needed to determine whether these new methods employed for defluorosilylation are capable of cleaving

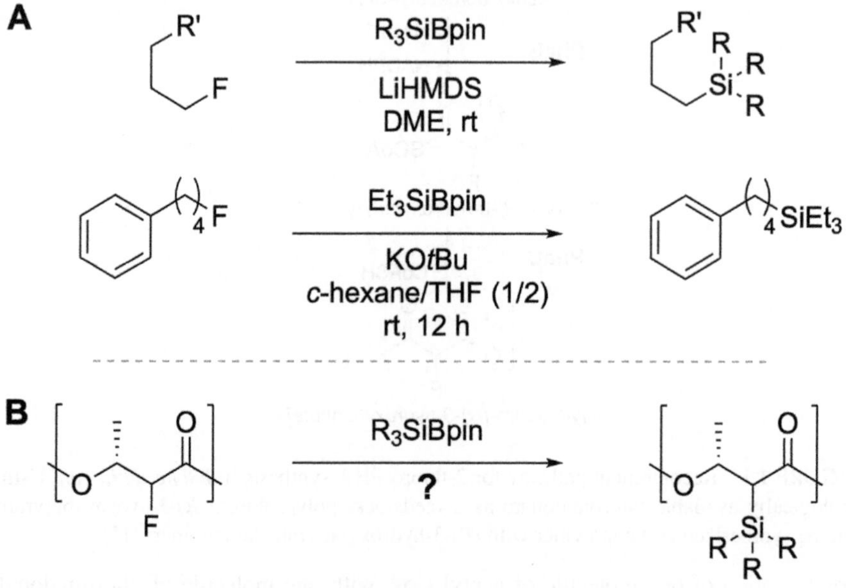

FIGURE 1.2 Recent methods for base-mediated defluorosilylation. A) Two examples of independently devised methods for defluorosilylation of alkyl fluorides [26, 28]. B) It may be possible to adapt these methods for the modification of 2-fluoro-PHA.

the C-F bond without destruction of the polymer. Chemical modification of the PHA α-carbon, if possible, would offer an interesting alternative to the current portfolio of functionalizations, most of which utilize reactive groups on the branch chain.

1.3 DEVELOPMENT OF AZIDO-PHA

Click reactions are comprised of a group of synthetic, chemoselective strategies that have no side products, occur under mild reaction conditions, and proceed rapidly with high yields [30]. Considered the paradigm among click reactions is the so-called copper alkyne-azide cycloaddition reaction (CuAAC), which reliably yields 1,2,3-triazole moieties under mild conditions (see Figure 1.3) [31]. The reaction involves a thermodynamically favored non-concerted cycloaddition featuring Cu(I) as the reactive coordination center by which alkynes and organoazides rearrange to regioselectively form triazole heterocycles. The extraordinary chemoselectivity of the CuAAC and its recognized value as a model reaction in pharmaceutical applications has led

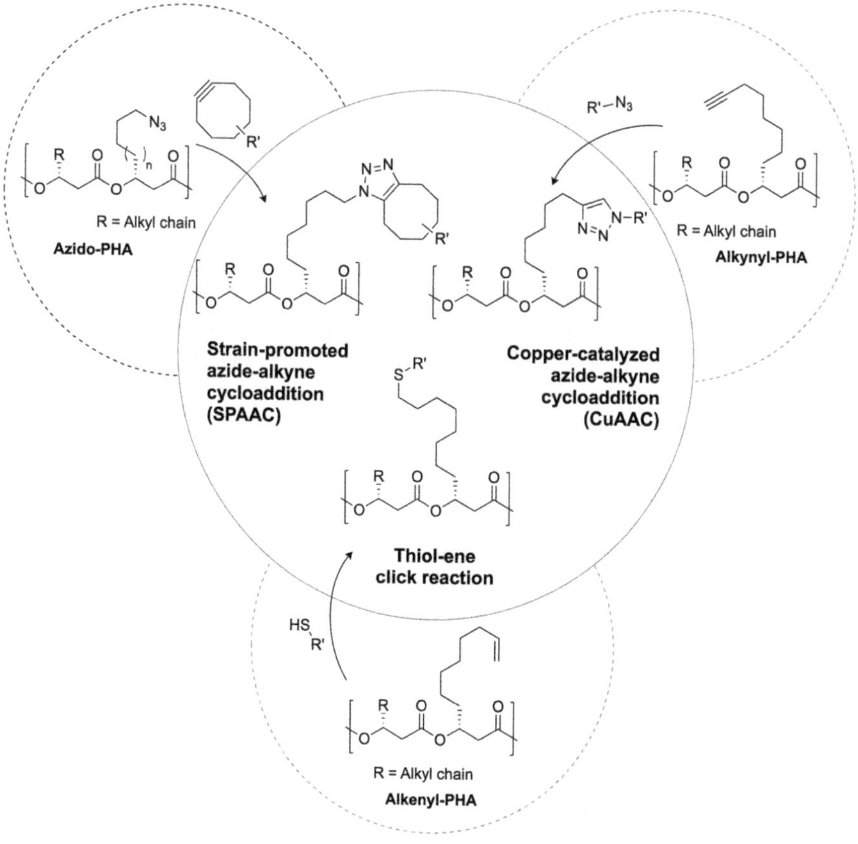

FIGURE 1.3 Click reactions can be used to expand functionality of PHA polymers. In addition to the copper catalyzed azide-alkyne cycloaddition reaction, thiol-ene click reactions, copper-free cycloadditions with strained alkynes are also possible.

to an expanded availability of commercial sources of azido- or alkynyl-conjugated biomolecules. More recently, strain-promoted variants of the alkyne-azide cycloaddition (SPAAC) were developed, with substituted cyclooctynes as their centerpiece, born out of the desire to employ these popular reactions in organisms, avoid the use of deleterious copper, and preserve the reaction's kinetic parameters *in vivo* [32, 33]. Progress in the budding field of orthogonal and bioorthogonal chemical transformations has been rapidly developing and has produced examples which include, but are not limited to: UV-promoted thiol-ene click reactions between thiols and alkenes/alkynes [34, 35], (4+2) cycloadditions between dienes and dienophiles such as normal and reverse-electron-demand Diels–Alder reactions [36, 37], and the Staudinger ligation between organoazides and organophosphorus reagents [38, 39].

In view of the need to develop simple and efficient solutions to the chemical modification of PHA, Pinto *et al.* led an effort to produce *mcl*-PHA polymers containing sidechains with reactive organoazide functional groups [40]. This modification would allow researchers to bridge the latent reactivity of these polymers with a vast library of click-chemistry based reagents, commercial and otherwise, and offers the potential to expand the reach of PHA applications into high-value areas such as biomedicine, advanced materials, and others.

Tappel *et al.* employed a synthetic/biosynthetic approach to the controlled incorporation of medium-chain-length azidofatty acid monomers into PHA [41]. A shortcoming of previous studies that used fatty acids as substrates for PHA production was the random, uncontrolled mixture of repeating units within the resultant PHA polymers. Since control of repeating unit composition is critical to producing polymers with desirable, reproducible properties, the Nomura lab engineered *E. coli* to produce PHA polymers with defined repeating unit compositions from fatty acids [41]. The parental strain used was *E. coli* LS5218 [42], which carries two significant mutations that make it ideal for producing PHA polymers from fatty acids: (1) a *fadR* mutation for enhanced expression of genes encoding enzymes involved in β-oxidation; and (2) an *atoC(Con)* mutation which results in the constitutive expression of the Ato short-chain-fatty acid uptake enzymes in *E. coli*. This strain was further engineered to control the production of enoyl-CoA by deleting the *fadB* [43] and *fadJ* [44] genes (*E. coli* LSBJ), both of which can catalyze the (*S*)-specific hydration of enoyl-CoA, resulting in a roadblock in β-oxidation. This resulted in a strain where any exogenous fatty acid supplied pooled as its enoyl-CoA intermediate with no loss of carbons from the substrate. For example, supplementation with 4-carbon fatty acid (butyrate) results in a butenoyl-CoA intermediate, and supplementation with 8-carbon fatty acid (octanoate) results in pooling of an octenoyl-CoA intermediate. Thus, specific fatty acid feeding regimens, coupled with expression of the (*R*)-specific enoyl-CoA hydratase (PhaJ4) and PhaC1(STQK) engineered synthase, results in production of PHA polymers with defined repeating unit compositions [45, 46].

The implementation of a bacterial strain like *E. coli* LSBJ, able to uptake and redirect fatty acid feedstocks to production of PHA, suggested the need to develop a short synthetic route to access chemically tractable azidofatty acids. For this purpose, and as shown in Figure 1.4A, 8 and 12-carbon α,ω-diols were converted in three steps to the corresponding ω-azidofatty acids with approximately 35% overall yields. In addition to the ω-azidofatty acids, 10-azidodecanoic acid was synthesized from commercially

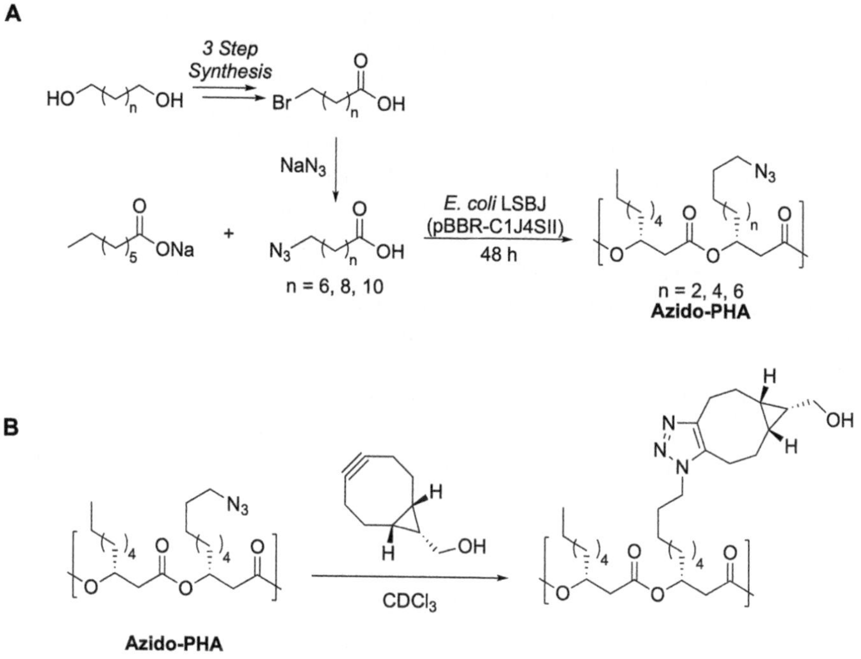

FIGURE 1.4 Azido-PHA and SPAAC reactions. A) A mixed synthetic/biosynthetic route was employed to produce Azido-PHA copolymers. B) Azido-PHA was tested in a SPAAC reaction with a commercially available cyclooctyne.

available 10-bromodecanoic acid in a single step and with 87% yield. The metabolic incorporation of the synthetic azidofatty acids into PHA copolymers was assessed by fermentation studies employing *E. coli* LSBJ and octanoic acid as a primary monomer. The study revealed the bacterial strain to be amenable to the uptake of all three azidofatty acids and leading to the production of copolymers of variable percentage azide compositions. Polymers yields with respect to cell dry mass (CDM) were obtained in the range of 20–30%, their structures characterized by ^{1}H NMR and IR spectroscopy. ^{1}H NMR showed the characteristic PHA signals at 5.18 ppm and 2.50 ppm associated with the β-proton and diastereotopic α-protons respectively. In addition, the presence of a signal 3.27 ppm was also observed that matched the chemical shift of the sidechain methylene (-CH_2-N_3) present in the azidofatty acid feedstocks. As added confirmation of the presence of the organoazide functionality in PHA, IR spectroscopy showed the characteristic azide sharp asymmetric stretch at 2096 cm^{-1}. The thermal properties of the isolated copolymer samples were analyzed by DSC and TGA and, unsurprisingly, the samples showed to be highly amorphous materials as per the absence of T_d and T_c and low T_g values ranging between −60°C and −40°C. Of the materials obtained, poly-[(*R*)-3-hydroxyoctanoate-*co*-10-azido-(*R*)-3-hydroxydecanoate] was employed by the researchers in SPAAC conjugation experiments with the commercially available cyclooctyne (1*R*,8*S*,9*s*)-bicyclo[6.1.0]non-4-yn-9-ylmethanol (Figure 1.4B). The experiment resulted in a clean cycloaddition reaction as confirmed by the downfield shift of the organoazide sidechain methylene from 3.27 ppm to 4.19 ppm in the ^{1}H NMR spectrum.

Pinto *et al.*'s exploratory study provided a clear answer about the feasibility of introducing organoazides into PHA copolymers and using strain-promoted click chemistry conditions to modify them [40]. More recently, Nkrumah-Agyeefi and Scholz expanded on this work by providing examples of how copper catalyzed click chemistry conditions could be applied to the modification of PHA [47]. Their study centered on the evaluation of two distinct yet functionally complementary copolymers with nonanoate as the primary monomer: one displaying an azide, and another an alkyne functional group (Figure 1.5A). Optimized production of the alkyne-containing polyester was performed by fermentation with *P. oleovorans* ATCC 29347 in the presence of a 1:1 molar ratio of sodium nonanoate and 10-undecynoic acid (a 40 mM total concentration), and 10 mM sodium acetate over 20 h. Rather than employing synthetic azidofatty acid feedstocks to fermentatively incorporate into PHA, Nkrumah-Agyeefi and Scholz reported an alternative approach that involved the production of a brominated PHA intermediate to be subsequently converted to Azido-PHA by nucleophilic substitution with NaN_3 [47]. Incorporation of brominated fatty

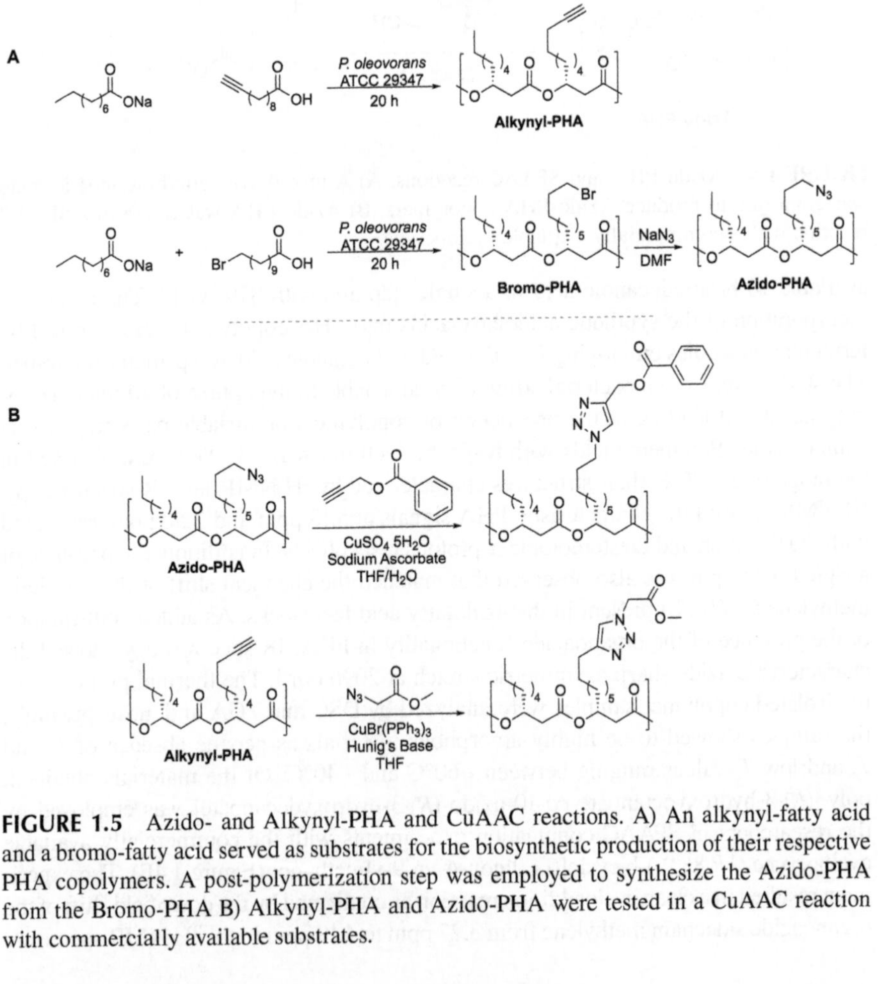

FIGURE 1.5 Azido- and Alkynyl-PHA and CuAAC reactions. A) An alkynyl-fatty acid and a bromo-fatty acid served as substrates for the biosynthetic production of their respective PHA copolymers. A post-polymerization step was employed to synthesize the Azido-PHA from the Bromo-PHA B) Alkynyl-PHA and Azido-PHA were tested in a CuAAC reaction with commercially available substrates.

acids into PHA was confirmed by ^{1}H NMR as represented by the methylene signal at 3.41 ppm (−CH_2Br). The brominated sidechains were converted by substitution to organoazides by reaction with sodium azide in DMF and their presence confirmed by their characteristic FTIR stretch at 2093 cm^{-1}. In addition, a slight upfield shift of the aforementioned methylene was observed to 3.28 ppm.

The azido-PHA copolymers obtained were modified by copper-catalyzed click chemistry (Figure 1.5B). The authors emphasized the need to test methodologies for this step, as they found difficulties when attempting to use a single copper-catalyzed method across polymer systems. The alkynyl-PHA copolymer was conjugated with methyl-2-azidoacetate in the presence of $CuBr(PPh_3)_3$, a copper catalyst soluble in tetrahydrofuran (THF) [48]. The reaction progress was monitored by the disappearance of the stretch at 3292 cm^{-1}, representative of terminal alkynes in FTIR spectroscopy. Formation of the triazole adduct was confirmed by ^{1}H NMR spectroscopy. An analogous method was employed in the click chemistry reaction of azido-PHA copolymer with propargyl benzoate. The key difference was the use of the more common click chemistry conditions of $CuSO_4$ $5H_2O$ as catalyst in an aqueous/THF solvent mixture containing sodium ascorbate. Under similar reaction conditions, the conjugation of azido-PHA with propargyl acetate failed to yield product.

1.4 CHEMICAL MODIFICATIONS OF UNSATURATED PHA

One type of chemically modifiable monomer unit that has been incorporated into PHA are ω-unsaturated fatty-acids of varying chain lengths (e.g., 10-undecenoic acid and 7-octenoic acid). The terminal alkenes featured in these polymers have become an attractive chemical handle for their modification post-polymerization, mainly by oxidative cleavage reactions (Figure 1.6). These strategies have been the subject of much scrutiny and have been featured and well described elsewhere [49].

This notwithstanding, a few salient examples have surfaced in recent years such as that by Sparks and Scholz which showed that PHA containing 10-undecenoate repeating units can be modified to produce ionizable and hydrophilic PHA [50]. In their study, the terminal alkene moieties in poly-[(*R*)-3-hydroxyoctanoate-*co*-(*R*)-3-hydroxyundecenoate] were epoxidized by treatment with 3-chloroperbenzoic acid and ring-opened by nucleophilic reaction with diethanolamine. The resulting ionizable PHA polymer was characterized by 1D and 2D NMR spectroscopy and solubility tests confirmed its miscibility in water (Figure 1.7A). The polycationic character of the modified PHA polymer served as the basis for investigating its value as a plasmid DNA delivery system. The authors demonstrated the polymer's ability to tightly bind and protect plasmid DNA from endonuclease degradation [51].

Other novel approaches for the modification and application of unsaturated PHA polymers have been performed by employing thiol-ene click chemistry. This type of click reaction involves the radical-driven covalent coupling of alkene and thiol functional groups in the presence of ultraviolet light (UV) and in the presence of a photoinitiator [52]. Examples of thiol-ene reactions without photoinitiators have also been studied [53]. Thiol-ene reactions have been demonstrated as robust reactions for organic synthesis, bio-functionalization, polymerization, and polymer modification. In a recent study, a PHA containing 10-undecenoate repeating units (55 mol %)

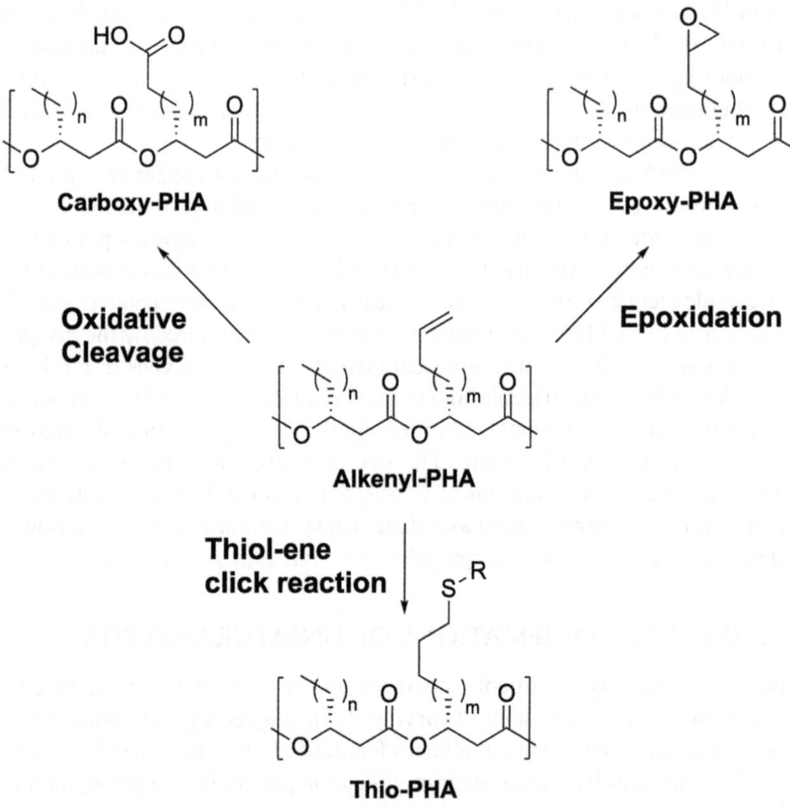

FIGURE 1.6 Common strategies employed in the chemical modification of unsaturated PHA. Three modes of reactivity have been exploited thus far and they include oxidative cleavage, epoxidation, and thiol-ene click reactions.

was both conjugated and cross-linked via thiol-ene click reaction to a mixture of polyhedral oligomeric silsesquioxane (POSS) containing a pendant thiol group, and pentaerythritol tetrakis(3-mercaptopropionate) (Figure 1.7B), resulting in polymers with excellent shape-memory properties [54].

1.5 CONCLUSIONS AND OUTLOOK

Polyhydroxyalkanoates (PHA) are attractive biopolymers owing to their biodegradable and biocompatible profiles and varied structural and physical properties. By simply controlling the material properties, PHA polymers can be used in a wide variety of applications ranging from replacements for non-biodegradable bulk-commodity plastics to biomedical materials. The potential use of PHA in biomedical applications such as drug delivery and tissue engineering has generally been limited by the methodologies available to chemically modify and fine-tune their properties.

Recent advances in the biosynthesis of azide-, fluorine-, and alkyne-containing PHA are an excellent first step towards the production of useful chemically modifiable materials. Creative avenues need to be explored to assay the polymerization of

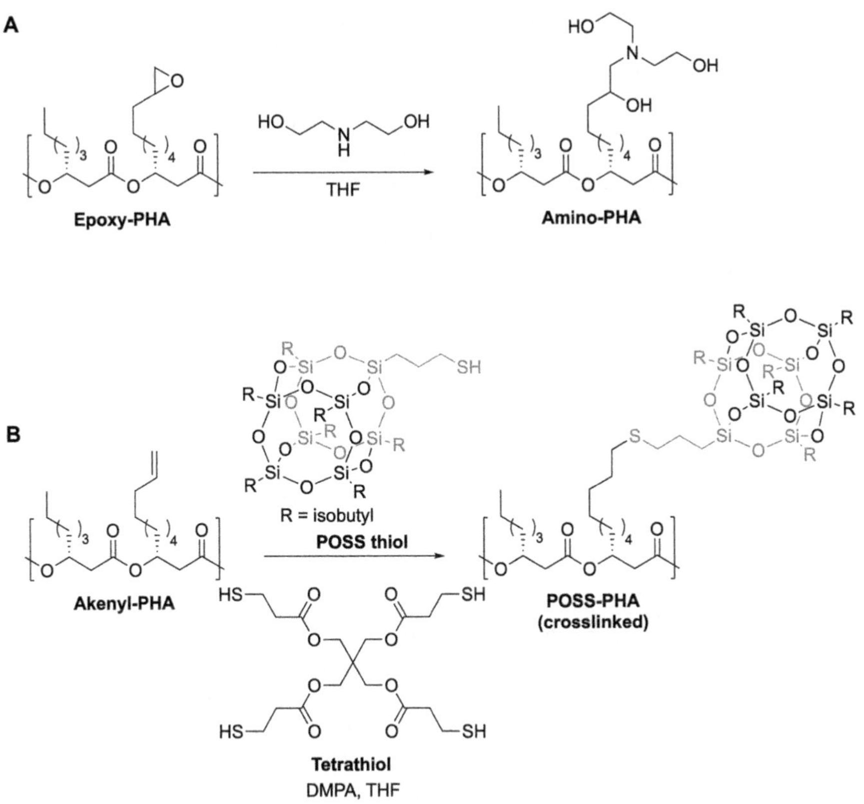

FIGURE 1.7 Salient examples in the modification of unsaturated PHA. A) Ring-opening of epoxide moiety produced an ionizable Amino-PHA with application. B) A thiol-ene click reaction with POSS thiol and tetrathiol led to the production of a shape memory polymer.

unnatural substrates and continue to expand the portfolio of functionalized PHA. Employing the various tools of metabolic engineering, synthetic chemistry, and materials science will provide new strategies for modification, such as the activation and substitution of recalcitrant C-F bonds, the application of more advanced cross-coupling reactions to terminal alkynes, the characterization of new biosynthetic pathways capable of introducing alternative functional groups, and engineering PHA synthases that can accept a broader range of substrates. As highlighted in this chapter, it is evident that no single approach stands out and a combination of multidisciplinary strategies and methodologies will be necessary to advance the current state of the science.

REFERENCES

1. Lu J, Tappel RC, Nomura CT. Mini-review: Biosynthesis of poly(hydroxyalkanoates). *Polym Rev* 2009; 49(3): 226–248.
2. Barham PJ, Barker P, Organ SJ. Physical properties of poly(hydroxybutyrate) and copolymers of hydroxybutyrate and hydroxyvalerate. *FEMS Microbiol Lett* 1992; 103(2): 289–298.

3. Matsumoto K, Takase K, Yamamoto Y, *et al.* Chimeric enzyme composed of polyhydroxyalkanoate (PHA) synthases from *Ralstonia eutropha* and *Aeromonas caviae* enhances production of PHAs in recombinant *Escherichia coli. Biomacromolecules* 2009; 10(4): 682–685.
4. Takase K, Taguchi S, Doi Y. Enhanced synthesis of poly(3-hydroxybutyrate) in recombinant *Escherichia coli* by means of error-prone PCR mutagenesis, saturation mutagenesis, and *in vitro* recombination of the type II polyhydroxyalkanoate synthase gene. *J Biochem* 2003; 133(1): 139–145.
5. Li Z-J, Qiao K, Shi W, *et al.* Biosynthesis of poly(glycolate-co-lactate-co-3-hydroxybu tyrate) from glucose by metabolically engineered *Escherichia coli. Metab Eng* 2016; 35: 1–8.
6. Li Z-J, Qiao K, Che X-M, *et al.* Metabolic engineering of *Escherichia coli* for the synthesis of the quadripolymer poly(glycolate-co-lactate-co-3-hydroxybutyrate-co-4-hyd roxybutyrate) from glucose. *Metab Eng* 2017; 44: 38–44.
7. Plunkett RJ. *Tetrafluoroethylene polymers.* US Patent 2230654A, 1941.
8. O'Hagan D. Understanding organofluorine chemistry. An introduction to the C–F bond. *Chem Soc Rev* 2008; 37(2): 308–319.
9. Purser S, Moore PR, Swallow S, Gouverneur V. Fluorine in medicinal chemistry. *Chem Soc Rev* 2008; 37(2): 320–330.
10. Kirsch, P. Applications of organofluorine compounds. In: Kirsch, P (ed.), *Modern Fluoroorganic Chemistry.* John Wiley & Sons, Ltd: Weinheim. 2005; pp. 203–277.
11. Walker JRL, Lien Bong Chui. Metabolism of fluoroacetate by a soil *Pseudomonas* sp. and *Fusarium solani. Soil Biol Biochem* 1981; 13(3): 231–235.
12. Twigg LE, Socha LV. Defluorination of sodium monofluoroacetate by soil microorganisms from central Australia. *Soil Biol Biochem* 2001; 33(2): 227–234.
13. Dong C, Huang F, Deng H, *et al.* Crystal structure and mechanism of a bacterial fluorinating enzyme. *Nature* 2004; 427(6974): 561–565.
14. Leong LEX, Khan S, Davis CK, *et al.* Fluoroacetate in plants - A review of its distribution, toxicity to livestock and microbial detoxification. *J Anim Sci Biotechnol* 2017; 8(55): 1–11.
15. Thuronyi BW, Privalsky TM, Chang MCY. Engineered fluorine metabolism and fluoropolymer production in living cells. *Angew Chem Int Ed* 2017; 56(44): 13637–13640.
16. Schaffitzel C, Berg M, Dimroth P, *et al.* Identification of an Na^+-dependent malonate transporter of *Malonomonas rubra* and its dependence on two separate genes. *J Bacteriol* 1998; 180(10): 2689–2693.
17. Crosby HA, Rank KC, Rayment I, *et al.* Structure-guided expansion of the substrate range of methylmalonyl coenzyme A synthetase (MatB) of *Rhodopseudomonas palustris. Appl Environ Microbiol* 2012; 78(18): 6619–6629.
18. Okamura E, Tomita T, Sawa R, *et al.* Unprecedented acetoacetyl-coenzyme A synthesizing enzyme of the thiolase superfamily involved in the mevalonate pathway. *Proc Natl Acad Sci USA* 2010; 107(25): 11265–11270.
19. Walker MC, Thuronyi BW, Charkoudian LK, *et al.* Expanding the fluorine chemistry of living systems using engineered polyketide synthase pathways. *Science* 2013; 341(6150): 1089–1094.
20. Peoples OP, Sinskey AJ. Poly-beta-hydroxybutyrate biosynthesis in *Alcaligenes eutrophus* H16. Characterization of the genes encoding beta-ketothiolase and acetoacetyl-CoA reductase. *J Biol Chem* 1989; 264(26): 15293–15297.
21. Peoples OP, Sinskey AJ. Poly-beta-hydroxybutyrate (PHB) biosynthesis in *Alcaligenes eutrophus* H16. Identification and characterization of the PHB polymerase gene (*phbC*). *J Biol Chem* 1989; 264(26): 15298–15303.

22. Perutz RN, Braun T. 1.26 - Transition metal-mediated C–F bond activation. In: Mingos DMP, Crabtree RH, Eds. *Comprehensive Organometallic Chemistry III.* Oxford: Elsevier 2007; pp. 725–758.
23. Fuchibe K, Kaneko T, Mori K, *et al.* Expedient synthesis of N-fused indoles: A C-F activation and C-H insertion approach. *Angew Chem Int Ed* 2009; 48(43): 8070–8073.
24. Pigeon X, Bergeron M, Barabé F, *et al.* Activation of allylic C-F bonds: Palladium-catalyzed allylic amination of 3,3-difluoropropenes. *Angew Chem Int Ed* 2010; 49(6): 1123–1127.
25. Tanaka J, Suzuki S, Tokunaga E, *et al.* Asymmetric desymmetrization via metal-free C-F bond activation: Synthesis of 3,5-diaryl-5-fluoromethyloxazolidin-2-ones with quaternary carbon centers. *Angew Chem Int Ed* 2016; 55(32): 9432–9436.
26. Cui B, Jia S, Tokunaga E, *et al.* Defluorosilylation of fluoroarenes and fluoroalkanes. *Nat Commun* 2018; 9(1): 1–8.
27. Mallick S, Xu P, Würthwein E-U, *et al.* Silyldefluorination of fluoroarenes by concerted nucleophilic aromatic substitution. *Angew Chem Int Ed* 2019; 58(1): 283–7.
28. Liu X-W, Zarate C, Martin R. Base-mediated defluorosilylation of C(sp^2)–F and C(sp^3)–F bonds. *Angew Chem Int Ed* 2019; 58(7): 2064–2068.
29. Komiyama T, Minami Y, Hiyama T. Recent advances in transition-metal-catalyzed synthetic transformations of organosilicon reagents. *ACS Catal* 2017; 7(1): 631–651.
30. Kolb HC, Sharpless KB. The growing impact of click chemistry on drug discovery. *Drug Discov Today* 2003; 8(24): 1128–1137.
31. Rostovtsev VV, Green LG, Fokin VV, *et al.* A stepwise Huisgen cycloaddition process: Copper(I)-catalyzed regioselective "ligation" of azides and terminal alkynes. *Angew Chem Int Ed* 2002; 41(14): 2596–2599.
32. Agard NJ, Prescher JA, Bertozzi CR. A strain-promoted [3 + 2] azide–alkyne cycloaddition for covalent modification of biomolecules in living systems. *J Am Chem Soc* 2004; 126(46): 15046–15047.
33. Baskin JM, Prescher JA, Laughlin ST, *et al.* Copper-free click chemistry for dynamic *in vivo* imaging. *Proc Natl Acad Sci USA* 2007; 104(43): 16793–16797.
34. Dondoni A. The emergence of thiol-ene coupling as a click process for materials and bioorganic chemistry. *Angew Chem Int Ed* 2008; 47(47): 8995–8997.
35. Hoyle CE, Bowman CN. Thiol-ene click chemistry. *Angew Chem Int Ed* 2010; 49(9): 1540–1573.
36. Shi M, Shoichet MS. Furan-functionalized co-polymers for targeted drug delivery: Characterization, self-assembly and drug encapsulation. *J Biomater Sci Polym Ed* 2008; 19(9): 1143–1157.
37. Pipkorn R, Waldeck W, Didinger B, *et al.* Inverse-electron-demand Diels-Alder reaction as a highly efficient chemoselective ligation procedure: Synthesis and function of a bioShuttle for temozolomide transport into prostate cancer cells. *J Pept Sci* 2009; 15(3): 235–241.
38. Saxon E, Bertozzi CR. Cell surface engineering by a modified staudinger reaction. *Science* 2000; 287(5460): 2007–2010.
39. Schilling CI, Jung N, Biskup M, *et al.* Bioconjugation via azide–Staudinger ligation: An overview. *Chem Soc Rev* 2011; 40(9): 4840–4871.
40. Pinto A, Ciesla JH, Palucci A, *et al.* Chemically intractable no more: *In vivo* incorporation of "click"-ready fatty acids into poly-[(*R*)-3-hydroxyalkanoates] in *Escherichia coli. ACS Macro Lett* 2016; 5(2): 215–219.
41. Tappel RC, Kucharski JM, Mastroianni JM, *et al.* Biosynthesis of poly[(*R*)-3-hydroxyalkanoate] copolymers with controlled repeating unit compositions and physical properties. *Biomacromolecules* 2012; 13(9): 2964–2972.

42. Spratt SK, Ginsburgh CL, Nunn WD. Isolation and genetic characterization of *Escherichia coli* mutants defective in propionate metabolism. *J Bacteriol* 1981; 146(3): 1166–1169.
43. Yang SY, He XY, Schulz H. Glutamate 139 of the large alpha-subunit is the catalytic base in the dehydration of both D- and L-3-hydroxyacyl-coenzyme A but not in the isomerization of delta 3, delta 2-enoyl-coenzyme A catalyzed by the multienzyme complex of fatty acid oxidation from *Escherichia coli*. *Biochemistry* 1995; 34(19): 6441–6447.
44. Campbell JW, Morgan-Kiss RM, Cronan JE. A new *Escherichia coli* metabolic competency: Growth on fatty acids by a novel anaerobic beta-oxidation pathway. *Mol Microbiol* 2003; 47(3): 793–805.
45. Tappel RC, Wang Q, Nomura CT. Precise control of repeating unit composition in biodegradable poly(3-hydroxyalkanoate) polymers synthesized by *Escherichia coli*. *J Biosci Bioeng* 2012; 113(4): 480–486.
46. Takase K, Taguchi S, Doi Y. Enhanced synthesis of poly(3-hydroxybutyrate) in recombinant *Escherichia coli* by means of error-prone PCR mutagenesis, saturation mutagenesis, and *in vitro* recombination of the type II polyhydroxyalkanoate synthase gene. *J Biochem* 2003; 133(1): 139–145.
47. Nkrumah-Agyeefi S, Scholz C. Chemical modification of functionalized polyhydroxyalkanoates via "click" chemistry: A proof of concept. *Int J Biol Macromol* 2017; 95, 796–808.
48. Malkoch M, Schleicher K, Drockenmuller E, *et al.* Structurally diverse dendritic libraries: A highly efficient functionalization approach using click chemistry. *Macromolecules* 2005; 38(9): 3663–3678.
49. Kai D, Loh XJ. Polyhydroxyalkanoates: Chemical modifications toward biomedical applications. *ACS Sustain Chem Eng* 2014; 2(2): 106–119.
50. Sparks J, Scholz C. Synthesis and characterization of a cationic poly(β-hydroxyalkanoate). *Biomacromolecules* 2008; 9(8): 2091–2096.
51. Sparks J, Scholz C. Evaluation of a cationic poly(β-hydroxyalkanoate) as a plasmid DNA delivery system. *Biomacromolecules* 2009; 10(7): 1715–1719.
52. Kade MJ, Burke DJ, Hawker CJ. The power of thiol-ene chemistry. *J Polym Sci Part A Polym Chem* 2010; 48(4): 743–750.
53. Uygun M, Tasdelen MA, Yagci Y. Influence of type of initiation on thiol-ene "click" chemistry. *Macromol Chem Phys* 2010; 211(1): 103–110.
54. Ishida K, Hortensius R, Luo X, *et al.* Soft bacterial polyester-based shape memory nanocomposites featuring reconfigurable nanostructure. *J Polym Sci Part B Polym Phys* 2012; 50(6): 387–393.

2 The Design of Functionalized PHA-Based Polymeric Materials by Chemical Modifications

Estelle Renard, Agustin Rios, Davy-Louis Versace, and Valérie Langlois

CONTENTS

2.1 INTRODUCTION

Novel functionalized PHA can be designed using two complementary approaches, direct biosynthesis of functionalized PHA and chemical modifications of PHA after biosynthesis. Although the first approach is very attractive, it remains limited because of the toxicity, the cost of certain substrates and the low polymer yields generally observed. In this chapter, we focused on the versatile chemical modifications after PHA biosynthesis. Chemical modifications of the skeleton or the side chains of PHA, restricted to homogeneous conditions, i.e., in solution, are presented first. The chemical modifications of the PHA side chains are mainly realized on unsaturated PHA. The most common unsaturated *mcl*-PHA directly obtained by fermentation is the poly(3-hydroxyoctanoate-*co*-hydroxyundecenoate) (PHOU), which contains terminal unsaturations on the side chains. The production of PHOU containing between 0 and 100% of unsaturated repetition, is now relatively well controlled [1, 2]. Crosslinking of PHA to enhance physical and mechanical properties and applicability of PHA are then reported. The last part will be devoted to the modification of the PHA surface, with a particular focus on photochemical processes.

2.2 CHEMICAL MODIFICATIONS IN HOMOGENEOUS CONDITIONS

2.2.1 Introduction of Reactive Groups

By their nature, natural *scl*-PHA do not carry chemical functionalities and consequently modification reactions are therefore limited. One method to functionalize *scl*-PHA consists of a substitution reaction through chlorine gas bubbling in a PHA solution [3, 4]. Depending on the amount of chlorine introduced, different structures were obtained with a variable chlorination rate (between 5.4 and 23.8% by mass) (Figure 2.1). In addition, postmodifications are possible with the partial conversion of the chlorine functions into quaternary ammonium, sulfate salt or phenyl derivative. The chlorination [3–5] or bromination [6] reactions of PHOU lead to a better-defined structure since the chlorine atoms were added to the double bonds.

Thanks to the presence of double bonds in the side chains, PHOU is more easily subject to chemical modifications, which have been conducted and have resulted in new materials (Figure 2.2). The conversion of alkene groups to hydroxyl groups has been reported by several authors. Hydroxylation was performed with 25% borane unsaturation followed by the addition of $NaOH/H_2O_2$ [7] with a slight decrease in molar mass. Eroglu *et al.* [8] reported a similar procedure using 9-bicyclononane (9-BBN) on poly(3-hydroxyundecenoate) (PHU). A conversion close to 100% was observed, with a significant decrease of the molar mass from 32 000 to 8 100 $g.mol^{-1}$. The hydrophilicity increases and the polymer becomes soluble in polar solvents such as methanol or ethanol. The introduction of vicinal diols on PHOU containing 45 to 93% of unsaturated bonds has been demonstrated with the use of $KMnO_4/Na_2CO_3$ at 20°C [9]. No severe decrease in molar mass is observed, and conversion to diol does not exceed 60%, regardless of the PHOU or experimental conditions used. However, the hydrophilicity of the polymer is greatly increased, making it soluble in polar

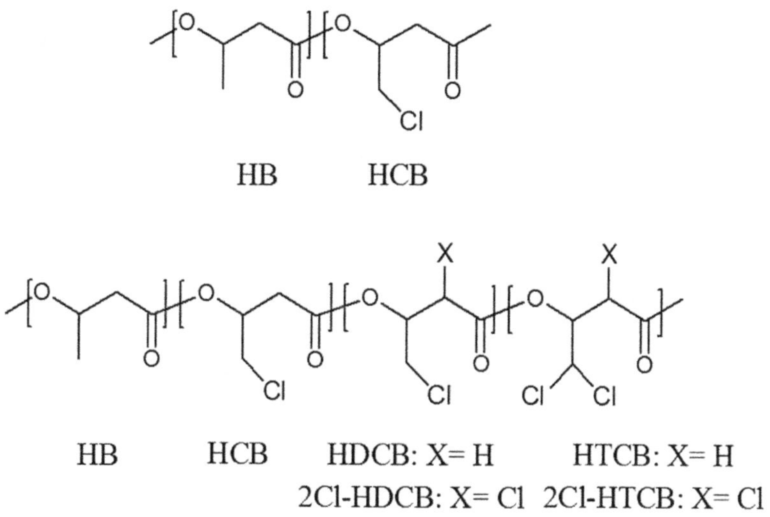

FIGURE 2.1 Chemical structure obtained by chlorination of PHB.

solvents such as methanol, DMSO or an 80/20 (v/v) acetone/water mixture. More recently, hydroxylated PHA has been obtained through thiol-ene photo-click reactions [10] without chain scission. The production of epoxy functionalized PHA by bioconversion is possible in the presence of metachloroperbenzoic acid (mCPBA). This method allows the quantitative epoxidation of PHOU without secondary reaction or decrease in molar mass [11, 12]. Oxidation of PHOU to carboxylic acids was first reported by Lee *et al.* [12] using a method derived from the formation of diols. Indeed, the use of $KMnO_4/KHCO_3$ at 55°C allows the formation of carboxylic acid function terminal. However, this method has some limitations, such as a maximum conversion of 50% and a significant decrease in molar mass from 137 000 to 19 000 $g.mol^{-1}$.

The quantitative conversion of a PHOU including 25% of unsaturations has been demonstrated by Kurth *et al.* [14] using $KMnO_4$ in the presence of 18-crown-6 crown ether in an acetic acid/dichloromethane mixture. The conversion of the alkene groups is enhanced without significant decrease of the molar mass. In addition, the introduction of carboxylic acid functions allows for a significant increase in the hydrolysis rate of the polymer; it is fully hydrolyzed at pH = 11 in 24 hours. Osmium tetroxide (OsO_4) also allows quantitative oxidation without decreasing the molar mass of the polymer [15]. The introduction of a carboxylic group can be carried out by a thiol-ene addition, which has been extensively exploited for the chemical modification of PHOU. The radical addition of 11-mercaptoundecanoic acid, or mercaptopropionic acid induced terminal carboxylic groups inside the chain [10, 16]. Knowing that the hydrophobic character of PHA limits their use for biomedical applications, many efforts have been focused on enhancing the hydrophilic character of PHA. For example, the epoxide ring of poly(3-hydroxyoctanoate-*co*-3-hydroxy-10-epoxyundecenoate) (PHOE) is opened with diethanolamine to form poly(3-hydroxyoctanoate-*co*-3-hydroxy-11-(bis(2-hydroxyethyl)amino)-10-hydroxyundecanoate) (PHON).

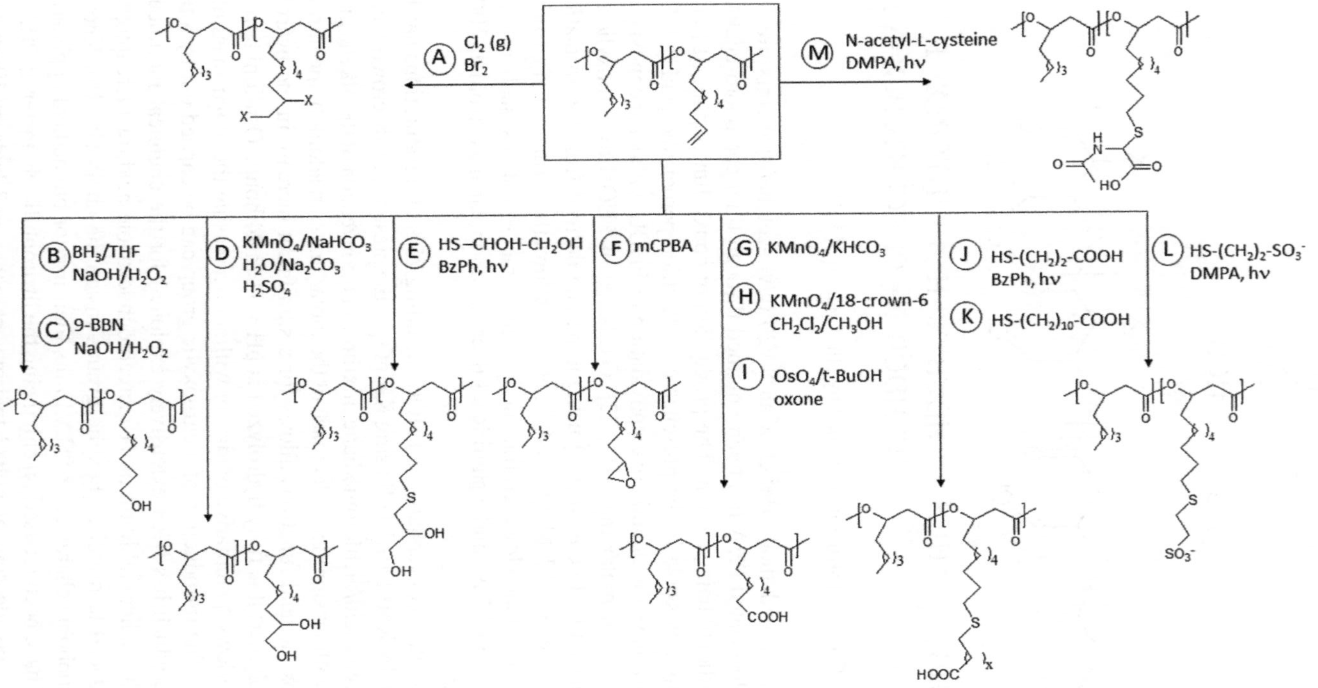

FIGURE 2.2 Chemical modifications of PHOU by halogenation (A) [3–6]; hydroxylation (B) [7], (C) [8]; diol formation (D) [9], (E) [10]; epoxidation (F) [11, 12]; oxidation of alkene to carboxyl group (G, E, I) [13–15], thiol-ene reaction (J, K, L, M) [10, 16–18].

This process causes a decrease in the molar mass of the polymer due to aminolysis of the skeleton but above all allows the synthesis of a water-soluble PHA; indeed, the amine, whose pKa is estimated at between 8 and 8.5, transforms this polymer into a water-soluble polycation with neutral pH (see Figure 2.3) [17].

Radical mediated thiol-ene reaction was investigated to functionalize unsaturated PHA using 3-mercapto-1-ethanesulfonate under photochemical initiation. When the molar ratio of sulfonate groups is about 27%, the ionic polyester is totally soluble in water for molar masses ranging from 5 000 to 223 000 g mol^{-1}. Under 27% of sulfonate groups content, particles were obtained with different diameters from 118 to 1746 nm, depending on the nature of the PHA [18]. The thiol-ene chemistry was also recently used to produced highly efficient rare-earth-modified fluorescent material. N-acetyl-L-cysteine was grafted by a UV thiol-ene process and the rare earth metal ions (Eu^{3+} and Tb^{3+}) were chelated through coordination reactions [19].

Progress in bioconversion has made it possible to obtain new functional PHA with reactive groups in side chains other than double bonds that open novel access to post-chemical modification. Brominated PHA was converted into an azido-terminated precursor for copper-catalyzed azide-alkyne cycloaddition (CuAAC) to graft propargyl benzoate or propargyl acetate (Figure 2.4). This highly efficient chemistry was applied to alkynyl PHA to link methyl-2-azidoacetate to the terminal side groups [20]. Except for previous anionic PHA, the introduction of reactive functions on the skeleton of PHA is not enough to dramatically modify the properties of PHA. Therefore, these reactive groups were then advantageously used to synthesize graft or block copolymers.

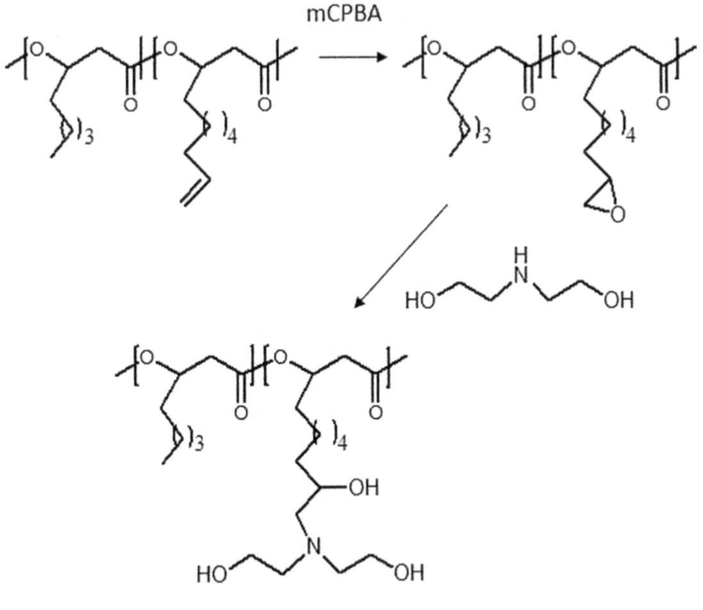

FIGURE 2.3 Synthesis of cationic PHA under physiological pH [17].

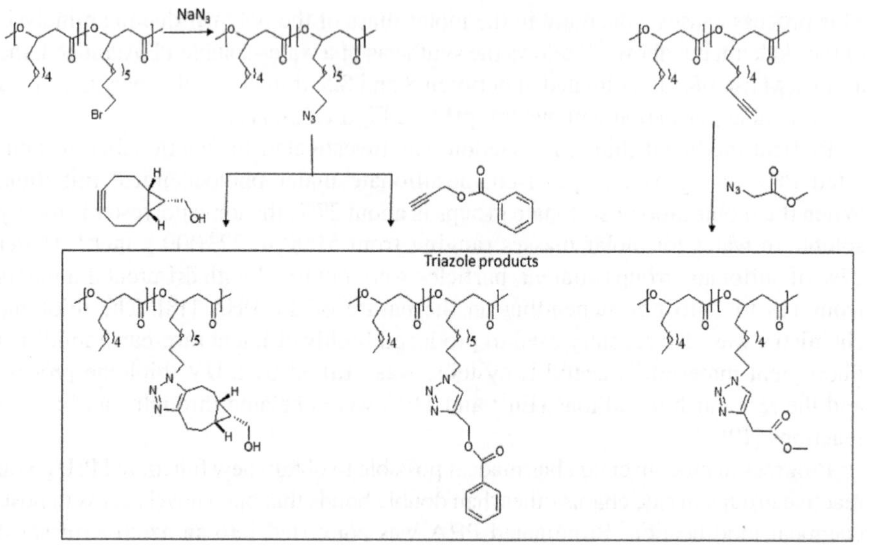

FIGURE 2.4 Chemical modification of bromo and alkyne functionalized PHA via copper-catalyzed azide-alkyne cycloaddition (CuAAC).

2.2.2 Preparation of Well-Defined Oligomers

The copolymer synthesis required the preparation of oligomers from native PHA. Depending on the method used, different types of oligomers are obtained. These methods include thermal treatment, [21–23] transesterification reactions, [24–26] hydrolysis [27] and methanolysis [28, 29]. Figure 2.5 provides an overview of the structures obtained by these different methods.

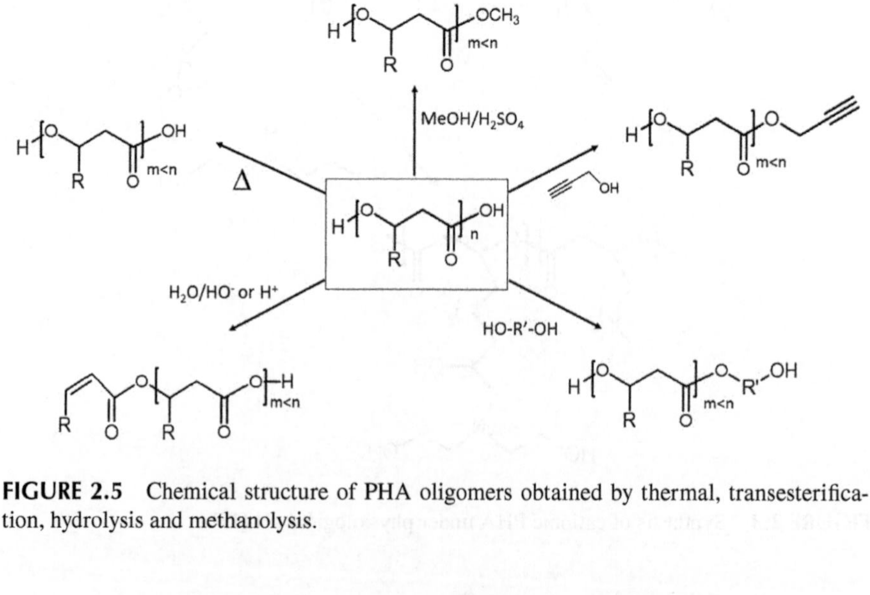

FIGURE 2.5 Chemical structure of PHA oligomers obtained by thermal, transesterification, hydrolysis and methanolysis.

Oligomers of PHB, PHV and PHBV have been synthesized by heat treatment, generating molar masses from 1 200 to 38 300 g.mol^{-1} via a β-elimination mechanism (McLafferty) [30]. The oligomers formed have well-defined terminal groups. A crotonate function of configuration *E* at one terminal end and carboxylic acid function at the other terminal end. Recently, Chan Sin *et al.* observed that the thermal degradation mechanism of the *mcl*-PHA was different from that of the *scl*-PHA [31]. Thermal degradation processes and decomposition mechanisms of PHBV were recently investigated by using thermal gravity analysis (TGA) [32]. The preparation of oligomers by microwave heating has been studied [23] to produce PHA oligomers of low molar masses (310<Mn<1 350 g.mol^{-1}) in very short reaction times. However, it should be noted that the polydispersities obtained are more important than with conventional heating. This is attributed to the heterogeneity of microwave heating, generating a gradient of temperature and therefore different degradation rates depending on the areas concerned. Hydroxy-telechelic oligomers of PHB, PHBV and PHO have been prepared by transesterification of native polymers with ethylene glycol [25, 33] or other diols [28, 34, 35]. Reactions with ethylene glycol have been the most studied and three types of extremities have been identified depending on the catalyst, solvent and temperature used [24]. The optimized reaction mainly produces the hydroxy-telechelic derivative in diglyme at 140°C, by using dibutyltin dilaurate as a catalyst. Other conditions tested lead to a mixture of oligomers with unsaturated end groups (due to the secondary alcohol dehydration), carboxylic acid and diol. Recently, Lemechko *et al.* developed a method for the transesterification of PHA in presence of propargyl alcohol in order to produce PHA oligomers that can be further modified by click chemistry [26]. Under hydrolysis conditions, the products obtained consist of one end hydroxyl and a carboxylic acid end [27]. For the PHO, several conditions were studied by Timbart *et al.* [29] In the case of basic hydrolysis, the mass distribution is bimodal, whereas in the case of catalysis (APTS), the formation of macrocycles has been highlighted by MALDI-TOF. Finally, methanolysis carried out in the presence of methanol and sulfuric acid was used for the synthesis of PHB oligomers [28, 36]. This method makes it possible to obtain well-defined oligomers with one hydroxyl end and one methyl ester. It is also a method of choice for the synthesis of oligomers of well-defined *mcl*-PHA, without formation of by-products [29].

2.3 SYNTHESIS OF COPOLYMERS

2.3.1 Copolymers Based on *scl*-PHA

2.3.1.1 Block Copolymers

The synthesis of block copolymers can be achieved by direct condensation between oligomers or by polymerization from a macroinitiator [37, 38] (see Figure 2.6). Different diblock copolymers PHB-*b*-PCL [39] or triblock copolymers PCL-*b*-PHBV-*b*-PCL [40] have been synthesized by the ring opening polymerization of ε-caprolactone (ROP) from PHA macromonomer. In the same way, diblock copolymers PHB-*b*-PLA, PHBV-*b*-PLA, PHBHHx-*b*-PLA [39, 41, 42] or triblock PHB-*b*-PLA-*b*-PCL [43] have been synthesized by ROP of lactide. This method allows the production of well-defined copolymers with a unimodal distribution. The molar

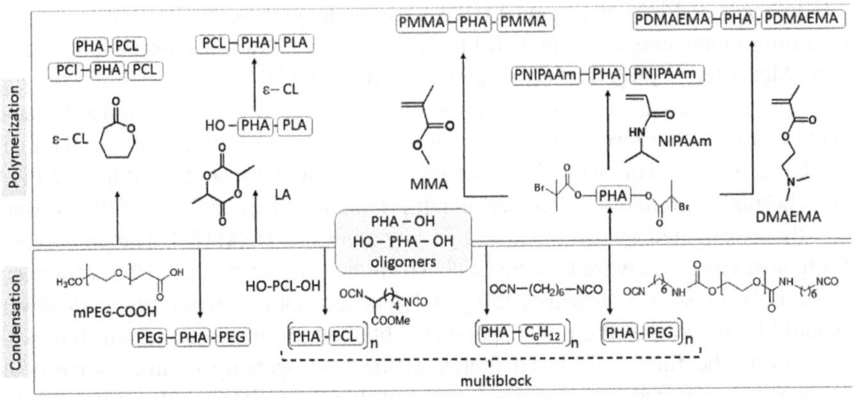

FIGURE 2.6 Synthesis of PHA *scl*-block copolymers by several methods including polymerization and condensation reactions.

masses of PLA or PCL blocks modulate the mechanical properties and/or degradation properties of the copolymers. Other authors have reported the use of ATRP for the synthesis of triblock copolymers based on PHB. [44, 45] After derivatization of PHB oligomer in di-brominated macroinitiator, Loh *et al.* reported the synthesis of a poly(N-isopropylacrylamide)-*b*-PHB-*b*-poly(N-isopropylacrylamide) triblock copolymer (PNIPAAm-*b*-PHB-*b*-PNIPAAm) by ATRP [46]. Through PNIPAAm blocks, this water-soluble copolymer is self-associated in micelles from 139 to 550 nm above its lower critical solubility temperature of 28–29°C (Temperature, LCST). Amphiphilic triblock copolymers were obtained by copolymerization of 2-(dimethylamino)ethylmethacrylate (DMAEMA) [47, 48]. A similar approach was used to synthesize PMMA-*b*-PHBV block copolymers with versatile mechanical properties depending on the amorphous component [37].

In addition, diblock [36] and triblock [49] copolymers have been prepared by condensation of PHB oligomer with methoxy-PEG-monocarboxylic acid, in the presence of N-N'-dicyclohexylcarbodiimide (DCC) and 4-(dimethylamino)pyridine (DMAP). However, this method offers low yields and requires steps of purification [50]. Block copolymers PHBV-*b*-PEG and P(3HB)4HB-*b*-PEG have been synthesized by a transesterification method developed by Ravenelle *et al.* [50–53]. However, this method involves two reactions: thermal degradation of the PHA by β-elimination and transesterification of the PHA by PEG. As a result, the final length of the PHA block is not very controllable and copolymers are poorly defined. The preparation of multi-block copolymers by forming urethane bonds has largely been studied. Hirt *et al.* reported the synthesis of multi-block copolymers including "rigid" blocks of PHB and "soft" blocks of poly(ε-caprolactone) (PCL) [33]. Macromonomers of PHB-diol and PCL-diol are coupled using diisocyanate without any catalyst. These materials present interesting mechanical properties, with an increase in the tensile strength and a decrease in elongation at break as a function of the proportion of PHB. Multi-block copolymers based on *scl*-PHA and PCL [33, 54], PLA [55, 56], poly(butylene glycol adipate) [35] or PEG [57–61] have been synthesized. In the latter case, the hydrophilicity and degradation properties of polymer are also modified in addition to the mechanical properties. Their hemocompatibility and their biodegradability have been demonstrated, making these new polymers promising candidates as novel implantable biomaterials [62, 63]. Finally, alternating multi-block copolymers have been synthesized by Pan *et al.* [64]. 1,6-hexamethylene diisocyanate is first coupled with poly(ethylene glycol), then the generated PEG-diisocyanate is reacted with PHB-diol; this strategy therefore ensures the synthesis of perfectly alternating block copolymers. The control of the material architecture is therefore more precise, allowing for a better understanding of structure–property relationships. More recently, the strain forward of click coupling was used to prepare a well-defined PHA-PEG di-block with alkyne-terminated PHA and azide terminated PEG. Alkyne functionalized PHA were prepared through condensation with propargyl amine. Ligation with azide terminated PEG was accomplished using the copper (I) catalyzed azide alkyne cycloaddition (CuAAC) [65].

2.3.1.2 Graft Copolymers

A first approach to prepare graft copolymers from *scl*-PHA is to remove the tertiary proton by radical reaction. PHB graft copolymers and PEG were thus synthesized in solution by reaction between PHB and PEG diacrylate, generating a more hydrophilic crosslinked polymer [66, 67]. Finally, other types of copolymers involve the use of PHA as grafts on polysaccharide skeletons such as cellulose, chitosan or their derivatives. Arslan *et al.* grafted PHBV oligomers onto chitosan by condensation between the carboxylic acid end of the PHA and the amine functions of chitosan [68]. This grafting modifies the solubility properties compared to chitosan, the copolymer becoming insoluble in 2% acetic acid. Methacrylic PHB has been synthesized from PHB oligomers with terminal carboxylic group obtained through thermal degradation. The esterification with HEMA leads to methacrylic macromonomer. Then, controlled radical copolymerization with other methacrylate monomers as MMA, ethylene glycol methyl ether methacrylate and PEG methyl ether methacrylate allows the synthesis of graft copolymers [69–71]. N-vinylpyrrolidone was grafted on PHBV with AIBN as initiator. The hydrophilic PVP grafting enhanced the swelling properties and active antibacterial effect of the PHBV-g-PVP copolymers [72]. P(3HB)4HB oligomers terminated with hydroxyl groups were functionalized with acrylate function. These acrylated monomethoxy PHA oligomers were coupling by the Michael addition with branched poly(ethylene imine). These grafted-branched copolymers showed high transfection efficiency [73].

2.3.2 Copolymers Based on *mcl*-PHA

Since *mcl*-PHA are not commercial, the available literature on copolymers based on *mcl*-PHAl is less plentifully available than *scl*-PHA. Nevertheless, several strategies for synthesizing block or graft copolymers with *mcl*-PHA have been reported.

2.3.2.1 Block Copolymers

Andrade *et al.* reported the synthesis of multi-block copolymers PHB-PHOHHx [74]. Macromonomers of PHB-diol (“rigid” segment) and PHOHHx-diol (“soft” segment) have been coupled using a diisocyanate in the presence of dilaurate of dibutyl tin. The resulting material consists of approximately 2 PHOHHHx-diol units, 2 PHB-diol units and 5 units L-lysine methyl ester diisocyanate (LDI); it thus has mechanical properties and thermal properties intermediate between PHB and pure PHOHHx. The same copolymer has also been prepared by coupling PHA diols with an acid dichloride (PHA chloride) [75]. Finally, Dai *et al.* reported a "green" synthesis of copolymers PHB-PHO by condensing PHO-diol and PHB-diol [76] with Novozym® 435 (*Candida antarctica* immobilized lipase B). As it was described above for *scl*-PHA, Timbart *et al.* reported the synthesis of a PHO-*b*-PCL block copolymer by ring opening of ε-caprolactone using macromonomers based on functionalized *mcl*-PHA oligomers [77]. Similar copolymers have been synthesized in the same way from PHOU macromonomers, bearing alkene groups into side chains, allowing their oxidation into acid functions and the formation of copolymers P((HO-COOH)-*b*-CL) [29] It has been shown that P((HO-COOH)-*b*-CL) films hydrolyzed more quickly than P(HO-*b*-CL) at pH 7.

2.3.2.2 Graft Copolymers

The synthesis of graft copolymers can be achieved either by chain grafting performed on the main chain ("grafting onto" method) or by polymerization of monomers from the main chain ("grafting from" method). However, "grafting onto" does not allow the coupling of large polymer chains. Arslan *et al.* reported the synthesis of a PHO modified by ATRP [78] with chlorinated PHO on the main chain. The chlorine atoms were then used to polymerize methyl methacrylate (MMA), styrene (S) or n-butyl methacrylate (BMA). The double bonds carried by *mcl*-PHA have been used to perform reactions radical grafting by thermal initiation or gamma or UV irradiation [79, 80]. These reactions are relatively easy to implement, but the products obtained are poorly defined because reactions often lead to products being insoluble due to crosslinking reactions. PMMA has been polymerized by grafting from the unsaturation of a *mcl*-PHA [81]. Moreover, functionalized groups such as cinnamic acid or sulfate groups, alkyl chains and a cyclic polysilsesquioxane (POSS) have been grafted by a thiol-ene addition to the alkene groups of PHA [82–84]. Cakmakli *et al.* have grafted PS or PMMA oligomers with a terminal peroxide function [85]. The PS, PMMA and alkyl chains are used to increase the mechanical strength of the PHA while the POSS grafting brings a shape memory effect [86]. Finally, graft copolymers have also been obtained by esterification reactions between the reactive COOH functions first introduced on the PHOU side chains and PLA and PEG oligomers [87–89]. These couplings are performed using dicyclohexylcarbodiimide (DCC) and dimethylaminopyridine (DMAP). The introduction of 2000 $g.mol^{-1}$ PEG improves the solubility of the copolymer in water. These copolymers allow the production of stable nanoparticles without the addition of surfactants. In order to achieve higher conversion rates, Babinot *et al.* used Huisgen's cycloaddition between the lateral alkyne functions of modified *mcl*-PHA and a PEG having a terminal function [90]. PHOU-g-PEGs with 550 and 5000 $g.mol^{-1}$ PEG grafts are obtained with conversions of 94% and 84% respectively. Thermoresponsive graft copolymers PHA were synthesized either via a thiol-ene reaction with thiol-terminated Jeffamine [91], or thiol-terminated PNIPAM [92]. These amphiphilic thermoresponsive copolymers could be promising materials for biomedical applications.

2.4 NETWORKS BASED ON PHA

The formation of a crosslinked structure is of much interest since it offers the possibility to increase both the mechanical and thermal properties. Networks can be directly obtained networks in the presence of a radical initiator, a multifunctional co-agent as thiol or the formation of semi-interpenetrating network in which the polyester is embedded in a tridimensional network or by sol-gel condensation (see Figure 2.7).

2.4.1 Networks Obtained by Using Radical Initiator

Crosslinking reactions of polyesters in the presence of radical initiators gave branched or crosslinked materials with better physical and mechanical characteristics [93, 94, 95]. In these conditions, the chain scissions are considered a drawback, because they are responsible for the loss of mechanical properties. As a consequence, the reactive extrusion process of polyesters using organic peroxides in the melt state is known to cause severe chain scissions [96, 97]. To counteract this negative effect, crosslinking was conducted in the presence of a multifunctional co-agent [98]. Peroxide treatment was also successfully employed to enhance the compatibilization between polymers by obtaining grafted copolymers and crosslinked networks. Reactive blending of PHO/PLA has been performed to produce *in situ* compatibilized blends with improved properties [98]. The modified PHO/PLA

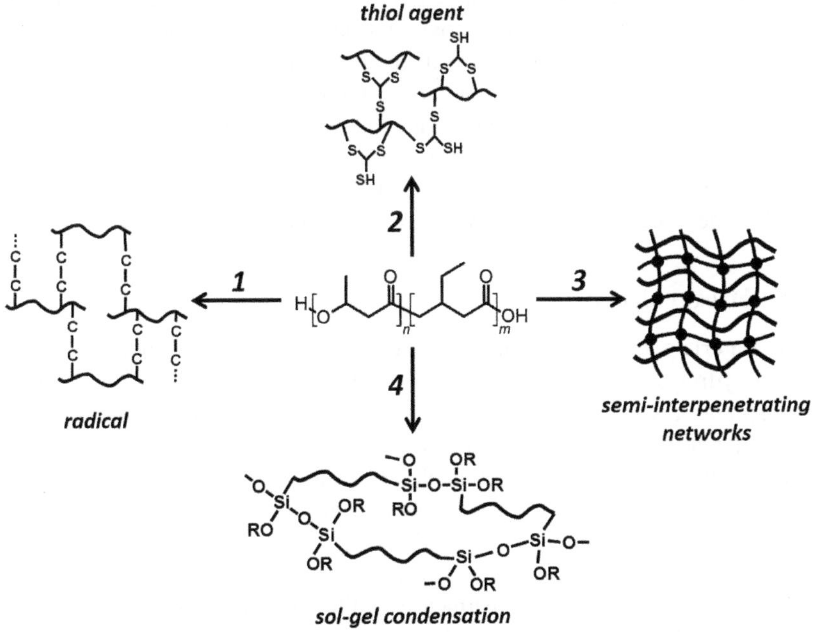

FIGURE 2.7 Chemical methods used for the crosslinking of polyesters (1) in the presence of peroxide initiator, (2) in the presence of multifunctional thiol, (3) by forming semi-interpenetrating network, (4) by sol-gel condensation.

blends had a finer morphology suggesting a compatibilization effect possibly arising from copolymer formation at the interface. A compatibilization between PHBV, poly(butylene adipate-*co*-terephthalate) (PBAT) and ENR was also achieved during melt processing [99]. Using this process, green composites based on these polymers and Miscanthus fibers were prepared by reactive extrusion to ensure good interfacial adhesion in the biocomposites. Moreover, thiol-ene chemistry [100] has been employed in various applications. Ishida *et al.* prepared a novel material by combining the PHOU, a thiol functionalized polyhedral oligomeric silsesquioxane (POSS) and a tetrathiol crosslinker [101] to obtain an elastomeric material with excellent shape fixing and recovery. The crosslinking of poly(3-hydroxybutyrate-*co*-3-hydroxyundecenoate) (PHBU) was also accomplished by using a pentaerythritol tetrakis(3-mercaptopropionate) to obtain a unique combination of improved strength and flexibility of the polyester [102]. According to a similar process, organo/hydrogel were synthesized using unsaturated PHA and poly(ethylene glycol dithiol) as crosslinker [103]. The swelling ratio, pore size and mechanical properties of the gels depend on the molar ratio of PHA to dithiol. These new biomaterials present promising biocompatibility properties.

2.4.2 Networks Based on Telechelic Oligoesters

The preparation of these networks required two steps: first the synthesis of the well-defined oligoesters and then the radical polymerization of these oligoesters (terminated with meth(acrylated) or alkene end groups) or their condensation (oligomers terminated with acid, isocyanate, epoxy or trialkyl siloxane end groups). Various methods have been developed for preparing, in one step, telechelic PHA having low molar mass [22–24, 29, 103–106]. Whatever the type of reaction used to carry out the crosslinking, the common point for these networks lies in the fact that the length of the crosslinking is directly linked to the molar masses of the oligoesters used. Although these processes are longer than the radical reactions in the presence of peroxides, the structures of the network are better defined.

2.4.3 Hybrid Networks

Hybrid materials were successfully produced by sol-gel process that consists of a two-step hydrolysis-condensation reaction of metal alkoxides. The interesting part of those materials is the possibility of tailoring the solid-state properties in relation to the nature and content of the constitutive components. Inorganic–organic hybrid networks based on biodegradable polyesters such as PHA have been prepared [107]. First, PHA was end-capped with triethoxysilane TEOS and then it was used to design a hybrid network using the sol-gel process. A biodegradable PHA-silica hybrid network was synthesized by sol-gel technique under acidic, basic conditions or UV-curing. The UV treatment was the more efficient and rapid route to prepare the PHA-silica network hybrid and promote a higher degree of condensation compared to the conventional sol-gel processes using acidic or basic catalysis. This material may be envisioned as a novel biodegradable material for tissue engineering applications.

2.4.4 (Semi-)Interpenetrating Polymer Networks (Semi-IPNs)

IPNs are defined as the combination of two independently crosslinked polymers. Such structures have been widely studied as they may develop microphase separated co-continuous morphologies due to their interlocking framework. IPNs provide an alternative response for the compatibilization of incompatible polymers and to improve the thermomechanical properties of the materials.

2.4.4.1 Semi-IPNs via Thiol-Ene Reactions

One of the main drawbacks of PHA when compared to another to other functional polymers is their mechanical brittleness. To overcome this issue, PHA can be readily plasticized with bio-sourced molecules [108–112]. However, as these molecules are only physically mixed with the polymer, they tend to exudate from it in time. To curtail this phenomenon, semi-interpenetrated networks (IPNs) are considered a suitable solution. Indeed, by inducing the formation of chemical links between functional additives mixed in a polymer matrix, two situations are solved: the matrix-additive compatibilization and the additive exudation from the matrix. This is achieved by the development within the material of co-continuous microphases interlocking the additive and the matrix. In this regard, semi-IPNs have been substantially and extensively studied in the case of PCL and PLA-based materials [113]. However, very limited studies on this field have been conducted on PHA, specifically the obtaining of semi-IPNs via thiol-ene reactions. Semi-IPNs are obtained by standard radical polymerization of a double bond in the presence of a radical initiator or by gamma radiation [114]. With the objective of increasing the mechanical properties of PHU as well as ameliorating its hydrolysis behavior, PHU-PLGA networks were prepared by mixing both polymers with benzophenone in chloroform and under UV irradiation. Rhee *et al.* observed that PLGA was successfully interpenetrated in PHU without any phase separation. Moreover, tensile tests showed that the Young modulus of PHU in the semi-IPNs increased threefold from 462 to 1 130 kPa when compared to that of the pure PHU polymer. Finally, the hydrolytic degradation of these PHU-PLGA semi-IPNs was increased by a factor of 7 for these materials when compared to the neat PHU polymer. Hao and Deng [115] studied PHB-PEG semi-IPNs. Acrylate functions were grafted at the end of PEG chains and these modified PEGs were mixed with PHB in dichloroethane and irradiated under UV light. The mechanical modulus of PHB increased tenfold for the PHB/PEG 75/25%wt. semi-IPN when compared to the neat PHB. Gamma radiation was also used to produce semi-IPNs [116, 117]. These investigations showed a good compatibilization between PHA and the co-reactants as well as an improvement of the functional properties of these polymers. A later comprehensive study was carried out by Mangeon *et al.* on PHBV using a thiol-ene reaction to produce semi-IPNs [118]. In this case, the way to get around inconveniences such as petro-sourced co-materials, sensitivity to water and chemical environment, while keeping all of the advantages offered by semi-IPN structures, is to use bio-sourced molecules that have unsaturated bonds in which reticulation reactions can be done. In this regard, PHBV semi-IPNs with sunflower oil (SO) were studied. These semi-IPNs were obtained by dissolving SO, PHBV, trimethylolpropane tris (3-mercaptopropionate) (tri-thiol) and dimethoxy-2-phenylacetophenone (DMPA) in dichloromethane. Raman analyses concluded that all unsaturated functions in SO reacted for all the prepared semi-IPNs, prompting the conclusion that the proposed thiol-ene reaction is effective in yielding such materials. DSC and DMA experiments showed that the PHBV glass transition temperature T_g diminished by a maximum of 35°C. These results are an indication of the plasticization of the matrix, which is sought to increase the ductility of a polymer. Moreover, the drop in T_g is thus an indication of a good compatibility between PHBV and semi-IPN SO. Finally, DMA and tensile mechanical tests showed that the elastic modulus *E'* diminished but most importantly the strain ratio at break λ_R (i.e., the ratio between the initial and the final length of the sample) increased from 1.1 to 2.7 in the presence of 40%wt. semi-IPN SO. This would mean that the presence of SO increases the mechanical toughness of PHBV (see Figure 2.8). Hence, it can be concluded that sunflower oil-based semi-IPNs embedded in PHBV are a promising solution to increase the toughening properties of PHA.

2.5 MODIFICATION OF PHA SURFACE

PHA can be considered to be promising biopolymers and a variety of biological applications have been proposed. However, the inertness of their surface limits their applications. Many physical or chemical modifications [119–130] have therefore been used to tune PHA surface properties, but mild grafting conditions are mandatory to preserve the integrity of the PHA. Among all the proposed techniques, photografting polymerization appears as a useful "green technique" due to its significant advantages [131], such as the low cost of operations or mild experimental conditions without the use of any hazardous solvent. The photopolymerization process is a substrate-independent technique which is well-adapted for both the covalent deposition of a broad range of polymers and the tuning of the surface properties of polymer matrices. These technical aspects make photopolymerization a useful method for the

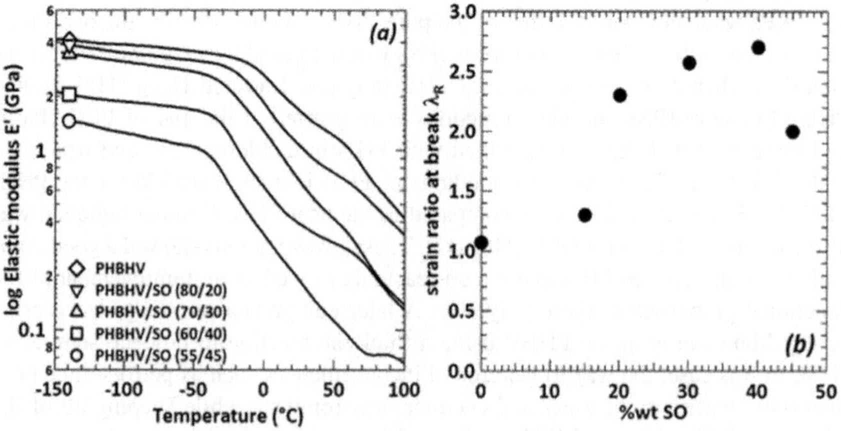

FIGURE 2.8 (a) Elastic modulus *E'* as a function of temperature obtained by DMA and (b) strain ratio at break λ_R obtained by tensile mechanical testing for various PHBHV/SO semi-interpenetrated networks.

surface modification of native PHA which do not exhibit any recognized biological properties. Few research groups have investigated so far the potentialities offered by the photoinduced grafted polymerization method for the surface modification of PHA films under light irradiation for medical applications. Its feasibility was demonstrated according to a "grafting-from" polymerization with the use of benzophenone [132–134], hydrogen peroxide [135], photosensitive system based on aryl azides [136] or water soluble photoinitiators [137, 138].

2.5.1 Photochemical Modification of PHA Surface for Cell Adhesion/Proliferation

For instance, Renard *et al.* [135] use benzophenone (BP) as a photosensitizer for the covalent grafting of poly(hydroxymethacrylate) (PHEMA) onto PHBV film surface via UV treatment. Because of the high crystallinity of PHBV films [139], a pretreatment of the surface with BP was realized to enhance the tertiary hydrogen absorption from the PHBV film under UV light exposure. Results demonstrated that the PHBV-derived films containing PHEMA exhibited a slower degradability rate than the native PHBV [135]. Another interesting feature of this new type of photo-grafting process relies on the 2D spatial control of the polymerization; indeed, the covalent grafting of PHEMA only occurs at the surface of the PHBV surface films with BP and H_2O_2 under UV light exposure, whereas the bulk material is partially modified when benzoyl peroxide is used as thermal initiator for HEMA grafting. Versace *et al.* have developed a novel photo-induced strategy to covalently graft polymer onto PHBV film surface in a one-step process [137]. A cationic photoinitiator, i.e., triarylsulfonium hexafluoroantimonate salt, was used as a phenyl radical generator to abstract protons from the PHBV surface, thus forming radical species able to initiate the free-radical polymerization of 2-hydroxyethyl methacrylate (HEMA) and methacrylic acid (MAA) from the surface. Interestingly, this method was used to barcode PHEMA microstructures onto PHBV surfaces; the great challenge was to develop micro-devices for promoting cell growth at specific locations. Another interesting way of functionalizing PHA was reported by Renard *et al.* [139] for promoting cell compatibility in mild experimental conditions. The modifications of nanofibrous PHA was implemented through two different ways for introducing epoxy groups: 1) preliminary chemical conversion of the double bonds of unsaturated PHA into epoxy groups following by electrospinning of epoxy-functionalized PHA blended with non-functionalized PHA, and 2) electrospinning of non-functionalized PHA, followed by the photo-grafting of glycidyl methacrylate under UV irradiation. Interestingly, biomimetic scaffolds were obtained via the covalent attachment of a peptide sequence (RGD) within the epoxy groups, thus favoring better adhesion of the human mesenchymal stromal cells (hMSCs) in comparison with native PHA. More recently, Versace *et al.* have described a new method for the chemical immobilization of polysaccharides on the surface of a PHBV electrospun fiber surface [140]. The objective was to design a new functional bio-based material with significant cell adhesion/proliferation properties. For this purpose, a new photoactivable dextran with anthraquinone moieties was synthesized and grafted under UV light irradiation to the PHBV fiber surface. Surprisingly, the proliferation rate of hMSCs increased with a higher extra-cellular matrix production than that observed with native fibers after only five days of culture.

2.5.2 Photochemical Modification of PHA Surface: An Efficient Way to Promote Anti-Adhesion or Bacterial Death

Infections by virulent micro-organisms constitute a major public health concern which is responsible for thousands of deaths worldwide every year as well as an incredible increase of healthcare costs. Nosocomial infections have been therefore classified by the World Health Organization as one of the main public health issues. Moreover, the abusive use of antibiotics has also accelerated the emergence of multi-drug-resistant bacteria strains. An alarming research study has proposed that the projected mortality rate in 2050 could reach 10 million annually due to nosocomial infections, which is much higher than the mortality rate for cancer [141]. To solve the problem of the resistance of bacteria strains toward antibiotics, researchers have focused their attention on developing new antimicrobial systems which can confer either biocide or antifouling properties [142]. However, few studies concern the applications of PHA as antibacterial materials using photochemistry technology. A straightforward process for immobilizing polymers and silver nanoparticles (NPs) onto the PHBV fiber surface has been recently established to develop highly efficient biocide materials [143]. The method is based on a two-step procedure using butan-2-one as a photoinitiator: first, the photolysis of butan-2-one has led to the formation of both alkyl and alkoxy radicals. The latter can abstract H from the PHBV surface, thus generating radicals on the surface of the PHBV fiber scaffolds and initiating the polymerization of methacrylic acid (MAA); in the second step, silver NPs were *in situ* synthesized under UV-light irradiation according to the reduction of a silver salt precursor by anthraquinone. Once the silver NPs were synthesized, they immediately complexed to the carboxylate groups of PMAA. SEM and TEM investigations have confirmed the well-dispersion of Ag-NPs at the surface of the modified PHBV materials. The antibacterial tests have demonstrated the synthesis of a very high-performing antibacterial material against *Escherichia coli* and *Staphylococcus aureus.* In line with the previous results, PHBV-derived polymer films with highly antifouling properties have been engineered under UV light activation in aqueous medium. A photoinduced free-radical technique employing butan-2-one has been successfully used to anchor fluorine and PEG moieties onto PHBV surfaces [143]. PEGylated surfaces have showed higher anti-adhesion properties than the fluorinated ones with tremendous inhibition of the *Escherichia coli* and *Staphylococcus aureus* adhesion (>98%). Interestingly, these surfaces have not exhibited any toxicity against human dermal fibroblasts. Finally, a green photoinduced method for the modification of PHBV [144] has been successfully carried out by grafting two potential antibacterial acrylate monomers, i.e., ampicillin-derived monomer and 2-[(methacryloyoxy)-ethyl]trimethylammonium chloride (META). The photografting was conducted in an aqueous medium employing a thiocarbamate-based photoinitiator previously grafted to the PHBV surface. Under appropriate experimental conditions, dithiocarbamyl and carbon-centered radicals are generated, thus initiating the polymerization of both monomers from the modified

PHBV surface. Such PHBV-derived films have led to a tremendous reduction by more than 90% of the *E. coli* adherence, but weaker anti-adherence properties against *S. aureus* have been observed, reflecting the higher resistance of *S. aureus* over *E. coli* to antibacterial treatment. Interestingly, no toxicity of these modified films against NIH-3T3 fibroblastic cells was observed.

2.6 CONCLUSIONS AND OUTLOOK

The chemical modification reactions carried out on PHA allow access to new chemical structures that were not possible by direct fermentation and thus broaden the range of these biocompatible and biodegradable polyesters. Different chemical modification pathways have been developed to increase their properties, in particular, biological, thermal and mechanical properties. Chemistry can propose solutions by using chemical reactions with high yields and thus makes it possible to propose different synthesis strategies to access polymers whose properties meet very specific specifications. This chapter reviewed the synthesis of well-defined graft, block copolymers and three-dimensional networks by chemistry in homogeneous or heterogeneous phase by using more and more green chemistry. Nevertheless, the good biocompatibility and degradability of the structures obtained remain to be verified in order to keep the essential properties of the PHA.

REFERENCES

1. Bear MM, Renard E, Randriamahefa S, Langlois V, Guerin P. Preparation of a bacterial polyester with carboxy groups in side chains. *Comptes Rendus De L Academie Des Sciences Fascicule C-Chimie* 2001; 4: 289–293.
2. Park WH, Lee MY. Characterization of bacterial copolyesters with unsaturated pendant groups - Effect of epoxidation on thermal degradation. *Korea Polymer Journal* 1998; 6: 219–224.
3. Arkin AH, Hazer B, Borcakli M. Chlorination of poly(3-hydroxy alkanoates) containing unsaturated side chains. *Macromolecules* 2000; 33: 3219–3223.
4. Arkin AH, Hazer B. Chemical modification of chlorinated microbial polyesters. *Biomacromolecules* 2002; 3: 1327–1335.
5. Erduranli H, Hazer B, Borcakli M. Plastics from bacteria: Natural functions and applications. *Macromolecular Symposia* 2008; 69: 161–169.
6. Kilicay E, Hazer B, Çoban B, Scholz C. Synthesis and characterization of the poly(ethylene glycol) grafted unsaturated microbial polyesters. *Hacettepe Journal of Biology and Chemistry* 2010; 38(1): 9–17.
7. Renard E, Poux A, Timbart L, Langlois V, Guerin P. Preparation of a novel artificial bacterial polyester modified with pendant hydroxyl groups. *Biomacromolecules* 2005; 6: 891–896.
8. Eroglu MS, Hazer B, Ozturk T, Caykara T. Hydroxylation of pendant vinyl groups of poly(3-hydroxy undec-10-enoate) in high yield. *Journal of Applied Polymer Science* 2005; 97: 2132–2139.
9. Lee MY, Park WH, Lenz RW. Hydrophilic bacterial polyesters modified with pendant hydroxyl groups. *Polymer* 2000; 41: 1703–1709.
10. Hazer B. Simple synthesis of amphiphilic poly(3-hydroxyalkanoate)s with pendant hydroxyl and carboxylic groups via thiol-ene photo clock reactions. *Polymer Degradation and Stability* 2015; 119: 159–166.

11. Bear MM, *et al.* Bacterial poly-3-hydroxyalkenoates with epoxy groups in the side chains. *Reactive and Functional Polymers* 1997; 34: 65–77.
12. Park WH, Lenz RW, Goodwin S. Epoxidation of bacterial polyesters with unsaturated side chains. I. Production and epoxidation of polyesters from 10-undecenoic acid. *Macromolecules* 1998; 31: 1480–1486.
13. Lee MY, Park WH. Preparation of bacterial copolyesters with improved hydrophilicity by carboxylation. *Macromolecular Chemistry and Physics* 2000; 201: 2771–2774.
14. Kurth N, *et al.* Poly(3-hydroxyoctanoate) containing pendant carboxylic groups for the preparation of nanoparticles aimed at drug transport and release. *Polymer* 2002; 43: 1095–1101.
15. Stigers DJ, Tew GN. Poly(3-hydroxyalkanoate)s functionalized with carboxylic acid groups in the side chain. *Biomacromolecules* 2003; 4: 193–195.
16. Constantin M, Simionescu C, Carpov A, Samain E, Driguez H. Chemical modification of poly(hydroxyalkanoates). Copolymers bearing pendant sugars. *Macromolecular Rapid Communications* 1999; 20: 91–94.
17. Sparks J, Scholz C. Synthesis and characterization of a cationic poly(β-hydroxyalkanoate). *Biomacromolecules* 2008; 9: 2091–2096.
18. Modjinou T, Lemechko P, Versace DL, Langlois V, Renard E. Poly(3-hydroxyalkanoate) sulfonate: From nanoparticles toward water soluble polyesters. *European Polymer Journal* 2015; 68: 471–479.
19. Yu LP, Zhang X, Wei DX, Wu Q, Jiang XR, Chen GQ. Highly efficient fluorescent material based on rare-earth-modified polyhydroxyalkanoates. *Biomacromolecules* 2019; 209: 3233–3241.
20. Nkrumah-Agyeefi S, Scholtz C. Chemical modification of functionalized polyhydroxyalkanoates via "click" chemistry: A proof concept. *International Journal of Biological Macromolecules* 2017; 95: 796–808.
21. Kunioka M, Doi Y. Thermal degradation of microbial copolyesters: Poly(3-hydroxyb utyrate-*co*-3-hydroxyvalerate) and poly(3-hydroxybutyrate-*co*-4-hydroxybutyrate). *Macromolecules* 1990; 23: 1933–1936.
22. Nguyen S, Yu GE, Marchessault RH. Thermal degradation of poly(3-hydroxyalkanoates): Preparation of well-defined oligomers. *Biomacromolecules* 2002; 3: 219–224.
23. Ramier J, Grande D, Langlois V, Renard E. Toward the controlled production of oligoesters by microwave-assisted degradation of poly(3-hydroxyalkanoate)s. *Polymer Degradation and Stability* 2012; 97: 322–328.
24. Hirt TD, Neuenschwander P, Suter UW. Telechelic diols from poly *(R)*-3-hydroxybutyric acid and poly{ *(R)*-3-hydroxybutyric acid -co- *(R)*-3-hydroxyvaleric acid }. *Macromolecular Chemistry and Physics* 1996; 197: 1609–1614.
25. Andrade AP, Witholt B, Hany R, Egli T, Li Z. Preparation and characterization of enantiomerically pure telechelic diols from mcl-poly *(R)*-3- hydroxyalkanoates. *Macromolecules* 2002; 35: 684–689.
26. Lemechko P, *et al.* Functionalized oligoesters from poly(3-hydroxyalkanoate)s containing reactive end group for click chemistry: Application to novel copolymer synthesis with poly(2-methyl-2-oxazoline). *Reactive and Functional Polymers* 2012; 72: 160–167.
27. Lauzier C, Revol JF, Debzi EM, Marchessault RH. Hydrolytic degradation of isolated poly(beta-hydroxybutyrate) granules. *Polymer* 1994; 35: 4156- 4162.
28. Reeve MS, McCarthy S, Gross RA. The chemical degradation of bacterial polyesters for use in the preparation of new degradable block copolymers. *Polymer Preprints (American Chemical Society, Division of Polymer Chemistry)* 1990; 31: 437–438.
29. Timbart L, Renard E, Tessier M, Langlois V. Monohydroxylated poly(3-hydroxyoctanoate) oligomers and its functionalized derivatives used as macroinitiators in the synthesis of degradable diblock copolyesters. *Biomacromolecules* 2007; 8: 1255–1265.

30. Lehrle RS, Williams RJ. Thermal degradation of bacterial poly(hydroxybutyric acid): Mechanisms from the dependence of pyrolysis yields on sample thickness. *Macromolecules* 1994; 7: 3782–3789.
31. Chan Sin M, Gan SN, Mohd Annuar MS, Ping Tan IK. Thermodegradation of medium-chain-length poly(3-hydroxyalkanoates) produced by *Pseudomonas putida* from oleic acid. *Polymer Degradation and Stability* 2010; 95: 2334–2342.
32. Xiang H, Wen X, Miu X, Li Y, Zhou Z, Zhun M. Thermal depolymerization mechanisms of poly (3-hydroxybutyrate-co-3-hydroxyvalerate). *Progress in Natural Science: Materials International* 2016; 26: 58–64.
33. Hirt TD, Neuenschwander P, Suter UW. Synthesis of degradable, biocompatible, and tough block-copolyesterurethanes. *Macromolecular Chemistry and Physics* 1996; 197: 4253–4268.
34. Hori Y, *et al.* A novel biodegradable poly(urethane ester) synthesized from poly(3-hydroxybutyrate) segments. *Macromolecules* 1992; 25: 5117–5118.
35. Saad GR, Seliger H. Biodegradable copolymers based on bacterial poly(*(R)*- 3-hydroxy-butyrate): Thermal and mechanical properties and biodegradation behaviour. *Polymer Degradation and Stability* 2004; 83: 101–110.
36. Marchessault RH, Yu GE. Preparation and characterization of low molecular weight poly(3-hydroxybutyrate)s and their block copolymers with poly(oxyethylene)s. *Polymer Preprints (American Chemical Society, Division of Polymer Chemistry)* 1999; 40: 527–528.
37. Wans S, Chen W, Xiang H, Yang J, Zhou Z, Zhu M. Modification and potential application of short-chain-length polyhydroxyalkanoate (*scl*-PHA). *Polymer* 2016; 8: 273–300.
38. Li Z, Loh XJ. Water soluble polyhydroxyalkanoates: Future materials for therapeutic applications. *Chemical Society Reviews* 2015; 44(10): 2865–79.
39. Reeve MS, McCarthy SP, Gross RA. Preparation and characterization of *(R)*-poly(beta-hydroxybutyrate) poly(epsilon-caprolactone) and *(R)*- poly(betahydroxybutyrate) poly(lactide) degradable diblock copolymers. *Macromolecules* 1993; 26: 888–894.
40. Liu QS, Shyr TW, Tung CH, Deng BY, Zhu MF. Block copolymers containing poly (3-hydroxybutyrate-*co*-3-hydroxyvalerate) and poly(ε-caprolactone) units: Synthesis, characterization and thermal degradation. *Fibers and Polymers* 2011; 12: 848–856.
41. Schreck KM, Hillmyer MA. Block copolymers and melt blends of polylactide with Nodax microbial polyesters: Preparation and mechanical properties. *Journal of Biotechnology* 2007; 132: 287–295.
42. Ramier J, Renard E, Grande D. Microwave-assisted synthesis and characterization of biodegradable block copolyesters based on poly(3-hydroxyalkanoate)s and poly(D,L-lactide). *Journal of Polymer Science, Part A: Polymer Chemistry* 2012; 50: 1445–1455.
43. Wu L, Chen S, Li Z, Xu K, Chen GQ. Synthesis, characterization and biocompatibility of novel biodegradable poly (*(R)*-3-hydroxybutyrate)-*block*-(D,L-lactide)-block-(ε-caprolactone) triblock copolymers. *Polymer International* 2008; 7: 939–949.
44. Zhang XQ, *et al.* Synthesis and characterization of biodegradable triblock copolymers based on bacterial poly *(R)*-3-hydroxybutyrate by atom transfer radical polymerization. *Journal of Polymer Science, Part A: Polymer Chemistry* 2005; 3: 4857–4869.
45. Loh XJ, Cheong WCD, Li J, Ito Y. Novel poly(*N*-isopropylacrylamide)-poly[(*R*)-3-hydroxybutyrate]-poly(*N*-isopropyl acrylamide) triblock copolymer surface as a culture substrate for human mesenchymal stem cells. *Soft Matter* 2009; 5: 2937–2946.
46. Loh XJ, Zhang ZX, Wu YL, Lee TS, Li J. Synthesis of novel biodegradable thermo-responsive triblock copolymers based on poly[*(R)*-3- hydroxybutyrate] and poly(N-isopropylacrylamide) and their formation of thermoresponsive micelles. *Macromolecules* 2009; 2: 194–202.
47. Loh XJ, Ong SJ, Tung YT, Choo HT. Dual responsive micelles based on poly[*(R)*-3-hydroxybutyrate] and poly(2-(di-methylamino)ethyl methacrylate) for effective doxorubicin delivery. *Polymer Chemistry* 2013; 4: 2564–2574.

48. Loh XJ, Ong SJ, Tung YT, Choo HT. Incorporation of poly[(*R*)-3-hydroxybutyrate] into cationic copolymers based on poly(2-(dimethylamino)ethyl methacrylate) to improve gene delivery. *Macromolecular BioScience* 2013; 13: 1092–1099.
49. Li J, Li X, Ni XP, Leong KW. Synthesis and characterization of new biodegradable amphiphilic poly(ethylene oxide)-*b*-poly[(R)-3-hydroxy butyrate-*b*-poly(ethylene oxide) triblock copolymers. *Macromolecules* 2003; 6: 2661–2667.
50. Ravenelle F, Marchessault RH. One-step synthesis of amphiphilic diblock copolymers from bacterial poly([R]-3-hydroxybutyric acid). *Biomacromolecules* 2002; 3: 1057–1064.
51. Shah M, Naseer MI, Choi MH, Kim MO, Yoon, SC. Amphiphilic PHA-mPEG copolymeric nanocontainers for drug delivery: Preparation, characterization and *in vitro* evaluation. *International Journal of Pharmaceutics* 2010; 400: 165–175.
52. Shah M, Choi MH, Ullah N, Kim MO, Yoon SC. Synthesis and characterization of PHV-block-mPEG diblock copolymer and its formation of amphiphilic nanoparticles for drug delivery. *Journal of Nanoscience and Nanotechnology* 2011; 11: 5702–5710.
53. Shah M, Ullah N, Choi MH, Kim MO, Yoon SC. Amorphous amphiphilic P(3HV-co-4HB)-*b*-mPEG block copolymer synthesized from bacterial copolyester via melt trans-esterification: Nanoparticle preparation, cisplatin loading for cancer therapy and *in vitro* evaluation. *European Journal of Pharmaceutics and Biopharmaceutics* 2012; 80: 518–527.
54. Liu QY, Cheng ST, Li ZB, Xu KT, Chen GQ. Characterization, biodegradability and blood compatibility of poly (R)-3-hydroxybutyrate based poly(ester-urethane)s. *Journal of Biomedical Materials Research Part A* 2009; 90A: 1162–1176.
55. Lendlein A, Neuenschwander P, Suter UW. Tissue-compatible multiblock copolymers for medical applications, controllable in degradation rate and mechanical properties. *Macromolecular Chemistry and Physics* 1998; 199: 2785–2796.
56. Lendlein A, Colussi M, Neuenschwander P, Suter UW. Hydrolytic degradation of phase-segregated multiblock copoly(ester urethane)s containing weak links. *Macromolecular Chemistry and Physics* 2001; 202: 2702–2711.
57. Li X, Loh XJ, Wang K, He CB, Li. J. Poly(ester urethane)s consisting of poly (R)-3-hydroxybutyrate and poly(ethylene glycol) as candidate biomaterials: Characterization and mechanical property study. *Biomacromolecules* 2005; 6: 2740–2747.
58. Loh XJ, Tan KK, Li X, Li J. The *in vitro* hydrolysis of poly(ester urethane)s consisting of poly[(*R*)-3-hydroxybutyrate] and poly(ethylene glycol). *Biomaterials* 2006; 27: 1841–1850.
59. Zhao Q, Cheng G, Li H, Ma X, Zhang L. Synthesis and characterization of biodegradable poly(3-hydroxybutyrate) and poly(ethylene glycol) multiblock copolymers. *Polymer* 2005; 46: 10561–10567.
60. Loh XJ, Wang X, Li HZ, Li X, Li J. Compositional study and cytotoxicity of biodegradable poly(ester urethane)s consisting of poly (R)-3-hydroxybutyrate and poly(ethylene glycol). *Materials Science & Engineering C-Biomimetic and Supramolecular Systems* 2007; 27: 267–273.
61. Liu QS, Zhu MF, Chen YM. Synthesis and characterization of multiblock copolymers containing poly (3-hydroxybutyrate)-co-(3-hydroxyvalerate) and poly(ethylene glycol). *Polymer International* 2010; 59: 842–850.
62. Loh XJ, Goh SH, Li J. Hydrolytic degradation and protein release studies of thermogelling polyurethane copolymers consisting of poly[(*R*)-3-hydroxybutyrate], poly(ethylene glycol), and poly(propylene glycol). *Biomaterials* 2007; 28: 4113–4123.
63. Li ZB, *et al.* Novel amphiphilic poly(ester-urethane)s based on poly[(*R*)-3-hydroxyalkanoate]: Synthesis, biocompatibility and aggregation in aqueous solution. *Polymer International* 2008; 7: 887–894.
64. Pan JY, *et al.* Alternative block polyurethanes based on poly(3-hydroxybutyrate-*co*-4-hydroxybutyrate) and poly(ethylene glycol). *Biomaterials* 2009; 30: 2975–2984.

65. Babinot J, Guigner JM, Renard E, Langlois V. A micellization study of medium chain length poly(3-hydroxyalkanoate)-based amphiphilic diblock copolymers. *Macromolecular Chemistry and Physics* 2012; 375: 88–93.
66. Zhijiang C, Zhihong W. Preparation of biodegradable poly(3-hydroxybutyrate) (PHB) and poly(ethylene glycol) (PEG) graft copolymer. *Journal of Materials Science* 2007; 42: 5886–5890.
67. Cai ZJ, Hou CW, Yang G. Crystallization behavior, thermal property and biodegradation of poly(3-hydroxybutyrate)-poly(ethylene glycol) grafting copolymer. *Polymer Degradation and Stability* 2011; 96: 1602–1609.
68. Arslan H, Hazer B, Yoon SC. Grafting of poly(3-hydroxyalkanoate) and linoleic acid onto chitosan. *Journal of Applied Polymer Science* 2007; 103: 81–89.
69. Neugbauer D, Rydz J, Goebel I, Dacko P, Kowalzuck M. Synthesis of graft copolymers containing biodegradable poly(3-hydroxybutyrate) chains. *Macromolecules* 2007; 40(5): 1767–1773.
70. Nguyen S, Marchessault RH. Synthesis and properties of graft copolymers based on poly(3-hydroxybutyrate) macromonomers. *Macromolecular Bioscienc*e 2007; 4(3): 262–268.
71. Nguyen S, Marchessault RH. Atom transfer radical copolymerization of bacterial poly(3-hydroxybutyrate) macromonomer and methyl methacrylate. *Macromolecules* 2005; 38(2): 290–296.
72. Saad G, Elsawy M, Elsabee M. Preparation, characterization and antimicrobial activity of poly(3-hydroxybutyrate-co-3-hydroxyvalerate)-g-poly(N-vinylpyrrolidone) Copolymers. *Polymer-Plastics Technology and Engineering* 2012; 51: 1113–1121.
73. Zhou L, Chen Z, Chi W, Yang X, Wang W, Zhang B. Mono methoxy poly(3-hydoxybu tyrate-co-4-hydroxybutyrate)-graft- hyperbranched polyethylenimine copolymers for siRNA delivery. *Biomaterials* 2012; 33(7): 2334–2344.
74. Andrade AP, *et al.* Synthesis and characterization of novel copoly(ester-urethane) containing blocks of poly-(*R*)-3-hydroxyoctanoate and poly-(*R*)-3-hydroxybutyrate. *Macromolecules* 2002; 35: 4946–4950.
75. Andrade AP, Witholt B, Chang DL, Li Z. Synthesis and characterization of novel thermoplastic polyester containing blocks of poly (R)-3- hydroxyoctanoate and poly[(*R*)-3-hydroxybutyrate]. *Macromolecules* 2003; 36: 9830–9835.
76. Dai S, Xue L, Zinn M, Li Z. Enzyme-catalyzed polycondensation of polyester macrodiols with divinyl adipate: A green method for the preparation of thermoplastic block copolyesters. *Biomacromolecules* 2009; 10: 3176–3181.
77. Timbart L, Renard E, Langlois V, Guerin P. Novel biodegradable copolyesters containing blocks of poly(3-hydroxyoctanoate) and poly(ε-caprolactone): Synthesis and characterization. *Macromolecular Bioscience* 2004; 4: 1014–1020.
78. Arslan H, Yesilyurt N, Hazer, B. Brush type copolymers of poly(3hydroxybutyrate) and poly(3-hydroxyoctanoate) with same vinyl monomers via "grafting from" technique by using atom transfer radical polymerization method. *Macromolecular Symposia* 2008; 269: 23–33.
79. Gagnon KD, Lenz RW, Farris RJ, Fuller RC. Chemical modification of bacterial elastomers peroxide cross-linking. *Polymer* 1994; 35: 4358–4367.
80. Hazer B, Lenz RW, Cakmakli B, Borcakli M, Kocer H. Preparation of poly(ethylene glycol) grafted poly(3-hydroxyalkanoate) networks. *Macromolecular Chemistry and Physics* 1999; 200: 1903–1907.
81. Ilter S, Hazer B, Borcakli M, Atici O. Graft copolymerisation of methyl methacrylate onto a bacterial polyester containing unsaturated side chains. *Macromolecular Chemistry and Physics* 2001; 202: 2281–2286.
82. Hany R, Bohlen C, Geiger T, Hartmann R, Kawada J, Schmid M, Zinn M, Marchessault RH. Chemical synthesis of crystalline comb polymers from olefinic medium-chain-length poly[3-hydroxyalkanoates]. *Macromolecules* 2004; 37: 385–389.

83. Hany R, Bohlen C, Geiger T, Schmid M, Zinn M. Toward non-toxic antifouling: Synthesis of hydroxy-, cinnamic acid-, sulfate-, and zosteric acid-labeled poly[3-hydroxyalkanoates]. *Biomacromolecules* 2004; 5(4): 1452–1456.
84. Hany R, Hartmann R, Bohlen C, Brandenberger S, Kawada J, Lowe C, Zinn M, Witholt B, Marchessault RH. Chemical synthesis and characterization of POSS-functionalized poly[3-hydroxyalkanoates]. *Polymer* 2005; 46(14): 5025–5031.
85. Cakmakli B, Hazer B, Borcakli M. Poly(styrene peroxide) and poly(methyl methacrylate peroxide) for grafting on unsaturated bacterial polyesters. *Macromolecular Bioscience* 2001; 1(8): 348–354.
86. Ishida K, Hortensius R, Luo XF, Mather P. Soft bacterial polyester-based shape memory nanocomposites featuring reconfigurable nanostructure. *Journal of Polymer Science Part B Polymer Physics* 2012; 50(6): 387–393.
87. Renard E, Ternat C, Langlois V, Guerin P. Synthesis of graft bacterial polyesters for nanoparticles preparation. *Macromolecular Bioscience* 2003; 3: 248–252.
88. Renard E, Tanguy PY, Samain E, Guerin P. Synthesis of novel graft polyhydroxyalkanoates. *Macromolecular Symposia* 2003; 197: 11–18.
89. Domenek S, Langlois V, Renard E. Bacterial polyesters grafted with poly(ethylene glycol): Behaviour in aqueous media. *Polymer Degradation and Stability* 2007; 92: 1384–1392.
90. Babinot J, Renard E, Langlois V. Preparation of clickable poly(3-hydroxyalkanoate) (PHA): Application to poly(ethylene glycol) (PEG) graft copolymers synthesis. *Macromolecular Rapid Communications* 2010; 31(7): 619–624.
91. Le Fer G, Babinot J, Versace DL, Langlois V, Renard E. An efficient thiol-ene chemistry for the preparation of amphiphilic PHA-based graft copolymers. *Macromolecular Rapid Communications* 2012; 33: 2041–2045.
92. Ma YM, Wei DX, Yao H, Wu LP, Chen GQ. Synthesis, characterization and application of thermoresponsive polyhydroxyalkanoate-graft-poly(N-isopropylacrylamide). *Biomacromolecules* 2016; 17: 2680–2690.
93. Ke Y, Zhang XY, Ramakrishna R, He LM, Wu G. Reactive blends based on polyhydroxyalkanoates: Preparation and biomedical application. *Materials Science and Engineering* 2017; C70: 1107–1119.
94. Wei L, McDonald AG. Peroxide induced cross-linking by reactive melt processing of two biopolyesters: Poly(3-hydroxybutyrate) and poly(L-lactic acid) to improve their melting processability. *Journal of Applied Polymer Science* 2015; 132: 1–15.
95. Fei B, Chen C, Chen S, Peng SW, Zhuang YG, An YX, Dong LS. Crosslinking of poly[(3-hydroxybutyrate)-co-(3-hydroxyvalerate)] using dicumyl peroxide as initiator. *Polymer International* 2004; 53: 937–943.
96. Kolahchi AR, Kontopoulou M. Chain extended poly(3-hydroxybutyrate) with improved rheological properties and thermal stability, through reactive modification in the melt state. *Polymer Degradation and Stability* 2015; 121: 222–229.
97. D'Haene P, Remsen EE, Asrar J. Preparation and characterization of a branched bacterial polyester. *Macromolecules* 1999; 32: 5229–5235.
98. Nerkar M, Ramsay J, Kontopoulou M, Ramsay B. Improvement of melt strength and crystallization rate of polylactic acid and its blends with medium-chain-length polyhydroxyalkanoate through reactive modification. *Plastics Engineering* 2014; 1: 339–342.
99. Zhang Z, Misra M, Mohanty AK. Toughened sustainable green composites from poly(3-hydroxybutyrate- co -3-hydroxyvalerate) based ternary blends and miscanthus biofiber. *ACS Sustainable Chemistry and Engineering* 2014; 2(10): 2345–2354.

100. Higham AK, Garber LA, Latshaw DC, Hall CK, Pojman JA, Khan SA. Gelation and cross-linking in multifunctional thiol and multifunctional acrylate systems involving an in situ comonomer catalyst. *Macromolecules* 2014; 47(2): 821–829.
101. Levine AC, Heberlig GW, Nomura CT. Use of thiol-ene click chemistry to modify mechanical and thermal properties of polyhydroxyalkanoates (PHAs). *International Journal of Biological Macromolecules* 2016; 83: 358–365.
102. Zhang X, Li Z, Che X, Yu L, Jia W, Shen R, Chen J, Ma Y, Chen GQ. Synthesis and characterization of polyhydroxyalkanoate organo/hydrogels. *Biomacromolecules* 2019; 20(9): 3303–3312.
103. Gross RA, McCarthy SP, Reeve MS. *Biodegradable and hydrodegradable diblock copolymers composed of poly(beta).* U.S. Patent No. 5,439,985. Washington, DC: U.S. Patent and Trademark Office, 1995.
104. Andrade AP, Witholt B, Hany R, Egli T, Li Z. Preparation and characterization of enantiomerically pure telechelic diols from *mcl*-poly[(R)-3-hydroxyalkanoates]. *Macromolecules* 2002; 35: 684–689.
105. Lav TX, Lemechko P, Renard E, Amiel C, Langlois V, Volet G. Development of a new azido-oxazoline monomer for the preparation of amphiphilic graft copolymers by combination of cationic ring-opening polymerization and click chemistry. *Reactive & Functional Polymers* 2013; 73: 1001–1008.
106. Marchessault RH, Yu G. Preparation and characterization of low molecular weight poly(3-hydroxybutyrate)s and their block copolymers with poly(oxyethylene)s. *ACS Division of Polymer Chemistry* 1999; 40: 527–528.
107. Lorenzini C, Versace DL, Babinot J, Renard E, Langlois V. Biodegradable hybrid poly(3-hydroxyalkanoate)s networks through silsesquioxane domains formed by efficient UV-curing. *Reactive & Functional Polymers* 2014; 84: 53–59.
108. Audic J-L, Lemieègre L, Corre Y-M. Thermal and mechanical properties of a polyhydroxyalkanoate plasticized with biobased epoxidized broccoli oil. *Journal of Applied Polymer Science* 2014; 131(6): 1–7.
109. Choi JS, Park WH. Effect of biodegradable plasticizers on thermal and mechanical properties of poly(3-hydroxybutyrate). *Polymer Testing* 2004; 23(4): 455–460.
110. Seydibeyoğğlu MOÖ, Misra M, Mohanty A. Synergistic improvements in the impact strength and elongation of polyhydroxybutyrate-co-valerate copolymers with functionalized soybean oils and POSS. *International Journal of Plastics Technology* 2010; 14(1): 1–16.
111. Hazer B. The properties of PLA/oxidized soybean oil polymer blends. *Journal of Polymers and the Environment* 2014; 22(2): 200–208.
112. Anderson KS, Schreck KM, Hillmyer MA. Toughening polylactide. *Polymer Reviews* 2008; 48(1): 85–108.
113. Mangeon C, Renard E, Thevenieau F, Langlois V. Networks based on biodegradable polyesters: An overview of the chemical ways of crosslinking. *Materials Science and Engineering C* 2017; 80(1): 760–770.
114. Kim HW, Chung CW, Kim YB, Rhee YH. Preparation and hydrolytic degradation of semi-interpenetrating networks of poly(3-hydroxyundecenoate) and poly(lactide-co-glycolide). *International Journal of Biological Macromolecules* 2005; 37: 221–226.
115. Hao J, Deng X. Semi-interpenetrating networks of bacterial poly(3-hydroxybutyrate) with net-poly(ethylene glycol). *Polymer* 2001; 42(9): 4091–4097.
116. Martellinia F, Innocentini Meia LH, Lorab S, Carenzam M. Semi-interpenetrating polymer networks of poly(3-hydroxybutyrate) prepared by radiation-induced polymerization. *Radiation Physics and Chemistry* 2004; 71: 255–260.

117. Paredes Zaldivar M, Galego Fernández N, Gastón Peña C, Rapado Paneque M, Antanés Valentín S. Synthesis and characterization of a new semi-interpenetrating polymer network hydrogel obtained by gamma radiations. *Journal of Thermal Analysis and Calorimetry* 2011; 106: 725–730.
118. Mangeon C, Modjinou T, Rios de Anda A, Thevenieau F, Renard E, Langlois V. Renewable semi-interpenetrating polymer networks based on vegetable oils used as plasticized systems of poly(3-hydroxyalkanoate)s. *ACS Sustainable Chemistry and Engineering* 2018; 6: 5034–5042.
119. Ying TH, Ishii D, Mahara A, Murakami S, Yamaoka T, Sudesh K, Samian R, Fujita M, Maeda M, Iwata T. Scaffolds from electrospun polyhydroxyalkanoate copolymers: Fabrication, characterization, bioabsorption and tissue response. *Biomaterials* 2008; 29: 1307–1317.
120. Cheng ML, Chen PY, Lan CH, Sun YM. Structure, mechanical properties and degradation behaviors of the electrospun fibrous blends of PHBHHx/PDLLA. *Polymer* 2011; 52: 1391–1401.
121. Li X, Liu KL, Wang M, Wong SY, Tjiu WC, He CB, Goh SH, Li J. Improving hydrophilicity, mechanical properties and biocompatibility of poly[(*R*)-3-hydroxybutyrat e-*co*-(*R*)-3-hydroxyvalerate] through blending with poly[(*R*)-3-hydroxybutyrate]-*alt*-poly(ethylene oxide). *Acta Biomaterialia* 2009; 5: 2002–2012.
122. Xu S, Luo R, Wu L, Xu K, Chen G-Q, Blending and characterizations of microbial poly(3-hydroxybutyrate) with dendrimers. *Journal of Applied Polymer Science* 2006; 102: 3782–3790.
123. Innocentini-Mei LH, Bartoli JR, Baltieri RC. Mechanical and thermal properties of poly(3-hydroxybutyrate) with starch and starch derivatives. *Macromolecular Symposia* 2003; 197: 77–88.
124. Li J, Yun H, Gong Y, Zhao N, Zhang X. Effects of surface modification of poly(3-hydroxybutyrate-*co*-3-hydroxyhexanoate) (PHBHHx) on physicochemical properties and on interactions with MC3T3-E1 cells. *Journal of Biomedical Materials Research Part A* 2005; 75A: 985–998.
125. Zhang DM, Cui FZ, Luo ZS, Lin YB, Zhao K, Chen GQ. Wettability improvement of bacterial poly(hydroxyalkanoates) via ion implantation. *Surface and Coatings Technology* 2000; 131: 350–354.
126. Kang IK, Choi SH, Shin DS, Yoon SC. Surface modification of polyhydroxyalkanoate films and their interaction with human fibroblasts. *International Journal of Biological Macromolecules* 2001; 28: 205–212.
127. Garrido L, Jiménez I, Ellis G, Cano P, García-Martínez JM, López L, de la Peña E. Characterization of surface-modified polyalkanoate films for biomedical applications. *Journal of Applied Polymer Science* 2011; 119: 3286–3296.
128. Wang YY, Luü LX, Shi JC, Wang HF, Xiao ZD, Huang N-P. Introducing RGD peptides on PHBHB films through PEG-containing cross-linkers to improve the biocompatibility. *Biomacromolecules* 2011; 12: 551–559.
129. Grøndahl L, Chandler-Temple A, Trau M. Polymeric grafting of acrylic acid onto poly(3-hydroxybutyrate-*co*-3-hydroxyvalerate): Surface functionalization for tissue engineering applications. *Biomacromolecules* 2005; 6: 2197–2203.
130. Hu SG, Jou CH, Yang MC. Antibacterial and biodegradable properties of polyhydroxyalkanoates grafted with chitosan and chitooligosaccharides via ozone treatment. *Journal of Applied Polymer Science* 2003; 88: 2797–2803.
131. Fouassier JP, Lalevee J. *Photoinitiators for Polymer Synthesis: Scope, Reactivity, and Efficiency*. John Wiley and Sons, Weinheim, Germany, 2012.

132. Lao HK, Renard E, Fagui A, Langlois V, Vallée-Rehel K, Linossier I. Functionalization of poly(3-hydroxybutyrate-*co*-3-hydroxyvalerate) films via surface-initiated atom transfer radical polymerization: Comparison with the conventional free-radical grafting procedure. *Journal of Applied Polymer Science* 2011; 120: 184–194.
133. Ma H, Davis RH, Bowman CN. A novel sequential photoinduced living graft polymerization. *Macromolecules* 1999; 33: 331–335.
134. Ke Y, Wang Y, Ren L, Lu L, Wu G, Chen X, Chen J. Photografting polymerization of polyacrylamide on PHBV films. *Journal of Applied Polymer Science* 2007; 104: 4088–4095.
135. Lao HK, Renard E, Langlois V, Vallée-Rehel K, Linossier I. Surface functionalization of PHBV by HEMA grafting via UV treatment: Comparison with thermal free radical polymerization. *Journal of Applied Polymer Science* 2010; 116: 288–297.
136. Rupp B, Ebner C, Rossegger E, Slugovc C, Stelzer F, Wiesbrock F. UV-induced crosslinking of the biopolyester poly(3-hydroxybutyrate)-*co*-(3-hydroxyvalerate). *Green Chemistry* 2010; 12: 1796–1802.
137. Versace DL, Ramier J, Grande D, Abbad Andaloussi S, Dubot P, Hobeika N, Malval JP, Lalevee J, Renard E, Langlois V. Natural biopolymer surface of poly(3-hydroxyb utyrate-*co*-3-hydroxyvalerate)-photoinduced modification with triarylsulfonium salts. *Advanced Healthcare Materials* 2013; 2: 1008–1018.
138. Ke Y, Wang Y, Ren L, Lu L, Wu G, Chen X, Chen J. Photografting polymerization of polyacrylamide on PHBV films. *Journal of Applied Polymer Science* 2007; 104: 4088–4095.
139. Ramier J, Boubaker M, Guerrouache M, Langlois V, Grande D, Renard E. Novel routes to epoxy functionalization of PHA-based electrospun scaffolds as ways to improve cell adhesion. *J. Polym. Sci. Part A: Polym. Chem.* 2014; 52(6): 816–824.
140. Versace DL, Ramier J, Babinot J, Lemechko P, Soppera O, Lalevee J, Albanese P, Renard E, Langlois V. Photoinduced modification of the natural biopolymer poly(3-hydroxybutyrate-*co*-3-hydroxyvalerate) microfibrous surface with anthraquinone-derived dextran for biological applications. *Journal of Materials Chemistry B* 2013; 1: 4834–4844.
141. Ducel G, Fabry J, Nicolle L. *Prevention of Hospital-Acquired Infections: A Practical Guide*. 2nd ed. World Health Organization, Geneva, Switzerland, 2002 © World Health Organization 2002.
142. Pace JL, Rupp ME, Finch RG. (eds.) *Biofilms, Infection, and Antimicrobial Therapy*. Taylor & Francis, CRC Press, Boca Raton, London, New York, Singapore. 2005.
143. Condat M, Helary C, Coradin T, Dubot P, Babinot J, Faustini M, Abbad Andaloussi S, Renard E, Langlois V, Versace DL. Design of cytocompatible bacteria-repellent bio-based polyester films via an aqueous photoactivated process. *Journal of Materials Chemistry B* 2016; 4: 2842–2850.
144. Poupart R, Haider A, Babinot J, Kang IK, Malval JP, Lalevée J, Abbad Andaloussi S, Langlois V, Versace DL. Photoactivable surface of natural poly(3-hydroxybutyrate-*co*-3-hydroxyvalerate) for antiadhesion applications. *ACS Biomaterials Science and Engineering* 2015; 1: 525–538.

3 Amphiphiles from Poly(3-hydroxyalkanoates)

Baki Hazer

CONTENTS

3.1 POLY(3-HYDROXYALKANOATES) (PHA)

Very widely produced and used commercial packaging plastic materials cause severe problems in the natural environment. Plastic pollution has been seen in the ground, in oceans, and in lakes because of its non-degradability. Mainly fuel fossil-based monomers have also been used in the production of these polymers. This is another disadvantage of the commercial non-degradable polymers, because fossil fuel reserves are limited and decreasing all the time. One of the solutions for plastic pollution is to use biodegradable polymers obtained from renewable resources. Microbial polyesters are very good candidates of these type of polymers; they can be obtained from natural products such as sugar, fatty acids, alkanoic acids, etc. via green chemistry [1, 2]. PHA exhibit valuable characteristics, such as biodegradability, biocompatibility, and thermoplasticity, and therefore can be used for medical, agricultural, and marine applications. PHA accumulate to high levels in bacteria (95% of the cellular dry mass), and their structures can be manipulated by genetic or physiological strategies [3–10]. Large number of bacteria, including *Ralstonia eutropha* (today: *Cupriavidus necator*) and *Pseudomonas putida*, produce PHA during metabolic stresses, such as a limitation of nitrogen, oxygen, or other essential nutrients in the presence of an excess of carbon source [11–16].

The general formula of the PHA is shown in Figure 3.1. The pendent alkyl group is decisive for the physical properties and the classification of the PHA. PHA are divided into three groups according to the pendant R group; short-chain-length PHA (*scl*-PHA) comprising 1–2 carbon atoms, medium-chain-length PHA (*mcl*-PHA) comprising 3–11 carbon atoms, and long-chain-length PHA (*lcl*-PHA) comprising more than 11 carbon atoms. The types of PHA are listed in Table 3.1. These PHA show different thermal and physical properties. Generally, *mcl*-PHA show lower T_m and T_g and more flexibility compared with *scl*-PHA. These *mcl*-PHA which can be produced using renewable resources are biocompatible, biodegradable, and thermoprocessable. They have low crystallinity, low glass transition temperature, low tensile strength, and high elongation at break, making them elastomeric polymers. *Mcl*-PHA and their copolymers are suitable for a range of biomedical applications where flexible biomaterials are required, such as heart valves and other cardiovascular applications, as well as matrices for controlled drug delivery. *Mcl*-PHA are more structurally diverse than short-chain-length PHA and hence can be more readily tailored for specific applications.

3.2 AMPHIPHILIC POLYMERS, GENERAL INTRODUCTION

Amphiphilic copolymers contain both hydrophilic and hydrophobic blocks. Amphiphilic polymers can be synthesized by introducing hydrophilic groups such as hydroxyl, carboxyl, amine, glycol, and hydrophilic polymers such as PEG, poly(vinyl alcohol), poly(acryl amide), poly(acrylic acids), hydroxyethylmethacrylate, poly(vinyl pyridine), and poly(vinyl pyrrolidone) onto a hydrophobic moiety. Such amphiphilic copolymers find numerous applications as emulsifiers, dispersants, foamers, thickeners, rinse aids, and compatibilizers [17–24].

3.3 AMPHIPHILIC PHA VIA CHEMICAL MODIFICATION REACTIONS

Amphiphilic PHA are hydrophobic, which is a disadvantage for some medical applications. Therefore, the hydrophobic natural polyesters, PHA, need to have

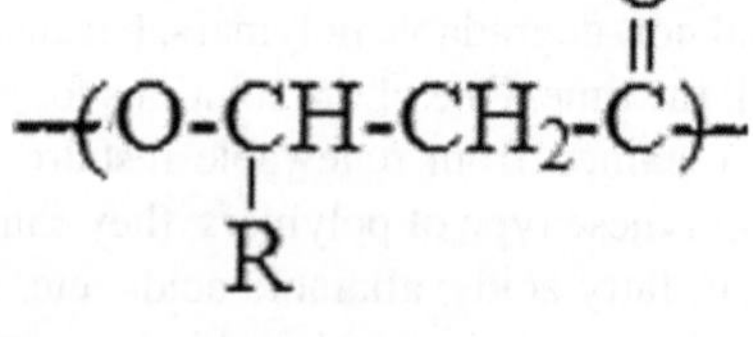

FIGURE 3.1 General chemical formula of PHA.

TABLE 3.1
Types of PHA

		Poly(3-hydroxyalkanoate), PHA				
Type of PHA accumulated	**Bacterium (production strain)**	**Side chain *(R)***	**Name of PHA***	**T_g [°C]**	**T_m [°C]**	**Elongation at break [%]**
scl-PHA	*R. eutropha* (today: *C. necator*)	methyl	PHB	0–15	137–170	5–30
		ethyl	PHV			
mcl-PHA	*P. oleovorans* (today: *P. putida*)	propyl	PHHx			
		butyl	PHHp	–40––20	45–90	600–800
		valeryl (pentyl)	PHO			
		hexyl	PHN			
		heptyl	PHD			
lcl-PHA	*P. oleovorans* (today: *P. putida*)	higher than octyl		–50	40	soft, sticky

*HB: 3-hydroxybutyrate, HV: 3-hydroxyvalerate, HHx: 3-hydroxyhexanoate, HHp: 3-hydroxyheptanoate, HO: 3-hydroxyoctanoate, HN: 3-hydroxynonanoate, HD: 3-hydroxydecanoate. T_g and T_m are glass transition and melting temperatures, respectively.

a hydrophilic character, particularly for tissue engineering and drug delivery systems [25].

PHA are grafted with some hydrophilic groups in order to enhance hydrophilicity. These hydrophilic moieties are listed in Table 3.2.

3.4 AMPHIPHILIC PHA

Microbial polyesters are excellent biodegradable hydrophobic biopolymers. Because of their limited mechanical, thermal, and non-hydrophilic properties, they are needed to diversify in reasonable chemical reactions to be able to be competitive against the convenient petroleum-based biomaterials [26].

There are three types of PHA: short-chain-length (e.g., PHB, PHBV), medium-chain-length (e.g., P(3HHx), P(3HHp), P(3HO), P(3HN), PHD, and P(3HU)), and long-chain-length (PHA derived from fatty acids) [13, 14]. Despite the many PHA that are produced by many different type of bacteria and identified in detail, only a few types of PHA have undergone the hydrophilic modification reactions: poly(3-hydroxyoctanoate-*co*-3-hydroxyundecenoate) (PHOU), PHO, PHN, poly(3-hydroxybutyrate) (PHB), poly(3-hydroxybutyrate-*co-3*-hydroxyvalerate), PHBV, and unsaturated *mcl*-PHA obtained from polyunsaturated plant oily acids. Here, we study the chemical reactions of the specific PHA one by one to gain hydrophilicity.

TABLE 3.2
The Hydrophilic Moieties Used in the Preparation of the Amphiphilic PHA

Hydrophilic moieties
-COOH
-OH
-NH2
Amino acid
Poly (ethylene glycol)
Acrylic acid
Methacrylic acid
N-isopropyl acryl amide
Dimethyl amino ethyl methacrylate

3.4.1 Amphiphilic PHOU Derivatives

Double bond functionality is readily open for the modification reactions. Therefore, the first step was the production of unsaturated *mcl*-PHA by Lenz's group. In an early experiment, *Pseudomonas oleovorans* was grown separately on 3-hydroxy-6-octenoic acid and 3-hydroxy-7-octenoic acid as the only carbon source and under ammonium nutrient-limiting conditions to produce storage polyesters. The polyesters produced contained mainly unsaturated C8 units [27].

However, the poly (3-hydroxyoctenoate) was rarely used in the derivatization reactions. The first attempt of the hydrophilic modification of the PHA started with the production of Poly(3-hydroxyoctanoate-*co*-3-hydroxyundecenoate), (PHOU) [28]. For this purpose, *Pseudomonas oleovorans* was grown on 10-undecenoic acid and on a mixture of sodium octanoate and 10-undecenoic acid. In the case of 10-undecenoic acid as the sole nutrient, the microorganism produced a polymer containing only repeating units with unsaturated side chains. Both the melting temperature (T_m) and glass transition temperature (T_g) were observed to decrease with increasing conversion of olefinic bonds. PHOU is a soft elastomer that is easy to handle while PHU is sticky and waxy. Interestingly, the fermentation of a mixture of nonanoic acid and 10-undecenoic acid gave a homopolymer mixture of PHN and PHU [29].

PHOU containing pendant hydroxyl and carboxylic acid groups can be obtained by varying oxidation conditions. Pendant double bonds can be oxidized by strong oxidized agents such as KMnO4 at 20°C in order to obtain hydroxylated PHOU. The polymers which were 40–60% hydroxylated were completely soluble in polar solvents including an 80/20 acetone/water mixture, methanol, and DMSO, indicating a considerably enhanced hydrophilicity of the modified PHA [30, 31].

The increase in percentage of hydroxyl groups of PHU and its improved water solubility are important for preparing drug delivery systems, artificial organs, and tissue engineering applications. Nearly 100% hydroxylation of double bonds of PHU was achieved by a hydroboration–oxidation reaction using 9-borobicyclononane. The hydroxylated PHU was fully soluble in methanol and almost soluble in water [32].

The pendant double bonds can be converted to pendant carboxylic acids using a biphasic CH_2C_{12}/CH_3COOH medium in the presence of KMnO4 and 18-crown-6-ether [33].

PHA containing pendant carboxyl groups was prepared by the chemical modification of unsaturated PHA using KMnO4 at 55°C, although there is a severe loss in molecular weight of PHA during the reaction. The degree of carboxylation increases to approximately 50% after 2 h of reaction time, but there is no further increase with prolonged reaction times. The polymers with a degree of carboxylation of 40–50% are completely soluble in water/Na2CO3, indicating a considerably enhanced hydrophilicity of the modified PHA [34].

With the use of osmium tetroxide and oxone (a triple salt mixture of $2KHSO_5$, K_2SO_4, and $KHSO_4$), double bonds of PHOU are converted to carboxylic groups, and the oxidation proceeds to completion with little backbone degradation [35]. It is worth saying that these modification reactions cause a severe decrease in molar masses of the PHA.

PHO and PCL films are not degraded at 37°C at pH 7.3 over 275 days. In the case of PHO75COOH25 films and copolymers P((HOCOOH)-*b*-CL), the presence of the carboxyl group promotes water penetration into the polymer and participates in ester group hydrolysis through better water penetration and catalysis. Results showed that the adhesion of cells was better on PHO75COOH25 films than in PHO films. This result can be explained by the presence of carboxyl groups at the films' surface which promoted cell adhesion [36].

The presence of carboxyl groups at the films' surface promotes cell adhesion. The biodegradation of PHO and PCL films are not degraded in aqueous solution at 37°C at pH 7.3 for nine months while PHOU-COOH is degradable because water penetration into polymer increases. Figure 3.2 shows the comparison between biodegradability of PHOU-COOH with some polymers.

3.5 EPOXIDATION

Epoxidation of the double bonds is very useful tool to obtain amphiphilic PHA. PHU epoxidation with m-chloroperbenzoic acid yields to quantitative conversions of the unsaturated groups into epoxy groups. There is no side reaction on the macromolecular chain by molecular weight measurements. It has also been possible to produce new functional bacterial polyesters containing terminal epoxy groups in the side chains, in variable proportions up to 37% by growing *P. oleovorans* on a 10-epoxyundecanoic acid and sodium octanoate culture feed mixture. But the chemical epoxidation reaction was widely used and shown that these compounds were totally epoxidized by using meta-chloroperbenzoic acid (MCPBA) [37–39].

3.6 THE POLYCATIONIC PHA

Epoxide groups can be opened by diethanol amine in order to obtain a water-soluble hydroxyl derivative of the PHOU, resulting in the polymer

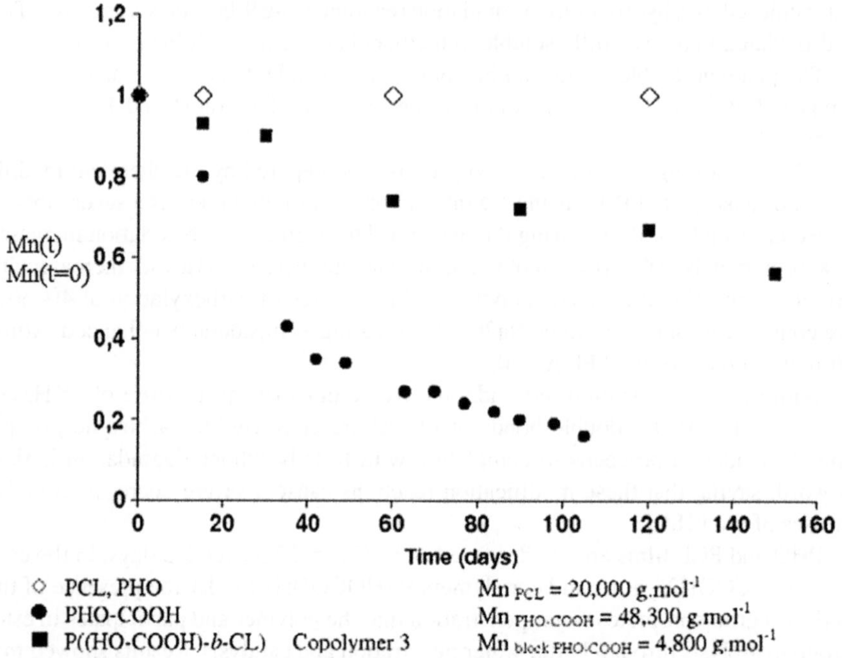

FIGURE 3.2 Plots of the molar masses versus time in aqueous solution. Hydrolytic degradation of the different polymers at pH 7.3 at 37°C [36].

poly(3-hydroxy-octanoate)-*co*-(3-hydroxy-11-(bis(2-hydroxyethyl)-amino)-10-hydroxyundecanoate) (PHON) [40]. PHON binds and condenses the DNA into positively charged particles smaller than 200 nm (Figure 3.3). In this manner, PHON protects plasmid DNA from nuclease degradation for up to 30 min. In addition, treatment of mammalian cells *in vitro* with PHON/DNA complexes results in a luciferase expression as the result of the delivery of the encoded gene [41].

Epoxy groups are chemically modified via the attachment of a peptide sequence such as Arg-Gly-Asp (RGD), to obtain biomimetic scaffolds. For tissue engineering, immobilizing the RGD sequence with aliphatic polyesters, amine groups of RGD were grafted onto the epoxidized PHOU through epoxy-amine chemistry.

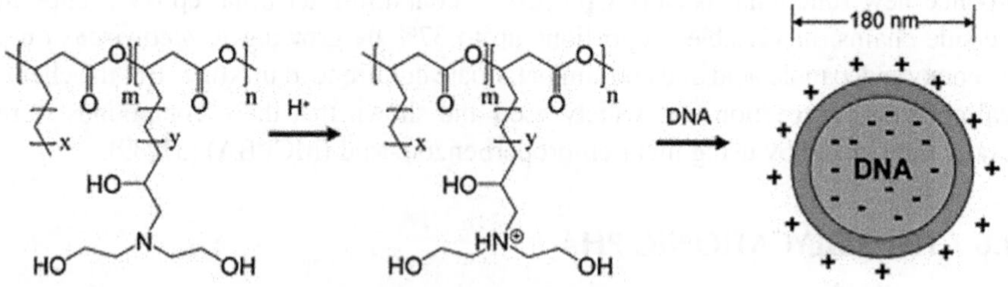

FIGURE 3.3 Formation of PHON/DNA complex [41].

The biological response of these new functional PHA scaffolds helps have more effect on cellular response [42].

3.7 PHOU WITH PENDANT PEG UNITS

PHA-PEG amphiphilic polymers find a wide variety of medical applications. For a medical application, adsorption of blood proteins and platelets on the modified PHA, PEG-g-PHU networks, are studied by Chung *et al.* [43]. Poly(ethylene glycol)-grafted poly(3-hydroxyundecenoate) (PEG-g-PHU) networks are prepared by irradiating the solution of PHU and the monoacrylate of poly(ethylene glycol) with UV light. The obtained cross-linked PHU and PEG-g-PHU result in lower adhesion of blood proteins and platelets. Blood compatibility increases as the PEG block increases.

Unsaturated *mcl*-PHA are also obtained by growing *Pseudomonas oleovorans* from the unsaturated fatty acids of the polyunsaturated plant oils [44]. Figure 3.4 shows the synthesis of the two types of unsaturated copolymers.

The side chain olefinic groups of the PHA can be converted to cross-linked PHA-g-PEG amphiphilic graft copolymers. For this purpose, a macroazo initiator derived from PEG and azo-bis-cyano pentanoyl chloride, was reacted with this unsaturated PHA [45]. This free radical addition of the PEG blocks on the unsaturated PHA results in the cross-linked PHA-g-PEG conjugates amphiphilic graft copolymers.

Synthesis of a PHOU-PEG comb-type graft copolymer can be carried out by the polyesterification reaction between carboxylic acid derivative of PHOU and PEG-OH using the catalyst system dicyclohexyl carbodiimide and dimethyl amino

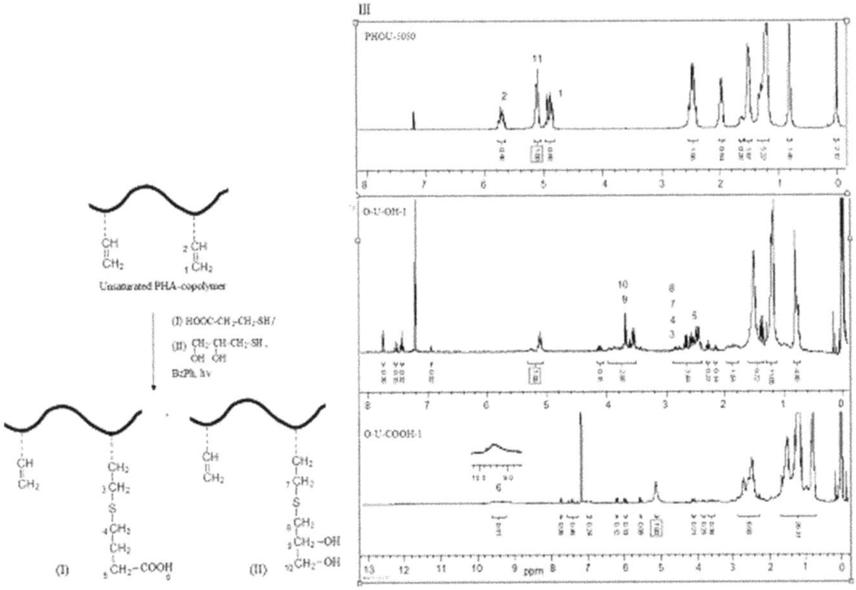

FIGURE 3.4 Synthesis of two types of unsaturated PHA from *Pseudomonas oleovorans* (I) grown on soybean oil (PHA-Sy) and (II) 10-undecenoic acid and octanoic acid (PHOU) [46].

pyridine. The graft copolymer is much more resistant to hydrolysis, at physiological pH, than PHOU-COOH. Water contact angles of PHOU-COOH and PHOU-PEG are very close to water contact angle of PHO [46].

3.8 CLICK REACTIONS

3.8.1 THIOL-ENE CLICK REACTIONS

Thiol-ene click reactions to double bonds are very attractive for producing carboxyl and hydroxyl derivatives of the unsaturated PHA without the polyester degradation. To do this, a methylene chloride solution of mercapto propionic acid or 3-thioglycerol in the presence of a photocatalyst in a Pyrex tube is irradiated by a mercury lamp. The same click reaction is also carried out by using the same thiols with the *mcl*-PHA obtained from soybean oil [47]. The PHOU-PEG amphiphilic copolymer leads to the formation of multi-compartment micelles [48]. Figure 3.5 shows the reaction design of the carboxylic acid (I) and hydroxyl (II) derivatives of PHOU and their 1H NMR spectra.

Glycopolymers are emerging as a novel class of neoglycoconjugates useful for biological studies. Thiosugar, a maltose-containing thiol group, is a potent tool in glycobiology. In order to obtain the anti-Markovnikov adduct, the thiol end of the thiol sugar is added to the double bond of the PHOU in the presence of diethylamine [49].

Poly(3-hydroxyoctanoate-*co*-3-hydroxyundecenoate) (PHOU) is methanolyzed and its unsaturated side chains are quantitatively oxidized to carboxylic acid. Alkyne-containing "clickable" PHA is obtained by the esterification with propargyl

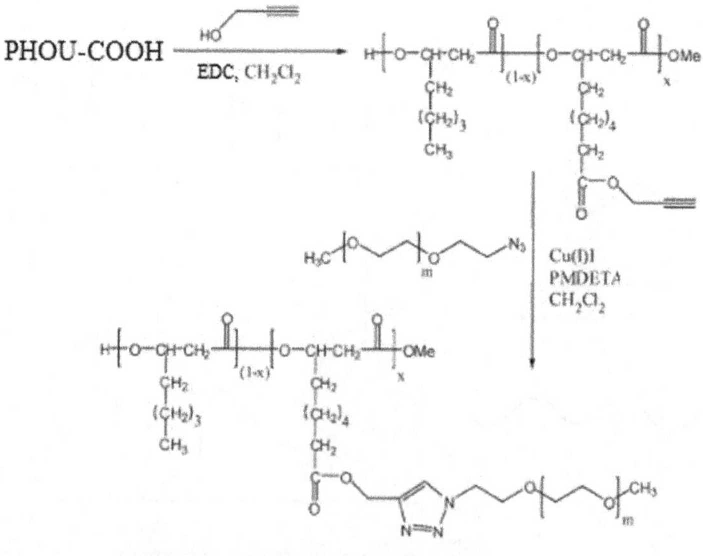

FIGURE 3.5 Reaction design of the carboxylic acid (I) and hydroxyl (II) derivatives of PHOU and their ^{1}H NMR spectra [47].

alcohol [50]. A click reaction of propargyl terminated PHOU with the azide terminated PEG leads to a PHOU-PEG graft copolymer. Figure 3.6 renders the synthesis of a PHOU-g-PEG graft copolymer using an azide-acetylene click reaction.

An unsaturated *mcl*-PHA, poly(3-hydroxyoctanoate-*co*-3-hydroxynonenoate), PHONe, was sulfonated by a radical thiol-ene or thiol-yne reaction in the presence of sodium mercapto ethane sulfonate. Figure 3.7 shows the reaction steps to prepare a sulfonate derivative of the unsaturated PHA [51].

Amphiphilic graft copolymers composed of PHA and thiol-appended PEG were synthesized by a thiol-ene addition [52]. Nanoscale polymersomes with a diameter of 63 nm and a membrane thickness of 8–9 nm were fabricated. Such a nanoparticulate system was assessed in encapsulation properties and biodegradability for potential use as delivery carriers.

A series of diblock copolymers based on a fixed poly(ethylene glycol) (PEG) block (5 000 g mol^{-1}) and a varying poly(3-hydroxyoctanoate-*co*-3-hydroxyhexanoate) (PHOHHx) segment (1 500–7 700 g/mol were synthesized using "click" chemistry. These copolymers self-assembled to form micelles in aqueous media. With increasing PHOHHx length, narrowly distributed spherical micelles with diameters ranging from 44 to 90 nm were obtained, with extremely low critical micelle concentration (CMC) of up to 0.85 mg/L [53].

A glycosaminoglycan-like marine exopolysaccharide, EPS HE800, was grafted to PHA in order to enhance cell adhesion. Novel graft copolymer HE800-g-PHA

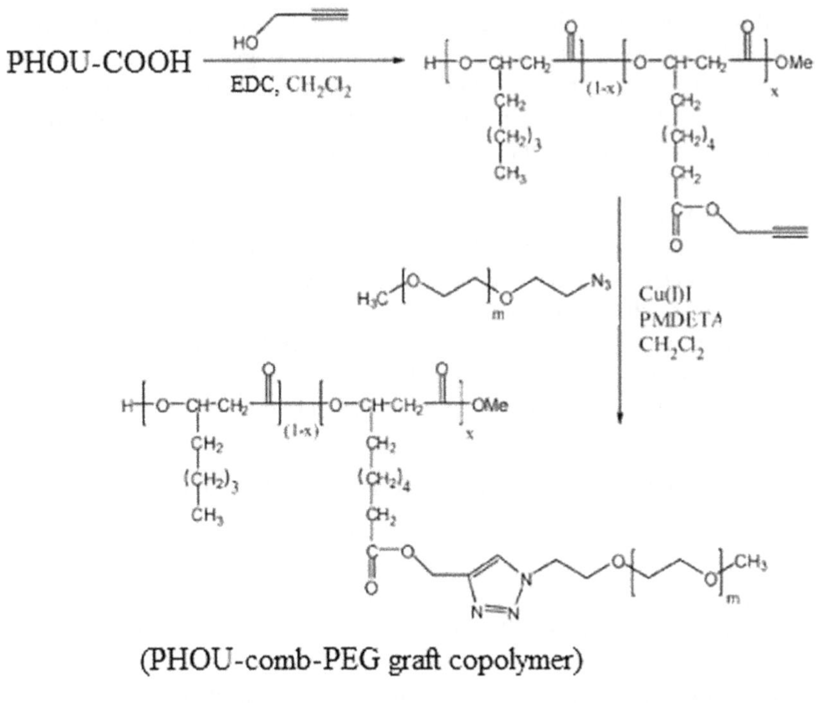

FIGURE 3.6 Synthesis of PHOU-g-PEG graft copolymer using azide-acetylene click reaction [50].

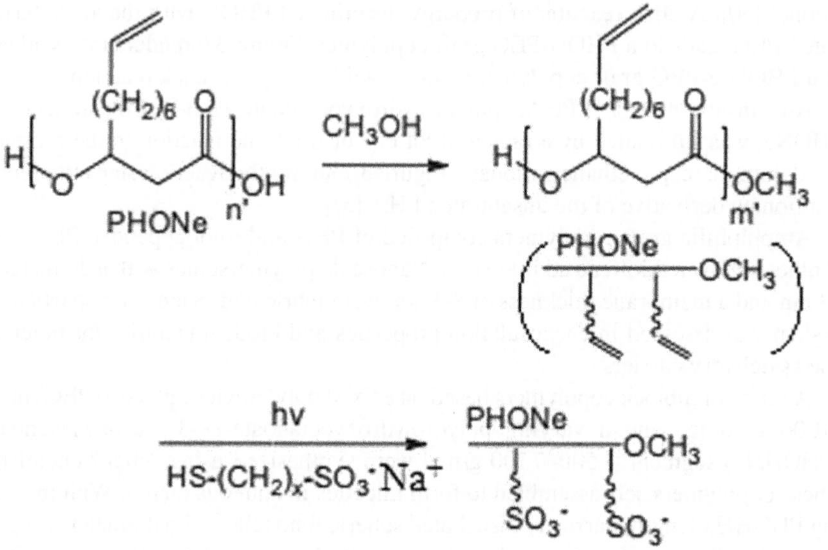

FIGURE 3.7 Synthesis of sulfonate derivative of PHOU using thiol-ene reaction with 3-mercapto-1-ethane sulfonate (redrawn from Ref. [51]).

was prepared to improve the compatibility between hydrophobic PHA and hydrophilic HE800. The carboxylic end groups of PHA oligomers were first activated with acyl chloride functions, allowing coupling to hydroxyl groups of HE800. Fibrous scaffolds were prepared by a modified electrospinning system which simultaneously combined PHA electrospinning and HE800-g-PHA copolymer electrospraying. Adhesion and growth of human mesenchymal stem cells on the HE800-g-PHA scaffolds showed a notable improvement over those on PHA matrices [54]. The interactions between the polymer and one of the main biomembrane components, 1,2-dioleoyl-sn-glycero-3-phosphocholine (DOPC) was studied using the Langmuir monolayer technique and Brewster angle microscopy. The addition of lipid to a polymer film does not change the monolayer phase behavior; however, the interactions between these two materials are repulsive and fall in two composition-dependent regimes [55]. The representative design of the synthesis of some modification reactions of the unsaturated *mcl*-PHA is summarized in Figure 3.8.

3.9 SATURATED PHA WITH HYDROPHILIC GROUPS

Saturated PHA are mainly PHB, PHBV, and PHO without any functional groups, such as double bonds, which can be easily modified. Therefore, the modification reactions reported of the saturated PHA are very limited. Among them is transesterification with poly(ethylene glycol), adding functional groups and copolymerization with hydrophilic monomers.

Some ester group(s) of the PHA is exchanged with an alcohol in the transesterification process. Transesterification is carried out in melt or in solution. Hydroxylation

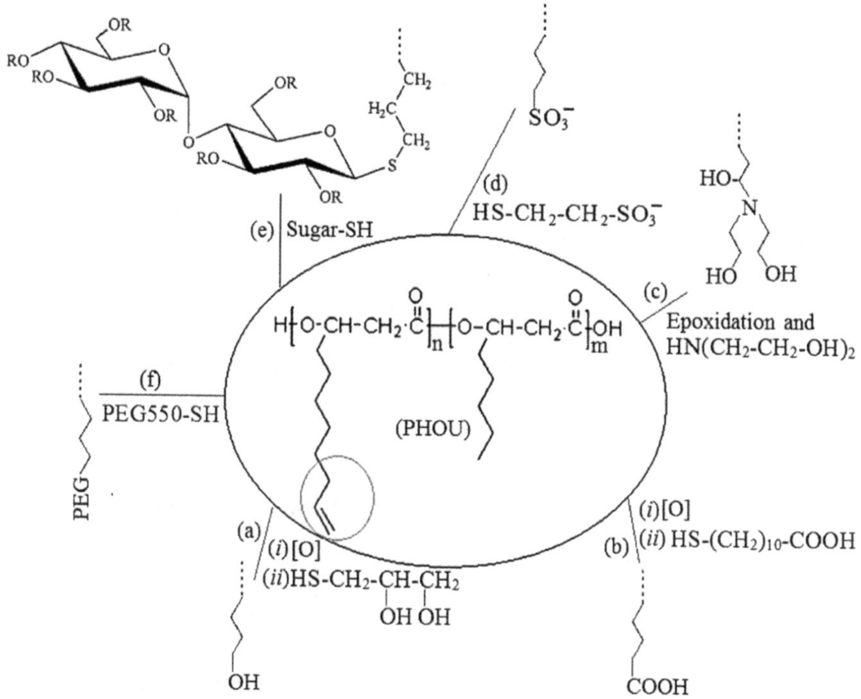

FIGURE 3.8 Representative design of the synthesis of some unsaturated *mcl*-PHA conjugates.

of the PHBs via chemical modification is usually achieved by the transesterification reactions to obtain diol ended PHB. Transesterification reactions in the melt between poly(ethylene glycol), mPEG, and PHB yield a diblock amphiphilic copolymer with a dramatic decrease in molecular weight. Catalyzed transesterification in the melt is used to produce diblock copolymers of poly([*R*]-3-hydroxybutyric acid), PHB, and monomethoxy poly(ethylene glycol), mPEG, in the presence of a catalyst, in a one-step process. The formation of diblocks is accomplished by the nucleophilic attack from the hydroxyl end group of the mPEG catalyzed by bis(2-ethylhexanoate) tin [56, 57].

PHB-*b*-mPEG diblock copolymers can self-assemble into nanoparticles which could find use as drug carriers, binders, and other specialty applications. Such drug carriers may show a longer lifetime in the bloodstream for they are robust versus dilution, that is, they have no critical micelle concentration because their core is hydrophobic and crystalline, and hence will not dissociate because of low concentration [58].

The refluxing of PHB and poly(ethylene glycol) bis (2-aminopropyl ether) with M_w 1 000 and 2 000 gave the block copolymers in a one-step transesterification reaction. According to the equivalency, AB and ABA type of block copolymers could be obtained.

PHB and PHU were microbially synthesized from *Alcaligenes eutrophus* (today: *C. necator*) fed with oleic acid and from *Pseudomonas oleovorans* fed with 10-undecenoic acid, respectively, according to the procedure cited in the literature. 1,2-Dichlorobenzene (DCB), 1,4-butanediol, poly(ethylene glycol) bis (2-aminopropyl ether) of average M_w 1 000 g/mol (PEG1KNH2) and M_w 2 000 g/mol (PEG2KNH2) were gifts from the Huntsman Co. (Switzerland) [59]. Transamidation reactions of PHB with primary amine-terminated poly(ethylene glycol) yield linear block copolymer of PHB with amine ends. PHO reacts with this amine-terminated PHB to give PHB-*b*–PEG-*b*–PHO block copolymers [60].

PEG is a polyether that is known for its exceptional blood and tissue compatibility. It is used extensively as biomaterial in a variety of drug delivery vehicles and is also under investigation as a surface coating for biomedical implants [25]. PHB-PEG amphiphilic polymers find wide variety of applications such as drug delivery systems [61] and adsorbents for metal cations [62].

3.10 POLYESTERIFICATION OF PHB-DIOL AND PEG-DIACID

Condensation polymerization of PHB-diol and PEG-diacid at equal molar amounts renders PHB-*alt*-PEG block copolymers. This polymerization is carried out in the presence of DMAP and DCC in dried methylene chloride [63]. PHB-*alt*-PEG alternating block copolymer can be seen in Figure 3.9.

Hydrogels can be prepared from cross-linked PEG-diacrylate and covered with PHB which is called a semi-interpenetrating network (IPN) hydrogel. IPN hydrogels are prepared by UV irradiation of the mixture of PEG-diacrylate and PHB. Net-PEG-based hydrogels all show higher equilibrium water contents (EWCs), while a remarkably decreased EWC is observed for the hydrogel containing 75% PHB. However, the crystallinity of PEG segments is noticeably decreased by cross-linking and would further drop with increasing amounts of PHB. Incorporation of a semi-IPN structure with PHB could significantly improve the mechanical properties of hydrogels when compared with those of pure net-PEG [64].

3.11 STIMULI RESPONSIVE PHA GRAFT COPOLYMERS

PHOU-g-PNIPAM comb-type thermoresponsive graft copolymer films show surface hydrophilicity improvement, thermoresponsive properties at different temperatures, and good biocompatibility for cell growth as well as thermoresponsive cell detachment ability [65, 66]. Chain transfer agent terminated PNIPAM is obtained by the

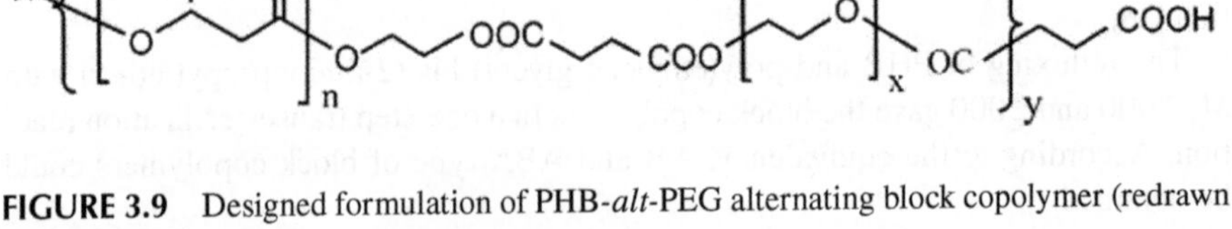

FIGURE 3.9 Designed formulation of PHB-*alt*-PEG alternating block copolymer (redrawn from Ref. [63]).

RAFT polymerization of NIPAM and then transformed to thiol-terminated PNIPAM, PNIPAM-SH, via aminolysis by the reaction with n-butyl amine. Then the unsaturated PHA, PHOU, is grafted with the PNIPAM-SH via a thiol-ene click reaction.

PNIPAAm-PHB-PNIPAAm triblock copolymers are non-toxic and soluble in water. Diol-terminated PHB is transformed to the dibromo-terminated PHB macroinitiator which is used in the atom transfer polymerization of NIPAM in order to obtain a thermoresponsive triblock copolymer. Dibromo-terminated PHB is prepared from PHB-diol by the reaction of the terminal hydroxyl end groups of PHB-diol with 2-bromoisobutyryl bromide [67]. Paclitaxel (PTX) is a common drug for cancer therapy. But it suffers from drug resistance. An amphiphilic cationic polyester PHB-PDMAEMA was designed to show better cell biocompatibility and comparable gene transfection efficiency. The bromo-terminated PHB can also be used in the atom transfer polymerization of DMAEMA in order to obtain PHB-*b*-PDMAEMA cationic polyester to encapsulate chemotherapeutic paclitaxel in a hydrophobic PHB domain [68–70]. Figure 3.10 shows the schematic illustration of amphiphilic cationic PHB-PDMAEMA copolymer with nanoparticulated polyplex formation ability to encapsulate chemotherapeutic paclitaxel (PTX) in its hydrophobic core.

PHB-PDMAEMA diblock copolymer with the ability to co-deliver PTX and Nur77 could effectively inhibit the growth of the drug-resistant cancer cells [71]. Mono brominated PHB is used in the ATRP of the DMAEMA to obtain amphiphilic cationic AB-type block copolymer. The reaction scheme is rendered in Figure 3.11.

3.12 ENHANCED HYDROPHILICITY VIA RADICAL FORMATION ONTO SATURATED PHA

Free radicals can be formed on the PHA films leading to the free radical polymerization of the hydrophilic monomers (e.g., acryl amide, hydroxyethylmethacrylate). For example, PHO films are treated with plasma of different discharge powers (10–50 W) and then treated with acrylamide solutions in order to prepare films with hydrophilic surfaces. The acrylamide-grafted PHO can be used as cell-compatible biomedical applications [72]. In this manner, the number of Chinese hamster ovary

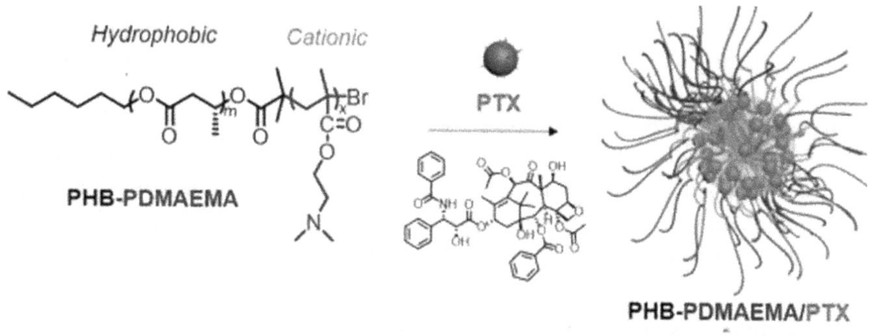

FIGURE 3.10 Schematic illustration of amphiphilic cationic PHB-PDMAEMA copolymer [69].

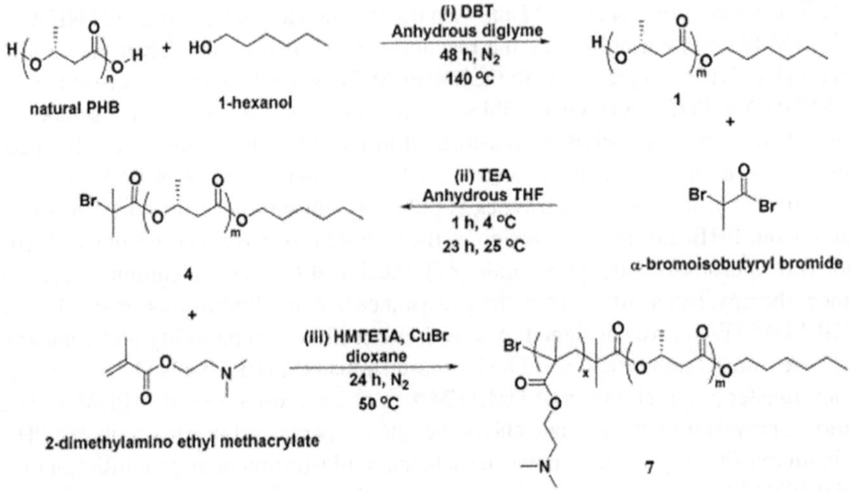

FIGURE 3.11 Chemical reaction route of PHB-*b*-PDMAEMA cationic copolymer: (i) transesterification of natural PHB with hexanol, (ii) bromo esterification, and (iii) ATRP reaction of DMAEMA monomer, in the presence of HMTETA and CuBr [71].

cells is investigated and it is observed that they adhere to and grow on the film surfaces depending on the hydrophilicity degree.

Radical formation on PHBV can be carried out by thermally heating with benzoyl peroxide in the presence of acrylamide. Graft polymerization of acrylamide onto PHBV leads to an amphiphilic PHBV-g-PAAm graft copolymer [73, 74]. The Langlois group grafted a hydrophilic monomer, HEMA, onto PHBV film at 80°C in the presence of benzoyl peroxide as a thermally radical producing source. Wettability has been obviously improved by grafting a hydrophilic monomer such as HEMA for a high graft yield (>130%) [75].

3.13 PEG GRAFTING ONTO SATURATED *MCL*-PHA

Radical terminated PEG was grafted onto medium-chain-length PHA with a double bond in order to obtain Poly(3-hydroxyalkanoate)-g-poly(ethylene glycol) cross-linked graft copolymers. PHA with low concentration of the double bonds lead to a branched graft copolymer instead of the cross-linked one [76].

In order to obtain biomaterials for packaging, PHB is grafted onto cellulose using a simple reactive extrusion process in the presence of dicumyl peroxide. The grafting reaction reduces the crystallinity because the grafting reaction occurs in the amorphous region and also slightly in crystalline regions of both cellulose and PHB. So the smaller crystal sizes cause the decrease of brittleness of PHB [77].

3.14 OZONIZATION OF PHB AND PHBV

Chitosan is natural biocompatible cationic polysaccharide. When sticking to the bacterial cell wall, chitosan can suppress the metabolism of bacteria. When PHA films are ozonized in acidic aqueous media, the film surface is peroxidized and then oxidized to carboxylic acid (Figure 3.12). The PHA film with carboxylic acid is grafted with chitosan in order to obtain antibacterial biomaterial against *Escherichia coli, Pseudomonas aeruginosa*, methicillin-resistant *Staphylococcus aureus*, and *S. aureus*. Acrylic acid grafting increases the biodegradability with *Alcaligens faecalis*, whereas chitosan grafting reduces the biodegradability [78].

3.15 CHLORINATION OF PHA

Chloride derivatives of the PHA open new modification reactions. Chlorine gas is passed into the carbon tetrachloride solution of the PHA moiety such as PHB, PHO, and also PHOU in order to obtain chloride derivatives (e.g., PHB-Cl, PHO-Cl, etc.) [79]. Decrease in methyl signals in ^{1}H NMR spectra can be attributed to the chlorination starts from the methyl groups of the PHA. Quaternization of the chlorinated PHA with triethyl amine makes them soluble in methanol and diethyl ether. For another amphiphilic derivative of the PHB-Cl, sodium thiosulfate is reacted with PHB-Cl to obtain the sulfonate derivative of bacterial polyester (PHB-SO3) [80].

3.16 PHB GRAFT COPOLYMERS WITH NATURAL HYDROPHILIC BIOPOLYMERS

Hyaluronic acid (HA) (Figure 3.13) end capped PHA graft copolymers are successfully used to encapsulate hydrophobic drugs. In order to obtain these graft copolymers, PHA oligomers were obtained by refluxing the PHA with a mixture of glacial acetic acid and distilled water at around 105°C. The carboxylic end of the oligo-PHA was esterified with the hydroxyl group of the hyaluronan: Poly(3-hydroxybutyrate)-HA, Poly(3-hydroxyoctanoate)-HA, Poly(3-hydroxyoctanoate-*co*-3-hydroxydecanoate)-HA, and Poly(3-hydroxyoctanoate-*co*-3-hydroxydecanoate-*co*-3-hydroxydodecanoate)-HA. HA and oligo (3-hydroxyalkanoates) copolymers were [81].

Carboxylic acid ends of the partially depolymerized PHB and PHO react with the amine groups of chitosan via the amidation reaction at 90°C for 5 h [82]. Chitosan-g-PHBV graft copolymers shown in Figure 3.14 exhibit different solubility behavior such as solubility, insolubility, or swelling in 2%wt. acetic acid and in water as a function of the degree of substitution of NH_2, while pure chitosan does not swell in water.

$$R \xrightarrow[\text{in HCl}]{O_3} \text{R-OOH} \longrightarrow \text{RCOOH}$$

R-COOH + HO-chitosan ⟶ R-COO-chitosan

or

R-COOH + HO-chitooligosaccharide ⟶ R-COO-chitooligosaccharide

where R = PHB or PHBV

FIGURE 3.12 Synthesis of PHA-grafted chitosan [78].

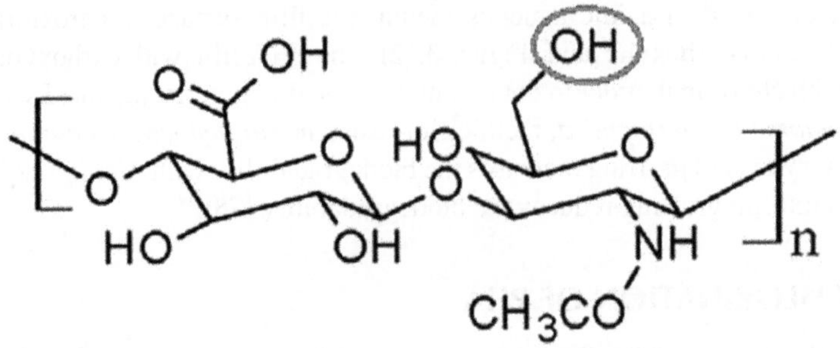

FIGURE 3.13 Hyaluronan (hyaluronic acid). The hydroxyl group in the circle reacts with the carboxylic acid end of the PHA oligomer.

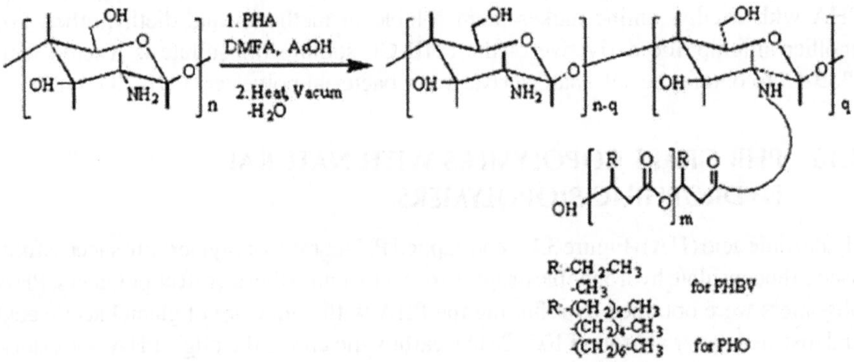

FIGURE 3.14 PHA-g-chitosan graft copolymer.

3.17 CONCLUSIONS AND OUTLOOK

PHA are excellent biodegradable materials for medical and industrial applications. In medical applications such as tissue engineering and drug delivery systems, their high hydrophobicity makes them limited usage. PHA-enhanced hydrophilicity is a very useful candidate for these medical applications. We have mentioned the hydrophilization process of the PHA which finds very different medical application areas. However, there are gradual increases in producing new amphiphilic PHA for these application areas. There is still a great challenge to diversifying PHA in view of the enhanced hydrophilicity including amphiphilic polymers, copolymers with hydrophilic monomers, and stimuli-responsive monomers.

ACKNOWLEDGMENTS

This work was supported by the Kapadokya University Research Fund (#KÜN.2018-BAGP-001) and Zonguldak Bülent Ecevit University, Faculty of Engineering.

REFERENCES

1. Anderson AJ, Dawes EA. Occurrence, metabolism, metabolic role, and industrial uses of bacterial polyhydroxyalkanoates. *Microbiological Reviews* 1990; 54:450–472.
2. Steinbuchel A, Valentin HE. Diversity of bacterial polyhydroxyalkanoic acids. *FEMS Microbiology Letters* 1995; 128:219–228.
3. Sudesh K, Abe H, Doi Y. Synthesis, structure and properties of polyhydroxyalkanoates: biological polyesters. *Progress in Polymer Science* 2000; 25:1503–1555.
4. Lenz RW, Marchessault RH. Bacterial polyesters: biosynthesis, biodegradable plastics and biotechnology. *Biomacromolecules* 2005; 6:1–8.
5. Steinbüchel A. Perspectives for biotechnological production and utilization of biopolymers: metabolic engineering of polyhydroxyalkanoate biosynthesis pathways as a successful example. *Macromolecular Bioscience* 2001; 1:1–24.
6. Koçer H, Borcaklı M, Demirel S, Hazer B. Production of bacterial polyesters from some various new substrates by *Alcaligenes eutrophus* and *Pseudomonas oleovorans. Turkish Journal of Chemistry* 2003; 27:365–373.
7. Chen GQ. A microbial polyhydroxyalkanoates (PHA) based bio- and materials industry. *Chemical Society Reviews* 2009; 38:2434–2446.
8. Chiellini E, Solaro R. Biodegradable polymeric materials. *Advanced Materials* 1996; 8:305–313.
9. Solaiman DKY, Ashby RD, Foglia TA, Marmer WN. Conversion of agricultural feedstock and coproducts into poly(hydroxyalkanoates). *Applied Microbiology and Biotechnology* 2006; 71:783–789.
10. Koray O, Koksal MS, Hazer B. Simple production experiment of poly (3-hydroxy butyrate) for science laboratories and its importance for science process skills of prospective teachers. *Energy Education Science and Technology Part B: Social and Educational Studies* 2010; 2(1–2):39–54.
11. Gross RA, Kalra B. Biodegradable polymers for the environment. *Science* 2002; 297:803–807.
12. Ishii-Hyakutake M, Mizuno S, Tsuge T. Biosynthesis and characteristics of aromatic polyhydroxyalkanoates. *Polymers* 2018; 10:1267. doi:10.3390/polym10111267
13. Hazer B, Steinbüchel A. Increased diversification of polyhydroxyalkanoates by modification reactions for industrial and medical applications. *Applied Microbiology and Biotechnology* 2007; 74:1–12.
14. Hazer DB, Kilicay E, Hazer B. Poly (3-hydroxyalkanoate)s: diversification and biomedical applications. A state of the art review. *Materials Science and Engineering C* 2012; 32:637–647.
15. Kai D, Loh XJ. Polyhydroxyalkanoates: Chemical modifications toward biomedical applications. *ACS Sustainable Chemistry and Engineering* 2013; 2:106–119.
16. Rai R, Keshavarz T, Roether JA, Boccaccini AR, Roy I. Medium chain length polyhydroxyalkanoates, promising new biomedical materials for the future. *Materials Science and Engineering R: Reports* 2011; 72:29–47.
17. Förster S, Antonietti M. Amphiphilic block copolymers in structure-controlled nanomaterial hybrids. *Advanced Materials* 1998; 10(3):195–217.
18. Wesslen B, Wesslen KB. Preparation and properties of some water-soluble, comb-shaped, amphiphilic polymers. *Journal of Polymer Science Part A: Polymer Chemistry* 1989; 27:3915–3926.
19. Riess G. Micellization of block copolymers. *Progress in Polymer Science* 2003; 28:1107–1170.
20. Balcı M, Allı A, Hazer B, Guven O, Cavicchi K, Cakmak M. Synthesis and characterization of novel comb-type amphiphilic graft copolymers containing polypropylene and polyethylene glycol. *Polymer Bulletin* 2010; 64:691–705.

21. Hedrick JL, Trollsas M, Hawker CJ, *et al.* Dendrimer-like star block and amphiphilic copolymers by combination of ring opening and atom transfer radical polymerization. *Macromolecules* 1998; 31:8691–8705.
22. Minoda M, Sawamoto M, Higashimura T. Amphiphilic block copolymers of vinyl ethers by living cationic polymerization. 3. Anionic macromolecular amphiphiles with pendant carboxylate anions. *Macromolecules* 1990; 23:1897–1901.
23. Zhang Z, Hadjichristidis N. Temperature and pH-dual responsive aie-active core crosslinked polyethylene-poly(methacrylic acid) multimiktoarm star copolymers. *ACS Macro Letters* 2018; 7:886–891.
24. Erhardt R, Zhang M, Boker A, *et al.* Amphiphilic janus micelles with polystyrene and poly(methacrylic acid) hemispheres. *Journal of the American Chemical Society* 2003; 125:3260–3267.
25. Hazer B. Amphiphilic poly (3-hydroxy alkanoate)s: potential candidates for medical applications. *International Journal of Polymer Science* 2010; 423460. doi: 10.1155/2010/423460
26. Wang Y, Yin J, Chen GQ. Polyhydroxyalkanoates, challenges and opportunities. *Current Opinion in Biotechnology* 2014; 30:59–65.
27. Andrade AP, Witholt B, Hany R, Egli T, Li Z. Preparation and characterization of enantiomerically pure telechelic diols from mcl-poly[(*R*)-3-hydroxyalkanoates]. *Macromolecules* 2002; 35:684–689.
28. Bear M-M, Leboucher-Durand M-A, Langlois V, Lenz RV, Goodwin S, Guerin P. Bacterial poly-3-hydroxyalkenoates with in the side chains epoxy groups. *Reactive & Functional Polymers* 1997; 34:65–77.
29. Kim YB, Rhee YH, Lenz RW. Poly(3-hydroxyalkanoate)s produced by *Pseudomonas oleovorans* grown by feeding nonanoic and 10-undecenoic acids in sequence. *Polymer Journal* 1997; 29:894–898.
30. Lee MY, Park WH, Lenz RW. Hydrophilic bacterial polyesters modified with pendant hydroxyl groups. *Polymer* 2000; 41:1703–1709.
31. Renard E, Poux A, Timbart L, Langlois V, Guerin P. Preparation of a novel artificial bacterial polyester modified with pendant hydroxyl groups. *Biomacromolecules* 2005; 6:891–896.
32. Eroğlu MS, Hazer B, Öztürk T, Çaykara T. Hydroxylation of pendant vinyl groups of poly(3-hydroxy undec-10-enoate) in high yield. *Journal of Applied Polymer Science* 2005; 97:2132–2139.
33. Bear MM, Mallarde D, Langlois V, *et al.* Natural and artificial functionalized biopolyesters.II. Medium-chain length polyhydroxyoctanoates from *Pseudomonas* strains. *Journal of Environmental Polymer Degradation* 1999; 7:179–184.
34. Lee MY, Park WH. Preparation of bacterial copolyesters with improved hydrophilicity by carboxylation. *Macromolecular Chemistry and Physics* 2000; 201:2771–2774.
35. Stigers DJ, Tew GN. Poly(3-hydroxyalkanoate)s functionalized with carboxylic acid groups in the side chain. *Biomacromolecules* 2003; 4:193–195.
36. Renard E, Timbart L, Vergnol G, Langlois V. Role of carboxyl pendant groups of medium chain length poly(3-hydroxyalkanoate)s in biomedical temporary applications. *Journal of Applied Polymer Science* 2010; 117:1888–1896.
37. Park WH, Lenz RW, Goodwin S. Epoxidation of bacterial polyesters with unsaturated side chains. I. Production and epoxidation of polyesters from 10-undecenoic acid. *Macromolecules* 1998; 31:1480–1486.
38. Park WH, Lenz RW, Goodwin S. Epoxidation of bacterial polyesters with unsaturated side chains. II. Rate of epoxidation and polymer properties. *Journal of Polymer Science Part A: Polymer Chemistry* 1998; 36:2381–2387.
39. Ashby RD, Foglia TA, Solaiman DKY, *et al.* Viscoelastic properties of linseed oil-based medium chain length poly(hydroxyalkanoate) films: effects of epoxidation and curing. *International Journal of Biological Macromolecules* 2000; 27:355–361.

40. Sparks J, Scholz C. Synthesis and characterization of a cationic poly(β-hydroxy alkanoate). *Biomacromolecules* 2008; 9:2091–2096.
41. Sparks J, Scholz C. Evaluation of a cationic poly(β-hydroxyalkanoate) as a plasmid DNA delivery system. *Biomacromolecules* 2009; 10:1715–1719.
42. Ramier J, Boubaker MB, Guerrouache M, *et al.* Novel routes to epoxy functionalization of pha-based electrospun scaffolds as ways to improve cell adhesion. *Journal of Polymer Science Part A: Polymer Chemistry* 2014; 52:816–824.
43. Chung CW, Kim HW, Kim YB, *et al.* Poly(ethylene glycol)-grafted poly(3-hydroxy undecenoate) networks for enhanced blood compatibility. *International Journal of Biological Macromolecules* 2003; 32:17–22.
44. Hazer B, Torul O, Borcaklı M, *et al.* Bacterial production of polyesters from free fatty acids obtained from natural oils by *Pseudomonas oleovorans*. Bacterial production of polyesters from free fatty acids obtained from natural oils by *Pseudomonas oleovorans. Journal of Environmental Polymer Degradation* 1998; 6:109–113.
45. Hazer B, Lenz RW, Çakmaklı B, *et al.* Preparation of poly(ethylene glycol) grafted poly(3-hydroxy alkanoate)s. *Macromolecular Chemistry and Physics* 1999; 200:1903–1907.
46. Domenek S, Langlois V, Renard E. Bacterial polyesters grafted with poly(ethylene glycol): behaviour in aqueous media. *Polymer Degradation and Stability* 2007; 92:1384–1392.
47. Hazer B. Simple synthesis of amphiphilic poly(3-hydroxy alkanoate)s with pendant hydroxyl and carboxylic groups via thiol-ene photo click reactions. *Polymer Degradation and Stability* 2015; 119:159–166.
48. Babinot J, Renard E, Le Droumaguet B, *et al.* Facile synthesis of multicompartment micelles based on biocompatible poly(3-hydroxyalkanoate). *Macromolecular Rapid Communications* 2013; 34:362–368.
49. Constantin M, Simionescu CI, Carpov A, *et al.* Chemical modification of poly(hydroxy alkanoates). Copolymers bearing pendant sugars. *Macromolecular Rapid Communications* 1999; 20:91–94.
50. Babinot J, Renard E, Langlois V. Preparation of *clickable* poly(3-hydroxyalkanoate) (PHA): Application to poly(ethylene glycol) (PEG) graft copolymers synthesis. *Macromolecular Rapid Communications* 2010; 31:619–624.
51. Modjinou T, Lemechko P, Babinot J, *et al.* Poly(3-hydroxyalkanoate) sulfonate: From nanoparticles toward water soluble polyesters. *European Polymer Journal* 2015; 68:471–479.
52. Babinot J, Guigner JM, Renard E, *et al.* Hide researcherID and ORCID poly(3-hydroxyalkanoate)-derived amphiphilic graft copolymers for the design of polymersomes. *Chemical Communications* 2012; 48:5364–5366.
53. Babinot J, Guigner JM, Renard E, *et al.* A micellization study of medium chain length poly(3-hydroxyalkanoate)-based amphiphilic diblock copolymers. *Journal of Colloid and Interface Science* 2012; 375:88–89.
54. Lemechko P, Ramier J, Versace DL, *et al.* Designing exopolysaccharide-graft-poly(3 -hydroxyalkanoate) copolymers for electrospun scaffolds. *Reactive & Functional Polymers* 2013; 73:237–243.
55. Jagoda A, Ketikidis P, Zinn M, *et al.* Interactions of biodegradable poly([R]-3-hydroxy-10-undecenoate) with 1,2-dioleoyl-*sn*-glycero-3-phosphocholine lipid: A monolayer study. *Langmuir* 2011; 27:10878–10885.
56. Hirt TD, Neuenschwander P, Suter UW. Telechelic diols from poly[(*R*)-3-hydroxy butyric acid] and poly{[(*R*)-3-hydroxy butyric acid]-*co*-[(*R*)-3-hydroxyvaleric acid]}. *Macromolecular Chemistry and Physics* 1996; 197:1609–1614.
57. Ravenelle F, Marchessault RH. One-step synthesis of amphiphilic diblock copolymers from bacterial poly([*R*]-3-hydroxybutyric acid). *Biomacromolecules* 2002; 3(5):1057–1064.

58. Ravenelle F, Marchessault RH. Self-assembly of poly ([R]-3-hydroxybutyric acid)-*block*-poly(ethylene glycol) diblock copolymers. *Biomacromolecules* 2003; 4:856–858.
59. Erduranlı H, Hazer B, Borcaklı M. Post polymerization of saturated and unsaturated poly(3-hydroxy alkanoate)s. *Macromolecular Symposia* 2008; 269:161–169.
60. Hazer B, Akyol E, Şanal T, *et al.* Synthesis of novel biodegradable elastomers based on poly [3 - hydroxy butyrate] and poly [3 - hydroxy octanoate] via transamidation reaction. *Polymer Bulletin* 2019; 76:919–932.
61. Kilicay E, Erdal E, Hazer B, *et al.* Antisense oligonucleotide delivery to cancer cell lines for the treatment of different cancer types. *Artificial Cells Nanomedicine Biotechnology: An International Journal* 2016; 44:1938–1948.
62. Wadhwa SK, Tuzen M, Kazi TG, *et al.* Polyhydroxybutyrate-*b*- polyethylene glycol block copolymer for the solid phase extraction of lead and copper in water, baby foods, tea and coffee samples. *Food Chemistry* 2014; 152:75–80.
63. Li X, Liu KL, Li J, *et al.* Synthesis, characterization, and morphology studies of biodegradable amphiphilic poly[(R)-3-hydroxybutyrate]-*alt*-poly(ethylene glycol) multiblock copolymers. *Biomacromolecules* 2006; 7:3112–3119.
64. Hao JY, Deng XM. Semi-interpenetrating networks of bacterial poly(3-hydroxybutyrate) with net-poly(ethylene glycol). *Polymer* 2001; 42:4091–4097.
65. Ma Y-M, Wie D-X, Yao H, *et al.* Synthesis, characterization and application of thermoresponsive polyhydroxyalkanoate-graft-poly(N-isopropylacrylamide). *Biomacromolecules* 2016; 17:2680–2690.
66. Toraman T, Hazer B. Synthesis and characterization of the novel thermoresponsive conjugates based on poly(3-hydroxy alkanoates). *Journal of Polymers and the Environment* 2014; 22:159–166.
67. Loh XJ, Zhang Z-X, Wu Y-L, *et al.* Synthesis of novel biodegradable thermoresponsive triblock copolymers based on poly[(R)-3-hydroxybutyrate] and poly(N-isopropylacrylamide) and their formation of thermoresponsive micelles. Amorphous amphiphilic PHB. *Macromolecules* 2009; 42:194–202.
68. Loh XJ, Ong SJ, Tung YT, *et al.* Dual responsive micelles based on poly[(*R*)-3-hydroxybutyrate] and poly(2-(di-methylamino)ethyl methacrylate) for effective doxorubicin delivery. *Polymer Chemistry* 2013; 4:2564–2574.
69. Cheng H, Wu Z, Wu C, *et al.* Overcoming STC2 mediated drug resistance through drug and gene codelivery by PHB-PDMAEMA cationic polyester in liver cancer cells. *Materials Science & Engineering C* 2018; 83:210–217.
70. Loh XJ, Ong SJ, Tung YT, *et al.* Incorporation of poly[(R)-3-hydroxybutyrate] into cationic copolymers based on poly(2-(dimethylamino)ethyl methacrylate) to improve gene delivery. *Macromolecular Bioscience* 2013; 13:1092–1099.
71. Wang X, Liow SS, Wu Q, *et al.* Codelivery for paclitaxel and Bcl-2 conversion gene by PHB-PDMAEMA amphiphilic cationic copolymer for effective drug resistant cancer therapy. *Macromolecular Bioscience* 2017; 17:1700186.
72. Kim HW, Chung CW, Kim SS, *et al.* Preparation and cell compatibility of acrylamide-grafted poly(3-hydroxyoctanoate). *International Journal of Biological Macromolecules* 2002; 30:129–135.
73. Hu SG, Jou CH, Yang MC. Protein adsorption, fibroblast activity and antibacterial properties of poly(3-hydroxybutyric acid-co-3-hydroxyvaleric acid) grafted with chitosan and chitooligosaccharide after immobilized with hyaluronic acid. *Biomaterials* 2003; 24:2685–2693.
74. Lee HS, Lee TY. Graft polymerization of acrylamide onto poly(hydroxybutyrate-co-h ydroxy-valerate) films. *Polymer* 1997; 38(17):4505–4511.
75. Lao H-K, Renard E, Linossier I, *et al.* Modification of poly(3-hydroxybutyrate-*co*-3-hydroxyvalerate) film by chemical graft copolymerization. *Biomacromolecules* 2007; 8:416–423.

76. Hazer B, Lenz RW, Çakmaklı B, *et al.* Preparation of poly(ethylene glycol) grafted poly(3-hydroxyalkanoate)s. *Macromolecular Chemistry and Physics* 1999; 200:1903–1907.
77. Wie L, McDonald AG, Stark NM. Grafting of bacterial polyhydroxybutyrate (PHB) onto cellulose via in situ reactive extrusion with dicumyl peroxide. *Biomacromolecules* 2015; 16:1040–1049.
78. Hu, SG; Jou CH, Yang MC. Antibacterial and biodegradable properties of polyhydroxy alkanoates grafted with chitosan and chito oligo saccharides via ozone treatment. *Journal of Applied Polymer Science* 2003; 88:2797–2803.
79. Arkin AH, Hazer B, Borcakli M. Chlorination of poly-3-hydroxy alkanoates containing unsaturated side chains. *Macromolecules* 2000; 33:3219–3223.
80. Arkin AH, Hazer B. Chemical modification of chlorinated microbial polyesters. *Biomacromolecules* 2002; 3(6):1327–1335.
81. Huerta-Angeles G, Brandejsová M, Nigmatullin R, *et al.* Synthesis of graft copolymers based on hyaluronan and poly(3-hydroxyalkanoates). *Carbohydrate Polymers* 2017; 171:220–228.
82. Arslan H, Hazer B, Yoon SC. Grafting of poly(3-hydroxyalkanoate) and linoleic acid onto chitosan. *Journal of Applied Polymer Science* 2007; 103(1):81–89.

4 Bioactive and Functional Oligomers Derived from Natural PHA and Their Synthetic Analogs

Anabel Itohowo Ekere, Iza Radecka, Piotr Kurcok, Grazyna Adamus, Tomasz Konieczny, Magdalena Zięba, Paweł Chaber, Fideline Tchuenbou-Magaia, and Marek Kowalczuk

CONTENTS

4.1 INTRODUCTION

Polyhydroxyalkanoates (PHA) are biopolyesters with good biodegradability and biocompatibility with broad application potential. The composition of monomeric building blocks, microstructures, and supra-macromolecular architecture determine the chemo-mechanical properties of PHA, and thus their suitability for defined technological applications [1, 2]. PHA were first discovered in the cytosols of certain microorganisms as high molar mass polymers stored in the form of granules [1]. These high molar mass PHA have more than 1 000 3-hydroxyacid residue units (approximately 50–5 000 kDa) and are generally referred to as storage PHA (sPHA) [3]. In certain prokaryotic bacteria, PHA are synthesized intracellularly in the form of inclusion bodies in their cytosols with their main function being for carbon and energy [4]. Much later, a low molar mass form of PHA referred to as PHA oligomers was discovered and reported by Reusch and Sadoff [5]. Consequently, PHA have now been classified into high molar mass PHA and low molar mass PHA (PHA oligomers), based on their molar mass.

PHA oligomers (oligo-PHA) are low molar mass PHA consisting of a small number of 3-hydroxyacids repeat units, usually three or more 3-hydroxyacids residue units, but not more than 200 residue units [6]. They are built from the same monomer units as sPHA, but possess a much shorter chain. While sPHA are primarily formed in many prokaryotes (Eubacteria and Archaea), PHA oligomers are components of all prokaryotes and eukaryotes and are thus thought to be present in all living organisms [3].

Over the last few years, storage PHA have been extensively studied and considered a more attractive biomaterial in research with high future impact due to their desirable properties and extensive applications. Nevertheless, the insufficient yields and poor mechanical properties of PHA limit its widespread commercialization. This is partly due to their hydrophobic nature (a long-chain fatty hydroxyl acid molecule) that results in fewer functional groups [7]. As a result, their applications in other advanced areas of importance are limited. Thus, to expand their widespread commercialization and improve their potential applications, PHA would need to have appropriate hydrolytic stability and enhanced chemical functionalities. This would improve their mechanical properties, amphiphilic character, and surface structure to meet the requirements of their tailored application [7, 8]. To achieve this, PHA oligomers containing reactive functional end groups with controlled molar mass can be used as building blocks of new block biopolymers with enhanced properties [9]. Furthermore, the high biocompability, bioactive function, and block copolymerisation of PHA oligomers with other polymers have been discovered to exhibit desirable properties for valued-added biomaterial applications, especially in medicine for therapeutic use, cosmetology, and agrichemistry [2, 10, 11]. Moreover, several researchers have also identified and confirmed the various range of PHA oligomers with high potential for being a good source of reactive oligoesters [12, 13, 14, 15, 16]. Thus, with these developments, it is expected that there would be an amplified interest in the functionalization of PHA biomaterials for other novel applications.

The physical properties of PHA oligomers are chiefly dependent on the length of the chain [17]. Aside from being biodegradable, PHA oligomers also exhibit excellent biocompatibility as confirmed by their lack of toxicity. This is well demonstrated by the natural presence of relatively large amounts of low-molar mass poly(3-hydroxybutyrate) (PHB) and other related oligomers in the bloodstream of humans [18]. In addition, PHA oligomers are components of all living organisms including eukaryotic cells and mammals where they are widely distributed in various intracellular fluids and fractions of these cells. In most cases, a short-chain form of low molecular mass (oligo-PHA) PHA is complexed to other cellular macromolecules such as proteins and inorganic polyphosphates and thus referred to as complexed PHA (cPHA).

4.2 OLIGOMERS DERIVED FROM NATURAL STORAGE

4.2.1 Oligo-PHB

Oligo-PHB was first discovered by [5] in the cytoplasmic membrane of *Azotobacter vinelandii* and *Bacillus subtilis*, comprising about 100–200 monomeric units. Oligo-PHB is known to often form complexes with calcium ions and polyphosphate (polyP) and is suggested to be responsible for the uptake of DNA during transformation [3]. Oligo-PHB is also part of low-density lipoproteins in human plasma and thus is suggested to be involved in arteriosclerosis [19].

4.2.2 Complexed PHB

cPHB was first discovered in *Escherichia coli*, but it has ever since then been found to be a ubiquitous constituent of both prokaryotic and eukaryotic cells. They usually consist of a low number of 3HB units (<~30) [3, 20] and are widely distributed in numerous cell fractions mostly in complex with other bio-macromolecules including inorganic polyPs and proteins [21]. They are usually water-soluble and chloroform-insoluble as long as the protein molecule to which it is attached to is water-soluble [3]. The majority of cPHB is complexed covalently with proteins, while a small fraction of them are non-covalently complexed with calcium polyP [22]. There is evidence from several reporters that these polyP complexes play a major role in the acquisition of competence in *Escherichia coli*, where they form ion channels in their plasma membranes [23, 24]. In addition, the covalent modification of proteins by cPHB, referred to as PHBylation, could have an important effect on the physiological properties of proteins. A significant amount of cPHB in *E. coli* is dependent on the presence of a novel type of PHB synthase called YdcS, different from sPHB synthase [25].

4.2.3 Synthesis of Natural PHA Oligomers

Oligomeric PHA could be synthesized either through intracellular degradation of PHA yielding natural oligomers, extracellular degradation or chemical modifications to yield synthetic oligomers. These pathways have been identified and confirmed by several researchers [26–29]. The extracellular degradation and chemical modifications for oligomer synthesis have been reported by many studies [9, 30–32], but only a few studies have been published on the intracellular degradation of PHA [33, 34].

4.2.4 Intracellular Degradation of Storage PHA

Intracellular degradation is the natural means by which PHA oligomers are being synthesized; it is, however, the less common route by which oligomers are synthesized, as gathered from the literature. This is because of the minute quantity of oligomeric PHA in cells, and thus the cellular mechanism behind the intracellular synthesis of PHA oligomers is yet to be fully understood.

So far, PHB has been extensively used to study the intracellular metabolism of PHA. Intracellular sPHB when accumulated with some intact surface layer and some PHB binding proteins in bacteria cells are referred to as native PHB (nPHB) or amorphous PHB (aPHB) [35]. If, however, these PHB-producing cells die (e.g. after cell lysis or solvent extraction) with or without a damaged surface layer, they are released into the environment, denature, and turn out to be more or less crystalline (semicrystalline/paracrystalline state) [36]. Although the extracellular degradation of denatured crystalline, PHB (dPHB), has been amplified in numerous bacteria, only a little is known about the intracellular degradation of native PHB [36].

Based on reported *in vivo* studies, sPHBs are usually degraded intracellularly by PHB depolymerase (PhaZ1) to obtain natural PHA oligomers [34]. PhaZ1 has been found to occur only as a form bound to PHA granules in cells, with various 3-hydroxybutyrate (3HB) oligomers being its major hydrolytic products from the enzymatic degradation of amorphous PHB [35]. This *in vivo* degradation process as described for *Ralstonia eutropha* (today: *Cupriavidus necator*) starts in the chains of amorphous PHB molecules. In this process, several links are created in the chain by intracellular PHB depolymerase (iPHB-PhaZ1) bound to exclusively inclusion bodies [33, 37]. Consequently, medium-sized 3HB-oligomers (that still bind to the granules) are produced mostly due to their hydrophobicity. In addition to this, some loosened 3HB ends of PHB chains protruding from the granules are produced, as well as small amounts of short-chain 3HB-oligomers (3–5 units) which diffuse from the granules [33, 34]. However, as reported by several researchers, this is only applicable to native or amorphous PHB granules but not crystalline PHB granules [27, 38].

Asides *R. eutropha*, there have been other reports on the intracellular degradation of sPHB in other bacteria such as *Escherichia coli* [35, 39, 40]. In most of these reported studies, PhaZ1 of *Ralstonai eurtopha* H16 was cloned and just a few novel intracellular PhaZ genes were identified and characterized [27, 40, 41]. In a report by Saegusa and his colleagues [40], an intracellular poly[D(-3)-3-hydroxybutyrate] depolymerase gene (PhaZ) was cloned from *R. eutropha* and expressed in *E. coli.* The crude extract of *E. coli* containing PhaZ gene digested amorphous PHB granules and released majorly oligomeric D(-)-3-hydroxybutyrate with the addition of some monomers [40]. Furthermore, it was reported in recent studies that an increase in PhaZ activity during culturing of *A. vinelandii* led to a decrease in the average molecular weight of sPHB and an increase in the fraction of low molecular weight PHB [27, 38, 42]. Besides these studies, no other study to the best of our knowledge has fully elucidated the metabolism behind the intracellular accumulation of PHA oligomers.

4.2.5 Extracellular Degradation of Storage PHA

Storage PHA that were synthesized by humans or released from PHA-accumulating microorganisms can be easily cleaved to by extracellular PHB depolymerases and many researchers have described numerous examples of extracellular PHB depolymerases over the last few decades [43–46]. A good number of microbes, particularly bacteria and fungi in soil, sludge, and seawater have been reported to excrete extracellular PHA-degrading enzymes able to hydrolyze sPHA into water-soluble oligomers [29]. Subsequently, these organisms make use of the resulting products as nutrients within their cells. In one of these studies carried out by Sugiyama *et al.* [31], PHB depolymerases (PHBDPs) were purified from two different bacteria (*Rasltonia pickettii* T1 and *Acidovorax* sp. SA1) to investigate their degradation products, kinetic properties, and substrate specificity. It was discovered that for *Ralstonia pickettii* T1, an extracellular PHB depolymerases enzyme degrades extracellular PHB to numerous sized 3HB-oligomers, and for *Acidovorax* sp. SA1, an extracellular PHBDP hydrolyzes extracellular PHB to small 3HB-oligomers (dimer and trimer) [31]. In another study by Gebauer and Jendrossek [36], protease-treated PHA granules were degraded with extracellular PHA depolymerases isolated from *P. lemoignei* to obtain 3HB and 3HV oligomers. In addition, Utsunomia and his colleagues [30] also recently reported on the production of D-lactate-3HB oligomers (D-LAOs) by engineered *Escherichia coli* expressing a D-specific lactate polymerizing enzyme. The extracted oligomers in this study contained 45 mol-% LA and 55 mol-% 3HB, suggesting that the obtained D-LAOs-DEG are random copolymers of LA and 3HB. The secreted D-LAOs also possessed two hydroxyl end group that can be used as building blocks for the production of numerous LA-based polymers including LA-based poly(ester-urethane) (PEU) [30].

4.2.6 Extraction of PHA Oligomers

The detailed extraction of PHA oligomers from cells have been previously described by Reusch and Sadoff [47], which has also been modified by Seebach *et al.* and Suzuki *et al.* [48, 49].

Due to the low concentrations of cPHB in certain solvents comprising of either lipids only, or lipids with proteins, its physical properties are usually modified by these solvents. Thus, the solubility of cPHB is majorly dependent on the properties of its complexant, where cPHB may be found in the residue part of its complexant, in any of the wash solutions or in the aqueous hypochlorite supernatant [19]. Even when cPHB is isolated from samples with chloroalkanes, cPHB would most often fail to dissolve in boiling chloroform. Consequently, designing a general procedure for the isolation of pure cPHA has become impossible. Unlike in bacteria, where the isolation of PHB is mostly based on the relative stability of the polymer to alkaline hypochlorite, the isolation of cPHB using the same protocol is highly unpredictable [19, 50]. Thus, as suggested by Reusch [19], it is generally advisable to quantify cPHB before attempting to separate or isolate it.

Reusch and Sadoff [47] first reported the isolation of cPHB from competent *E. coli*, where total PHB was extracted with cold $CHCl_3$ to yield 9.6 μμg/g (wt/wt) wet mass of cPHB. Since then, only three pure cPHA have been successfully isolated. Seebach *et al.* [48] was able to extract cPHB from competent *E. coli* cells using an enhanced isolation procedure different from that used by Reusch and Sadoff [47]. Here, the pelleted cells were freeze-dried and the residue was extracted with dry $CHCl_3$, saturated with cold (4°C) SDS. The remaining insoluble material was boiled for a long period in $CHCl_3$. The extracts (both hot and cold) obtained afterward were purified by a Bio-Rad S-X3 bead treatment and further analyzed in a high-field NMR spectrometer. Both extracts showed signals of P(3-HB) with the hot extract containing significantly higher amounts. The hot extract was further purified by its continuous injection into a GPC system. Through this, chain lengths of cPHB were roughly assigned [48]. Seebach *et al.* [48] researched further to isolate cPHB component from spinach, beef-heart, human aortae, and mitochondria to obtain H-NMR detectable amounts. Through this study, it was established that cPHB extracted from *E. coli* and spinach leaves is composed of *(R)*-3-hydroxybutanoic acid moieties [48].

Further isolation of eukaryotic cPHA in its pure state from sugar beet (*Beta vulgaris L.*) has been described [49]. In this study two different methods were used to yield 500 μg (80 g beet powder) and 600 μg (100 g beet powder) of pure cPHA respectively (see Figure 4.1).

The highest cPHA yield was observed using method 2. This was probably due to the addition of a partition procedure where the methanol precipitate was partitioned between chloroform and water (2:1) and this must have led to further reduction of impurities.

It is important to note that for the isolation of low concentrations of cPHB from these samples, large amounts of samples were required.

4.2.7 PHA Oligomers in Eukaryotes

It is well known that sPHB is a ubiquitous biopolymer confined to certain prokaryotic bacteria. Low molecular weight PHB found complexed to other macromolecules (cPHB) has also been shown to be widely distributed in representative organisms of nearly all biological cells and phyla [51]. While it's been reported that PHA oligomers are present in all living compartments of prokaryotic cells, it has, however, only been extensively examined in the Prokaryotic cell, *Escherichia coli*. Wild type *E. coli* is itself not capable of producing sPHA under normal growth conditions, but it has been found capable of producing small amounts of low molecular weight PHB complexed to cell membranes within its plasma membranes and not within its cytosols [47].

Reusch [5] was the first to isolate this complex form of PHB from the cell membrane of competent *E. coli* while examining transitions in lipid phase membranes of *Azotobacter vinelandii* and *Bacillus subtilis* cells during the development of genetic competence. The increase in transformability was followed by the occurrence of a

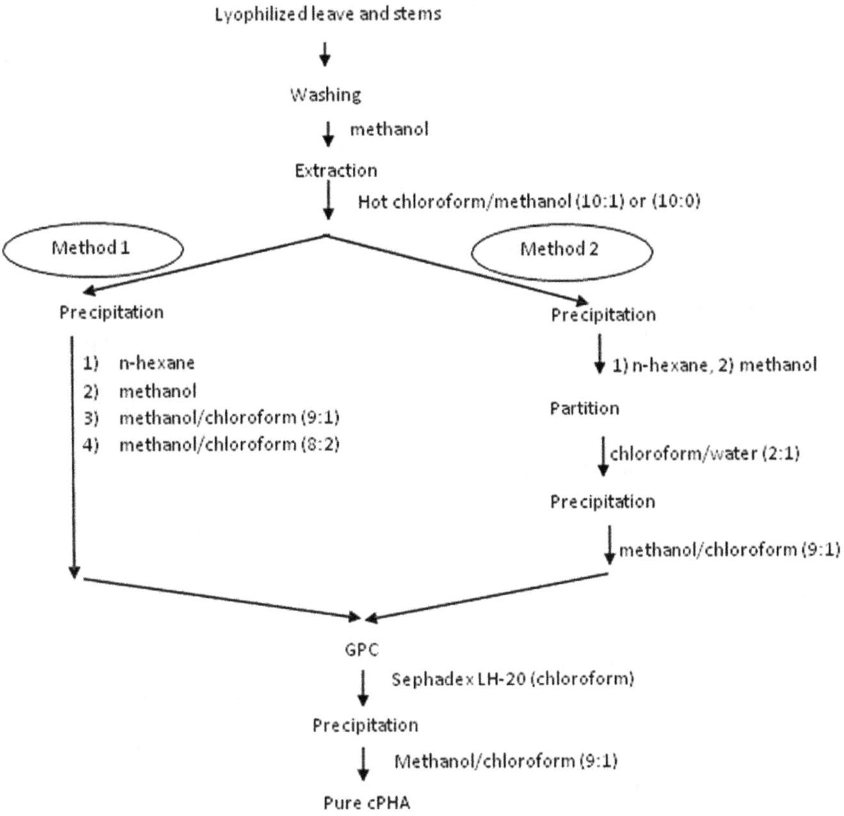

FIGURE 4.1 Isolation and purification outline of cPHA from sugar beet [49].

sharp and irreversible fluorescence peak corresponding to an increase in the concentration of PHB in the plasma membranes of *E. coli*. Naturally, PHB is not known to form intramolecular bonds and thus not capable of forming the quasi-crystalline structure on its own as indicated by the sharp fluorescence peak. As a result, the PHB was thought to be complexed and its complexant was assumed to be water-soluble due to a significant decrease in peak intensity when the cells were washed, especially with chelating buffers [5, 23]. PHB extracted from *E. coli* contained the anionic polymer PPi and Ca^{2+} which were surmised to be complexants of PHB since these hydrophilic moieties are greatly insoluble in chloroform [51]. Furthermore, it was discovered in Reusch's study [51], that this oligomer is located in the inner membrane of *E. coli*, where it occurs together with the calcium salt of inorganic polyP. The structure, composition, and distribution of the cPHB suggest that it may play a role in the regulation of intracellular calcium and in calcium signaling, as well as permitting calcium, DNA, or phosphate transport across its inner membrane [52]. Since then, *E. coli* has become the most preferred organism for cPHB studies. This is because it does not form inclusion bodies with sPHB that would otherwise have been difficult to separate from low molar mass PHB.

Huang and Reusch [20] further studied the presence of PHB in *E. coli* and discovered that the PHB in this cell is found all through the cells complexed to $CaPolyP_i$ and primarily associated with proteins. Furthermore, it was detected in this study that the properties of the protein-associated PHB were different from those of cellular inclusion sPHB. More than 80% of the PHB detected in *E. coli* were found in the cytoplasm, with the greatest concentration being observed in the ribosomal fraction [20, 53].

In another study by Tsuge *et al.* [54], low molar mass PHA was synthesized by *R. eutropha* engineered with recombinant plasmid pBBREE32d13 carrying PHA synthase of *Aeromonas caviae*. Marine *Bacillus subtilis* and *Streptomyces lividans* were also detected to produce oligo-PHB in associations with calcium ions and polyP [55].

Since the regulation of calcium transport is being viewed as an important physiological function in bacteria, and $PolyP_i$ synthesis is also known to be widely distributed in eukaryotic systems, it was therefore important to investigate if cPHB also exists in eukaryotic membranes. To this end, there have been quite a few studies (49, 50, 52] on detecting the presence of cPHB in eukaryotic cells to which a number of eukaryotic cells have been investigated to synthesize and accumulate cPHB. These cells include yeast (*Saccharomyces cerevisiae*), sugar beet (*Beta vulgaris* L.), peanut, cat muscles, sheep (intestine), and spinach, further showing that oligomeric PHB has excellent biocompatibility in animals [22, 55, 56]. In these studies, the PHBs form complexes with polyphosphate, various lipids, or proteins and play a role in membrane transport, thus making its solubility performance unpredictable [48, 52, 57].

A survey carried out on plant and animal tissues by Reusch [52] showed that PHB is indeed synthesized by a wide variety of eukaryotic cells, further illustrating that PHB is nearly ubiquitous in all cells. The long-chain-length PHB with high dispersity that forms inclusion bodies in some bacteria was not found in these studied eukaryotic cells. Meanwhile, it was short-chain-length PHB associated with $CapolyP_i$ that was omnipresent in these cells. Just like in *E.coli* and other reported bacteria known to accumulate cPHB, cPHB was also found to be in membrane fractions of these eukaryotes, as indicated by the intracellular distribution of this complex in bovine liver [52], although the complex formed was primarily in the mitochondria and microsomes. The eukaryotic $polyP_i$ chain length was a little greater (170–220) than that of bacterial $PolyP_i$ (130–170), although the eukaryotic cPHB had the same broad range of chain lengths (120–200 subunits) as those in bacterial membranes. As suggested by the ubiquitous nature of the complex, the structure of PHB-Ca-$polyP_i$ has an important physiological function. Furthermore, the intracellular distribution of the complex, its composition, and its structure also suggest its involvement in the storage and transport of Ca^{2+} and PO_4^{2-}, giving it the function of regulating intracellular Ca^{2+} and transmitting Ca^{2+} signals [52].

Previous work by Reusch [19] also showed that cPHB is present in minute quantities in human blood plasma (predominantly associated with low-density lipoproteins (LDL), human aorta, bovine heart, liver mitochondria, and bovine serum albumin [19, 48, 58]. Seebach *et al.* [48] also reported their study on the isolation of cPHB from several eukaryotic organisms including yeast, spinach leaves, celery leaves, celery stalk, chicken liver, beef liver, lamb heart, beef heart, and beef brain.

To obtain more information on eukaryotic cPHA-biosynthesis mechanisms, Suzuki *et al.* [49] isolated a cPHA-component obtained from a complex with Ca-polyP from sugar beet (*Beta vulgaris* L.) and further determined its structure. This complex was discovered to be a homopolymer comprised of 3-hydroxybuyrate. Using MALDI MS, the number average molar mass and dispersity index of the isolated cPHA were determined to be M_n = 9 100 Da and $Đ$ = 1.01 respectively, indicating that beet cPHA has a somewhat lower molar mass than the already known cPHA of *Escherichia coli.* Beet cPHB had a shorter chain PHB of n = 106 with a much lower 3HV content than cPHA from *E. coli.* The structural analysis of the obtained cPHA in this study also revealed that 100 mol-% of the carboxyl end is free, about 30 mol-% of the hydroxyl end is also free with about 70 mol-% of this end being masked, and the hydroxyl end group is also masked by at a minimum six identified short-chain alkanedioic and alkanoic acids [49]. These findings confirmed the organism-dependent structural diversity of cPHA (see also Figure 4.2).

Suzuki *et al.* [21] also reported his findings from the isolation of cPHA from commercial baker's yeast. In his study, 2.6 mg of pure cPHA (referred to as yeast cPHA-1) was isolated from 850 g dry weight of the yeast cells, making it the second cPHA being purified from eukaryotic organisms. The cPHA-1 obtained was further characterized using ^{1}H-NMR which revealed four comonomers; 3-HV, 4-hydroxybutyrate (4-HB), 3-hydroxypropanoate (3-HP), and crotonate (CA). The presence of 3-HB, 3-HV, and very small units of CA was also confirmed by further GC-MS of cPHA-1 ethanolyzates. The amount of each comonomer in cPHA-1 was determined to be 1.03 mol-% of 3HV, 0.06 mol-% of 4-HB, 0.07 mol-% of 3-HP, and 0.02 mol-% of CA. Notably, as observed in this study, yeast cPHA-1 had an almost identical monomer content to beet cPHA (0.1 mol-% 4-HB, 0.01 mol-% CA) but had a ca. 10-fold greater 3HV content than beet cPHA [21]. The M_n of cPHA-1 was observed to be 57.4, which is much shorter than the chain length observed for beet cPHA. Significant signals indicating free-end 3-hydroxyl and carboxyl groups of cPHA-1 were also identified by the ^{31}P NMR spectrum [21]. The content of these groups corresponded to 17.5% of the molecules for the hydroxyl groups and 3.6% of the molecules for the carboxyl groups. These values also suggested the presence of cyclic molecules. It was also discovered in this study that changing the culture medium could directly influence the molecular weight and not the polydispersity of cPHA from baker's yeast.

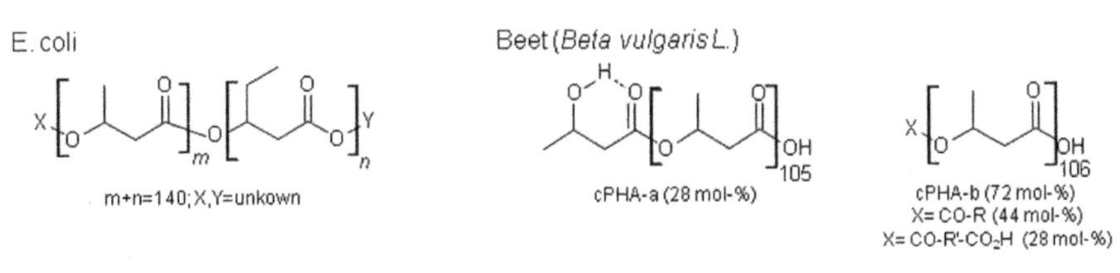

FIGURE 4.2 Chemical structures of beet cPHA and *E. coli* cPHA [49].

4.2.8 PHA Oligomers for the Structural Studies of Their Precursors by Mass Spectrometry

The microstructure of high-molecular-weight bacterial PHA has been originally determined by ^{13}C NMR based on diad and triad analysis. However, the significant information on PHA copolymers microstructure was provided by mass spectrometry. For this purpose, the theoretical mass spectra of PHA copolymers containing various compositions of repeat units were generated, and compared with those experimentally acquired with the aid of such "soft" MS ionization techniques as matrix-assisted laser desorption-ionization (MALDI), fast atom bombardment (FAB), desorption-chemical ionization (DCI), and electrospray-ionization mass spectrometry (ESI-MS) [59]. Moreover, the electrospray ionization multistage mass spectrometry (ESI-MS^n) has been successfully applied for verification of PHA microstructure. By means of these methodologies, the random comonomer arrangements have been confirmed in poly(3-hydroxybutyrate-*co*-3-hydroxyvalerate) (PHBV), poly(3-hydroxybutyrate-*co*-3-hydroxyhexanoate) (PHBH), copolymers, and recently for PHA with more than two different repeat units [60].

The ESI-MS/MS structural studies of bacterial PHA copolymers were based on the analysis of their oligomers obtained either by alkaline hydrolysis or controlled moderate-temperature degradation induced by carboxylate moieties. Using both of these approaches, oligomers containing carboxylic and olefinic end groups and the same composition and sequence distribution as the starting materials were obtained as revealed by ^{1}H-NMR, ^{13}C NMR, and ESI-MS/MS analysis [61].

4.2.9 Functionalization of PHA Oligomers Contained Unsaturated End Groups

PHA can be chemically modified to produce oligomers with the introduction of functional groups depending on the tailored functionality. So far, there have been several publications related to the various chemical modifications that can be employed for the production of synthetic oligomers [7, 26, 28].

4.2.10 Thermal Degradation

This is the most widely employed degradation process used for the production of synthetic PHA oligomers. It is the degradation of polymer chains due to overheating, where at high temperatures, this process results in volatile monomers, dimers, and trimers [7]. At moderately low temperatures (170–200°C), the degradation of PHA results in the production of a well-defined oligomer (usually 500–10000 g/mol macromolecule) with an unsaturated group (predominantly a trans-alkenyl end group) and a carboxylic group. In Yu's [62] patent, a controlled thermal degradation method was described to produce desired and well defined low molar mass PHA with functional end groups. In this study, a chemically stable substance (e.g., diethylene glycol, poly(ethylene glycol) or tetraethylene glycol) with a high boiling point of about 250°C was added to the reaction process to serve as a solvent or diluter during the thermal degradation process of high molar mass PHA. With this method, the window to obtain PHA oligomers with controlled molar mass is large, so that the degradation process occurs in a better controlled manner, with improved heat transfer and more efficient stirring. Due to the large reaction period, a desirable molar mass can be achieved [62].

4.2.11 Controlled Moderate-Temperature Degradation

Moderate-temperature degradation of poly(3-hydroxyalkanoate)s (3-PHA) can be induced by carboxylate groups according to ElcB mechanism (see Figure 4.3). In the case of 3-PHB with end groups in the form of carboxylic acid salts with Na^+, K^+, and Bu_4N^+ counterions, the degradation induced by intermolecular α-deprotonation by carboxylate was suggested. It was assumed that this process is the main 3-PHA decomposition pathway at moderate temperatures. Oligomers containing carboxylic and unsaturated end groups were formed in this way [63].

Moreover, the degradation of 3-PHA with selected salts of organic and mineral acids was investigated. The significant decrease in 3-PHA thermal stability in the presence of salts of weak Bronsted-Lowry acids was observed due to an anionic degradation reaction proceeding via an ElcB mechanism. Furthermore, continuous poly(3-hydroxybutyrate) controlled degradation was developed by a moderate-temperature process using carbonic acid salts as "initiators" of moderate-temperature degradation. Foamed 3-PHA oligomeric macromonomers, were obtained by a reactive extrusion process [64]

Furthermore, bacterial (P(3HB4HB)) degradation under mild reaction conditions induced by carboxylate salts leads to the thermal degradation of poly(3-hydroxybutyrate-*co*-4-hydroxybutyrate) and allows the obtaining of uniform linear oligomers. Moderate reaction temperature conditions and the presence of carboxylate salt allow for avoiding of the degradation process through a back-biting reaction which occurs even when a small amount of 4HB repeating units is present [65].

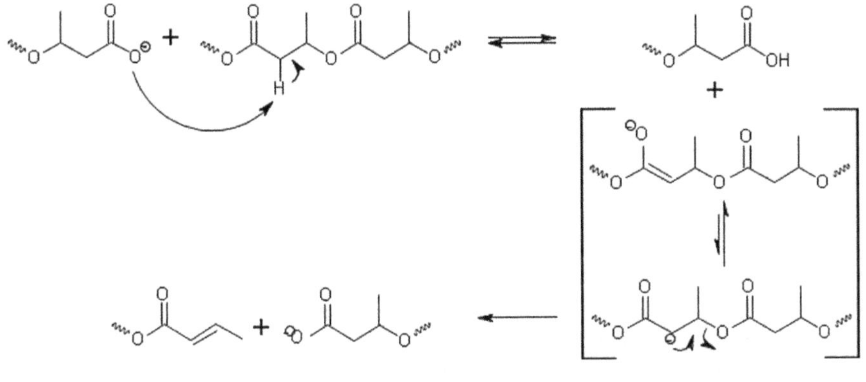

FIGURE 4.3 ElcB PHB degradation mechanism [63].

4.2.12 Thermooxidative Degradation of PHB

The thermal treatment of PHB in an oxygen/ozone mixture resulted in an increased rate of polymer backbone scission. The non-volatile degradation products contained macromolecules with several types of terminal groups, but also a part of the 3-hydroxybutyrate repeating units was transformed into 3-malic acid units. NMR and multistage MS characterization revealed the random distribution of 3-malic acid units in the oligomeric products as well as the content of the malic acid units being dependent on oxidation conditions [66].

4.2.13 Microwave-Assisted Degradation and UV Irradiation

Recently, microwave-assisted degradation of PHA has been employed as an inexpensive degradation route to PHA oligomers, since this method generates an efficient internal heating system by direct coupling of microwave energy with molecules [67, 68]. Here, sPHA samples are degraded in a microwave synthesis reactor operated at a desired power and temperature levels [9]. Ramier *et al.* [9] used this approach at 200°C–220°C to obtain high yields of oligomers (Mn ≤1 000 g/mol) characterized by a carboxyl group at one chain end and terminal crotonic ones on the other side. The obtained oligomeric PHA was also characterized by fluctuating molar masses in the shortest of times possible, about a 100 times faster rate than those obtained with conventional thermal degradation. This study further established that two microwave-assisted procedures could be successfully used: the first is irradiation with a constant temperature to obtain PHA oligomers with high yields and molar masses lower than 1 000 g/mol and the second is irradiation under constant power to afford oligoesters with higher molar masses [9]. Shangguan *et al.* [69] made use of UV irradiation to obtain oligomers with reactive radical groups after the quick controlled degradation of poly(3-hydroxybutyrate-*co*-3-hydroxyhexanoate) (PHBHHx) [69, 70].

4.2.14 Acid/Base Catalyzed Hydrolysis/Methanolysis of PHA

In this method, PHA are subjected to hydrolysis in either acid or base, leading to a slow and well-controlled depolymerization of the aliphatic polyester chains. Timbart *et al.* [16] have reported the synthesis of well-defined *mcl*-PHA oligomers (Poly(3-hydroxyoctanoate) oligomers – PHO) from natural *mcl*-PHA using a basic hydrolysis, acid-catalyzed reaction with para-toluenesulfonoic acid monohydrate (APTS) and methanolysis catalyzed by H_2SO_4. The ^{1}HNMR spectrum showed that the PHO oligomers produced by basic degradation at pH = 14 contained an unsaturated end group. However, more efficient production of PHO oligomers was observed with acid-catalyzed reaction and methanolysis. A MALDI-TOF spectra of PHO oligomers obtained with acid-catalyzed reaction showed the formation of linear oligomers having a hydroxyl group at one end and a carboxylic group at another end, with a very low proportion of cyclic structure. When Lukasiewicz and his colleagues [26] recently used an acid hydrolysis method, oligomeric PHA was obtained from natural *mcl*-PHA. The plasticizing effect of the obtained oligomeric PHA on natural PHB was also studied via characterization of mechanical and thermal properties of the blends during the course of aging at varying ambient conditions. The oligomeric PHA obtained in this study was further confirmed as a suitable biodegradable and biocompatible plasticizer for natural PHB, where plasticizing of PHB with oligomeric PHA resulted in a softer and more flexible PHB biomaterial [26].

4.2.15 Functionalization of PHA to Oligomers Containing Hydroxyl End Groups

Degradation of PHA via a reduction reaction with lithium borohydride leads to oligo(hydroxyalkanoate) diols (see Figure 4.4):

The structural characterization of the oligomers conducted using NMR and ESI-MS^n analyses confirmed that oligomers were terminated by two hydroxyl end groups. The reduction of the PHA occurred in a statistical way, regardless of the chemical structure of the comonomer units or of the microstructure of the polyester chain [71, 72]. This method can be used to synthesize various PHA oligodiols that were useful in the further synthesis of tailor-made biodegradable materials [73, 74].

4.2.16 Bioactive Oligomers of PHA

In a transesterification reaction, an alcohol molecule and an ester molecule react in the presence of acid; basically, one ester is transformed into another ester. This process leads to the synthesis of oligomers with either carboxyl or hydroxyl functionalities. Using this transesterification reaction, Kwiecień *et al.* [75] were able to develop two methods for the preparation of pesticide-oligomer conjugates. In the first method (one-pot), an MCPA-oligo (3HB-*co*-4HB) conjugate was synthesized using a transesterification reaction of poly(3-hydroxybutyrate-*co*-4-hydroxybutyrate) and P(3HB-*co*-4HB) biopolyester with (4-chloro-2-methylphenoxy) acetic acid (MCPA) in the presence of 4-toluenesulfonic acid monohydrate (see Figure 4.5).

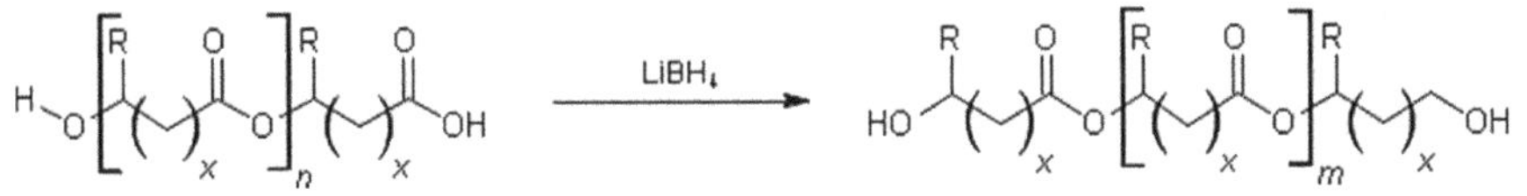

FIGURE 4.4 Reductive degradation of PHA [71].

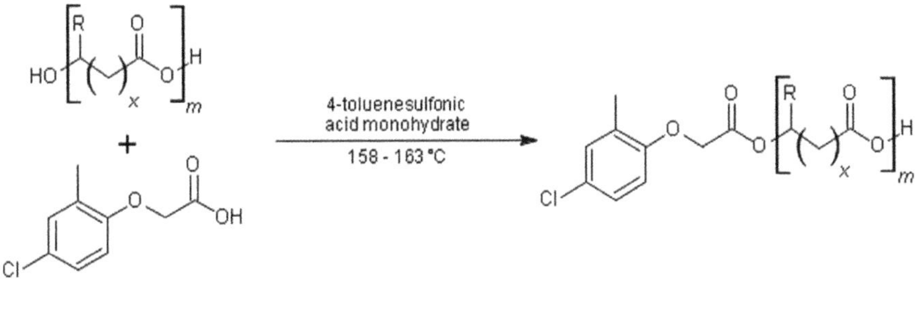

FIGURE 4.5 Transesterification reaction of P(3HB-*co*-4HB) by MCPA [75].

Oligo (3HB)-tyrosol conjugate was synthesized using a two-step approach. In the first approach, cyclic oligomers were obtained via a ring-closing depolymerization reaction and the cyclic oligomers were reacted with 4-(2-hydroxyethyl) phenol (tyrosol) in the presence of lipase from *Candida antarctica* (see Figure 4.6).

The synthetic route to prepare a PHB-amine conjugate containing a hydrolyzable imine bond was also reported. A short-chain PHB crotonate obtained by moderate-temperature degradation of natural polyester was converted into PHB glyoxylate via ozonolysis followed by reductive decomposition of peroxidic products with dimethylsulfide. Aldehyde-functionalized PHB was obtained quantitatively without polymer backbone degradation. The aldehyde-functionalized PHB can be a valuable biocompatible carrier for novel drug delivery systems [76].

4.2.17 Copolymers Containing PHA Building Blocks

Low-molar mass macroinitiators derived from natural PHA, containing unsaturated and activated by 18-crown-6 ether carboxylate end groups, were used in anionic ring-opening polymerization (ROP) of racemic β-butyrolactone, and new diblock copolymers of selected PHA (PHB, PHBV, PHO) with atactic poly[(R,S)-3-hydroxybutyrate] (aPHB) were obtained. The suitability of these polymeric materials for cardiovascular engineering and as blend compatibilizers was demonstrated [77].

Block copolyesters containing PHA and poly(D,L-lactide) structural units were obtained with a microwave-assisted process [9]. The PHA oligomers used for this purpose were obtained by acid-catalyzed methanolysis of corresponding native PHA.

The amphiphilic hyaluronan (HA) grafted with poly(3-hydroxyalkanoates) (PHA) oligomers, were obtained by a "grafting to" strategy [28]. PHA oligomers containing carboxylic terminal moieties were prepared by partial hydrolysis of PHA in acetic acid/water system. Such graft copolymers can be physically loaded with hydrophobic drugs and may serve as drug delivery systems.

Nguyen [78] had earlier suggested that PHB oligomers with carboxylic acid terminal group could be further subjected to modifications and graft polymerization

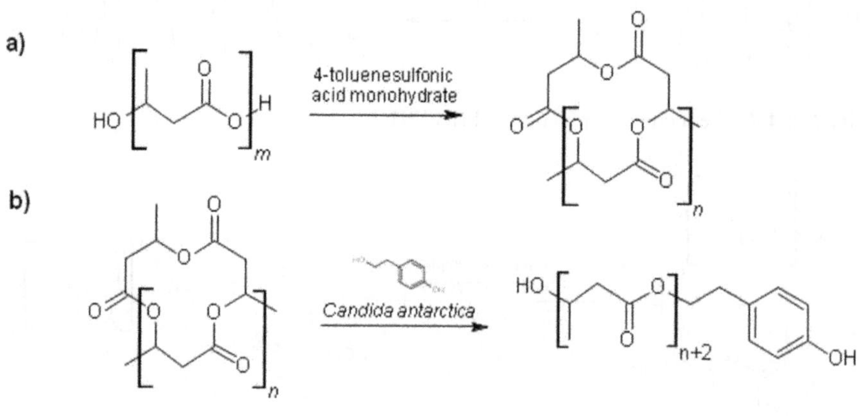

FIGURE 4.6 (a) Ring-closing depolymerization of PHB, (b) reaction between cyclic oligo (3-PHB) and tyrosol [75].

through a readily available radical mechanism in acrylic backbone and PHB side chain copolymers.

4.3 OLIGOMERS OF SYNTHETIC ANALOGS OF NATURAL PHA

4.3.1 Monodisperse Synthetic Analogs of PHA

Monodisperse oligomers of PHA, up to the 128 repeat units (molecular mass ca. 11 000 Da, as revealed by MALDI MS) were synthesized by fragment coupling (see Figure 4.7).

With these oligomers it was possible to calibrate the standards used for molar mass determinations of PHB by gel-permeation chromatography (GPC).

Using the same methodology as for the *isotactic* monodisperse PHB the *syndiotactic*, *atactic*, and *block* oligomers of PHB, with a given sequence of (*R*) and (*S*)-configuration of the 3HB units along the chain were prepared. Such non-natural PHA oligomers are important for studying the stereoselectivity of enzymatic PHB depolymerization [22].

4.3.2 Functional Oligomers of PHA Analogs

The facile synthesis of biomimetic predominantly isotactic oligo-(*R*)-3-hydroxybutyrate by region-selective anionic polymerization of *(S)*-β-butyrolactone initiated with 3-hydroxyacid sodium salt/crown ether complexes was described [79].

Such a synthetic analog of natural PHB was used for the preparation of the models of artificial channels in cell membranes. Das *et al.* [80] found that complexes of biomimetic OHBs with M_n 1670 and *Đ* 1.2 and with inorganic polyPs can be used to

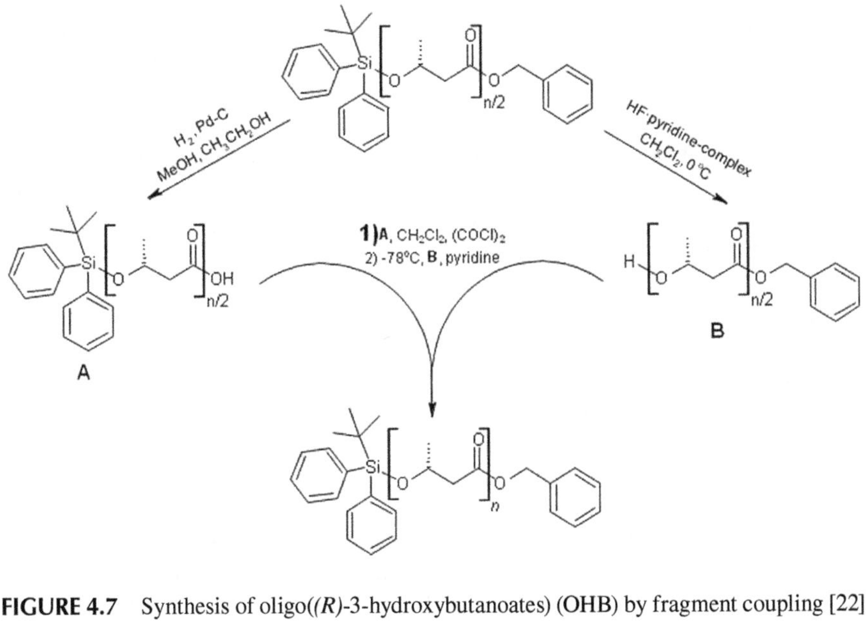

FIGURE 4.7 Synthesis of oligo((*R*)-3-hydroxybutanoates) (OHB) by fragment coupling [22].

create serviceable artificial cation channels of limited divalent cation selectivity in planar bilayers.

Piddubniak *et al.* [81] described the chemical synthesis and toxicity studies of well-defined tailor-made oligo-[R,S]-3-hydroxybutyrates (OHBs). The results indicate potential applicability of these oligomers as drug delivery carriers. Several OHBs with a number average molar mass (M_n) ranging from 800 to 2400 have been synthesized and tested on transformed hamster V79 fibroblasts and murine melanoma B16(F10) cells using the 3-[4,5-dimethylthiazol-2-yl]2,5-diphenyltetrazolium bromide (MTT)-based drug resistance and clonogenic survival assays. It was shown that 96 h incubation of cells with 1–9 mg/ml of OHBs did not affect cell viability. Moreover, incubation of OHBs with rat hepatoma FTO-2B cells stably transfected with a chloramphenicol acetyltransferase (CAT) gene ligated to a heat-inducible hsp70i gene promoter demonstrated that OHBs did not induce a cellular stress response. Furthermore, they demonstrate that doxorubicin conjugated with OHB is effectively taken up by murine melanoma B16(F10) cells *in vitro* and localizes in the cytoplasm. These data show for the first time that tailor-made biodegradable and biocompatible oligomers of 3-hydroxybutyric acid can be taken into consideration as effective, non-toxic vectors for the delivery of drugs in a conjugated form [81].

Elustondo *et al.* [82] investigated the ability of aPHB to interact with living cells and isolated mitochondria and the effects of these interactions on membrane ion transport using a fluorescein derivative of aPHB. The obtained results indicated that PHB demonstrates ionophoretic properties in biological membranes, and this effect is most profound in mitochondria due to the selective accumulation of the polymer in this organelle [82].

The transformation method of low molar mass crotonate-terminated poly(3-hydroxybutyrate)s (PHB), obtained by anionic ROP of β-butyrolactone [83] and controlled degradation of high-molar mass bacterial PHB, into mono- and di-epoxy-functionalized reactive poly(3-hydroxybutyrate)s via oxidation of crotonate end groups to α-3-methyloxirane-2-carboxylates, while carboxylate terminal groups functionalized by simple alkylation with epibromohydrin were described. Combining these two methods resulted in telechelic structures: α-3-methyloxirane-2-carboxylat e-ω-glicydyl PHB. Moreover, the reactivity of 3-methyloxirane groups of a functional aPHB was confirmed in experiments with primary amine and primary alcohol, revealing very high yield in reaction with alcohol, while degradation of polymer backbone was noticed during reaction with amine [84].

4.3.3 Bioactive Oligomers of PHA Analogs

Synthetically prepared PHA oligomers were found to be non-toxic and they may be used as carriers covalently bounded to appropriate bioactive compounds suitable for medical, cosmetic, agrichemical, and functional packaging applications.

The drug delivery systems were focused on penicillin G, acetylsalicylic acid, and ibuprofen [85, 86, 87]. Novel conjugates of the non-steroidal anti-inflammatory drug ibuprofen were synthetized by anionic ring-opening polymerization of (R,S)-β-butyrolactone initiated with an alkali metal salt of *(S)*-(+)-2-(4-isobutylphenyl) propionic acid (ibuprofen). Using the MTT cell proliferation assay, it was demonstrated that such conjugates exhibited significantly increased potential to inhibit proliferation of HT-29 and HCT 116 colon cancer cells when compared to free ibuprofen. Moreover, the conjugates of ibuprofen and OHB are less toxic as was shown in oral acute toxicity test in rats [87].

Incorporation of bioactive compounds into the β-lactones structure may lead to homo- and co-oligoesters with a bioactive moiety covalently linked as pendant groups along an oligomer backbone. This synthetic strategy was applied for preparation of the PHA synthetic analogs with ibuprofen pendant groups [88], pesticide moieties [32], and antioxidants used in cosmetics [89, 90, 91].

4.3.4 Copolymers Containing PHA Analogs

A synthetic aPHB oligomer possessing hydroxyl end groups, obtained in anionic polymerization of (*R*,*S*)-*β*-butyrolactone, was applied as a precursor for preparation of a coordination macroinitiator to be used for *ε*-caprolactone polymerization. The respective block copolymers were prepared in this way [92].

aPHB with a bishydroxy chain terminus was obtained through ring-opening polymerization of (*R*,*S*)-*β*-butyrolactone mediated by the activated anionic initiator 2,2-bis(hydroxymethyl)butyric acid (BHBA) tetrabutylammonium salt. Then, the polyester was modified to obtain a macroinitiator bifunctional species applicable in an ATRP process. Next, in an ATRP of methacrylic PEG macromonomer a water-soluble brush copolymer was synthesized [93].

Block copolymers of (*R*,*S*)-β-butyrolactone with pivalolactone (PVL) are prepared in order to define the effect of crystalline domains provided by poly(pivalolactone) on the biodegradability of atactic poly(β-butyrolactone), aPHB. The aPHB was synthesized from racemic β-butyrolactone, in the presence of a potassium alkoxide/18-crown-6 complex, and such a living polymer was applied for polymerization of PVL, yielding block copolymers, aPHB-*b*-PPVL, of tailored molecular weight and composition. While plain aPHB does not biodegrade, the biodegradation rate of aPHB-*b*-PPVL copolymers increases along with the increase of crystalline PPVL domains [94].

Graft copolymers were synthesized via an anionic grafting reaction of β-butyrolactone on poly(methyl methacrylate) (PMMA). Partially saponified PMMA bearing carboxylate anions complexed by 18-crown-6 potassium counterion acts as a macroinitiator of β-butyrolactone polymerization. As a result, graft polymers of PMMA with grafted poly(butyrolactone) side chains are produced in high yield over a wide graft composition range [95]. aPHB graft copolymers were also prepared via a macromonomer method, i.e., grafting through copolymerization [96]. Poly[(R,S)-3-hydroxybutyrate] telechelics were used for the synthesis of poly(methyl methacrylate)-*b*-poly(3-hydroxybutyrate) block copolymers [97]

Branched, aliphatic polyurethanes (PURs) were synthesized and compared to linear analogs. The influence of polycaprolactonetriol and synthetic aPHB oligomers in soft segments on structure, thermal, and sorptive properties of PURs was determined. It was found that increasing the aPHB amount in the structure of branched PURs reduced a tendency of urethane groups to hydrogen bonding. Thus, by controlling the number of branches and the amount of a-PHB in soft segments, thermal and absorptive properties of aliphatic PURs could be controlled [98].

4.4 CONCLUSIONS AND OUTLOOK

Polyhydroxyalkanoates (PHA) are a group of polyesters with excellent biodegradable and biocompatible properties. Additionally, their degradation products after hydrolysis are non-toxic, therefore making this biopolymer an attractive biomaterial for various end use applications. However, due to several undesirable physical properties of storage PHA (e.g., poor mechanical properties, poor solubility) and limited functionalities, the application of PHA in the other areas of importance is still limited. In recent years, significant efforts have been made to investigate a relationship between structure, properties, and function of advanced biodegradable polymeric materials in order to fulfill requirements for their specific novel applications in the future.

One way to increase the mechanical properties of PHA and expand their potential applications is to use PHA oligomers containing reactive functional end groups which could be used as building blocks for the preparation of new biopolymers with enhanced properties and high-value applications. PHA oligomers are low molecular weight PHA with a limited number of 3-hydroxyacids repeated units (max. 200 residual units). They are present in numerous intracellular fluids of all types of cells and in some cases, complexed to other cellular macromolecules including inorganic polyPs. They can also be synthesized either naturally in prokaryotic and eukaryotic cells or via several chemical modifications such as basic hydrolysis or transesterification. The chemical modification of storage PHA yields synthetic oligomers with the additional functional end groups, depending on the tailored functionality. These chemical modifications, such as controlled moderate-temperature degradation, thermal degradation, and thermo-oxidative degradation of PHB often leads to the synthesis of oligomers with unsaturated, carboxylic, or hydroxyl end groups.

The PHA oligomers derived either from natural aliphatic polyesters or their synthetic analogs could be used as carriers covalently bound to bioactive compounds. Polymeric controlled-release delivery systems of bioactive compounds can provide novel opportunities suitable for various applications in medicine, cosmetic industry, food packaging, or agri-chemistry.

REFERENCES

1. Lemoigne M. Production d'acide β-oxybutyrique par certaines bactéries du groupe du *Bacillus subtilis. Comptes Rendus de l'Académie des Sciences.* 1973; 176(1923): 1760–1761.
2. Koller M. Biodegradable and biocompatible polyhydroxy-alkanoates (PHA): Auspicious microbial macromolecules for pharmaceutical and therapeutic applications. *Molecules* 2018; 23(2): 362.
3. Jendrossek D, Pfeiffer D. New insights in the formation of polyhydroxyalkanoate granules (carbonosomes) and novel functions of poly (3-hydroxybutyrate). *Environ Microbiol* 2014; 16(8): 2357–2373.
4. Elustondo P, Zakharian E, Pavlov E. Identification of the polyhydroxybutyrate granules in mammalian cultured cells. *Chem Biodiv* 2012; 9(11): 2597–2604.
5. Reusch RN, Sadoff HL. D-(-)-poly-3-hydroxybutyrate in membranes of genetically competent bacteria. *J Bacteriol* 1983; 156(2): 778–788.
6. Martin DP, Peoples OP, Williams SF, *et al.* Nutritional and therapeutic uses of 3-hydroxyalkanoate oligomers. U.S. Patent 6,380,244. 2002.

7. Raza ZA, Riaz S, Banat IM. Polyhydroxyalkanoates: Properties and chemical modification approaches for their functionalization. *Biotechnol Prog* 2018; 34(1): 29–41.
8. Li Z, Loh XJ. Water soluble polyhydroxyalkanoates: Future materials for therapeutic applications. *Chem Soc Rev* 2015; 44(10): 2865–2879.
9. Ramier J, Grande D, Langlois V, *et al.* Toward the controlled production of oligoesters by microwave-assisted degradation of poly(3-hydroxyalkanoate)s. *Polym Degr Stab* 2012; 97(3): 322–328.
10. Adamus G, Kurcok P, Radecka I, *et al.* Bioactive oligomers from natural polyhydroxyalkanoates and their synthetic analogues. *Polimery* 2017; 62(4): 317–322.
11. Maksymiak M, Debowska R, Bazela K, *et al.* Designing of biodegradable and biocompatible release and delivery systems of selected antioxidants used in cosmetology. *Biomacromolecules* 2015; 16(11): 3603–3612.
12. Babinot J, Renard E, Langlois V. Controlled synthesis of well-defined poly (3-hydroxyalkanoate) s-based amphiphilic diblock copolymers using click chemistry. *Macromol Chem Phys* 2011; 212(3): 278–285.
13. Ajioka M, Suizu H, Higuchi C, *et al.* Aliphatic polyesters and their copolymers synthesized through direct condensation polymerization. *Polym Degr Stab* 1998; 59(1–3): 137–143.
14. Hiki S, Miyamoto M, Kimura Y. Synthesis and characterization of hydroxy-terminated [RS]-poly (3-hydroxybutyrate) and its utilization to block copolymerization with l-lactide to obtain a biodegradable thermoplastic elastomer. *Polymer* 2000; 41(20): 7369–7379.
15. Andrade AP, Neuenschwander P, Hany R, *et al.* Synthesis and characterization of novel copoly (ester-urethane) containing blocks of poly-[(R)-3-hydroxyoctanoate] and poly-[(R)-3-hydroxybutyrate]. *Macromolecules* 2002; 35(13): 4946–4950.
16. Timbart L, Renard E, Tessier M, *et al.* Monohydroxylated poly (3-hydroxyoctanoate) oligomers and its functionalized derivatives used as macroinitiators in the synthesis of degradable diblock copolyesters. *Biomacromolecules* 2007; 8(4): 1255–1265.
17. Klaerner G, Padmanabhan R. Multi-step/step-wise polymerization of well-defined oligomers. In: Hashmi S, Smithers G, Eds., *Reference Module in Materials Science and Materials Engineering*. Elsevier, 2018.
18. Freier T, Kunze C, Nischan C, *et al. In vitro* and *in vivo* degradation studies for development of a biodegradable patch based on poly (3-hydroxybutyrate). *Biomaterials* 2002; 23(13): 2649–2657.
19. Reusch RN. Biological complexes of poly-β-hydroxybutyrate. *FEMS Microbiol Rev* 1992; 9(2–4): 119–129.
20. Huang R, Reusch RN. Poly (3-hydroxybutyrate) is associated with specific proteins in the cytoplasm and membranes of *Escherichia coli*. *J Biol Chem* 1996; 271(36): 22196–22202.
21. Suzuki Y, Esumi Y, Koshino H, *et al.* Characterization of short-chain poly3-hydroxybutyrate in baker's yeast. *Phytochemistry* 2008; 69(2): 491–497.
22. Seebach D, Fritz MG. Detection, synthesis, structure, and function of oligo (3-hydroxyalkanoates): Contributions by synthetic organic chemists. *Int J Biol Macromol* 1999; 25(1–3): 217–236.
23. Reusch RN, Hiske TW, Sadoff HL. Poly-β-hydroxybutyrate membrane structure and its relationship to genetic transformability in *Escherichia coli*. *J Bacteriol* 1986; 168(2): 553–562.
24. Seebach D, Brunner A, Büger HM, *et al.* Channel-forming activity of 3-hydroxybutanoic-acid oligomers in planar lipid bilayers. *Helv Chim Acta* 1996; 79(2): 507–517.
25. Dai D, Reusch RN. Poly-3-hydroxybutyrate synthase from the periplasm of *Escherichia coli*. *Biochem Biophys Res Commun* 2008; 374(3): 485–489.
26. Lukasiewicz B, Basnett P, Nigmatullin R, *et al.* Binary polyhydroxyalkanoate systems for soft tissue engineering. *Acta Biomater* 2018; 71: 225–234.

27. Adaya L, Millán M, Peña C, *et al.* Inactivation of an intracellular poly-3-hydroxybutyrate depolymerase of *Azotobacter vinelandii* allows to obtain a polymer of uniform high molecular mass. *Appl Microbiol Biotechnol* 2018; 102(6): 2693–2707.
28. Huerta-Angeles G, Brandejsová M, Nigmatullin R, *et al.* Synthesis of graft copolymers based on hyaluronan and poly (3-hydroxyalkanoates). *Carbohydr Polym* 2017; 171: 220–228.
29. Shrivastav A, Kim HY, Kim YR. Advances in the applications of polyhydroxyalkanoate nanoparticles for novel drug delivery system. *BioMed Research International* 2013; 2013: 581684.
30. Utsunomia C, Saito T, Matsumoto KI, *et al.* Synthesis of lactate (LA)-based poly (ester-urethane) using hydroxyl-terminated LA-based oligomers from a microbial secretion system. *J Polym Res* 2017; 24(10): 167.
31. Sugiyama A, Kobayashi T, Shiraki M, *et al.* Roles of poly (3-hydroxybutyrate) depolymerase and 3HB-oligomer hydrolase in bacterial PHB metabolism. *Curr Microbiol* 2004; 48(6): 424–427.
32. Kwiecień I, Bałakier T, Jurczak J, *et al.* Molecular architecture of novel potentially bioactive (co) oligoesters containing pesticide moieties established by electrospray ionization multistage mass spectrometry. *Rapid Commun Mass Spectrom* 2015; 29(6): 533–544.
33. Guérin P, Renard E, Langlois V. Degradation of natural and artificial poly [(R)-3-hydroxyalkanoate]s: From biodegradation to hydrolysis. In: Chen G, Ed., *Plastics from Bacteria*. Berlin, Heidelberg: Springer 2010; pp. 283–321.
34. Kobayashi T, Uchino K, Abe T, *et al.* Novel intracellular 3-hydroxybutyrate-oligomer hydrolase in Wautersia eutropha H16. *J Bacteriol* 2005; 187(15): 5129–5135.
35. Tseng CL, Chen HJ, Shaw GC. Identification and characterization of the Bacillus thuringiensis phaZ gene, encoding new intracellular poly-3-hydroxybutyrate depolymerase. *J Bacteriol* 2006; 188(21): 7592–7599.
36. Gebauer B, Jendrossek D. Assay of poly (3-hydroxybutyrate) depolymerase activity and product determination. *Appl Environ Microbiol* 2006; 72(9): 6094–6100.
37. Kobayashi T, Shiraki M, Abe T, *et al.* Purification and properties of an intracellular 3-hydroxybutyrate-oligomer hydrolase (PhaZ2) in Ralstonia eutropha H16 and its identification as a novel intracellular poly(3-hydroxybutyrate) depolymerase. *J Bacteriol* 2003; 185(12): 3485–3490.
38. Arikawa H, Sato S, Fujiki T, *et al.* A study on the relation between poly (3-hydroxybutyrate) depolymerases or oligomer hydrolases and molecular weight of polyhydroxyalkanoates accumulating in *Cupriavidus necator* H16. *J Biotechnol* 2016; 227: 94–102.
39. Handrick R, Reinhardt S, Kimmig P, *et al.* The "intracellular" poly (3-hydroxybutyrate)(PHB) depolymerase of *Rhodospirillum rubrum* is a periplasm-located protein with specificity for native PHB and with structural similarity to extracellular PHB depolymerases. *J Bacteriol* 2004; 186(21): 7243–7253.
40. Saegusa H, Shiraki M, Kanai C, *et al.* Cloning of an intracellular poly [d (-)-3-hydroxybutyrate] depolymerase gene from *Ralstonia eutropha* H16 and characterization of the gene product. *J Bacteriol* 2001; 183(1): 94–100.
41. Abe T, Kobayashi T, Saito T. Properties of a novel intracellular poly (3-hydroxybutyrate) depolymerase with high specific activity (PhaZd) in *Wautersia eutropha* H16. *J Bacteriol* 2005; 187(20): 6982–6990.
42. Agus J, Kahar P, Hyakutake M, *et al.* Unusual change in molecular weight of polyhydroxyalkanoate (PHA) during cultivation of PHA-accumulating *Escherichia coli. Polym Degrad Stab* 2010; 95(12): 2250–2254.
43. Zaheer MR, Kuddus M. PHB (poly-β-hydroxybutyrate) and its enzymatic degradation. *Polym Adv Technol* 2018; 29(1): 30–40.

44. Volova TG, Prudnikova SV, Vinogradova ON, *et al.* Microbial degradation of polyhydroxyalkanoates with different chemical compositions and their biodegradability. *Microb Ecol* 2017; 73(2): 353–367.
45. Jendrossek D. Microbial degradation of polyesters. *Adv. Biochem. Eng. Biotechnol.* 2001; 71: 293–325.
46. Handrick R, Reinhardt S, Jendrossek D. Mobilization of poly (3-hydroxybutyrate) in *Ralstonia eutropha. J Bacteriol* 2000; 182(20): 5916–5918.
47. Reusch RN, Sadoff HL. Putative structure and functions of a poly-β-hydroxybutyrate/calcium polyphosphate channel in bacterial plasma membranes. *PNAS* 1988; 85(12): 4176–4180.
48. Seebach D, Brunner A, Bürger HM, *et al.* Isolation and 1H-NMR Spectroscopic identification of poly(3-hydroxybutanoate) from prokaryotic and eukaryotic organisms: Determination of the absolute configuration (R) of the monomeric unit 3-hydroxybutanoic acid from *Escherichia coli* and Spinach. *Eur J Biochem* 1994; 224(2): 317–328.
49. Suzuki Y, Esumi Y, Koshino H, *et al.* Isolation and structure determination of complexed poly (3-hydroxyalkanoate) from beet (*Beta vulgaris* L.). *Macromol Biosci* 2005; 5(9): 853–862.
50. Law JH, Slepecky RA. Assay of poly-β-hydroxybutyric acid. *J Bacteriol* 1961; 82(1): 33–36.
51. Reusch RN. Low molecular weight complexed poly (3-hydroxybutyrate): A dynamic and versatile molecule *in vivo. Can J Microbiol* 1995; 41(13) Supplement 1: 50–54.
52. Reusch RN. Poly-β-hydroxybutyrate/calcium polyphosphate complexes in eukaryotic membranes. *Proc Soc Exp Biol Med* 1989; 191(4): 377–381.
53. Nobes GA, Maysinger D, Marchessault RH. Polyhydroxyalkanoates: Materials for delivery systems. *Drug Deliv* 1998; 5(3): 167–177.
54. Tsuge T, Watanabe S, Sato S, *et al.* Variation in copolymer composition and molecular weight of polyhydroxyalkanoate generated by saturation mutagenesis of *Aeromonas caviae* PHA synthase. *Macromol Biosci* 2007; 7(6): 846–854.
55. Reusch RN. Streptomyces lividans potassium channel contains poly-(R)-3-hydroxybutyrate and inorganic polyphosphate. *Biochemistry* 1999; 38(47): 15666–15672.
56. Kavitha G, Rengasamy R, Inbakandan D. Polyhydroxybutyrate production from marine source and its application. *Int J Biol Macromol* 2018; 111: 102–108.
57. Abd-El-haleem D, Amara A, Zaki S, *et al.* Biosynthesis of biodegradable polyhydroxyalkanotes biopolymers in genetically modified yeasts. *Int J Environ Sci Technol* 2007; 4(4): 513–520.
58. Müller HM, Seebach D. Poly (hydroxyalkanoates): A fifth class of physiologically important organic biopolymers? *Angew Chem Int Ed* 1993; 32(4): 477–502.
59. Adamus G, Sikorska W, Kowalczuk M, *et al.* Sequence distribution and fragmentation studies of bacterial copolyester macromolecules: Characterization of PHBV macroinitiator by electrospray ion-trap multistage mass spectrometry. *Macromolecules* 2000; 33(16): 5797–5802.
60. Johnston B, Radecka I, Chiellini E, *et al.* Mass spectrometry reveals molecular structure of polyhydroxyalkanoates attained by bioconversion of oxidized polypropylene waste fragments. *Polymers* 2019; 11(10): 1580.
61. Kowalczuk M, Adamus G. Mass spectrometry for the elucidation of the subtle molecular structure of biodegradable polymers and their degradation products. *Mass Spectrom Rev* 2016; 35(1): 188–198.
62. Yu GE. Process of producing low molecular weight poly (hydroxyalkanoate) s from high molecular weight poly (hydroxyalkanoate)s. U.S. Patent 7,361,725. 2008.
63. Kawalec M, Adamus G, Kurcok P, *et al.* Carboxylate-induced degradation of poly (3-hydroxybutyrate)s. *Biomacromolecules* 2007; 8(4): 1053–1058.

64. Kawalec M, Sobota M, Scandola M, *et al.* A convenient route to PHB macromonomers via anionically controlled moderate-temperature degradation of PHB. *J Polym Sci Part A: Polym Chem* 2010; 48(23): 5490–5497.
65. Kwiecień M, Kawalec M, Kurcok P, *et al.* Selective carboxylate induced thermal degradation of bacterial poly (3-hydroxybutyrate-co-4-hydroxybutyrate)–Source of linear uniform 3HB4HB oligomers. *Polym Degrad Stab* 2014; 110: 71–79.
66. Michalak M, Kwiecień M, Kawalec M, *et al.* Oxidative degradation of poly (3-hydroxybutyrate). A new method of synthesis for the malic acid copolymers. *RSC Adv* 2016; 6(16): 12809–12818.
67. Kappe CO. Controlled microwave heating in modern organic synthesis. *Angew Chem Int Ed* 2004; 43(46): 6250–6284.
68. Galema SA. Microwave chemistry. *Chem Soc Rev* 1997; *26*(3): 233–238.
69. Shangguan YY, Wang YW, Wu Q, *et al.* The mechanical properties and *in vitro* biodegradation and biocompatibility of UV-treated poly (3-hydroxybutyrate-co-3-hydrox yhexanoate). *Biomaterials* 2006; 27(11): 2349–2357.
70. Gumel, AM, Aris, MH, Annuar, MSM. Modification of polyhydroxyalkanoates (PHAs). In: Roy I, Visakh PM, Eds., *Polyhydroxyalkanoate (PHA) Based Blends, Composites and Nanocomposites*. RSC, 2014. pp. 141–182.
71. Kwiecień M, Adamus G, Kowalczuk M. Selective reduction of PHA biopolyesters and their synthetic analogues to corresponding PHA oligodiols proved by structural studies. *Biomacromolecules* 2013; 14(4): 1181–1188.
72. Chaber P, Kwiecień M, Zięba M, *et al.* The heterogeneous selective reduction of PHB as a useful method for preparation of oligodiols and surface modification. *RSC Adv* 2017; 7(56): 35096–35104.
73. Michalak M, Kwiecien I, Kwiecien M, *et al.* Diversifying polyhydroxyalkanoates–end-group and side-chain functionality. *Curr Org Synth* 2017; 14(6): 757–767.
74. Kwiecień M, Kwiecień I, Radecka I, *et al.* Biocompatible terpolyesters containing polyhydroxyalkanoate and sebacic acid structural segments- synthesis and characterization. *RSC Adv* 2017; 7: 20469–20479.
75. Kwiecień I, Radecka I, Kowalczuk M, *et al.* Transesterification of PHA to oligomers covalently bonded with (bio)active compounds containing either carboxyl or hydroxyl functionalities. *PLoS One* 2015; 10(3): e0120149.
76. Michalak M, Marek AA, Zawadiak J, *et al.* Synthesis of PHB-based carrier for drug delivery systems with pH-controlled release. *Eur Polym J* 2013; 49(12): 4149–4156.
77. Adamus G, Sikorska W, Janeczek H, *et al.* Novel block copolymers of atactic PHB with natural PHA for cardiovascular engineering: Synthesis and characterization. *Eur Polym J* 2012; 48(3): 621–631.
78. Nguyen S. Graft copolymers containing poly (3-hydroxyalkanoates)—A review on their synthesis, properties, and applications. *Can J Chem* 2008; 86(6): 570–578.
79. Jedliński Z, Kurcok P, Lenz RW. First facile synthesis of biomimetic poly-(R)-3-hydroxybutyrate via regioselective anionic polymerization of (S)-β-butyrolactone. *Macromolecules* 1998; 31(19): 6718–6720.
80. Das S, Kurcok P, Jedlinski Z, *et al.* Ion channels formed by biomimetic oligo-(R)-3-hydroxybutyrates and inorganic polyphosphates in planar lipid bilayers. *Macromolecules* 1999; 32(26): 8781–8785.
81. Piddubnyak V, Kurcok P, Matuszowicz A, *et al.* Oligo-3-hydroxybutyrates as potential carriers for drug delivery. *Biomaterials* 2004; 25(22): 5271–5279.
82. Elustondo PA, Angelova PR, Kawalec M, *et al.* Polyhydroxybutyrate targets mammalian mitochondria and increases permeability of plasmalemmal and mitochondrial membranes. *PLoS One* 2013; 8(9): e75812.

83. Kurcok P, Matuszowicz A, Jedliński Z. Anionic polymerization of β-lactones initiated with potassium hydride. A convenient route to polyester macromonomers. *Macromol Rapid Commun* 1995; 16(3): 201–206.
84. Michalak M, Kawalec M, Kurcok P. Reactive mono-and di-epoxy-functionalized poly (3-hydroxybutyrate) s. Synthesis and characterization. *Polym Degrad Stab* 2012; 97(10): 1861–1870.
85. Adamus G, Kowalczuk M. Electrospray multistep ion trap mass spectrometry for the structural characterisation of poly [(R, S)-3-hydroxybutanoic acid] containing a β-lactam end group. *Rapid Commun Mass Spectrom* 2000; 14(4): 195–202.
86. Juzwa M, Rusin A, Zawidlak-Węgrzyńska B, *et al.* Oligo(3-hydroxybutanoate) conjugates with acetylsalicylic acid and their antitumour activity. *Eur J Med Chem* 2008; 43: 1785–1790.
87. Zawidlak-Węgrzyńska B, Kawalec M, Bosek I, *et al.* Synthesis and antiproliferative properties of ibuprofen–oligo (3-hydroxybutyrate) conjugates. *Eur J Med Chem* 2010; 45(5): 1833–1842.
88. Kowalczuk M, Adamus G, Kwiecień I, *et al.* New biodegradable (co)polyesters from β-lactone monomers containing biologically active substances. Patent No. PL-225906. 2016.
89. Maksymiak M, Bałakier T, Jurczak J, *et al.* Bioactive (co) oligoesters with antioxidant properties–synthesis and structural characterization at the molecular level. *RSC Adv* 2016; 6(62): 57751–57761.
90. Maksymiak M, Dębowska R, Jelonek K, *et al.* Structural characterization of biocompatible lipoic acid – oligo(3-hydroxybutyrate) conjugates by ESI-mass spectrometry. *Rapid Commun. Mass Spectrom.* 2013: 27(7): 773–783.
91. Maksymiak M, Kowalczuk M, Adamus G. Electrospray tandem mass spectrometry for the structural characterization of p-coumaric acid - oligo(3-hydroxybutyrate) conjugates. *Int J Mass Spectrom* 2014; 359: 6–11.
92. Kurcok P, Dubois P, Sikorska W, *et al.* Macromolecular engineering of lactones and lactides. 24. Controlled synthesis of (R, S)-β-butyrolactone-*b*-ε-caprolactone block copolymers by anionic and coordination polymerization. *Macromolecules* 1997; 30(19): 5591–5595.
93. Koseva NS, Novakov CP, Rydz J, *et al.* Synthesis of aPHB-PEG brush co-polymers through ATRP in a macroinitiator–macromonomer feed system and their characterization. *Des Monomers Polym* 2010; 13(6): 579–595.
94. Scandola M, Focarete ML, Gazzano M, *et al.* Crystallinity-induced biodegradation of novel [(R, S)-β-butyrolactone]-*b*-pivalolactone copolymers. *Macromolecules* 1997; 30(25): 7743–7748.
95. Kowalczuk M, Adamus G, Jedlinski Z. Synthesis of new graft polymers via anionic grafting of β-butyrolactone on poly(methyl methacrylate). *Macromolecules* 1994; 27(2): 5720–5725.
96. Neugebauer D, Rydz J, Goebel I, *et al.* Synthesis of graft copolymers containing biodegradable poly(3-hydroxybutyrate) chains. *Macromolecules* 2007; 40(5): 1767–1773.
97. Alli A, Hazer B, Adamus G, *et al.* Polyhydroxyalkanoates/polyhydroxybutyrates (PHAs/PHBs). In: Guillaume SM, Ed., *Handbook of Telechelic Polyesters, Polycarbonates, and Polyethers*. Singapore: Pan Stanford Publishing 2017; pp. 65–113.
98. Brzeska J, Elert A, Morawska M, *et al.* Branched polyurethanes based on synthetic polyhydroxybutyrate with tunable structure and properties. *Polymers* 2018; 10(8): 826.

Part II

Processing of PHA

5 Processing and Thermomechanical Properties of PHA

Vito Gigante, Patrizia Cinelli, Maurizia Seggiani, Vera A. Alavarez, and Andrea Lazzeri

CONTENTS

5.1 INTRODUCTION

Polyhydroxyalkanoates (PHA) are biodegradable aliphatic polyesters that can be produced biotechnologically by numerous naturally occurring microorganisms, with very different types of substrates, from fossil resources (methane, mineral oil, lignite, hard coal) to organic (sucrose, starch, cellulose, etc.), chemicals (propionic acid, 4-hydroxybutyric acid), and food by-products (molasses, whey, glycerol) [1]. The overfed bacteria produce PHA as an energy reserve, which is then consumed when the bacteria are short of a feeding substrate. The price of PHA is currently quite high (€4–5/kg). Some researchers seek to optimize production by using waste-based substrates for the growth of PHA and extraction processes with innovative technologies such as enzymes, ionic liquids, carbon dioxide, etc.

PHA find many uses in biomedical applications, but have been proposed even for single use and commodities applications, such as packaging, because of low oxygen permeability and easy biodegradability in different environments [2, 3].

PHA applicability is limited not just by high production costs, which are progressively being reduced year by year, but even by processing and performance limitations due to brittleness and low temperature stability in the molten state, with consequent ease of degradability during processing.

Thus, for PHA processing and extended applications to commodities and single-use items, a sound knowledge of PHA's physical, thermal, and rheological properties are essential to achieve a proper management of this challenging, but appealing, polymeric family.

The aim of this chapter is to review some relevant PHA characteristics, with particular relevance to PHA shortcomings such as thermomechanical degradation and brittleness, and correlation between thermal degradation and rheological properties. In fact, for PHA to become a friendly commodity polymer, these shortcomings must be addressed, through the development of processing methods which avoid degradation, and through modifications which prevent the formation of micro-fractures.

5.2 POLYHYDROXYALKANOATES: PHYSICAL PROPERTIES

The typical structure of a PHA polymer is shown in Figure 5.1, where "R" refers to the side group, whereas "m" refers to the number of "CH_2" groups. Both "R" and "m" determine the HA unit constituting the polymer chain.

The physical and chemical characteristics of PHA vary as a function of their varied monomer structure. There is a wide possible variation in the length and composition of the side chains "R" groups, and over 150 different types of hydroxyalkanoate (HA) monomers have been identified [4]. Factors affecting the monomer content include microorganisms (gram-negative or gram-positive), modes of fermentation (batch, fed-batch, continuous), media ingredients, fermentation conditions, and recovery [1].

The most studied PHA are the homopolymer poly(3-hydroxybutyrate) (PHB, a.k.a. P(3HB)) and the copolymer poly(3-hydroxybutyrate-*co*-3-hydroxyvalerate) (PHBV) where R is respectively a methyl and an ethyl group for the 3-hydroxybutyrate (3HB), and the 3-hydroxyvalerate (3HV) monomers.

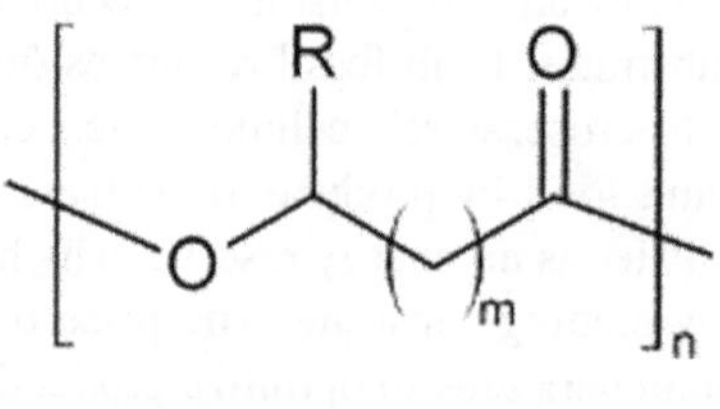

FIGURE 5.1 Chemical structure of PHA.

PHB is the linear polyester of D(-)-3-hydroxybutyric acid. It is a semi-crystalline isotactic stereo regular polymer with 100% *(R)*-configuration, which allows a high level of degradability. PHB is a thermoplastic polymer, with high crystallinity, high melting temperature, and good resistance to organic solvents. PHB possesses excellent mechanical strength and modulus, resembling that of poly(propylene), although its applications are hindered by several serious drawbacks, in particular, pronounced brittleness, very low deformability, high susceptibility to a rapid thermal degradation, and difficult processing by conventional methods.

Bacterial PHA can be broadly divided into three main types depending on the number of carbon atoms in the monomers. This category includes: short-chain-length PHA (*scl*-PHA), with monomers consisting of 3–5 carbon atoms, medium-chain-length PHA (*mcl*-PHA), with monomers consisting of 6–14 carbon atoms, and long-chain-length PHA, which are obtained from long chain fatty acids with more than 14 carbon atoms [5].

mcl-PHA are characterized by a much lower crystallinity, tensile strength, and melting temperature with respect to *scl*-PHAs, and are in general more flexible. So, the introduction of different HA monomers in the PHB chain, such as 3-hydroxyhexanoate (3HHx) can strongly influence and improve the material properties.

PHB is a linear polyester of D(-)-3-hydroxybutyric acid, thus is a semi-crystalline isotactic stereo regular polymer with 100% *(R)*-configuration that allows a high level of degradability. PHB molecular weight differs depending on the organism and conditions of growth and method of extraction, and for isolated PHB, it can vary from about 50000 to well over a million, with a polydispersity of about 2, and an average crystallinity rate of approximately 65% [6].

5.2.1 Properties Relevant for PHA Processing

Under crystallization from the melt, PHB crystal structure is the α-form, characterized by two antiparallel chains in the left-handed 2_1 helical conformation packed in an orthorhombic unit cell having axes a = 0.576 nm, b = 1.320 nm, and c (fiber axis) = 0.596 nm [7].

Mechanical properties of PHB are close to poly(propylene) (PP), and may outperform poly(ethylene) (PE) in some parameters. A main drawback for PHB is the low values for elongation at break, and the relevant brittleness. The reason for PHB brittleness arises mainly from the presence of large crystals in the form of spherulites that show radial and circumferential cracks, from which fractures can originate and propagate [8]. In addition, it has been reported that PHB undergoes further embrittlement during storage at room temperature, due to additional secondary crystallization [9]. Anyway, the high values of the strength and Young's modulus allow a large range of possible modifications to increase deformability and toughness.

This is detailed in Table 5.1, in which the mechanical properties of P(3HB), PP and low-density- and high-density-poly(ethylene), LDPE and HDPE, are also reported for comparison.

PHB presents several properties similar to PP such as Young's modulus and tensile strength while the elongation at break of PHB is markedly lower, resulting in PHB as a more brittle material than either PP, LDPE, and HDPE.

TABLE 5.1
Comparison of Some Thermal and Mechanical Properties of PHB, PLA, PP, LDPE, and HDPE (Adapted from [9, 10, 11])

	T_g [°C]	T_m [°C]	Average Crystallinity [wt.%]	Young's Modulus [GPa]	Tensile Strength [MPa]	Elongation at Break [%]
PHB	0	165	70	1.5–3.5	25–35	5–45
PLA	58	150	20	3.5	70	3
PP	-3	170	60	1.4	36	>500
LDPE	–36	110	50	0.2	10	>400
HDPE	–36	130	80	1.1	25	>500

T_g: Glass transition temperature, T_m: melting temperature, X_c: percentage of crystallinity.

Considering that the average crystallinity is quite high for all of these polymers, an influence on mechanical properties is ascribable even to organization and morphology of the crystals, and interconnection between the crystalline and the amorphous regions [10].

Moreover, PHB exhibits quite ductile behavior just after molding, but with storage at room temperature, it becomes very brittle. Since the PHB glass transition temperature is near to 0°C, at room temperature, polymer chains have sufficient mobility to organize into crystal structures. This process can last for many days, and the higher density of crystalline phases with respect to the amorphous phase induces cracks between the crystals. These structure modifications are thus responsible for the slow transition from a ductile freshly molded material to a very brittle material after storage. After one year of aging, de Koning and Lemstra observed a Young's modulus increase from about 1.5 GPa up to 3.5 GPa, whereas elongation at break decreased from about 45% to approximately 5% [9]. The increase was found to be linear with the logarithm of the aging time.

It is stated in the literature that one of the major drawbacks to most bio-based polymers, as PHA, is their potential thermal degradation under typical processing conditions [2, 11]. In fact, this thermal degradation typically occurs at lower temperatures than for conventional polymers [3, 12, 13]. PHBs are less thermally stable than other hydroxyl carboxylic acids, such as PLA and PCL [14].

5.3 PHA PROCESSING METHODS

The workability of PHA varies depending on the molecular weight, as well as on the primary structure. As reported before, PHB displays some properties similar to petroleum-based polymers, such as PP, as regards melting temperature (165°C) and relatively high tensile strength (25–35 MPa), but has low values of elongation at break and a narrow processing window, which is a main issue in PHB processing. Thus, the short permanence of PHB at a temperature just above its melting point induces degradation, with consequent production of degraded products from the olefinic and carboxylic acid compounds, such as crotonic acid and various oligomers through a random chain scission reaction that involves a cis-elimination reaction of β-CH and a six-member ring transition [15, 16]. A proposed mechanism of polyesters' degradation is reported in Figure 5.2. Degradation induces a decrease in PHB molecular weight and melt viscosity, as well as an increase of the crystallization time, since crystallization occurs at lower temperatures.

This very limited resistance to processing from the melt constitutes the main problem for PHB processing, which consequently requires the addition of some plasticizers, which allow lowering of the melting temperature and increasing in the processing windows, thus avoiding thermal degradation. Another approach for improving PHB processability has been the production of copolyesters of PHB with other 3-hydroxyalkanoates units, such as poly(3-hydroxybutyrate-*co*-hydroxyvalerate) (PHBV) and poly(3-hydroxybutyrate-*co*-hydroxyexanoate) [17].

5.3.1 Extrusion, Injection Molding, Thermoforming

5.3.1.1 Extrusion-Injection Molding

The low resistance of PHB to thermal degradation is the main issue related to processing in extrusion, as reported above. The degradability of PHB chains in processing is reduced by the addition of lubricants due to lowering of PHB melting temperature as reported in Section 5.5.1 on plasticizers. Fast processing speeds can be achieved through the sharp transition fluid/solid melting of PHB right behind the filling zone and lowering the temperature towards the die. Typical viscosity for this material is similar to PP with an MFI: 30–40 g/ 10 min [11]. In order to avoid degradation when processing PHA, it is recommended to dry the polymer before extrusion; this is usually performed by air dryers, e.g., for over 2 h at 80°C. The pellets regain the original humidity within 30 minutes after they are removed from the dryer if left in an open environment.

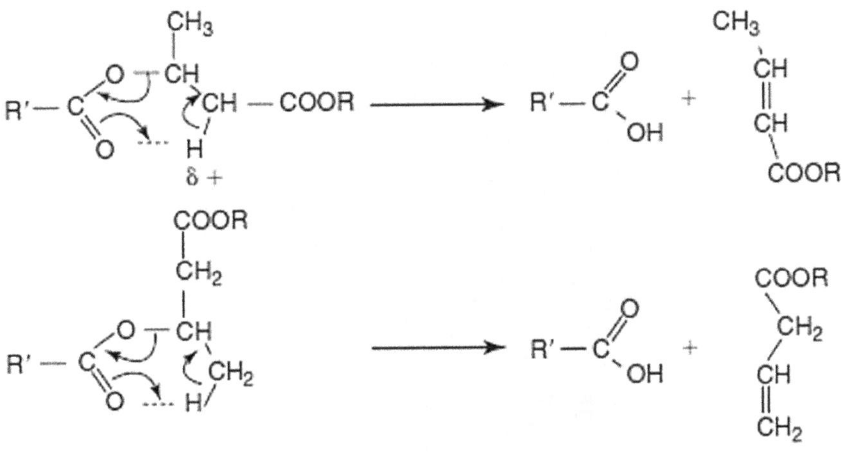

FIGURE 5.2 Cis-elimination in polyesters.

In order to reduce the cost of products based on PHA efforts, have been made to incorporate low-value materials, such as starch, or natural fillers [18, 19] using, in particular, agri-food by-products or waste lignocellulose fibers, highly available and at low cost, sourced from agricultural and industrial crops [13, 20, 21].

An additional benefit in the final products is due to the effect of natural fibers in bio-composites, which present a faster and higher biodegradation when subjected to outdoor applications compared to composites with synthetic fibers or to materials produced with the raw polymers [22]. This is because the biodegradation of a bio-composite occurs with the degradation of its individual components as well as with the loss of interfacial strength between them [23].

Composites based on PHBV and fibers of *Posidonia oceanica*, a dominant Mediterranean seagrass, were successfully produced by melt extrusion and their degradability was investigated in seawater. The results showed an increase in the impact resistance of the composites with increasing fiber content [3], while the presence of fibers favored the physical disintegration of the composite, increasing its biodegradation rate under simulated and real marine conditions.

Composites based on PHBV and waste sawdust fibers (Figure 5.3), derived from the wood industry, were also successfully produced at industrial scale by extrusion and injection molding [20]. Given their use in terrestrial applications, biodegradation tests were completed on the developed composites under simulated composting conditions in accordance with standard methods from EN 13427:2000. The preliminary evaluation of degradability in soil of PHBV/wood fibers-based molded specimens (pots with thickness of 1 mm) was positive [24].

For applications more demanding in performance than agricultural pots, it is important to consider that in composites of PHA with natural fibers, the weak interfacial bonding between the polar natural fiber and the non-polar matrix can lead to a reduction in final properties of the composite, ultimately hindering their industrial usage. Thus it is harder to obtain high degrees of alignment with natural fibers than

FIGURE 5.3 Injection molding of a PHA/wood fiber pots.

with synthetic fibers, since during the extrusion process long natural fibers tend to twist randomly [25, 26, 27, 28]. This behavior compromises the mechanical properties of the composites, as most of the natural fibers are not aligned parallel to the direction of the applied load. Different methods have been applied to improve the compatibility and interfacial bond strength, including the use of various surface modification techniques based on chemical and enzymatic approaches [29]. For example, in order to improve the mechanical response of the PHBV/potato pulp powder biocomposites, surface treatment of the potato pulp powder with bio-based and petroleum-based waxes was investigated and a good enhancement of the mechanical properties was achieved with natural carnauba and bee waxes [24]. This last approach is very promising and easy to achieve, and is thus under investigation even for other natural fibers such as bran and wood fibers.

5.3.1.2 Thermoforming

In thermoforming, a thermoplastic sheet is heated (see Figure 5.4) to its softening point and stretched over a mold and held in place while it cools and solidifies to form the desired part [30].

Thermoforming is a secondary process because it involves pre-processing of selected materials to manufacture semi-finished products such as sheets.

Very often thermoforming is confused with compression molding, in which, instead, thermoplastic material (in the form of granules, sheet, or prepregs) is melted in the mold cavity under heat and pressure, followed by cooling and part-removal after solidification. The other main differences are shown in Table 5.2 [31].

The thermoforming process of PHA is described in the literature both for PHA polymers [32], and also for their blends [33] and composites with the addition of natural fibers [34]. PHA sheets thermoformed with hemp, jute, or flax, for example, in which the composition of natural fibers ranged from 15 to 50%, were patented [35].

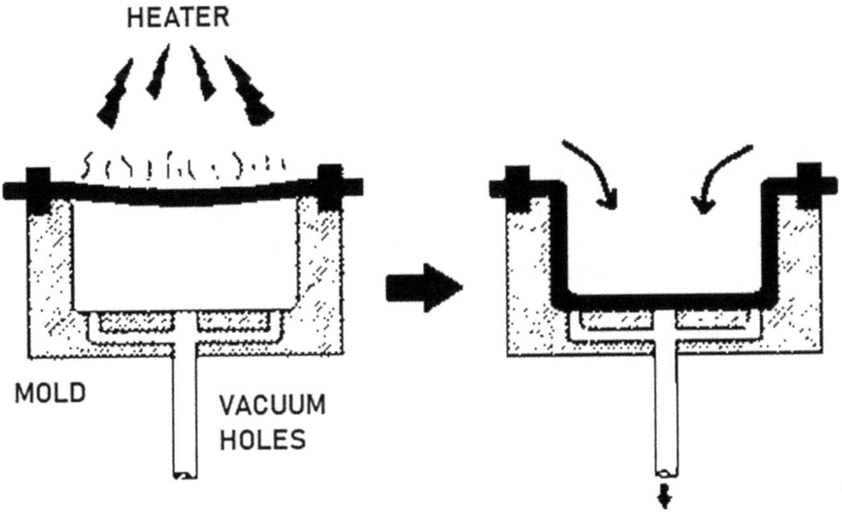

FIGURE 5.4 Scheme of thermoforming process.

TABLE 5.2
Differences between Compression Molding and Thermoforming Processes

Features	Compression Molding (CM)	Thermoforming (TF)
1. Type of process	Discontinuous	Continuous
2. Application type	Low-volume production and high-tolerance parts	High-volume production and low-tolerance parts
3. Acceptable raw material form	Sheets, composite granules, prepregs	Sheets only
4. Material waste	Low	High
5. Application	Automotive hood, fender, door panels, gears, etc.	Food and packaging containers, aircraft windscreens, interior panels, boat hulls, etc.

Properties such as melt viscosity, thermal conductivity, and crystallinity affect PHA processing [36]. High melt viscosity leads to problems such as undesirable molecular orientation and internal stress [37]. The thermoforming of PHA is similar to that of poly(styrene) in cycle time and mold release. Thermoforming grades have a heat-distortion temperature above 98°C, which allows creating microwaveable food containers, medical/pharmaceutical trays, and boxes.

PHA can be thermoformed with conventional equipment. The best softening temperature for thermoforming is around 180°C [38], which allows the material not to overheat, which could degrade it into toxic low-molecular-weight organic acids.

In continuous, in-line thermoforming, cooler molds can be used. In this case, a lower mold temperature is required to cool parts faster since the residence time in a continuous forming operation is generally shorter.

It should be noted that application of other biopolyesters such as PLA is very limited due to the rather fragile nature of this material in its pure form, making it not accessible for thermoforming in packaging film production. In contrast, PHA are more versatile, being based on a multitude of possible compositions at the monomer level where building blocks, such as 3HV, confer some flexibility to the polymer [39].

PHBV has been partnered with PLA to improve its gas barrier and mechanical properties [40]. In this study, two methods, classical 3-layer co-extrusion and extrusion of the original polymer blend, were used to combine PLA and PHBV to obtain films with different PLA-PHBV structures. The thermoformability of the different films was compared and the final structure was studied in relation to their gas barrier properties. The effect of the thermoforming step was studied by separating the effects of heating and stretching. The morphology of the mixture has been shown to cause superior mechanical and barrier properties than single PLA [41]. These experiences open the door to the design of flexible thermoformed films that will be used for food packaging.

5.3.2 Film Processing

Plastic films and multilayer systems can be manufactured using different conversion processes, such as blown film extrusion, flat die extrusion (see Figure 5.5), extrusion coating, extrusion lamination, and co-extrusion. These methods have advantages

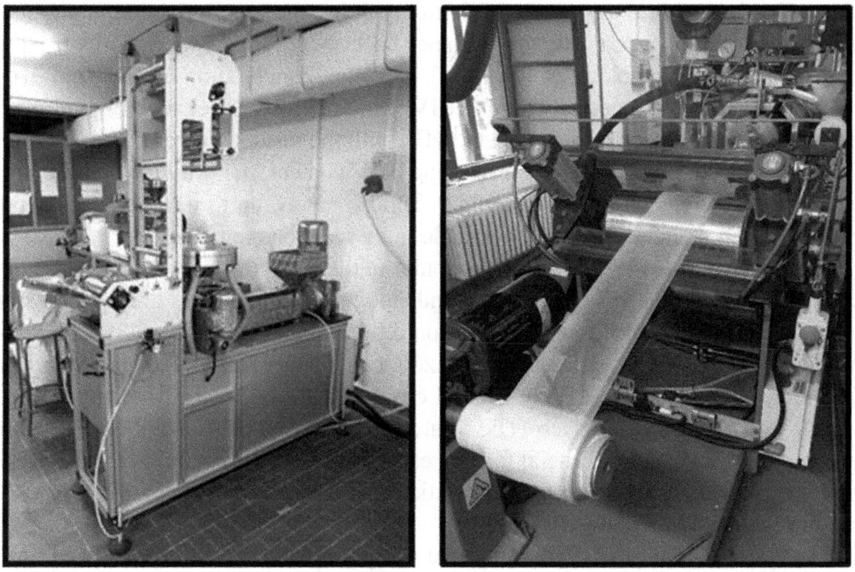

FIGURE 5.5 Equipment for film production by blowing and calendering.

and disadvantages depending on the type of material used, the width and thickness of the film, and the required properties of the film.

In particular, one of the most commonly used methods for preparing a plastic film is the extrusion of a blown film [42]. The film produced is tubular, therefore this process is generally used for the production of bags, industrial bags, or packaging film for shrink film. This technique consists in treating the polymer in a screw extruder, and pumping the melt through a mold. The spindle is axially elongated and radially expanded in the form of a thin-walled tube to obtain a continuous film (furlough). On the other hand, flat matrix extrusion consists of an extrusion through a linear matrix of adjustable thickness (matrix space) generally comprised between 3 and 1.4 mm. This technology allows the production of sheets and polymeric films (with thicknesses ranging from 50 microns to 1 millimeter) and consists of the extrusion of melted polymer through a matrix of rectangular geometry [43].

Film extrusions, such as the processing methods briefly described above, are also used to process PHA. A number of melt extrusion compositions have been developed based on PHB or PHBV mixed with additives, other polymers, and/or inorganic fillers, to improve the workability of the material and also to reduce costs. PHA formulations approved for contact with food are sold by Telles and TianAn [44].

The M_w of the PHA in the pellets used to produce the films is at least 470 000. It has to be at least 435 000 if PHA thermal stabilizers are used. In order to achieve a stable, not backed up film having desirable elongation and tensile properties, the M_w of PHA in the film has to be greater than about 420 000 [45].

However, films with suitable properties for various applications are difficult to obtain using PHA polymers because of their often-unacceptable characteristics such as poor melt strength, rapid aging, and brittleness. For example, biodegradable PHA cannot be formed into thin films while maintaining the required strength and tear resistance for applications such as a diaper back-sheet [35]. Several works tried to understand the morphology of PHA [46]. The copolymerization of 3HB and 3HV monomers leads to the copolymers PHBV, which exhibit lower melting and glass transition temperatures with respect to PHB, with the result that the processing window is extended. As an effect of increasing 3HV content, the brittleness of the processed PHBV films can be reduced [47]. Unfortunately, up to now, materials with different 3HV content are hardly available in kg scale on the market.

By incorporating the 3HV as co-monomer in the PHB, the mechanical properties can be improved in terms of flexibility and elongation. In this way, other processing techniques such as flat film extrusion should be enabled. Jost *et al.* [48] stated that a blend of PHBV with different plasticizers can further improve their properties. Glycerol, propylene glycol (PG), triethyl citrate, castor oil, epoxidized soybean oil (ESO), and poly(ethylene glycol) (PEG) are the most used plasticizers. Thus the use of a suitable plasticizer is crucial for processing of PHA, but Farris *et al.* proved the necessity to control the migration of plasticizer in order to avoid loss in viscosity of the material and loss of ductility [49].

Moreover, Cunha *et al.* [50] stated that it is possible to obtain PHBV-based films by blowing extrusion in scalable conditions for industrial production. The effects of blending with other biodegradable polymers, such as PBAT, is evident; better processability and better film properties compared to blown films were achieved. The mechanical characteristics (creep resistance at break) of these two-layer films correspond to those of commercial PBAT blown films intended for food packaging, which reached a strain of about 750%. However, the layers show low adhesion, and PHBV stratification can occur in strains of only 10%.

5.3.3 Electrospinning

Electrospinning (see Figure 5.6) has been widely used in laboratories to fabricate nanofibers due to its relative ease of setting up and requirements. However, this process presents several challenges when scaling up to mass-produce nanofibers. The unique challenge in electrospinning versus other fiber-spinning techniques is its use of electrical charges to stretch the polymer solution into fiber. Beyond this, it shares the same requirements as conventional fiber-spinning processes.

The selection of an appropriate non-hazardous solvent or solvent system is essential to determine the rheological properties and electrospinnability of the solution, the productivity, and the morphology of nanofibers (as described in Figure 5.7).

Electrospun nanofibers have shown significant potential in a number of applications, such as filtration systems, chemical and optical sensors, tissue engineering, wound healing, and release of drugs, due to their high surface area to volume ratio, small pore size, and high porosity [51, 52, 53, 54, 55]. The electrospinning process relies on the application of an electric field [56] between a needle-tipped syringe containing the polymeric solution and a collector for the deposition of nanofibers. Therefore, the polymeric solution is electrically charged and a conical droplet is formed at the needle tip. As the electric forces overcome the surface tension of the solution, a polymeric jet is generated from the surface of the

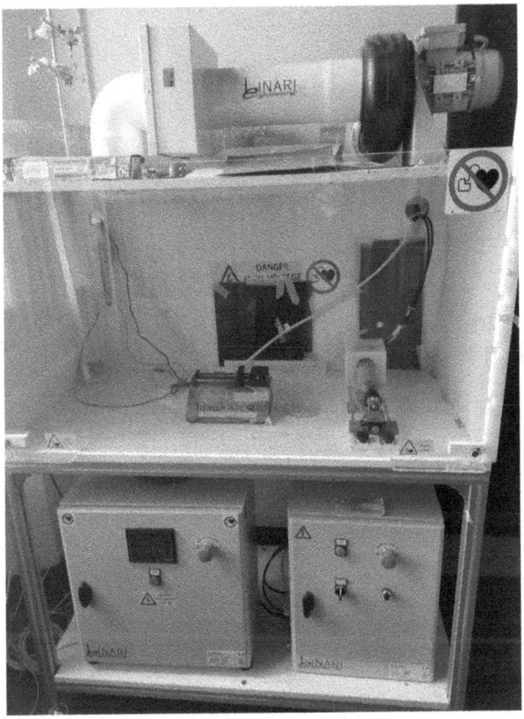

FIGURE 5.6 Example of an electrospinning equipment.

droplet and travels towards the collector. Between the needle and the collector, the solvent evaporates from the jet and consequently nanofibers can be collected [57]. The morphology and diameter of the resultant nanofibers depend on many parameters. Regarding the production of PLA nanofibers, the effects of polymer concentration and process parameters, such as electric field, feed rate, and distance between needle and collector, on nanofiber morphology have been investigated [58, 59, 60, 61], and are currently transferred and adapted to the processing of PHA in electrospinning.

5.4 PHA RHEOLOGY

5.4.1 Introduction

Rheological properties are of fundamental importance for the proper handling of PHA, and PHB in particular, especially when studying variations of properties at different temperatures. There are several basic tests available to characterize PHs' rheological behavior, such as shear stress as a function of shear rate, stress sweep, frequency sweep, and dynamic rheological tests [62].

In order to expand the use of PHA as commodity materials, this chapter aims to underline and examine the rheological, viscous, and elastic behavior of PHA, as a function of temperature, to improve processability.

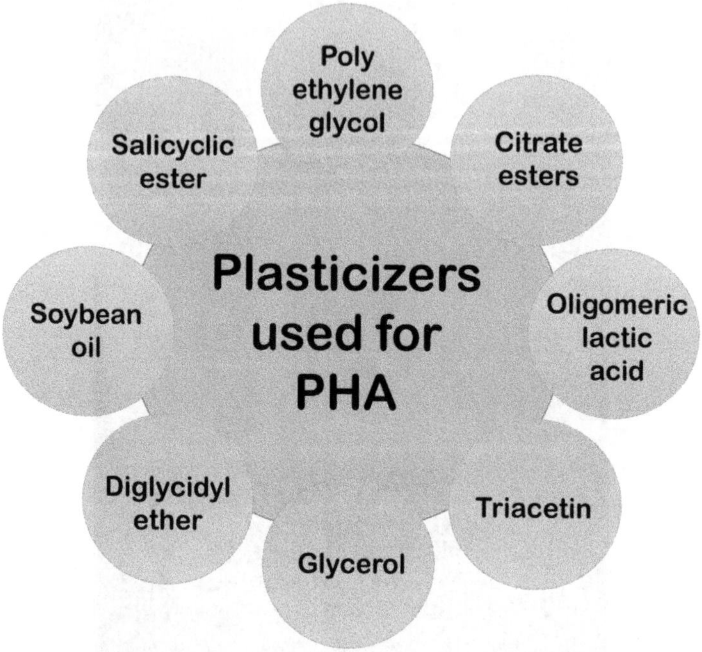

FIGURE 5.7 Electrospinning mass production requirements.

The viscosity of very viscous non-Newtonian fluids at low-velocity gradients is typically measured in rotational rheometers equipped with a cone-plate or plate-plate geometry [63].

The principle is to allow the fluid to move along closed trajectories, i.e., repeated indefinitely over time. Rotational rheometers are based on the relative rotary motion of two surfaces. The complex viscosity η^* is usually evaluated as a frequency response starting from viscoelasticity considerations [64]. The viscoelasticity of the fluid is therefore measured through mechanical-dynamic tests by subjecting the polymer melt to an oscillating flow motion with a sinusoidal displacement applied to one of the plates of the rheometer.

The viscoelastic response of the fluid to this oscillating deformation represents the tangential stress, which is the sum of two components, one elastic and the other viscose. Dividing everything by the maximum deformation, Equation (5.1) is obtained:

$$\frac{\tau}{\dot{\gamma}} = G'\sin(\omega t) + G''\cos(\omega t) \tag{5.1}$$

where G' and G'' represent the elastic and viscous response to the imposed oscillation having the dimensions of a module. They are called, respectively, conservative and dissipative modulus, and the modulus of the so-called complex viscosity can be obtained from these frequency viscoelastic measures such as in Equation (5.2):

$$\left|\eta^*\right| = \frac{1}{\omega}\sqrt{G'^2 + G''^2} \tag{5.2}$$

The capillary rheometer is the most common rheological equipment for the measurement of viscosity above all at high shear rates. Indeed, the straight trajectories and the absence of edge effects, present in all rotational geometries, allow the achievement of extremely high sliding gradients [65].

Operatively, the liquid is pushed (generally by means of a piston) through a relatively small cross-sectional channel. The pressure variation at the inlet and at the outlet of the conduit is measured (where the pressure is generally atmospheric). Knowing the flow-rate passing through the conduit (determined by the advancement speed of the plunger), it is possible to measure the viscosity of the fluid.

High strain rate rheological measurements are also carried out through an injection molding device operating as a rheometer [66]. Injection molding machines provide a good method for making measurements at these high strain rates because their large injection volume allows a steady state pressure to be reached and provides a direct method to measure rheological properties typically observed in this process [67].

The complex viscosity of a molten polymer can be easily and accurately determined from oscillatory shear, under a few conditions. First, the amplitude of the oscillating shear must be small enough to be within the linear viscoelastic region of the polymer, meaning small enough that the response to strain is not a function of past strain. Second, the complex viscosity under oscillating strain must follow the same trend as viscosity under steady strain at a shear rate numerically equal to the frequency of the oscillating strain. This second condition is also called the Cox-Merz rule, written more simply. The Cox-Merz empirical law states [68] that modulus of the complex viscosity is equal to the viscosity in stationary conditions when it is measured at a frequency (in rad/s) numerically equal to the shear rate (in s^{-1}) at which the shear viscosity in flow is measured. The Cox-Merz rule has previously also been validated for PHB by Park *et al.* [69].

As regards the elongational viscosity, also known as extensional or stretching flow, a fluid element is stretched in one or more directions [70]. It measures how the extension influences the way in which the fluid resists a deformation. Early experiments of Trouton [71] on uniaxial elongation by stretching a fiber or a filament of a Newtonian liquid stated that at low shear rates, the elongation viscosity was considered to be three times more than the corresponding shear viscosity. The ratio of the two values is called the Trouton ratio [72].

For a non-Newtonian fluid this statement is not true; in fact, the shear viscosity is a function of the shear rate and the elongation viscosity is a function of the rate of stretching.

Extensional flow of polymer melts has been extensively studied because of its importance in many technological processing operations and, from a more fundamental point of view, because the tensile properties of the polymer melts cannot be correlated directly with the shear viscosity behavior [73].

The basic technique for the measurement of the elongation viscosity of polymer melts consists of a tensile test carried out on a sample subjected to an isothermal uniform elongation flow, and the dependence of the extensional viscosity on the strain, under either constant strain rate or constant stress, is directly measured by means of an extensional rheometer [74].

Very important also are the relationships between viscosity of a molten aliphatic polymer and its average molecular weight; this can be modeled with a power-law relationship, shown in Equation (5.3), where the polymer is very often 3.4 for a linear polymer with no branching.

$$\eta = MW^{\alpha} \tag{5.3}$$

Degradation can be monitored in real time during processing, and this is a very important for polymers that easily degrade, such as the PHA. Melik and Schechtman [75] discussed a method to monitor degradation during compounding by inferring molecular weight from the compounding torque, which is correlated to the viscosity.

5.4.2 Rheology of PHA

Many researchers have developed models allowing PHB degradation to be estimated from typical processing parameters. In one of these studies, degradation in injection molding was found to be highly correlated to injection pressure and injector capillary diameter [76], while in another experiment, injection parameters were successfully correlated with tensile mechanical properties of PHB as well as with molecular weight [77].

Thellen *et al.* stated that blending of a long chain branched polymer and a linear polymer, combined with tailoring of composition and molecular weight of the base polymer, affects the rheology of PHA allowing it to achieve what is needed to impart high melt strength and melt elasticity. Once branching is introduced, the rheological and crystallization behavior of the branched PHA can be characterized and optimized for each application [78].

As reported above, the chain-scission degradation mechanism of PHA causes a reduction of the molecular weight of the polymer, which has a significant effect on the viscosity. By investigating the rheology of these polymers, it is possible to quantify the degradation rate. Several rheological studies have been performed which examine the degradation behavior of PHA [79, 80]. It was shown that by performing sequential dynamic frequency sweeps or steady shear rate experiments on the same PHA sample, a decrease in the measured viscosity occurred in successive experiments due to the thermal degradation of the material [81].

More specifically, Conrad *et al.* [82] measured shear stress as a function of shear rate (see Figure 5.8). At 180°C, the shear stress was found to increase exponentially with respect to the shear rate up to about 100 s^{-1}. For higher shear rate, the stress appeared almost constant, in agreement with a typical shear-thinning behavior.

To examine the rheological behavior of PHB as a function of temperature, dynamic tests were performed (as reported in Figure 5.9 [82]). The temperature sweep was conducted from 150°C to 185°C. Viscosity was found to be decreasing dramatically after 175°C G′ and G″ values were constant up to 175°C, but after that, they start to decrease and were found to cross each other at 180°C.

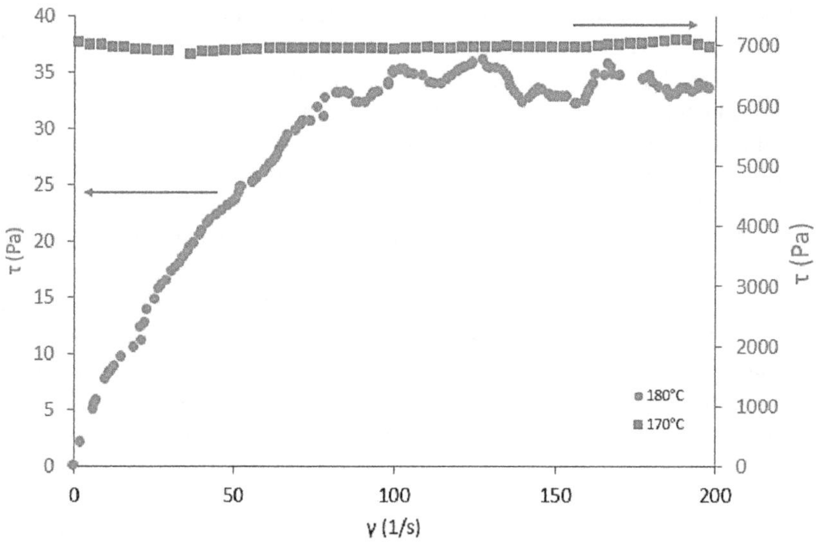

FIGURE 5.8 Shear stress of PHB as a function of shear rate at different temperatures (adjusted from [82]).

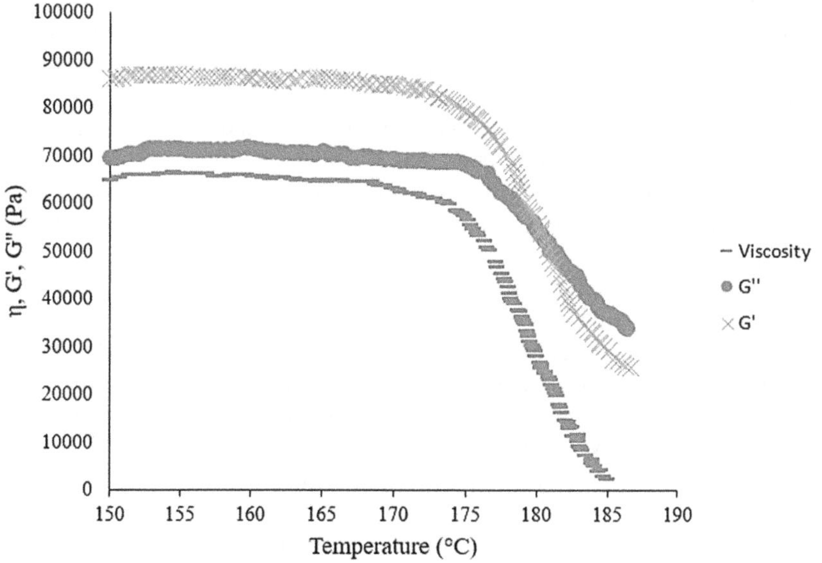

FIGURE 5.9 G′, G″, and Viscosity vs Temperature for PHB at 1 rad/sec (adjusted from [82]).

It is interesting to evidence how PHB rheological properties can be modified adding different percentages of valerate co-units. The shear rheology of PHBV copolymers shows that the zero-shear viscosity increases with increasing 3HV content.

The growing of viscosity is basically due to an increase in the molecular weight of the polymer. The weight average molecular weight of the PHA was determined based on the zero-shear viscosities using the D'Haene model [83] developed for a poly(3HB-*co*-5%-3HV) copolymer of several molecular weights. This model also guarantees a method to predict the change in molecular weight that occurred during extrusion and during the degradation rheology experiments [84, 85].

Another factor that may contribute to higher viscosity is that at increasing 3HV content, more rubbery and elastomeric properties are observed [86, 87]. The complex viscosity for the 5%, 8%, and 12% 3HV specimens reported in Figure 5.10 reports a viscosity decrease with increasing temperature. When the frequency is increased, the complex viscosities at the varying temperatures approach one another, because with increasing temperature, the onset frequency of shear thinning increases, whereas the relaxation time decreases. A minor decrease in viscosity can be observed at low frequencies for the 8% and 12% 3HV samples. This decrease can be correlated to thermal degradation occurring during the test at the lower frequencies [88].

As discussed before in all processing applications, a polymer will encounter both shear and extensional flow fields. Therefore, the extensional rheology of the materials is also significant in addition to the shear rheology [89].

The residence time in the rheometer results in minimal degradation during the experiments. Extensional viscosity for the poly(3HB-*co*-3HV) samples was measured at a temperature of 165°C and strain rates of 0.4, 0.6, and 1 s^{-1} (not much higher because otherwise it is impossible to evaluate the elongation viscosity).

Conrad also studied [82] the transient evolution of the extensional viscosity as a function of strain rate at a temperature of 165°C for the PHBV with different percentages of valerate units to those used previously. The extensional viscosity rapidly reaches a steady state plateau, which again indicates largely viscous behavior of the polymer melt, and the short relaxation times 400, 750, and 2000 Pa•s are, respectively the extensional viscosities found [83].

The time required to reach the plateau extensional viscosity correlates well with the relaxation times determined from the onset of shear thinning in the shear rheology.

For all three PHBVs, the time to reach the steady state plateau is approximately 0.07 seconds, whereas the relaxation times were 0.0067–0.02 s.

As the melt processing at higher temperatures affects the rheological behavior of PHB and PHBV, due to degradation processes, the incorporation of additives can be useful to improve the processing conditions. Appropriate additives can allow extrusion of PHB and PHBV at lower processing temperatures (170°C–180°C) without any change in viscosity, molar mass, and crystallization behavior [90]. Moreover, in the literature [91], it is found that with increasing temperatures of melt processing, the activation energy of blends increases in the opposite of raw PHB.

5.4.3 Rheology of PHA Blends

Another possible approach is the reactive modification of PHA in the presence of peroxide and co-agent, to generate branched molecular weights distributions, which results in significant variations in the rheological and thermal properties of the polymers [92]. A possible product of the reaction of PHB with peroxide and triallyl trimesate (TAM) as tri-functional co-agents is described by Kolanchi *et al.* [93].

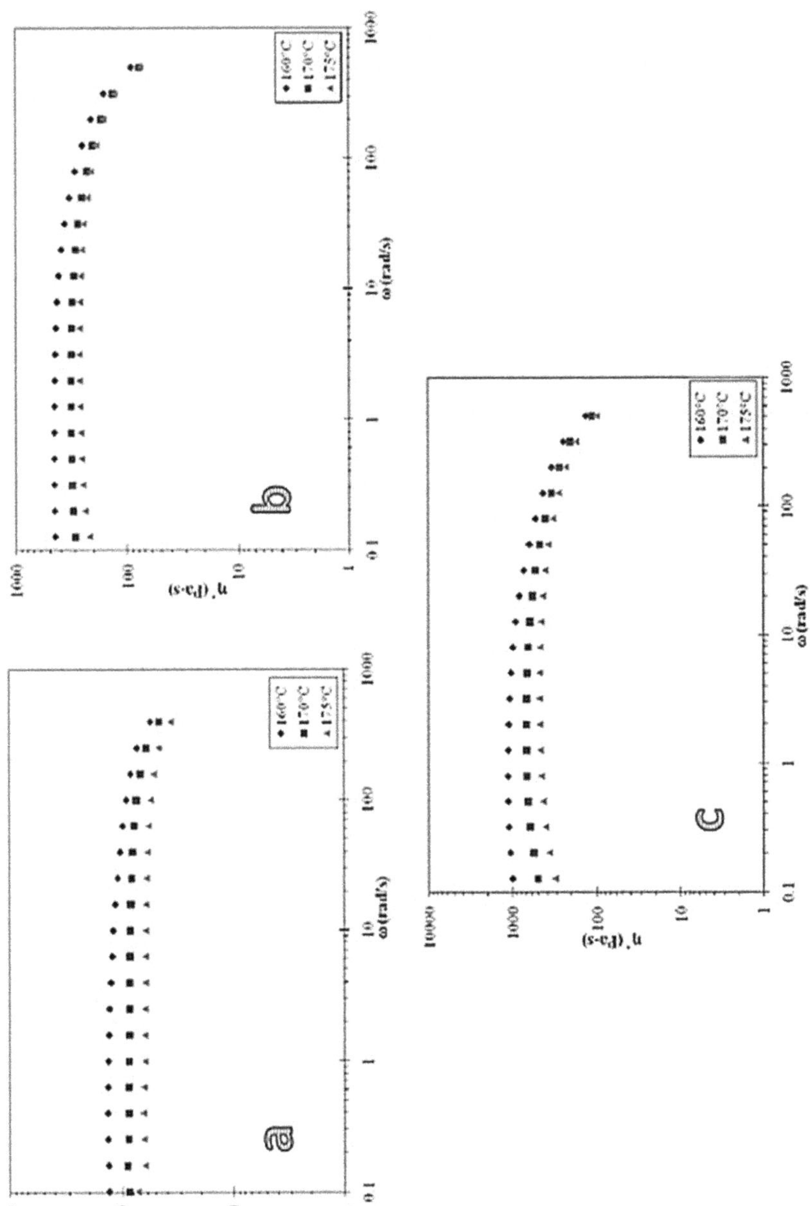

FIGURE 5.10 PHBV frequency sweep at different temperatures for a) 5, b) 8, and c) 12% 3HV (adapted from [84]).

The unmodified PHB exhibited very low viscosity, and a Newtonian plateau with minimal shear thinning. PHB specimens reacted with peroxides that displayed slightly lower viscosity, because of degradation attributed to chain scission. The addition of the TAM co-agent, while keeping the amount of peroxide constant at 0.2 wt.% resulted in progressively higher values of viscosity, with loss of the Newtonian plateau and increased shear thinning (Table 5.3).

Some very interesting work has been done by Bousfield [94] in which, starting from pure PHB, torque and temperature were plotted during compounding to understand the effect of degradation in real time (see Figure 5.11). It can be noticed that, after just 10 minutes, the degradation is complete; no appreciable differences can be assessed.

TABLE 5.3
Rheological Properties of Neat PHB, and Reactively Modified PHB at a Frequency of 1 rad/s (Adapted from [93])

Sample	Complex Viscosity [Pa • s]	Storage Modulus [Pa]	Phase Angle Degree	Tan Delta
PHB neat	607	13.3	88	21.0
PHB + 0.2% DCP	435	28.2	85	11.3
PHB + 0.3% DCP	444	30.8	84	10.0
PHB + 0.2% DCP + 0.2% TAM	784	227	73	3.0
PHB + 0.3% + 0.2% TAM	3970	2450	52	1.3

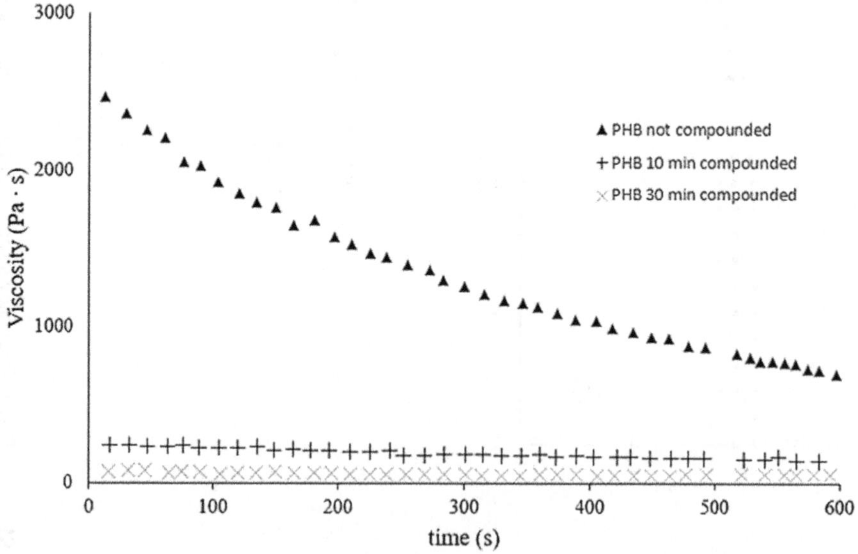

FIGURE 5.11 Effect of compounding on PHB complex viscosity (adapted from [94]).

The sample which was not exposed to compounding has a much higher complex viscosity than a sample of PHB after 10 minutes compounding. The change between the complex viscosity of PHB after 30 minutes is only slightly less than it was after 10 minutes, suggesting that the majority of degradation occurred in the first 10 minutes of compounding. Complex viscosity decay followed the same exponential trend for all three samples.

The second step of the work was on studying the effect of chain extenders on PHB. Four different chain extenders were added to PHB. Pyromellitic dianhydride and polycarbodiimide do not appear to yield any increase in complex viscosity, while addition of hexamethylene diisocyanate results in a small increase, and Joncryl substantially increases the polymers' complex viscosity. Significantly, none of the chain extenders alters the rate of degradation, although some effect on complex viscosity is apparent. From this information it is inferred that the successful chain extenders reacted to completion during compounding, but their action did not inhibit future degradation. Although complex viscosity responded in some cases to chain extenders, the degradation rate was not affected by chain extension.

5.5 ADDITIVES FOR PHA PROCESSING

Efforts in compounding PHA are focused on the search of plasticizers and nucleating agents capable of reducing the crystallization process and improving flexibility.

5.5.1 Plasticizers

The council of the IUPAC (International Union of Pure and Applied Chemistry) defined a plasticizer as "a substance or material incorporated in a material (usually a plastic or elastomer) to increase its flexibility, workability, or distensibility". These substances reduce the tension of deformation, hardness, density, viscosity, and electrostatic charge of a polymer, and at the same time increase the polymer chain flexibility, resistance to fracture, and dielectric constant. The low molecular size of a plasticizer allows it to occupy intermolecular spaces between polymer chains, reducing the secondary forces among them. In the same way, these molecules change the three-dimensional molecular organization of polymers, reducing the energy required for molecular motion and the formation of hydrogen bonding between the chains. As a consequence, an increase in the free volume and, hence, in the molecular mobility is observed. Plasticizers are widely used additives for polymeric materials to enhance their flexibility, processability, and ductility. Generally, an efficient plasticizer reduces the glass transition temperature (T_g) and melting temperature of the plasticized materials [95].

External plasticization is more efficient to conduct since it provides an easy route to advance the technical characteristics of PHA, and are also generally of low cost. Plasticizers for biodegradable polymers should preferably be selected among those that are biodegradable as well. In this respect, most of the plasticizers used in classical polymer processing are not appropriate for PHB [96].

Blending PHA and in particular PHB with plasticizers, may offer opportunities to improve processability by lowering the processing temperature and reducing the brittleness of PHA-based plastics. So far, many blends containing PHB/PHA have been studied and also many types of plasticizers have been proposed [97]. In particular, see Figure 5.12.

A lot of literature studies report mixing of PHB with poly(ethyleneglycol) (PEG) [98, 99, 100] and the better improvements achieved are detailed in Table 5.4.

The degree of plasticity of polymers is largely dependent on the chemical structure of the plasticizer, including chemical composition, molecular weight, and functional groups.

The selection for a specified system is normally based on different factors (see Figure 5.13).

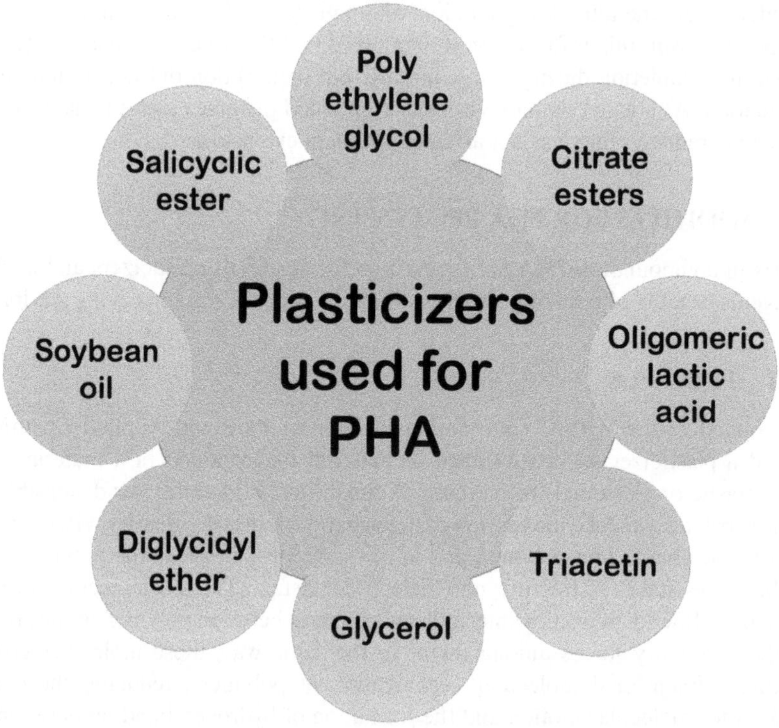

FIGURE 5.12 Plasticizers commonly used for PHA.

TABLE 5.4
Result of the PHB Plasticization with PEG (Adjusted from [100])

Use of PEG of different molecular mass	Increase in elongation at break four times vs pure PHB
	Weakening of intermolecular forces
	Lowering of viscosity
	Reduction of melting temperature
	Improving of processability

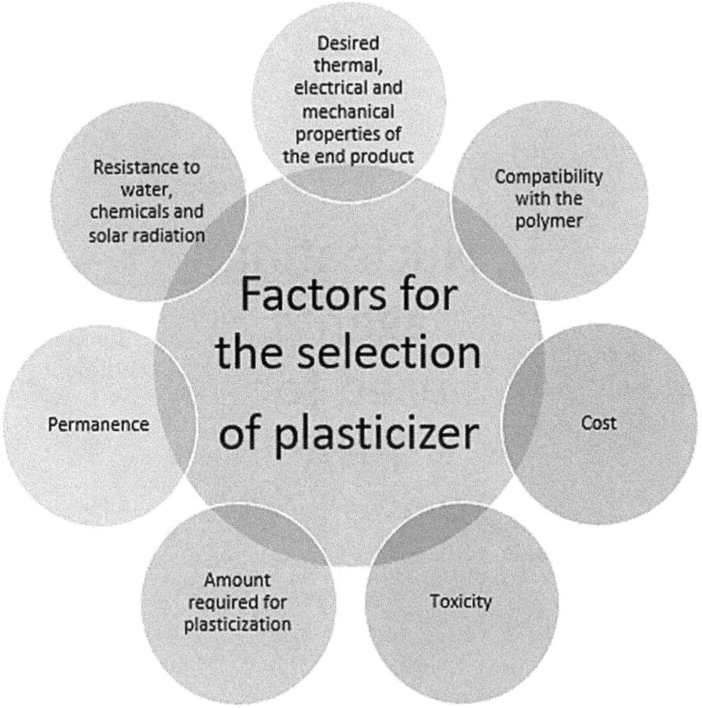

FIGURE 5.13 Factors taken into account for the selection of plasticizers.

5.5.2 Nucleating Agents

Nucleating agents are used in polymers that are able to crystallize, with the aim of accelerating the crystallization and consequently the speed cycle times in processing. After the melting, the rate of solidification of the plastic into a useful shape controls the processing and cycle time. Pure PHB exhibits a high melting temperature (T_m, 177°C), a low crystallization temperature (T_c, 79°C), and few heterogeneous nuclei, which results in a slow crystallization process for PHB. These characteristics make it difficult to use PHB to make products via injection-molding. P(3HB-*co*-3HV) copolymers also exhibit similar slow nucleation behavior comparable to the homopolymer PHB. In general, the addition of a nucleating agent accelerates the crystallization of polymers and provides polymers with heterogeneous nuclei allowing a more rapid crystallization during the cool-down period after the polymer melts [101], since the nucleation density increases and the spherulite size decreases so that the crystallization rate is increased [102]. Various nucleating agents, such as orotic acid [103], α-cyclodextrin [104], boron nitride, talc, terbium oxide, lanthanum oxide [105], saccharin, and phthalimide [106] have been tested for enhancing the crystallization of PHA at elevated temperatures (see Figure 5.14).

With the aim of reducing the spherulites size, studies on the addition of nucleants to P(3HB) were performed, but very little difference between the toughness of the nucleated and un-nucleated P(3HB) was reported [107].

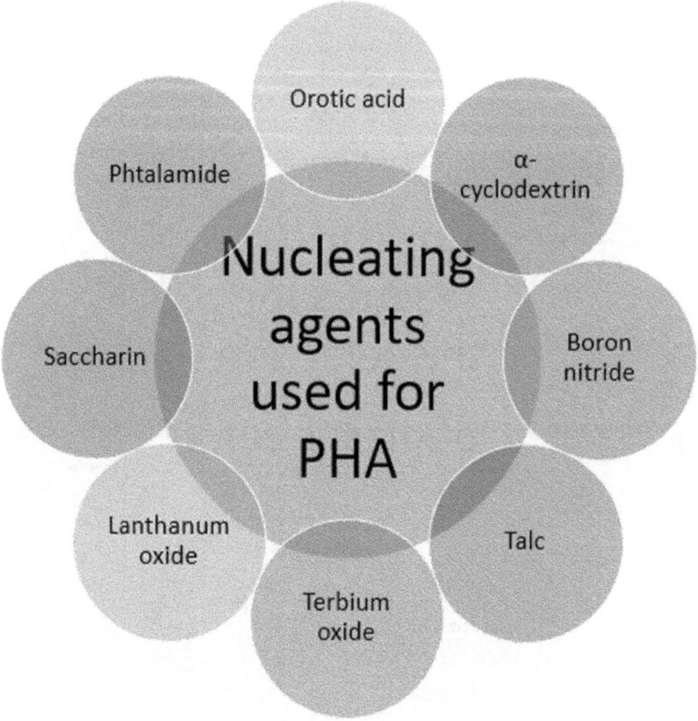

FIGURE 5.14 Nucleating agents commonly used for PHA.

5.6 CONCLUSIONS AND OUTLOOK

Polyhydroxyalkanoates (PHA) represent a valuable family of biopolyesters with assessed applications in the biomedical field, which can find applications even in commodities as alternatives to the non-degradable polymers currently used, with the benefit of being bio-based and highly biodegradable. Moreover, compounding with natural fiber fillers lowers the cost of the final products and promotes disintegration and subsequent biodegradation.

In order to expand their applications, PHA's drawbacks, such as narrow processing windows, brittleness, and changes in properties with aging, must be considered. Chemical and physical modification through the use of PHA copolymers, plasticizers, and nucleating agents, as well as knowledge of PHA behavior in the melt are very important to manage PHA processing, and to optimize PHA-based materials. Several studies on these topics are reported and many are ongoing, in particular on the rheology of PHA materials, preliminarily reported in this chapter, but the object of further investigation and analysis.

REFERENCES

1. C.S.K. Reddy, R. Ghai, Rashmi, V.C. Kalia, Polyhydroxyalkanoates: An overview, *Bioresour. Technol.* 87 (2003) 137–146. doi:10.1016/S0960-8524(02)00212-2.

2. T.G. Volova, A.N. Boyandin, A.D. Vasiliev, V.A. Karpov, S. V Prudnikova, O. V Mishukova, U.A. Boyarskikh, M.L. Filipenko, V.P. Rudnev, B. Bá Xuân, V. Việt Dũng, I.I. Gitelson, Biodegradation of polyhydroxyalkanoates (PHAs) in tropical coastal waters and identification of PHA-degrading bacteria, *Polym. Degrad. Stab.* 95 (2010) 2350–2359. doi:10.1016/j.polymdegradstab.2010.08.023.
3. M. Seggiani, P. Cinelli, E. Balestri, N. Mallegni, E. Stefanelli, A. Rossi, C. Lardicci, A. Lazzeri, Novel sustainable composites based on poly(hydroxybutyrate-co-hydroxyvalerate) and seagrass beach-CAST fibers: Performance and degradability in marine environments, *Materials (Basel).* 11 (2018) 772. doi:10.3390/ma11050772.
4. S. Khanna, A.K. Srivastava, Recent advances in microbial polyhydroxyalkanoates, *Process Biochem.* 40 (2005) 607–619. doi:10.1016/j.procbio.2004.01.053.
5. E. Prados, S. Maicas, Bacterial production of hydroxyalkanoates (PHA), *Univers. J. Microbiol. Res.* 4 (2016) 23–30.
6. M. Scandola, G. Ceccorulli, M. Pizzoli, M. Gazzano, Study of the crystal phase and crystallization rate of bacterial poly(3-hydroxybutyrate-co-3-hydroxyvalerate), *Macromolecules.* 25 (1992) 1405–1410. doi:10.1021/ma00031a008.
7. J.C. Worch, H. Prydderch, S. Jimaja, P. Bexis, M.L. Becker, A.P. Dove, Stereochemical enhancement of polymer properties, *Nat. Rev. Chem.* 3 (2019) 514–535. doi:10.1038/s41570-019-0117-z.
8. P.J. Barham, A. Keller, The relationship between microstructure and mode of fracture in polyhydroxybutyrate, *J. Polym. Sci. Part B Polym. Phys.* 24 (1986) 69–77. doi:10.1002/polb.1986.180240108.
9. G.J.M. De Koning, P.J. Lemstra, Crystallization phenomena in bacterial poly [(R)-3-hydroxybutyrate]: 2. Embrittlement and rejuvenation, *Polymer (Guildf).* 34 (1993) 4089–4094.
10. E. Bugnicourt, P. Cinelli, A. Lazzeri, V. Alvarez, The main characteristics, properties, improvements, and market data of polyhydroxyalkanoates, in: *Handbook of Sustainable Polymers Processing and Applications*, V.K. Thakur, M.K. Thaku, Eds., Panipat City, 2016, pp. 899–928.
11. E. Bugnicourt, P. Cinelli, A. Lazzeri, V. Alvarez, Polyhydroxyalkanoate (PHA): Review of synthesis, characteristics, processing and potential applications in packaging, *Express Polym. Lett.* 8 (2014) 791–808. doi:10.3144/expresspolymlett.2014.82.
12. M. Seggiani, P. Cinelli, N. Mallegni, E. Balestri, M. Puccini, S. Vitolo, C. Lardicci, A. Lazzeri, New bio-composites based on polyhydroxyalkanoates and *Posidonia oceanica* fibres for applications in a marine environment, *Materials (Basel).* 10 (2017) 326. doi:10.3390/ma10040326.
13. P. Cinelli, N. Mallegni, V. Gigante, A. Montanari, M. Seggiani, B. Coltelli, S. Bronco, A. Lazzeri, Biocomposites based on polyhydroxyalkanoates and natural fibres from renewable byproducts, *Appl. Food Biotechnol.* 6 (2019) 35–43.
14. Y. Aoyagi, K. Yamashita, Y. Doi, Thermal degradation of poly [(R) –3-hydroxybutyrate], poly [e -caprolactone], and poly [(S) -lactide], *Polym. Degrad. Stab.* 76 (2002) 53–59.
15. N. Grassie, E.J. Murray, P.A. Holmes, The thermal degradation of poly(–(d)-β-hydroxybutyric acid): Part 1—Identification and quantitative analysis of products, *Polym. Degrad. Stab.* 6 (1984) 47–61. doi:10.1016/0141-3910(84)90016-8.
16. N. Grassie, E.J. Murray, P.A. Holmes, The thermal degradation of poly(–(D)-β-hydroxybutyric acid): Part 2—Changes in molecular weight, *Polym. Degrad. Stab.* 6 (1984) 95–103. doi:10.1016/0141-3910(84)90075-2.
17. Y.-Z. Qiu, J. Han, J.-J. Guo, G.-Q. Chen, Production of poly(3-hydroxybutyrate-co-3-hy droxyhexanoate) from gluconate and glucose by recombinant *Aeromonas hydrophila* and *Pseudomonas putida*, *Biotechnol. Lett.* 27 (2005) 1381–1386. doi:10.1007/s10529-005-3685-6.

18. M. Zhang, N.L. Thomas, Preparation and properties of polyhydroxybutyrate blended with different types of starch, *J. Appl. Polym. Sci.* 116 (2010) 688–694. doi:10.1002/app.30991.
19. M. Seggiani, P. Cinelli, E. Balestri, N. Mallegni, E. Stefanelli, A. Rossi, C.L. Id, A.L. Id, Novel sustainable composites based on degradability in marine environments, *Materials.* (2018). doi:10.3390/ma11050772.
20. P. Cinelli, M. Seggiani, N. Mallegni, V. Gigante, A. Lazzeri, Processability and degradability of PHA-based composites in terrestrial environments, *Int. J. Mol. Sci.* 20 (2019). doi:10.3390/ijms20020284.
21. M. Seggiani, P. Cinelli, S. Verstichel, M. Puccini, I. Anguillesi, A. Lazzeri, Development of fibres-reinforced biodegradable composites, *Chem. Eng. Trans.* 43 (2015) 1813–1818. doi:10.3303/CET1543303.
22. J.K. Pandey, V. Nagarajan, A.K. Mohanty, M. Misra 1 - Commercial potential and competitiveness of natural fiber composites, in: *Woodhead Publishing Series in Composites Science and Engineering*, M. Misra, J.K. Pandey, A.K.B.T.-B. Mohanty Eds., Woodhead Publishing, Sawston, 2015, pp. 1–15. doi:10.1016/B978-1-78242-373-7.00001-9.
23. Z.N. Azwa, B.F. Yousif, A.C. Manalo, W. Karunasena, A review on the degradability of polymeric composites based on natural fibres, *Mater. Des.* 47 (2013) 424–442.
24. M.C. Righetti, P. Cinelli, N. Mallegni, A. Stäbler, A. Lazzeri, Thermal and mechanical properties of biocomposites made of poly(3-hydroxybutyrate-co-3-hydroxyvalerate) and potato pulp powder, *Polymers (Basel).* 11 (2019). doi:10.3390/polym11020308.
25. L. Aliotta, V. Gigante, M.B. Coltelli, P. Cinelli, A. Lazzeri, Evaluation of mechanical and interfacial properties of bio-composites based on poly (lactic acid) with natural cellulose fibers, *Int. J. Mol. Sci.* (2019). doi:10.3390/ijms20040960.
26. V. Gigante, L. Aliotta, V.T. Phuong, M.B. Coltelli, P. Cinelli, A. Lazzeri, Effects of waviness on fiber-length distribution and interfacial shear strength of natural fibers reinforced composites, *Compos. Sci. Technol.* 152 (2017) 129–138. doi:10.1016/j.compscitech.2017.09.008.
27. K.G. Satyanarayana, G.G.C. Arizaga, F. Wypych, Biodegradable composites based on lignocellulosic fibers—An overview, *Prog. Polym. Sci.* 34 (2009) 982–1021. doi:10.1016/j.progpolymsci.2008.12.002.
28. S.J. Eichhorn, C.A. Baillie, N. Zafeiropoulos, L.Y. Mwaikambo, M.P. Ansell, A. Dufresne, K.M. Entwistle, P.J. Herrera-Franco, G.C. Escamilla, L. Groom, M. Hughes, C. Hill, T.G. Rials, P.M. Wild, Current international research into cellulosic fibres and composites, *J. Mater. Sci.* 36 (2001) 2107–2131. doi:10.1023/A:1017512029696.
29. L.G. Angelini, M. Scalabrelli, S. Tavarini, P. Cinelli, I. Anguillesi, A. Lazzeri, Ramie fibers in a comparison between chemical and microbiological retting proposed for application in biocomposites, *Ind. Crops Prod.* 75 (2015) 178–184. doi:10.1016/j.indcrop.2015.05.004.
30. F. Erchiqui, F. Godard, A. Gakwaya, A. Koubaa, M. Vincent, H. Kaddami, Engineering investigations on the potentiality of the thermoformability of HDPE charged by wood flours in the thermoforming part, *Polym. Eng. Sci.* 49 (2009) 1594–1602. doi:10.1002/pen.21394.
31. K.C. Birat, M. Pervaiz, O. Faruk, J. Tjong, M. Sain, Green composite manufacturing via compression molding and thermoforming, in: *Manufacturing of Natural Fibre Reinforced Polymer Composites*, Springer, New York City, 2015, pp. 45–63.
32. E. Bugnicourt, P. Cinelli, A. Lazzeri, V. Alvarez, Polyhydroxyalkanoate (PHA): Review of synthesis, characteristics, processing and potential applications in packaging, *Express Polym. Lett.* 8 (2014) 791–808. doi:10.3144/expresspolymlett.2014.82.
33. M.P. Arrieta, J. López, A. Hernández, E. Rayón, Ternary PLA–PHB–Limonene blends intended for biodegradable food packaging applications, *Eur. Polym. J.* 50 (2014) 255–270. doi:10.1016/j.eurpolymj.2013.11.009.

34. M. Breulmann, A. Künkel, S. Philipp, V. Reimer, K.O. Siegenthaler, G. Skupin, M. Yamamoto, Polymers, biodegradable, *Ullmann*'s *Encycl.* Ind. Chem. (2009). doi: 10.1002/14356007.n21_n01.
35. S.L. Billington, C.S. Criddle, C.W. Frank, M.C. Morse, S.J. Christian, A.J. Pieja. A Patent on Bacterial poly(hydroxy alkanoate) polymer and natural fiber composites US2008/0160567, 2008.
36. R. Auras, L.-T. Lim, S.E.M. Selke, H. Tsuji, *Poly (Lactic Acid): Synthesis, Structures, Properties, Processing, and Applications*, John Wiley & Sons, Hoboken, New Jersey, 2011.
37. C. Menzel, E. Olsson, T.S. Plivelic, R. Andersson, C. Johansson, R. Kuktaite, L. Järnström, K. Koch, Molecular structure of citric acid cross-linked starch films, *Carbohydr. Polym.* 96 (2013) 270–276. doi:10.1016/j.carbpol.2013.03.044.
38. S.M. Sapuan, N. Bin Yusoff, The relationship between manufacturing and design for manufacturing in product development of natural fibre composite, In: Salit M., Jawaid M., Yusoff N., Hoque M. (eds), *Manufacturing of Natural Fibre Reinforced Polymer Composites.* Springer, Cham, 2015. doi:10.1007/978-3-319-07944-8_1.
39. Koller, M., Poly(hydroxyalkanoates) for food packagi, *Appl Food Biotechnol.* 1 (2014) 3–15.
40. T. Messin, N. Follain, A. Guinault, C. Sollogoub, V. Gaucher, N. Delpouve, S. Marais, Structure and barrier properties of multinanolayered biodegradable PLA/PBSA films: Confinement effect via forced assembly coextrusion, *ACS Appl. Mater. Interfaces.* 9 (2017) 29101–29112. doi:10.1021/acsami.7b08404.
41. A. Guinault, A.S. Nguyen, G. Miquelard-Garnier, D. Jouannet, A. Grandmontagne, C. Sollogoub, The effect of thermoforming of PLA-PHBV films on the morphology and gas barrier properties, *Key Eng. Mater.* 504–506 (2012) 1135–1138. doi:10.4028/www.scientific.net/KEM.504-506.1135.
42. K.K. Majumder, Molecular, rheological, and crystalline properties of low-density polyethylene in blown film extrusion. *Polym. Eng. Sci.* 47 (2007), 1983.
43. M.-B. Coltelli, V. Gigante, P. Cinelli, A. Lazzeri, Flexible food packaging using polymers from biomass, in: *Bionanotechnology to Save the Environment*, P. Morganti, Ed., MDPI, 2019, pp. 272–298. doi:10.3390/books978-3-03842-693-6.
44. S.A. Ashter, Introduction to bioplastics engineering, Andrew, W., Ed., William Andrew: London, 2016. ISBN 0323394078.
45. J. Asrar, J.R. Pierre, *PHA compositions and methods for their use in the production of PHA films*, 2003.
46. A.J. Domb, Degradable polymer blends. I. Screening of miscible polymers, *J. Polym. Sci. Part A Polym. Chem.* 31 (1993) 1973–1981. doi:10.1002/pola.1993.080310805.
47. H.-J. Endres, A. Siebert-Raths, Engineering biopolymers, *Eng. Biopolym.* 71148 (2011) 3–15.
48. V. Jost, H.C. Langowski, Effect of different plasticisers on the mechanical and barrier properties of extruded cast PHBV films, *Eur. Polym. J.* 68 (2015) 302–312. doi:10.1016/j.eurpolymj.2015.04.012.
49. G. Farris, P. Cinelli, I. Anguillesi, S. Salvadori, M.-B. Coltelli, A. Lazzeri, Effect of ageing time on mechanical properties of plasticized poly(hydroxybutyrate) (PHB), *AIP Conf. Proc.* 1599 (2014) 294–297. doi:10.1063/1.4876836.
50. M. Cunha, B. Fernandes, J.A. Covas, A.A. Vicente, L. Hilliou, Film blowing of PHBV blends and PHBV-based multilayers for the production of biodegradable packages, *J. Appl. Polym. Sci.* 133 (2016) 1–11. doi:10.1002/app.42165.
51. H.L. Schreuder-Gibson, P. Gibson, P. Tsai, Cooperative charging effects of fibers from electrospinning of electrically dissimilar polymers, *Int. Nonwovens J.* os-13(4) (2004). doi:10.1177/1558925004os-1300406.

52. X. Wang, Y.-G. Kim, C. Drew, B.-C. Ku, J. Kumar, L.A. Samuelson, Electrostatic assembly of conjugated polymer thin layers on electrospun nanofibrous membranes for biosensors, *Nano Lett.* 4 (2004) 331–334. doi:10.1021/nl034885z.
53. S. Agarwal, J.H. Wendorff, A. Greiner, Use of electrospinning technique for biomedical applications, *Polymer (Guildf)*. 49 (2008) 5603–5621. doi:10.1016/j.polymer.2008.09.014.
54. M.-S. Khil, D.-I. Cha, H.-Y. Kim, I.-S. Kim, N. Bhattarai, Electrospun nanofibrous polyurethane membrane as wound dressing, *J. Biomed. Mater. Res. Part B Appl. Biomater.* 67B (2003) 675–679. doi:10.1002/jbm.b.10058.
55. E.-R. Kenawy, G.L. Bowlin, K. Mansfield, J. Layman, D.G. Simpson, E.H. Sanders, G.E. Wnek, Release of tetracycline hydrochloride from electrospun poly(ethylene-co-vinylacetate), poly(lactic acid), and a blend, *J. Control. Release.* 81 (2002) 57–64. doi:10.1016/S0168-3659(02)00041-X.
56. J. Doshi, D.H. Reneker, Electrospinning process and applications of electrospun fibers, *J. Electrostat.* 35 (1995) 151–160. doi:10.1016/0304-3886(95)00041-8.
57. D.H. Reneker, A.L. Yarin, Electrospinning jets and polymer nanofibers, *Polymer (Guildf)*. 49 (2008) 2387–2425. doi:10.1016/j.polymer.2008.02.002.
58. S.-Y. Gu, J. Ren, Process optimization and empirical modeling for electrospun poly(D,L-lactide) fibers using response surface methodology, *Macromol. Mater. Eng.* 290 (2005) 1097–1105. doi:10.1002/mame.200500215.
59. F. Yang, C.Y. Xu, M. Kotaki, S. Wang, S. Ramakrishna, Characterization of neural stem cells on electrospun poly(L-lactic acid) nanofibrous scaffold, *J. Biomater. Sci. Polym. Ed.* 15 (2004) 1483–1497. doi:10.1163/1568562042459733.
60. M.P. Prabhakaran, L. Ghasemi-Mobarakeh, G. Jin, S. Ramakrishna, Electrospun conducting polymer nanofibers and electrical stimulation of nerve stem cells, *J. Biosci. Bioeng.* 112 (2011) 501–507. doi:10.1016/j.jbiosc.2011.07.010.
61. D. Ishii, T.H. Ying, A. Mahara, S. Murakami, T. Yamaoka, W. Lee, T. Iwata, In vivo tissue response and degradation behavior of PLLA and stereocomplexed PLA nanofibers, *Biomacromolecules.* 10 (2009) 237–242. doi:10.1021/bm8009363.
62. C. Macosko, R.G. Larson, *Rheology: Principles, Measurements, and Applications*, VCH, New York City, 1994.
63. T.G. Mezger, *The Rheology Handbook: For Users of Rotational and Oscillatory Rheometers*, Vincentz Network GmbH & Co KG, Hanover, Germany, 2006.
64. M. Guaita, F. Ciardelli, F.P. La Mantia, E. Pedemonte, Fondamenti di scienza dei polimeri, Università di Pisa, Catalogo Ricerca UNIPI (1998). http://hdl.handle.net/11568/52954
65. R. Chhabra, Non-Newtonian fluids: An introduction, in: Krishnan J., Deshpande A., Kumar P. (eds), *Rheology of Complex Fluids*, Springer, New York, 2010, pp. 3–34.
66. L.R. Schmidt, A special mold and tracer technique for studying shear and extensional flows in a mold cavity during injection molding, *Polym. Eng. Sci.* 14 (1974) 797–800.
67. B.J. Shah, *Reverse Temperature Profile Rheology and Recycle Study of Polyhydroxybutyrate Copolymer Within an Injection Molding Machine*, University of Massachusetts Lowell, 2012.
68. N. Grizzuti, Reologia dei materiali polimerici. Scienza ed ingegneria, Nuova Cultura, 2012. https://books.google.it/books?id=f4_KvWplJUMC.
69. S.H. Park, S.T. Lim, T.K. Shin, H.J. Choi, M.S. Jhon, Viscoelasticity of biodegradable polymer blends of poly(3-hydroxybutyrate) and poly(ethylene oxide), *Polymer (Guildf)*. 42 (2001) 5737–5742. doi:10.1016/S0032-3861(01)00071-4.
70. B.A. Morris, Rheology of polymer melts, in: B.A.B.T.-T.S. and T. of F.P. Morris (Ed.), *The Science and Technology of Flexible Packaging: Multilayer Films from Resin and Process to End Use*, William Andrew Publishing, Oxford, 2017, pp. 121–147. doi:10.1016/B978-0-323-24273-8.00005-8.

71. F.T. Trouton, On the coefficient of viscous traction and its relation to that of viscosity, *Proc. R. Soc. London. Ser. A, Contain. Pap. a Math. Phys. Character.* 1 (1906) 426–440.
72. C.J.S. Petrie, Extensional viscosity: A critical discussion, *J. Nonnewton. Fluid Mech.* 137 (2006) 15–23. doi:10.1016/j.jnnfm.2006.01.011.
73. F. Baldi, A. Franceschini, T. Riccò, Determination of the elongational viscosity of polymer melts by melt spinning experiments. A comparison with different experimental techniques, *Rheol. Acta.* 46 (2007) 965–978.
74. J. Meissner, J. Hostettler, A new elongational rheometer for polymer melts and other highly viscoelastic liquids, *Rheol. Acta.* 33 (1994) 1–21. doi:10.1007/BF00453459.
75. H.D. Melik, A.L. Schechtman, Biopolyester melt behavior by torque rheometry, *Polym. Eng. Sci.* 35 (1995) 1795–1806. doi:10.1002/pen.760352209.
76. E. Leroy, I. Petit, J.L. Audic, G. Colomines, R. Deterre, Rheological characterization of a thermally unstable bioplastic in injection molding conditions, *Polym. Degrad. Stab.* 97 (2012) 1915–1921.
77. R. Renstad, S. Karlsson, A. Albertsson, P. Werner, M. Westdahl, Influence of processing parameters on the mass crystallinity of poly (3-hydroxybutyrate-co-3-hydroxyvalerate), *Polym. Int.* 43 (1997) 201–209.
78. C. Thellen, M. Coyne, D. Froio, M. Auerbach, C. Wirsen, J.A. Ratto, A processing, characterization and marine biodegradation study of melt-extruded polyhydroxyalkanoate (PHA) films, *J. Polym. Environ.* 16 (2008) 1–11. doi:10.1007/s10924-008-0079-6.
79. B.S. Thorat Gadgil, N. Killi, G.V.N. Rathna, Polyhydroxyalkanoates as biomaterials, *Medchemcomm.* 8 (2017) 1774–1787. doi:10.1039/C7MD00252A.
80. S. Ponnusamy, S. Viswanathan, A. Periyasamy, S. Rajaiah, Production and characterization of PHB-HV copolymer by Bacillus thuringiensis isolated from *Eisenia foetida*, *Biotechnol. Appl. Biochem.* 66 (2019) 340–352. doi:10.1002/bab.1730.
81. M.M. Satkowski, D.H. Melik, J.P. Autran, P.R. Green, I. Noda, L.A. Schechtman, *Biopolymers*, Wiley-VCH Weinheim, Germany, 2001.
82. J.D. Conrad, G.M. Harrison, The rheology and processing of renewable resource polymers, *AIP Conf. Proc.* 1027 (2008) 114–116. doi:10.1063/1.2964497.
83. P. D'Haene, E.E. Remsen, J. Asrar, Preparation and characterization of a branched bacterial polyester, *Macromolecules.* 32 (1999) 5229–5235.
84. H.J. Park, K. Muthukumarappan, J.J. Julson, Characterization of rheological properties of poly (3-hydroxybutyric acid), *2005 ASAE Annu.* Int. Meet. 0300 (2005). doi:10.13031/2013.19645.
85. H.J. Choi, J. Kim, M.S. Jhon, Viscoelastic characterization of biodegradable poly (3-hydroxybutyrate-co-3-hydroxyvalerate), *Polymer (Guildf).* 40 (1999) 4135–4138.
86. S.W. Ko, R.K. Gupta, S.N. Bhattacharya, H.J. Choi, Rheology and physical characteristics of synthetic biodegradable aliphatic polymer blends dispersed with MWNTs, *Macromol. Mater. Eng.* 295 (2010) 320–328. doi:10.1002/mame.200900390.
87. H.J. Choi, S.H. Park, J.S. Yoon, H.S. Lee, S.J. Choi, Rheological study on biodegradable poly (3-hydroxybutyrate) and its copolymer, *J. Macromol. Sci. Part A.* 32 (1995) 843–852.
88. G.M. Harrison, D.H. Melik, Application of degradation kinetics to the rheology of poly (hydroxyalkanoates), *J. Appl. Polym. Sci.* 102 (2006) 1794–1802.
89. F.P. La Mantia, M. Morreale, L. Botta, M.C. Mistretta, M. Ceraulo, R. Scaffaro, Degradation of polymer blends: A brief review, *Polym. Degrad. Stab.* 145 (2017) 79–92. doi:10.1016/j.polymdegradstab.2017.07.011.
90. A. El-Hadi, R. Schnabel, E. Straube, G. Müller, M. Riemschneider, Effect of melt processing on crystallization behavior and rheology of poly (3-hydroxybutyrate)(PHB) and its blends, *Macromol. Mater. Eng.* 287 (2002) 363–372.
91. C.R. Arza, P. Jannasch, P. Johansson, P. Magnusson, A. Werker, F.H.J. Maurer, Effect of additives on the melt rheology and thermal degradation of poly [(R)-3-hydroxybutyric acid], *J. Appl. Polym. Sci.* 132 (2015).

92. J.S. Parent, S.S. Sengupta, M. Kaufman, B.I. Chaudhary, Coagent-induced transformations of polypropylene microstructure: Evolution of bimodal architectures and cross-linked nano-particles, *Polymer (Guildf).* 49 (2008) 3884–3891. doi:10.1016/j.polymer.2008.07.007.
93. A.R. Kolahchi, M. Kontopoulou, Chain extended poly(3-hydroxybutyrate) with improved rheological properties and thermal stability, through reactive modification in the melt state, *Polym. Degrad. Stab.* 121 (2015) 222–229. doi:10.1016/j.polymdegradstab.2015.09.008.
94. G. Bousfield, Effect of chain extension on rheology and tensile properties of PHB and PHB-PLA blends, 2014. (Doctoral dissertation, École Polytechnique de Montréal) doi:10.1024/0300-9831/a000127.
95. H. Younes, D. Cohn, Phase separation in poly(ethylene glycol)/poly(lactic acid) blends, *Eur. Polym. J.* 24 (1988) 765–773. doi:10.1016/0014-3057(88)90013-4.
96. J.S. Choi, W.H. Park, Effect of biodegradable plasticizers on thermal and mechanical properties of poly(3-hydroxybutyrate), *Polym. Test.* 23 (2004) 455–460. doi:10.1016/j.polymertesting.2003.09.005.
97. L.H. Innocentini-Mei, J.R. Bartoli, R.C. Baltieri, Mechanical and thermal properties of poly(3-hydroxybutyrate) blends with starch and starch derivatives, *Macromol. Symp.* 197 (2003) 77–88. doi:10.1002/masy.200350708.
98. H. Abe, Y. Doi, Y. Kumagai, Synthesis and characterization of Poly[(R,S)-3-hydroxy butyrate-b-6-hydroxyhexanoate] as a compatibilizer for a biodegradable blend of poly[(R)-3-hydroxybutyrate] and poly(6-hydroxyhexanoate), *Macromolecules.* 27 (1994) 6012–6017. doi:10.1021/ma00099a012.
99. J.A.F.R. Rodrigues, D.F. Parra, A.B. Lugao, Crystallization on films of PHB/PEG blends, *J. Therm. Anal. Calorim.* 79 (2005) 379–381. doi:10.1007/s10973-005-0069-z.
100. M.C. Righetti, A. Lazzeri, P. Cinelli, *Recent Advances in Biotechnology*, Vol. 2, Martin Koller, Bentham Science Publisher, Sharjah, UAE, 2016.
101. J. Qian, L. Zhu, J. Zhang, R.S. Whitehouse, Comparison of different nucleating agents on crystallization of poly(3-hydroxybutyrate-co-3-hydroxyvalerates), *J. Polym. Sci. Part B Polym. Phys.* 45 (2007) 1564–1577. doi:10.1002/polb.21157.
102. W. Kai, Y. He, Y. Inoue, Fast crystallization of poly(3-hydroxybutyrate) and poly(3-hyd roxybutyrate-co-3-hydroxyvalerate) with talc and boron nitride as nucleating agents, *Polym. Int.* 54 (2005) 780–789. doi:10.1002/pi.1758.
103. N. Jacquel, K. Tajima, N. Nakamura, H. Kawachi, P. Pan, Y. Inoue, Nucleation mechanism of polyhydroxybutyrate and poly(hydroxybutyrate-co-hydroxyhexanoate) crystallized by orotic acid as a nucleating agent, *J. Appl. Polym. Sci.* 115 (2010) 709–715. doi:10.1002/app.30873.
104. Y. He, Y. Inoue, Effect of α-cyclodextrin on the crystallization of poly(3-hydroxybutyrate), *J. Polym. Sci. Part B Polym. Phys.* 42 (2004) 3461–3469. doi:10.1002/polb.20213.
105. C. Zhu, C.T. Nomura, J.A. Perrotta, A.J. Stipanovic, J.P. Nakas, The effect of nucleating agents on physical properties of poly-3-hydroxybutyrate (PHB) and poly-3-hydroxyb utyrate-co-3-hydroxyvalerate (PHB-co-HV) produced by *Burkholderia cepacia* ATCC 17759, *Polym. Test.* 31 (2012) 579–585. doi:10.1016/j.polymertesting.2012.03.004.
106. R.E. Withey, J.N. Hay, The effect of seeding on the crystallisation of poly(hydroxybutyrate), and co-poly(hydroxybutyrate-co-valerate), *Polymer (Guildf).* 40 (1999) 5147–5152. doi:10.1016/S0032-3861(98)00732-0.
107. J.K. Hobbs, The fracture of poly(hydroxybutyrate). Part I Fracture mechanics study during ageing, *J. Mater. Sci.* 33 (1998) 2509–2514. doi:10.1023/A:1004384631218.

6 Additive Manufacturing of PHA

Dario Puppi and Federica Chiellini

CONTENTS

6.1 INTRODUCTION

Additive manufacturing (AM) techniques are based on a computer-aided design and manufacturing process for the layered fabrication of tridimensional (3D) objects with geometry and dimensions predefined by a digital model. The fabrication process involves a sequential delivery of energy and/or materials, typically starting from the bottom of the object and building layers up, with each newly formed layer adhering to the previous one. A wide range of materials (i.e., metals, polymers, ceramics, and composites) can be processed into specific final products by properly designed AM technology, processing conditions, and material formulation. The great interest raised in the last years in the industrial translation of AM is a consequence of the huge advantages of this class of techniques in terms of a reduced number of manufacturing steps between the virtual design and the ready-to-use part, optimization of materials employment, and possibility of fabricating complex shapes and porous structures not obtainable using traditional production processes (e.g., melt extrusion and molding) [1]. For all these reasons AM is currently seen as an emerging technology with high potential to impact the global consumer market and specialized industrial areas, such as the aerospace and medical device sectors.

Polyhydroxyalkanoates (PHA) are aliphatic polyesters synthesized by various Gram-positive and Gram-negative bacteria under unbalanced growth conditions [2, 3]. They generally consist of 1 000–10 000 monomeric units with different pendant groups at the beta-position, and they are classified as short- or medium-chain-length if they consist of 3 to 5, or 6 to 14 carbons in the hydroxyalkanoate unit. The chemical structure of the first PHA isolated in the 1920s, i.e., poly(3-hydroxybutyrate) (PHB) [4], and the other two PHA copolyesters investigated so far for AM, i.e., poly(3-hydroxybutyrate-*co*-3-hydroxyvalerate) (PHBV) and poly[3-hydroxybutyrate-*co*-3-hydroxyhexanoate) (PHBHHx), are reported in Figure 6.1.

The increasing interest in potential industrial applications of PHA is justified by the possibility of their production starting from abundantly available renewable resources, as well as by their biodegradability under aerobic and anaerobic conditions, in agreement with a modern concept of sustainable development [5]. In addition, PHA's thermoplastic behavior and mechanical properties make them reliable candidates as environmentally friendly alternatives to the so-called "commodity plastics" (e.g., polyolefins). For these reasons, great efforts have been spent on PHA production enhancement in terms of economic feasibility, as well as environmental sustainability: for instance, through its integration into biorefinery concepts and waste treatment facilities [6–8]. Numerous investigations to increase the competitiveness of PHA large-scale commercialization are based on new technologies via metabolic engineering, synthetic biology, and bioinformatics [9].

Packaging materials, e.g., for the food sector, is the main applicative field of PHA and its follow-up products, with the agricultural, biomedical, and pharmaceutical sectors as other industrial options [10, 11]. A tremendous amount of literature has been published on PHA investigation for the development of biodegradable devices such as screws and pins for orthopedic surgery, patches for wound treatment, implantable and injectable drug release systems, and tissue engineering scaffolds [12–14]. These efforts are attested to by various PHA-based medical devices currently available on the market, including wound and surgical sutures (e.g., TephaFlex®, MonoMax®, and Phantom Fiber™), as well as surgical meshes (e.g., Tornier® and GalaFLEX) [15].

The integration of the high versatility of AM, in terms of automation degree, design freedom, and customization of product composition, shape, and microstructure, with

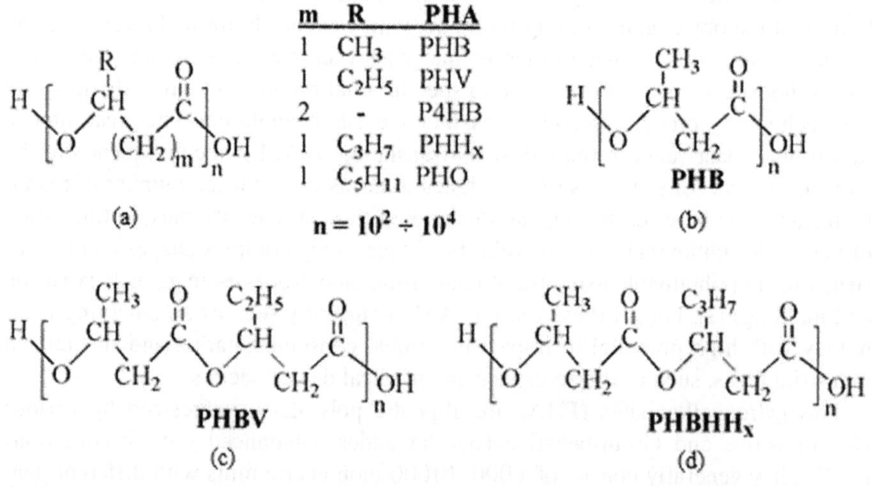

FIGURE 6.1 Chemical structure of polyhydroxyalkanoates (PHA) investigated for AM: (a) general structure of short-chain PHA; (b) poly(3-hydroxybutyrate) (PHB); (c) poly(3-hydroxybutyrate-*co*-3-hydroxyvalerate) (PHBV); and (d) poly[3-hydroxybutyrate-*co*-3-hydroxy hexanoate) (PHBHHx).

the sustainable development aspects of PHA production is a stimulating research aspect that could have a tremendous impact on the global consumer market and specialized industrial areas. This topic is particularly hot in the biomedical field in which AM is seen as a revolutionary approach to personalized medicine, thanks to the possibility of modeling devices on medical imaging data obtained from the analysis of anatomical parts. In addition, advanced AM technologies enable the customization of compositional and microstructural features that play a key role in therapeutic aspects, such as tissue regeneration and bioactive stimulation [16]. This chapter is aimed at providing an updated overview of published research on PHA processing by means of AM. Processing properties and requirements of PHA with particular reference to AM are discussed. Theoretical and applicative aspects of the different AM techniques, i.e., selective laser sintering (SLS), fused deposition modeling (FDM), and computer-aided wet-spinning (CAWS), applied so far to process PHA are outlined. In addition, investigations reporting on AM processing of different PHA, i.e., PHB, PHBV, and PHBHHx, possibly in combination with bioactive ceramics (i.e., calcium phosphate) are critically reviewed. An overview is given in Table 6.1.

6.2 PROCESSING PROPERTIES OF PHA

Thanks to its aliphatic polyester macromolecular structure, PHA generally show superior mechanical properties and much higher processing versatility in comparison to other classes of natural polymers (i.e., proteins and polysaccharides). For this reason, as previously mentioned, PHA are widely investigated and employed for applications requiring a structural and load-bearing role in different industrial sectors. The great number of homopolymers and copolymers belonging to this class of materials, as well as the possibility of engineering their composition and molecular weight by changing microbial fermentation conditions, offer great versatility in tuning properties relevant to their application, including biodegradation kinetics, mechanical behavior, and processability.

The length of the pendant groups in their monomeric unit and the distance between ester linkages remarkably affect the structural, processing, and mechanical properties of PHA. Due to its high stereochemical regularity, PHB has a high degree of crystallization (60–80%) and a stiff, brittle mechanical behavior with an elongation at break typically below 10%. Its glass transition temperature (T_g), melting temperature (T_m), and degradation temperature (T_{deg}) are around 0°C, 180°C, and 220°C, respectively [17]. The resulting processing temperature window is often too narrow to avoid polymer degradation with the consequent formation of trans-crotonic acids [15, 18]. PHB is soluble only in a few solvents, including halogenated solvents, such as chloroform and dichloromethane, and dimethyl formamide, typically requiring high temperatures and pressures to achieve its solubilization [19]. As a consequence of a different molecular structure, poly(4-hydroxybutyrate) (P4HB) has a lower degree of crystallization (~35%), T_g (~−50°C), and T_m (~60°C) than PHB. These differences are reflected in the higher flexibility and easier processing of P4HB [20]. Indeed, copolymerization between 3-hydroxybutyrate (3HB) and 4-hydroxybutyrate (4HB) is an effective way to tune morphological, processing, and mechanical

TABLE 6.1
State of the Art on AM of PHA

Material	AM Technique	Product	Highlight	Ref.
PHB	SLS	Porous construct with square base geometry, composed of 10 layers; pore size: 1 000 μm.	First investigation on SLS of PHA to fabricate a 12-layer scaffold.	[50]
PHB	SLS	Tetragonal construct (13 × 13 × 26 mm^3) with a dense base (4 mm height) and a 3D orthogonal periodic porous architecture (22 mm height); square pores (1 × 1 × 1 mm^3) and struts of 1 mm.	SEM analysis showed a rough and porous surface due to incomplete sintering.	[51]
PHB	SLS	Porous construct with approximate shape of a cube (10.4 × 10.4 × 10.1 mm^3) and internal architecture composed of 16 circular pins (diameter of 1.7 mm) arranged orthogonally with 32 pins (diameter of 1.6 mm), resulting in orthogonal channels measuring 0.8 mm in diameter.	^1H NMR and DSC analysis showed that any thermal degradation of polymer occurred upon laser sintering.	[52]
PHBV, PHBV/CaP	SLS	Tetragonal construct (8 × 8 × 15.5 mm^3) with 3D orthogonal periodic porous architecture; pore size 1 mm, strut size 0.5 mm, solid base (9 × 9 × 3 mm^3); theoretical CaP loading of 15 wt.%.	Developed a S/O/W emulsion/solvent evaporation method to prepare composite microspheres made of PHBV loaded with Ca-P nanoparticles.	[53, 54]
PHBV/CaP	SLS	Various complex porous structures, and an anatomical construct reproducing a scale-down to 40% of a digital model of a human proximal femoral condyle scaffold; theoretical CaP loading of 15 wt.%.	Developed a two-step functionalization method (i. physical entrapment of gelatin, ii. heparin immobilization) to enhance scaffold wettability and provide binding sites for rhBMP-2.	[55]
PHBV/PCL	FDM	Tetragonal scaffolds (6.0 × 6.0 × 2.5 mm^3), lay-down pattern of 0/90/45/135°; strand diameter in the range of 370–390 μm, pore size in the range 190–210 μm; PHBV/PCL weight ratio of 75/25, 50/50, or 25/75.	By increasing PHBV content in the blend and submitting the resulting scaffolds to a surface plasma treatment, samples' surface roughness and wettability were enhanced.	[69]

(Continued)

TABLE 6.1 (CONTINUED)
State of the Art on AM of PHA

Material	AM Technique	Product	Highlight	Ref.
PHB/PDLLA/ Citroflex®	FDM	Dog-bone shape samples; PHB/PLA/Citroflex® weight ratio of 60/25/15.	The plasticizer significantly increased the shape fidelity, thermal properties, and mechanical behavior of printed samples.	[70]
PHBHHx	CAWS	Tetragonal scaffolds with different external shape and dimensions, pore size (100–800 μm), and geometry (lay-down pattern 0–90° or 0–45–90–135°).	As a consequence of the phase inversion governing polymer solidification, a dual-scale porous structure was achieved.	[88]
PHBHHx	CAWS	Anatomical scaffolds modelled on a critical size (2 mm length) segment of a New Zealand rabbit's radius model, and endowed with a longitudinal macrochannel.	The scaffolds showed anisotropic mechanical behavior, and the longitudinal channel significantly increased scaffold mechanical properties under compression.	[89]
PHBHHx/PCL	CAWS	Cylindrical scaffolds (15 mm diameter, 5 mm height); PHBHHx/PCL weight ratio of 3:1, 2:1, or 1:1.	Blend ratio significantly affected material processing properties as well as the morphological, thermal, and mechanical properties of the resulting scaffolds.	[90]
PHBHHx	CAWS	Small caliber (φ~2 mm) vascular stents (length of 5 mm or 10 mm).	PHBHHx stents showed a dual-scale porosity as a consequence of phase inversion, great radial elasticity, as well as tunable pore sizes and wall thickness by varying fabrication parameters.	[91]

parameters of the resulting copolyester [21]. Inhibition of secondary crystallization in PHBV results in increased flexibility and ductility in comparison to PHB, as well as enhanced solubility in organic solvents and reduced T_m down to 130°C, broadening material processing versatilities [22]. Being characterized by a longer alkyl side group, PHBHHx has even lower T_m and degree of crystallization, down to 54°C and 18%, respectively, and much higher elongation at break – up to 850% [23–25].

The use of plasticizers, such as oxypropylated glycerin, glycerol, glycerol triacetate, 4-nonylphenol, 4,40-dihydroxydiphenylmethane, and acetyl tributyl citrate, has been reported to improve the processability of PHA by lowering their T_m and reducing their brittleness, especially in the case of PHB [26]. For instance, PHB properties can be improved through blending with additives, such as plasticizers and lubricants, by mixing in a Brabender (kneader) and subsequent melt molding. Additive blending has greatly broadened the range of techniques available for PHA processing, depending on the polymer molecular weight and macromolecular composition (e.g., comonomer content). Indeed, various technological approaches, including melt-spinning, blow molding, injection molding, thermoforming, lamination, and casting, have been optimized for PHB copolymers processing. More advanced processing techniques, such as electrospinning [27], laser microperforation [28], and AM [29], are typically adopted to fabricate PHA medical devices to meet the more stringent requirements of material composition, bioactivity, and morphology relevant to modern approaches in the biomedical field.

6.3 ADDITIVE MANUFACTURING (AM) OF PHA

To the best of the authors' knowledge, almost all the articles available in the literature reporting on AM application to PHA are aimed at the development of porous medical implants, typically for bone tissue engineering. The tremendous progress achieved in the last two decades on AM materials and technologies is propelling their translation to the medical industry [30, 31]. Indeed, different customized biodegradable devices have been implanted in humans, improving health conditions and saving lives of patients undergoing cranioplasty, oral surgery, or pediatric otolaryngologic intervention [32]. In this context, the interest in PHA for medical implants comes from the perspective of combining the load-bearing properties of aliphatic polyesters with the environmental, social, and economic benefits of using biodegradable materials from renewable and sustainable resources. However, the number of articles reporting the successful application of AM to PHA is still limited, due to the aforementioned processing drawbacks. As will be described in the following sections, SLS, FDM, and CAWS have been successfully employed over the past decade to develop experimental protocols tailored to the fabrication of PHA-based scaffolds with predefined shapes and internal porous architecture.

6.3.1 Selective Laser Sintering (SLS)

SLS was the first AM technique successfully applied to process different PHA, i.e., PHB and PHBV, possibly in combination with ceramic fillers to develop composite materials. SLS is industrially employed for prototyping devices, small components, and cost-effective consumer good parts in different sectors, including electronics, aerospace, and automotive. It has also found application in the fabrication of personalized medical implants for cranial, maxillofacial, and spinal surgery [33], as well as in surgical pediatric otolaryngology [34].

SLS is based on a computer-controlled translating laser beam as a power source to selectively sinter a powder bed made of polymer, ceramic, or hybrid particles [35]. A new powder bed with a thickness of around 100 μm is mechanically spread over each layer after its fabrication, by means of a roller or a blade system, to start the sintering of the subsequent layer (Figure 6.2a). This powder sintering/spreading cycle is repeated several times with a layer-by-layer process until the 3D object is built up. Sintering within a layer and between adjacent layers is achieved by using a high-intensity laser beam (e.g., CO_2 laser) which is absorbed by bed particles that raise their temperature to a value high enough to become soft and fuse together. The fabrication chamber is typically maintained at a temperature close to the melting or softening point to optimize the sintering process and rate. For this reason, an inert gas atmosphere in the chamber is often required to prevent polymer oxidative degradation. The optimization of various processing parameters, including laser power, scan spacing (i.e., distance between adjacent scan lines), layer/bed thickness, powder bed temperature, roller speed, and scan speed, is required when developing a novel fabrication protocol.

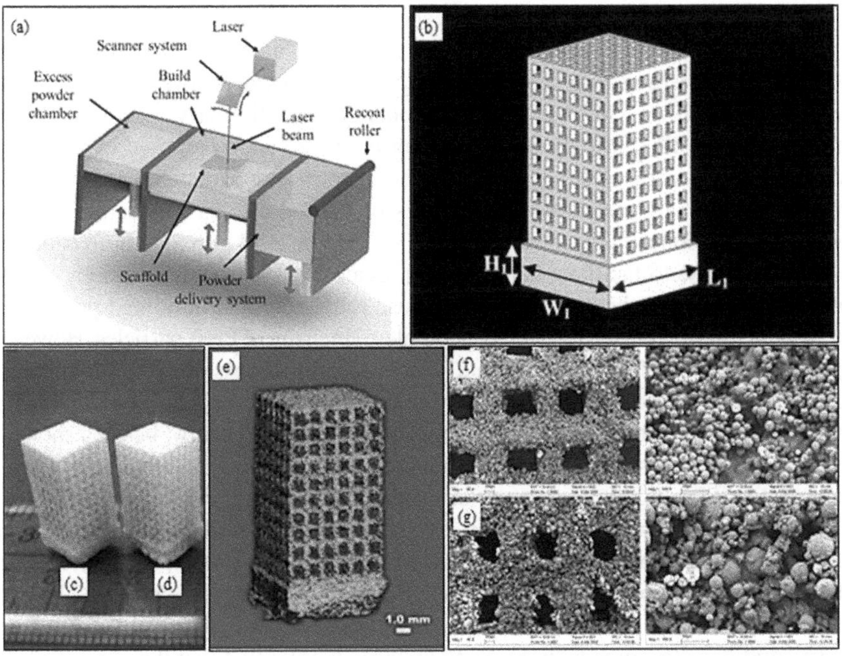

FIGURE 6.2 Selective laser sintering (SLS) of PHA: (a) schematic representation of SLS process (modified from [30]); (b) digital model of PHBV-based scaffolds; (c) photograph of PHBV and (d) PHBV-CaP scaffolds; (e) micro-CT image of PHBV-CaP scaffold; (f) SEM micrographs of PHBV; and (g) PHBV-CaP scaffolds (modified from [53]).

One of the advantages of SLS is that complex geometries can be manufactured without the implementation of support structures since non-sintered powder can support the subsequent layers. In addition, SLS avoids the use of solvents, diluents, monomers, or other reactive agents that can have harmful effects during fabrication and as residual components in the final item [36]. However, post-processing treatments are often required to eliminate non-sintered powder that may remain entrapped in porous architectures. Surface finishing treatments, such as milling and coating, are also necessary due to a poor control of SLS over surface topography. Indeed, a high surface roughness determined by partial particles binding is generally achieved as a consequence of the limited laser intensity applied to avoid polymer thermal degradation. Particles with a narrow size distribution in the range 20–100 μm and a spherical geometry are typically required to optimize powder flowability and density, as well as minimize particles agglomeration [37]. Depending on laser parameters and particle properties, a resolution in the range of 100 μm is typically achieved.

Owing to a complex combination of material requirements in terms of optical, thermal, viscosity, and surface tension properties, as well as particle shape and size, a limited number of polymers are industrially processed by SLS, mainly polyamides (e.g., PA12) and few other semicrystalline polymers, such as thermoplastic polyurethanes and poly(ether ether ketone) (PEEK) [38–41]. Amorphous polymers, such as polycarbonate, are poorly used for SLS due to their relatively high melt viscosity hindering polymer particles' full coalescence [40].

Together with poly(ε-caprolactone) (PCL) [42–46] and poly(d,l lactide) (PDLLA) [47–49], PHA were among the first biodegradable polymers investigated for SLS processing. In 2007, Oliveira *et al.* [50] published preliminary results about the fabrication of PHB scaffolds with 1 000 μm pore size and composed of 12 layers. Afterwards, a systematic study on the effect of processing parameters on the resulting morphology and mechanical properties of PHB scaffolds was carried out on tetragonal interconnected porous architectures (13 mm × 13 mm × 26 mm) endowed with a solid base to facilitate handling [51]. Scanning Electron Microscopy (SEM) analysis highlighted the presence of a diffuse roughness and unwanted porosity in the polymer matrix derived from the incomplete sintering of PHB particles. However, upon optimization of scan spacing and powder layer thickness, the scaffolds showed mechanical properties within the lower range of human trabecular bone. A further investigation confirmed PHB microporous and rough topography, besides showing by proton nuclear resonance (^{1}H NMR) and differential scanning calorimetry (DSC) analysis that SLS did not cause any thermal degradation of the polymer [52].

Duan *et al.* [53] employed a solid-in-oil-in-water (S/O/W) emulsion/solvent evaporation method to prepare composite microspheres made of PHBV loaded with calcium phosphate (CaP) nanoparticles. The prepared microspheres were processed by SLS to fabricate tetragonal scaffolds with a 3D orthogonal periodic porous architecture and a solid base (Figures 6.2b–e). SEM analysis highlighted that the surface of the scaffolds was composed of partially fused and even intact microspheres (ϕ~50 μm), resulting in a rough and irregular topography (Figures 6.2f,g). The presence of CaP resulted in a significantly smaller microsphere size, and required a higher input laser energy for sintering, with the overall effect of a different surface microtopography and a higher compressive stiffness [53, 54]. The developed technology was employed to fabricate an anatomical PHBV/CaP construct, representing a scale-down to 40% of a digital model of a human proximal femoral condyle scaffold, as well as to integrate a two-step scaffold functionalization method involving (i) physical entrapment of gelatin, and (ii) heparin immobilization [55]. The surface-modification improved scaffold wettability and provided specific binding sites between conjugated heparin and the growth factor recombinant human bone morphogenetic protein-2 (rhBMP-2) to enhance material osteo-inductive properties.

6.3.2 Fused Deposition Modeling (FDM)

FDM and other melt-extrusion AM (ME-AM) techniques are increasingly employed in different industrial sectors for fabricating functional prototypes, models, and small parts of vehicles or civil aircrafts. In addition, they are the most exploited AM techniques in the biomedical field for the development of porous scaffolds and customized biodegradable implants made of aliphatic polyesters, such as PCL [56], PDLLA [57], and poly(D,L-lactide-*co*-glycolide) (PLGA) [57, 58], thanks to their versatility in terms of material selection, design freedom, mechanical properties tuning, and achievable resolution.

ME-AM techniques generally involve the controlled deposition of a polymeric material extruded through a nozzle which is kept at a temperature above polymer T_g [59, 60]. The extruded material is deposited through a layer-by-layer process defined by the computer-controlled motion of the extrusion head and/or the building platform. In the case of FDM, also referred to as fused filament fabrication (FFF), the raw material is in the form of a continuous filament which is fed to the heated nozzle (Figure 6.3a). FDM allows an increased automation degree, as well as the minimization of residence time at high temperatures in comparison to other ME-AM techniques. On the other hand, technological solutions based on processing materials in the form of pellets or powder (e.g., screw extruder, pressurized nozzle, and force-controlled plunger) offer the possibility of directly processing physical mixtures of polymers and inorganic/organic fillers without the need of pre-processing into filament [60–62]. Cutting-edge advancements on technological solutions for polymer processing have led to the integration of FDM and electrospinning into a hybrid AM technique, i.e., melt electrospinning writing (MEW) [63]. By depositing a stable electrified molten jet onto a collector translating at a high speed, 3D layered structures composed by axially aligned submicrometric polymeric fibers can be fabricated in this way [64–66].

Polymers showing suitable viscoelastic properties and good thermal stability, such as amorphous thermoplastics like acrylonitrile butadiene styrene (ABS), are typically employed in ME-AM. In general, semicrystalline polymers are more challenging to process by FDM due to shrinkage and warpage during part building as a consequence of polymer crystallization [67]. A heated chamber or deposition platform, as well as polymer matrix loading with fillers characterized by high thermal conductivity (e.g., carbon fibers), are effective means to reduce warpage effects related to material thermal gradients in both semicrystalline and amorphous polymers [67, 68]. Together with poly(L-lactide) (PLLA) which is the most used

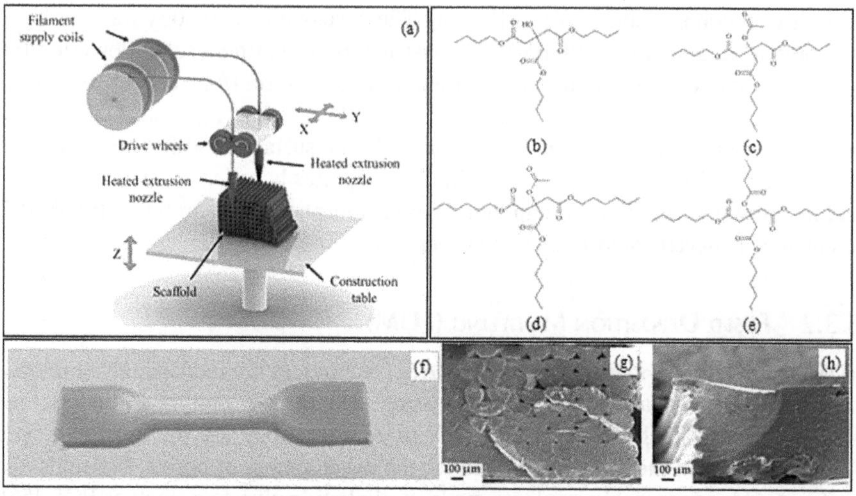

FIGURE 6.3 Fused deposition modeling (FDM) of PHA: (a) schematic representation of FDM process (modified from [30]); plasticizers based on esters of citric acid: (b) tributyl citrate, (c) acetyl tributyl citrate, (d) acetyl trihexyl citrate, and (e) n-butyryl tri-n-hexyl citrate; (f) picture of a FDM-printed dog-bone shape sample made of PHB/PLA/plasticizer; (g) SEM micrographs of cross-section of neat PHB/PDLLA; and (h) plasticized PHB/PDLLA samples (modified from [70]).

semicrystalline polymer for FDM, a wide range of polymers, e.g., polycarbonate, PEEK, and poly(methyl methacrylate), are commercially available nowadays as filaments for FDM (diameter of 1.75 or 2.85 ±0.05 mm), even if the relevant medical grade segment is still limited.

Owing to the aforementioned melt processing shortcomings, the successful application of FDM to PHA has been reported so far by only a few studies. In 2016, Kosorn *et al.* [69] melt-extruded filaments composed by blends of PHBV with PCL, one of the most investigated biodegradable polymers for FDM thanks to its melt rheological properties [32]. By employing a 0/90/45/135° lay-down pattern, filaments made of PHBV/PCL in different weight ratios (i.e., 75:25, 50:50, and 25:75) were processed by FDM at 170°C into 3D tetragonal scaffolds (6.0 × 6.0 × 2.5 mm^3). The scaffold's compressive strength was enhanced by increasing the weight percentage of PHBV in the blend. In addition, scaffolds were submitted to a low-pressure oxygen plasma treatment that was found to increase the surface roughness and wettability, without affecting sample mechanical properties. Porcine chondrocytes exhibited higher proliferative capacity and chondrogenic potential when being cultured on the scaffolds with greater PHBV contents and submitted to plasma treatment.

A recent study explored the influence of four commercial monomeric plasticizers based on esters of citric acid (trademark Citroflex®), i.e., tributyl citrate, acetyl tributyl citrate, acetyl trihexyl citrate, and n-butyryl tri-n-hexyl citrate, on the preparation and properties of PHB/PDLLA blend scaffolds by FDM (Figures 6.3b–e) [70]. In particular, a PHB/PDLLA/plasticizer blend (weight ratio of 60/25/15) was processed into filaments for FDM by means of single screw and co-rotating meshing twin screw extruders. Thermal and tensile mechanical properties of the plasticized blends, in the form of filaments or FDM-printed dog bone shape samples (Figure 6.3f), were improved by selecting the proper plasticizer. For instance, elongation at break of printed samples increased from 5% for neat non-plasticized PHB/PDLLA blend samples to 187% for the optimized plasticized blends. In addition, warping effects during printing at 190°C were minimized in plasticized blends allowing enhanced shape stability in comparison to neat PHB/PDLLA samples (Figure 6.3g,h).

A recent article reported on melt extrusion at 140°C of filaments made of a composite made from a maleic anhydride-grafted PHA (PHA-g-MA) loaded with palm fibers that were previously treated with a silane coupling agent as effective means to increase the interfacial adhesion [71]. Similar approaches were adopted to prepare filaments made of PHA-g-MA loaded with wood flour [72] or multi-walled carbon nanotubes [73]. However, although the diameter of these PHA-g-MA-based filaments was tailored to FDM requirements, their FDM processing was not reported in any of the mentioned articles. Filaments made of PDLLA blended with 12 wt.%. PHA are available on the market [74] and their FDM processing is reported in the literature [75, 76].

6.3.3 Computer-Aided Wet-Spinning (CAWS)

CAWS is a hybrid AM technique investigated over the last years for the fabrication of biomedical polymeric devices endowed with a dual-scale macro/microporous architecture. Either biodegradable [77] or biostable [78] polymers have been processed by CAWS into devices designed as bone or cardiovascular implants [79] as well as *in vitro* cancer models [80,81].

Ongoing research on the combination of AM with other materials processing techniques (e.g., electrospinning, salt-leaching, and freeze-drying) is mainly focused on increasing fabrication resolution, enhancing structural stability, or endowing the resulting manufactured items with microstructural features [82]. In this context, CAWS technique represents a successful example of hybrid AM enabling the resulting polymeric construct to be endowed with a local microporosity [79]. Indeed, the application of the AM principles to wet-spinning enables the controlled deposition of an extruded polymeric solution or suspension directly into a coagulation bath, typically composed by a non-solvent with respect to the polymer (Figure 6.4a–c). The synchronized motion of the needle and the coagulation bath control the deposition of the solidifying filament to build up a 3D construct with a layer-by-layer process. Under optimal processing conditions, a network of macropores with a size determined by the lay-down pattern can be integrated with a microporosity in the polymeric matrix formed as a consequence of the non-solvent-induced phase separation (NIPS) process governing polymer coagulation and solidification [78, 83–86]. The control and variation of various processing conditions related to polymer solution/suspension (e.g., concentration and additives), deposition process (e.g., lay-down velocity, solution feed rate, and coagulation bath composition), and environmental conditions (e.g., temperature and humidity) allow tailoring of microporosity dimensional features. This dual-scale porosity offers larger pores ensuring efficient nutrient supply and cell migration, together with micropores providing high surface area

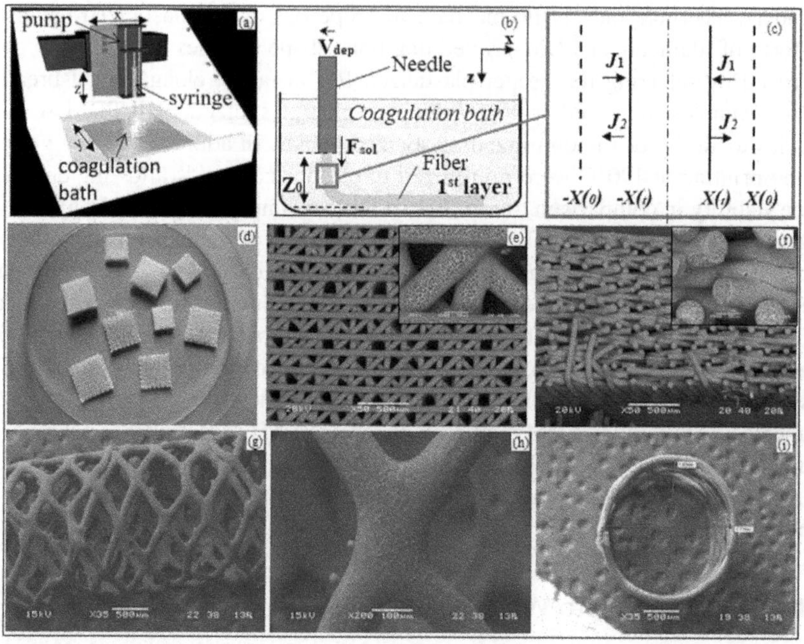

FIGURE 6.4 Computer-aided wet-spinning (CAWS) of PHA: (a) schematic representation of CAWS apparatus; (b) extruded polymer solution deposition process (Z_0: initial distance between needle-tip and deposition plane; V_{dep}: deposition velocity; F_{sol}: solution feed rate); and (c) counter diffusion between non-solvent (J1) and solvent (J2) at the coagulation interface (modified from [78]); (d) PHBHHx scaffolds by CAWS: photograph of scaffolds with different external dimensions, pore size, and pore geometry; SEM micrographs of (e) top-view and (f) cross-section of scaffolds fabricated with 0°–45°–90°–135° lay-down pattern (modified from [88]); PHBHHx stents by CAWS: SEM micrographs of (g) and (h) lateral-view, and top-view (i) of small caliber tubular constructs (φ~2 mm).

and roughness favoring cell adhesion and tissue integration. Another advantage of CAWS over other AM techniques is the possibility of biofunctionalizing the device by simply adding a drug to the solution before its extrusion [87].

The first article on PHA processing by CAWS was focused on chloroform suspensions of PHBHHx extruded in an ethanol bath [88]. The application of optimized processing conditions allowed the fabrication of 3D scaffolds with tunable pore size and geometry, as well as overall porosity, by varying the lay-down pattern and the distance between deposition lines (Figure 6.4d). These changes in architectural parameters were reflected in significant effects on scaffold compressive properties. In addition, SEM analysis of the fibers' external surface and cross-section clearly showed a "spongy" morphology as a result of a diffuse microporosity in the polymeric matrix (Figure 6.4e–f). Optimized PHBHHx scaffolds were able to sustain *in vitro* the proliferation of the MC3T3-E1 murine pre-osteoblast cell line.

A recent study resulted in the development of PHBHHx scaffolds with an anatomical shape and dimensions resembling those of a critical size segment (length of 2 mm) of a New Zealand rabbit's radius model [89]. In addition, the scaffolds were endowed with a predesigned longitudinal macrochannel to favor host tissue infiltration from the edges of the resected bone upon implantation. These scaffolds showed anisotropic behavior under either tension or compression loading, and the presence of the longitudinal channel resulted in increased compressive stiffness.

Another article demonstrated that the CAWS approach was suitable for processing PHBHHx/PCL blends dissolved in tetrahydrofuran. By varying the weight ratio between the two blend components (3:1, 2:1, 1:1) different scaffold prototypes with an interconnected macroporous structure and a microporosity in the polymer matrix were developed [90]. The variation of blend composition significantly influenced the scaffold morphological, thermal, and mechanical properties. *In vitro* biological characterization studies showed that the developed scaffolds were able to sustain the adhesion and proliferation of MC3T3-E1 murine pre-osteoblast cells.

An innovative CAWS approach involves the employment of a rotating mandrel immersed in an ethanol bath as a fiber collector [91]. In this way, 3D tubular constructs can be fabricated by winding the coagulating wet-spun polymeric fiber around the mandrel with a computer-controlled layering process (Figure 6.4g–i). PHBHHx devices developed adopting this fabrication strategy were investigated as potential biodegradable intravascular stents for small-caliber blood vessels, showing great radial elasticity and excellent results in terms of thromboresistivity when in contact with human blood. Although this layer-by-layer approach complies with generic descriptions of AM commonly adopted in literature, more specific definitions consider a complete digitalization, and elimination of tools and dies as distinctive features of AM compared with other manufacturing processes [92]. CAWS fabrication of polymeric stents without using an auxiliary cylindrical collector should therefore be tested in order for this approach to fully qualify as falling under rigorous AM definitions.

6.4 CONCLUSIONS AND OUTLOOK

An increasing amount of literature is reporting on the application of AM to PHA processing into predesigned constructs mainly tailored to biomedical applications. The widespread interest in this research trend relies on the possibility of combining the advantages of AM in controlling the composition and structural features at different length scale levels with the renewable and sustainable development principles driving PHA industrial research.

SLS, FDM, and CAWS approaches have been customized to fabricate a range of 3D porous constructs based on PHA. In particular, SLS enables the fabrication of 3D porous PHB and PHBV constructs, possibly in combination with osteoconductive CaP ceramics, even if homogeneous, dense topographies have not been achieved due to limited coalescence of polymeric powder. FDM was successfully employed to fabricate 3D PHBV-based scaffolds with a well-defined 3D porous architecture that supported the *in vitro* proliferation and differentiation of porcine chondrocytes. However, FDM approaches require blending with other polymers or plasticizers to enhance PHA processing properties, which are limited by a small melt temperature window. CAWS technique has shown great versatility in the customization of the macroshape and microarchitecture of 3D porous scaffolds and small-caliber stents made of PHBHHx. In particular, pore size and geometry, as well as device shape and dimensions, can be tuned by varying the design and deposition parameters, while the polymer matrix microporosity can be engineered by the actions of polymer coagulation variables.

Ongoing relevant research is particularly focused on developing novel PHA formulations suitable to be processed by AM technologies already employed in industrially relevant environments, as well as on the development of novel AM approaches tailored to meet the peculiar processing requirements of PHA. Future research in this area should be aimed at broadening the range of applications of AM applied to PHA in the biomedical area as well as in other industrial sectors.

REFERENCES

1. Bose S, Ke D, Sahasrabudhe H, *et al.* Additive manufacturing of biomaterials. *Prog Mater Sci* 2018; 93: 45–111.
2. Schellauf F, Grillo Fernandes E, Braunegg G, *et al.* Properties of PHAs and their correlation to fermentation conditions in biorelated polymers. In: Chiellini E, Gil H, Braunegg G, Buchert J, Gatenholm P, van der Zee M, Eds. *Sustainable Polymer Science and Technology.* New York: Kluwer Academic/Plenum Publishers 2001; pp. 115.
3. Zinn M, Witholt B, Egli T. Occurrence, synthesis and medical application of bacterial polyhydroxyalkanoate. *Adv Drug Del Rev* 2001; 53 (1): 5–21.
4. Lemoigne M. Produits de deshydration et de polymerisation de l'acide b-oxobutyrique. *Bull Soc Chem Biol* 1926; 8: 770–82.
5. Koller M Advances in polyhydroxyalkanoate (PHA) production. *Bioengineering* 2017; 4 (4): 88.
6. Shahzad K, Narodoslawsky M, Sagir M, *et al.* Techno-economic feasibility of waste biorefinery: Using slaughtering waste streams as starting material for biopolyester production. *Waste Manag* 2017; 67: 73–85.
7. Pittmann T, Steinmetz H. Polyhydroxyalkanoate production on waste water treatment plants: Process scheme, operating conditions and potential analysis for German and European municipal waste water treatment plants. *Bioengineering* 2017; 4 (2): 54. doi:10.3390/bioengineering4020054
8. Troschl C, Meixner K, Drosg B. Cyanobacterial PHA production—Review of recent advances and a summary of three years' working experience running a pilot plant. *Bioengineering* 2017; 4 (2): 26. doi:10.3390/bioengineering4020026
9. Koller M, Maršálek L, Miranda de Sousa Dias M, *et al.* Producing microbial polyhydroxyalkanoate (PHA) biopolyesters in a sustainable manner. *New Biotech* 2017; 37: 24–38.
10. Koller M. Switching from petro-plastics to microbial polyhydroxyalkanoates (PHA): The biotechnological escape route of choice out of the plastic predicament? *The EuroBiotech J* 2019; 3 (1): 32–44.
11. Morelli A, Puppi D, Chiellini F. Polymers from renewable resources. *J Renew Mat* 2013; 1 (2): 83–112.
12. Puppi D, Chiellini F, Dash M, *et al.* Biodegradable polymers for biomedical applications In: Felton GP, Ed. *Biodegradable Polymers: Processing, Degradation and Applications.* New York: Nova Science Publishers, Inc. 2011; pp. 545–604.
13. Nigmatullin R, Thomas P, Lukasiewicz B, *et al.* Polyhydroxyalkanoates, a family of natural polymers, and their applications in drug delivery. *J Chem Technol Biotechnol* 2015; 90 (7): 1209–21.

14. Koller M. Biodegradable and biocompatible polyhydroxy-alkanoates (PHA): Auspicious microbial macromolecules for pharmaceutical and therapeutic applications. *Molecules* 2018; 23 (2): 362.
15. Manavitehrani I, Fathi A, Badr H, *et al.* Biomedical applications of biodegradable polyesters. *Polymers* 2016; 8 (1): 20.
16. Puppi D, Chiellini F. Biofabrication via integrated additive manufacturing and electrofluidodynamics. In: Guarino V, Ambrosio L, Eds. *Electrofluidodynamic Technologies (EFDTs) for Biomaterials and Medical Devices.* Duxford, UK: Woodhead Publishing 2018; pp. 71–85.
17. Anbukarasu P, Sauvageau D, Elias A. Tuning the properties of polyhydroxybutyrate films using acetic acid via solvent casting. *Sci Rep* 2015; 5: 17884.
18. Leroy E, Petit I, Audic JL, *et al.* Rheological characterization of a thermally unstable bioplastic in injection molding conditions. *Polym Degrad Stab* 2012; 97 (10): 1915–21.
19. Madkour MH, Heinrich D, Alghamdi MA, *et al.* PHA recovery from biomass. *Biomacromolecules* 2013; 14 (9): 2963–72.
20. Martin DP, Williams SF. Medical applications of poly-4-hydroxybutyrate: A strong flexible absorbable biomaterial. *Biochem Eng J* 2003; 16 (2): 97–105.
21. Miranda De Sousa Dias M, Koller M, Puppi D, *et al.* Fed-batch synthesis of poly(3-hydroxybutyrate) and poly(3-hydroxybutyrate-co-4-hydroxybutyrate) from sucrose and 4-hydroxybutyrate precursors by *Burkholderia sacchari* strain DSM 17165. *Bioengineering* 2017; 4 (2): 36.
22. Radecka IK, Jiang G, Hill DJ, *et al.* Poly(hydroxyalkanoates) composites and their applications. In: Inamuddin, Ed. *Green Polymer Composites Technology: Properties and Applications.* Boca Raton: CRC Press 2016; pp. 163–75.
23. Padermshoke A, Katsumoto Y, Sato H, *et al.* Surface melting and crystallization behavior of polyhydroxyalkanoates studied by attenuated total reflection infrared spectroscopy. *Polymer* 2004; 45 (19): 6547–54.
24. Feng L, Watanabe T, Wang Y, *et al.* Studies on comonomer compositional distribution of bacterial poly(3-hydroxybutyrate-co-3-hydroxyhexanoate)s and thermal characteristics of their factions. *Biomacromolecules* 2002; 3 (5): 1071–7.
25. Yang Q, Wang J, Zhang S, *et al.* The properties of poly(3-hydroxybutyrate-co-3-hy droxyhexanoate) and its applications in tissue engineering. *Curr Stem Cell Res Ther* 2014; 9 (3): 215–22.
26. Bugnicourt E, Cinelli P, Lazzeri A, *et al.* Polyhydroxyalkanoate (PHA): Review of synthesis, characteristics, processing and potential applications in packaging. *eXPRESS Polym Lett* 2014; 8 (11): 791–808.
27. Sanhueza C, Acevedo F, Rocha S, *et al.* Polyhydroxyalkanoates as biomaterial for electrospun scaffolds. *Int J Biol Macromol* 2019; 124: 102–10.
28. Ellis G, Cano P, Jadraque M, *et al.* Laser microperforated biodegradable microbial polyhydroxyalkanoate substrates for tissue repair strategies: An infrared microspectroscopy study. *Anal Bioanal Chem* 2011; 399 (7): 2379–88.
29. Puppi D, Pirosa A, Morelli A, *et al.* Design, fabrication and characterization of tailored poly[(R)-3-hydroxybutyrate-co-(R)-3-hydroxyexanoate] scaffolds by computer-aided wet-spinning. *Rapid Prototyp J* 2018; 24 (2): 1–8.
30. Mota C, Puppi D, Chiellini F, *et al.* Additive manufacturing techniques for the production of tissue engineering constructs. *J Tissue Eng Regen Med* 2015; 9 (3): 174–90.
31. Ligon SC, Liska R, Stampfl J, *et al.* Polymers for 3D printing and customized additive manufacturing. *Chem Rev* 2017; 117 (15): 10212–90.
32. Youssef A, Hollister SJ, Dalton PD. Additive manufacturing of polymer melts for implantable medical devices and scaffolds. *Biofabrication* 2017; 9 (1): 012002.

33. Adamzyk C, Kachel P, Hoss M, *et al.* Bone tissue engineering using polyetherketoneketone scaffolds combined with autologous mesenchymal stem cells in a sheep calvarial defect model. *J Craniomaxillofac Surg* 2016; 44 (8): 985–94.
34. Morrison RJ, Hollister SJ, Niedner MF, *et al.* Mitigation of tracheobronchomalacia with 3D-printed personalized medical devices in pediatric patients. *Sci Transl Med* 2015; 7 (285): 285ra64.
35. Froyen L, Kruth JP, Laoui T, *et al.* Lasers and materials in selective laser sintering. *Assembly Autom* 2003; 23 (4): 357–71.
36. Fina F, Goyanes A, Gaisford S, *et al.* Selective laser sintering (SLS) 3D printing of medicines. *Int J Pharm* 2017; 529 (1): 285–93.
37. Drummer D, Rietzel D, Kühnlein F. Development of a characterization approach for the sintering behavior of new thermoplastics for selective laser sintering. *Phys Procedia* 2010; 5: 533–42.
38. Schmid M, Amado A, Wegener K. Polymer powders for selective laser sintering (SLS). *AIP Conf Proc* 2015; 1664 (1): 160009.
39. Schmid M, Amado A, Wegener K. Materials perspective of polymers for additive manufacturing with selective laser sintering. *J Mater Res* 2014; 29 (17): 1824–32.
40. Schmid M, Wegener K. Additive manufacturing: Polymers applicable for laser sintering (LS). *Procedia Eng* 2016; 149: 457–64.
41. Mazzoli A. Selective laser sintering in biomedical engineering. *Med Biol Eng Comput* 2013; 51 (3): 245–56.
42. Williams JM, Adewunmi A, Schek RM, *et al.* Bone tissue engineering using polycaprolactone scaffolds fabricated via selective laser sintering. *Biomaterials* 2005; 26 (23): 4817–27.
43. Wiria FE, Leong KF, Chua CK, *et al.* Poly-ε-caprolactone/hydroxyapatite for tissue engineering scaffold fabrication via selective laser sintering. *Acta Biomater* 2007; 3 (1): 1–12.
44. Eosoly S, Lohfeld S, Brabazon D. Effect of hydroxyapatite on biodegradable scaffolds fabricated by SLS. *Key Eng Mater* 2009; 396–398: 659–62.
45. Eshraghi S, Das S. Mechanical and microstructural properties of polycaprolactone scaffolds with one-dimensional, two-dimensional, and three-dimensional orthogonally oriented porous architectures produced by selective laser sintering. *Acta Biomater* 2010; 6 (7): 2467–76.
46. Lee P-H, Chang E, Yu S, *et al.* Modification and characteristics of biodegradable polymer suitable for selective laser sintering. *Int J Precis Eng Manuf* 2013; 14 (6): 1079–86.
47. Antonov EN, Bagratashvili VN, Whitaker MJ, *et al.* Three-dimensional bioactive and biodegradable scaffolds fabricated by surface-selective laser sintering. *Adv Mater* 2004; 17 (3): 327–30.
48. Antonov EN, Bagratashvili VN, Howdle SM, *et al.* Fabrication of polymer scaffolds for tissue engineering using surface selective laser sintering. *Laser Phys* 2006; 16 (5): 774–87.
49. Gayer C, Abert J, Bullemer M, *et al.* Influence of the material properties of a poly(D,L-lactide)/β-tricalcium phosphate composite on the processability by selective laser sintering. *J Mech Behav Biomed Mater* 2018; 87: 267–78.
50. Oliveira MF, Maia IA, Noritomi PY, *et al.* Construção de Scaffolds para engenharia tecidual utilizando prototipagem rápida. *Matéria* 2007; 12 (2): 373–82.
51. Pereira TF, Silva MAC, Oliveira MF, *et al.* Effect of process parameters on the properties of selective laser sintered poly(3-hydroxybutyrate) scaffolds for bone tissue engineering. *Virtual Phys Prototyp* 2012; 7 (4): 275–85.
52. Pereira TF, Oliveira MF, Maia IA, *et al.* 3D printing of poly(3-hydroxybutyrate) porous structures using selective laser sintering. *Macromol Symp* 2012; 319 (1): 64–73.

53. Duan B, Wang M, Zhou WY, *et al.* Three-dimensional nanocomposite scaffolds fabricated via selective laser sintering for bone tissue engineering. *Acta Biomater* 2010; 6 (12): 4495–505.
54. Duan B, Cheung WL, Wang M. Optimized fabrication of Ca–P/PHBV nanocomposite scaffolds via selective laser sintering for bone tissue engineering. *Biofabrication* 2011; 3 (1): 015001.
55. Duan B, Wang M. Customized Ca–P/PHBV nanocomposite scaffolds for bone tissue engineering: Design, fabrication, surface modification and sustained release of growth factor. *J R Soc Interface* 2010; 7 (Suppl 5): S615–S29.
56. Rohner D, Hutmacher Dietmar W, Cheng Tan K, *et al.* In vivo efficacy of bone-marrow-coated polycaprolactone scaffolds for the reconstruction of orbital defects in the pig. *J Biomed Mater Res B Appl Biomater* 2003; 66B (2): 574–80.
57. Park SH, Park DS, Shin JW, *et al.* Scaffolds for bone tissue engineering fabricated from two different materials by the rapid prototyping technique: PCL versus PLGA. *J Mater Sci Mater Med* 2012; 23 (11): 2671–8.
58. Jinku K, Sean M, Brandi T, *et al.* Rapid-prototyped PLGA/β-TCP/hydroxyapatite nanocomposite scaffolds in a rabbit femoral defect model. *Biofabrication* 2012; 4 (2): 025003.
59. Hutmacher DW. Scaffolds in tissue engineering bone and cartilage. *Biomaterials* 2000; 21 (24): 2529–43.
60. Wang F, Shor L, Darling A, *et al.* Precision extruding deposition and characterization of cellular poly-ε-caprolactone tissue scaffolds. *Rapid Prototyp J* 2004; 10 (1): 42–9.
61. Gloria A, Russo T, De Santis R, *et al.* 3D fiber deposition technique to make multifunctional and tailor-made scaffolds for tissue engineering applications. *J Appl Biomater Biomech* 2009; 7 (3): 141–52.
62. Mota C, Puppi D, Dinucci D, *et al.* Dual-scale polymeric constructs as scaffolds for tissue engineering. *Materials* 2011; 4 (3): 527–42.
63. Brown TD, Dalton PD, Hutmacher DW. Melt electrospinning today: An opportune time for an emerging polymer process. *Prog Polym Sci* 2016; 56: 116–66.
64. Dalton PD, Vaquette C, Farrugia BL, *et al.* Electrospinning and additive manufacturing: Converging technologies. *Biomater Sci* 2013; 1 (2): 171–85.
65. Brown TD, Dalton PD, Hutmacher DW. Direct writing by way of melt electrospinning. *Adv Mater* 2011; 23 (47): 5651–7.
66. Gernot H, Tomasz J, Toby DB, *et al.* Additive manufacturing of scaffolds with submicron filaments via melt electrospinning writing. *Biofabrication* 2015; 7 (3): 035002.
67. Fitzharris ER, Watanabe N, Rosen DW, *et al.* Effects of material properties on warpage in fused deposition modeling parts. *Int J Adv Manuf Technol* 2018; 95 (5): 2059–70.
68. Love LJ, Kunc V, Rios O, *et al.* The importance of carbon fiber to polymer additive manufacturing. *J Mater Res* 2014; 29 (17): 1893–8.
69. Kosorn W, Sakulsumbat M, Uppanan P, *et al.* PCL/PHBV blended three dimensional scaffolds fabricated by fused deposition modeling and responses of chondrocytes to the scaffolds. *J Biomed Mater Res B Appl Biomater* 2016; 105: 1141–1150.
70. Menčík P, Přikryl R, Stehnová I, *et al.* Effect of selected commercial plasticizers on mechanical, thermal, and morphological properties of poly(3-hydroxybutyrate)/poly(lactic acid)/plasticizer biodegradable blends for three-dimensional (3D). *Print Mater* 2018; 11 (10): 1893.
71. Wu C-S, Liao H-T, Cai Y-X. Characterisation, biodegradability and application of palm fibre-reinforced polyhydroxyalkanoate composites. *Polym Degrad Stab* 2017; 140: 55–63.
72. Wu C-S, Liao H-T. Fabrication, characterization, and application of polyester/wood flour composites. *J Polym Eng* 2017; 37 (7): 689.

73. Wu CS, Liao HT. Interface design of environmentally friendly carbon nanotube-filled polyester composites: Fabrication, characterisation, functionality and application. *Express Polym Lett* 2017; 11 (3): 187–98.
74. colorFabb BV. The Netherlands. https://colorfabb.com, accessed June 2020.
75. Gonzalez Ausejo J, Rydz J, Musioł M, *et al.* A comparative study of three-dimensional printing directions: The degradation and toxicological profile of a PLA/PHA blend. *Polym Degrad Stab* 2018; 152: 191–207.
76. Kaygusuz B, Özerinç S. Improving the ductility of polylactic acid parts produced by fused deposition modeling through polyhydroxyalkanoate additions. *J Appl Polym Sci* 2019; 136 (43): 48154.
77. Puppi D, Mota C, Gazzarri M, *et al.* Additive manufacturing of wet-spun polymeric scaffolds for bone tissue engineering. *Biomed Microdevices* 2012; 14 (6): 1115–27.
78. Puppi D, Morelli A, Bello F, *et al.* Additive manufacturing of poly(methyl methacrylate) biomedical implants with dual-scale porosity *Macromol Mater Eng* 2018; 303 (9): 1800247.
79. Puppi D, Chiellini F. Wet-spinning of biomedical polymers: From single-fibre production to additive manufacturing of three-dimensional scaffolds. *Polym Int* 2017; 66 (12): 1690–6.
80. Puppi D, Migone C, Morelli A, *et al.* Microstructured chitosan/poly(γ-glutamic acid) polyelectrolyte complex hydrogels by computer-aided wet-spinning for biomedical three-dimensional scaffolds. *J Bioact Compatible Polym* 2016; 31 (5): 531–49.
81. Chiellini F, Puppi D, Piras AM, *et al.* Modelling of pancreatic ductal adenocarcinoma in vitro with three-dimensional microstructured hydrogels. *RSC Adv* 2016; 6 (59): 54226–35.
82. Giannitelli SM, Mozetic P, Trombetta M, *et al.* Combined additive manufacturing approaches in tissue engineering. *Acta Biomater* 2015; 24: 1–11.
83. Mota C, Puppi D, Dinucci D, *et al.* Additive manufacturing of star poly(ε-caprolactone) wet-spun scaffolds for bone tissue engineering applications. *J Bioact Compat Polym* 2013; 28 (4): 320–40.
84. Puppi D, Migone C, Grassi L, *et al.* Integrated three-dimensional fiber/hydrogel biphasic scaffolds for periodontal bone tissue engineering. *Polym Int* 2016; 65 (6): 631–40.
85. Puppi D, Piras AM, Pirosa A, *et al.* Levofloxacin-loaded star poly(ε-caprolactone) scaffolds by additive manufacturing. *J Mater Sci Mater Med* 2016; 27 (3): 44.
86. Romagnoli C, Zonefrati R, Puppi D, *et al.* Human adipose tissue-derived stem cells and a poly(ε-caprolactone) scaffold produced by computer-aided wet spinning for bone tissue engineering. *J Biomater Tissue Eng* 2017; 7 (8): 622–33.
87. Puppi D, Piras AM, Pirosa A, *et al.* Levofloxacin-loaded star poly(ε-caprolactone) scaffolds by additive manufacturing. *J Mater Sci - Mater Med* 2016; 27 (3): 44.
88. Mota C, Wang SY, Puppi D, *et al.* Additive manufacturing of poly[(*R*)-3-hydroxybutyrate-*co*-(*R*)-3-hydroxyhexanoate] scaffolds for engineered bone development. *J Tissue Eng Regen Med* 2017; 11 (1): 175–86.
89. Puppi D, Pirosa A, Morelli A, *et al.* Design, fabrication and characterization of tailored poly[(R)-3-hydroxybutyrate-*co*-(R)-3-hydroxyexanoate] scaffolds by computer-aided wet-spinning. *Rapid Prototyp J* 2018; 24 (1): 1–8.
90. Puppi D, Morelli A, Chiellini F. Additive manufacturing of poly(3-hydroxybutyra te-co-3-hydroxyhexanoate)/poly(ε-caprolactone) blend scaffolds for tissue engineering. *Bioengineering* 2017; 4 (2): 49.
91. Puppi D, Pirosa A, Lupi G, *et al.* Design and fabrication of novel polymeric biodegradable stents for small caliber blood vessels by computer-aided wet-spinning. *Biomed Mater* 2017; 12 (3): 035011.
92. Gibson I, Rosen D, Stucker B. *Additive Manufacturing Technologies: Rapid Prototyping to Direct Digital Manufacturing.* New York/Heidelberg/Dordrecht/London: Springer 2010.

7 Mechanical and Permeation Properties of PHA-Based Blends and Composites

Verena Jost

CONTENTS

7.1 INTRODUCTION

The final properties of films from polyhydroxyalkanoates (PHA) are affected by various factors. The production conditions of PHA (as described in the previous chapters) affect, amongst other things, the chain length of the monomer(s), the molecular weight (M_w) of the homopolymer or copolymer and the monomer ratio of the copolymers; these factors then determine the final properties of the films. Other crucial factors are the degree of polymerisation and the use and amount of a nucleating agent as well as the processing, crystallisation and storage conditions.

The chain length of the monomers, namely of the hydroxycarboxylic acids, determines the final properties to a large extent. A short chain length – resulting in a short side chain – gives highly crystalline and brittle films. An example of this is the homopolymer poly(3-hydroxybutyrate) (PHB). By increasing the chain length, and thereby the side chain, the final films have a lower melting temperature (T_m) and glass transition temperature (T_g), reduced crystallinity and increased flexibility [1–3]. Examples are films of poly(3-hydroxyvalerate) (PHV) and poly(3-hydroxyhexanoate) (PHH). Commercially available grades are mostly copolymers such as poly(3-hydroxybutyrate-*co*-3-hydroxyvalerate) (PHBV). For these grades, the 3HV content is decisive. Also, the M_w is an important factor affecting, amongst other things, the T_m of the polymer [4]. PHB with a high M_w can have a similar spherulite morphology to low M_w PHB when it crystallises at a higher crystallisation temperature (T_c) [5].

Another decisive factor is the processing (the preparation method) of PHA. Due to their solubility in chloroform, PHA can be processed via solvent mixing and casting (solution casting). As this process is performed without (intense) thermal treatment, thermally induced changes to PHA can be minimised. However, this process is only feasible on a laboratory scale. Another possibility is a small-scale moulding process under heat and pressure (compression moulding). This process reduces the applied stress due to the lack of shear force. However, this process is only feasible for small amounts of PHA and low throughput rates. Commonly, PHA are processed by thermoplastic extrusion and subsequent shaping via injection moulding, flat film extrusion or blown film extrusion. These processes can be scaled up, enabling high throughput. Challenges of thermoplastic extrusion can be the residual moisture, low viscosity and low melt strength of PHA as well as the slow crystallisation rate. As the decomposition temperature (T_{dec}) is quite low for some PHA and therefore close to the T_m, thermally induced changes can even occur during processing (reduction of the M_w to approximately half of the original value for a residence time of 1 hour at 190°C [5]). This can lead to modification of, amongst other things, the colour, mechanical properties and permeation properties. The processing window can be extended by reducing the T_m via an increase of the side chain length.

PHA have comparably slow crystallisation rates and a high post crystallisation. Crystallisation can be affected by nucleating agents. The addition of the latter leads to the formation of more and smaller spherulites. This usually results in films with increased flexibility [6]. A further effect is an increased T_c, while the T_m is only a little affected [7]. PHB crystallises in two different forms: a helical (α-form) and a planar zigzag (β-form) conformation [8]. Annealing PHB at 100°C leads to more prevalent formation of the α-form crystals. However, β-form crystals give films with improved mechanical properties [8].

7.2 PROPERTIES OF PHA FILMS

PHA are biobased and biodegradable polyesters which can be processed thermally. Noteworthy are their high strength and elastic modulus, good water resistance and low permeability to oxygen and especially to water vapour compared to other biopolymers. However, challenges are their high crystallinity which leads to brittle and stiff films, low toughness and flexibility, poor ductility and poor resistance to high temperatures, acids and bases.

The low oxygen and water vapour permeability of several PHA (PHB, PHBV, poly(3-hydroxybutyrate-*co*-4-hydroxybutyrate) (PHBHB)) are illustrated in Figure 7.1 (based on [9]) and are compared to other thermoplastic biopolymers and polymers.

Besides possessing these attractive permeation properties, the challenging processing and the low flexibility of PHA films still limit their broad application. A major drawback is their low elongation at break (Figure 7.2, based on [9]).

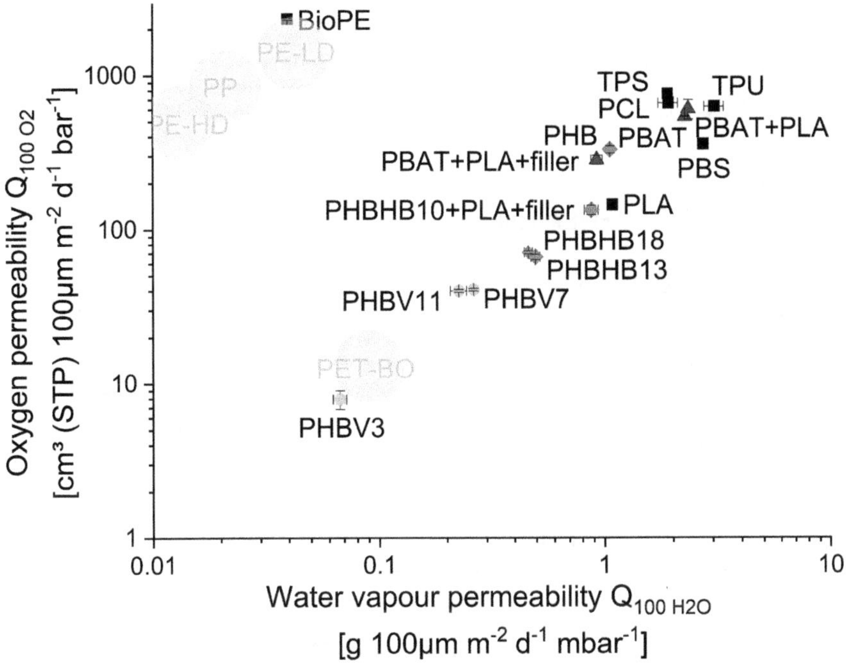

FIGURE 7.1 Oxygen permeability (23°C, 50% r.h.) versus water vapour permeability (23°C, 85 → 0% r.h.) of PHA: PHBV (grey spheres: number indicates the 3HV content) and PHBHB (grey rhombuses: number indicates the 3HB content) compared to PBAT-based materials (black triangles), other biopolymers (black squares) [9], and conventional polymers (large transparent spheres) [10].

7.3 ADDITIVES

In order to use PHA for applications such as packaging, the processing and the final structural and mechanical properties need to be improved. In order to improve the slow crystallisation, to reduce the high brittleness and to enhance the low flexibility, various approaches are feasible. This chapter focuses on the effect of additives on the properties of PHA. Additives which particularly affect the processing and mechanical properties include low molecular weight plasticisers, blend partners and fillers. Compatibilisers and crosslinkers can also be added to blends to further improve the miscibility of multi-component systems.

7.3.1 Low Molecular Weight Plasticisers

The addition of a plasticiser improves the flexibility of a polymer. Plasticisers are commonly substances with a low M_w and low steric hindrance which interact with the polymer chains to enhance the chain mobility. The free volume theory explains their effect on the polymer, such as the reduction of T_g and increase of the elongation by increasing the free volume, which induces increased motion of the polymer chain [11, 12]. The effect of a plasticiser is not only limited to structural properties. The

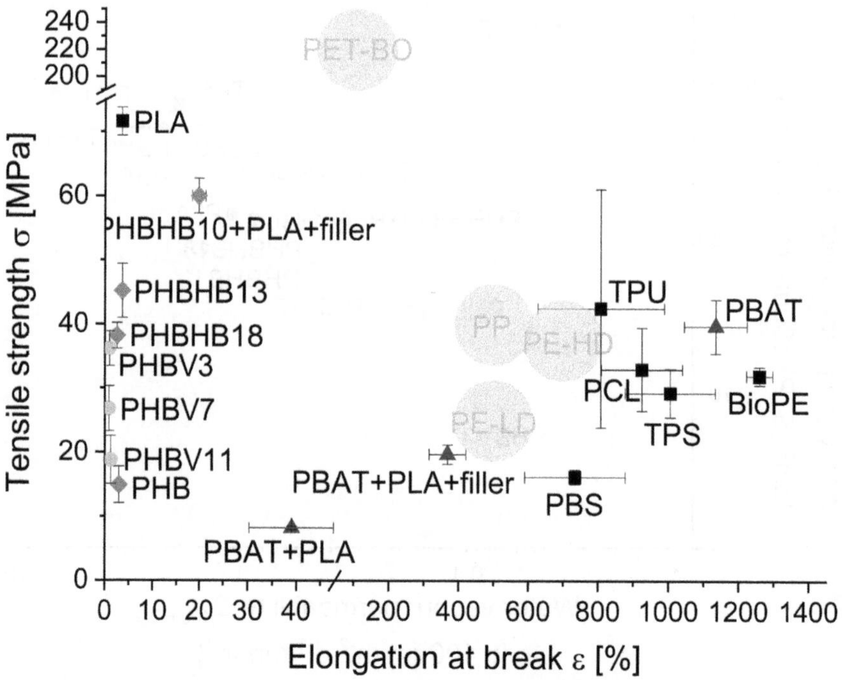

FIGURE 7.2 Tensile strength versus elongation at break of PHA: PHBV (grey spheres: number indicates the 3HV content) and PHBHB (grey rhombuses: number indicates the 3HB content) compared to PBAT-based materials (black triangles), other biopolymers (black squares) [9] and conventional polymers (large transparent spheres) [10].

final properties of the films, such as the mechanical properties and the permeability to various gases and water vapour, are also affected. Table 7.1 summarises the effects of different plasticisers on the structural, mechanical and permeation properties of PHA-based films as described in the literature. Information about the preparation method and the base material are additionally included due to their high importance for the polymer and the effect of the additive.

The effect of plasticisers is mainly evaluated by their effect on the structural and mechanical properties. The analysed polymers were PHB and PHBV and no studies on other PHA were found. Generally, the incorporation of an effective plasticiser leads to lower T_m and T_g and also lower crystallinity of PHA-based films. The films become more flexible as a result of the reduced elastic modulus and tensile strength and the enhanced elongation. Notable is that the preparation method seems to affect the impact of the plasticiser. Films processed via solvent mixing and casting have a greater increase in elongation than films processed via melt blending and extrusion. This underlines the thermal effect on the biopolymer during this method of processing and film forming. However, the effect of the plasticiser on the processability and in particular the final mechanical properties of PHA is limited due to a maximum concentration. Many plasticisers are liquids which is the reason for this maximum concentration in the melt blending process. The effect of plasticisers on permeability

TABLE 7.1
Effect of Different Plasticisers on PHA-Based Films

Applied plasticiser	Amount of applied plasticiser	Base material	Preparation method	Effect on structural properties	Effect on mechanical properties	Effect on permeation properties	Ref.
Butyryl-trihexyl citrate	30 wt.%	PHB	Solvent mixing and casting	Reduction of crystallinity (69→52%)	Reduction of EM and σ, increase of ε (1→28%)		[13]
Castor oil	13 wt.%	PHBV (3% 3HV)	Melt blending and cast film extrusion		Reduction of EM and σ, no effect on ε	No effect on WVP, slight increase of OP	[14]
Dibutyl phthalate	20 wt.%	PHBV (6% 3HV)	Solvent mixing and casting, press	Reduction of T_g (−7→−29°C)	Increase of ε (~130→~235%)		[15]
Epoxidised soybean oil	13 wt.%	PHBV (3% 3HV)	Melt blending and cast film extrusion	Reduction of T_m (173→169°C)	Reduction of EM and σ, no effect on ε	Slight increase of WVP and OP	[14]
Epoxidised soybean oil	20 wt.%	PHBV (6% 3HV)	Solvent mixing and casting, press	Reduction of T_g (−7→−19°C)	Increase of ε (~130→~175%)		[15]
Glycerol	13 wt.%	PHBV (3% 3HV)	Melt blending and cast film extrusion		Reduction of EM and σ, no effect on ε	Slight increase of WVP and OP	[14]
Oxypropylated glycerol	23 wt.%	PHB	Solvent mixing and casting		Increase of ε (~2→~193%)		[16]
Poly(ethylene glycol) (PEG) 300	20 wt.%	PHB	Solvent mixing and casting	Reduction of T_m (179→166°C) and crystallinity (31→26%)	Reduction of σ (28→13 MPa), increase of ε (9→32%)	Increase of WVTR (8.3→37 $*10^{-4}$ g mm m^{-2} d^{-1})	[17]
PEG 300	23 wt.%	PHB	Solvent mixing and casting		Increase of ε (~2→~121%)		[16]
PEG 1.000	20 wt.%	PHBV (3% 3HV)	Melt blending and cast film extrusion		Reduction of EM and σ, no effect on ε	Increase of WVP (0.21→1.22 g 100 µm m^{-2} d^{-1} $mbar^{-1}$), no significant effect on OP	[14]

(Continued)

TABLE 7.1 (CONTINUED)
Effect of Different Plasticisers on PHA-Based Films

Applied plasticiser	Amount of applied plasticiser	Base material	Preparation method	Effect on structural properties	Effect on mechanical properties	Effect on permeation properties	Ref.
PEG1.350-1.650	20 wt.%	PHBV (3.6% 3HV)	Solvent mixing and casting	Reduction of T_m (172→169°C) and crystallinity (~75→~60%)			[18]
PEG	50 wt.%	PHBV (12% 3HV)	Solvent mixing and casting	Reduction of T_m (160→153°C), acceleration of hydrolytic degradation			[19]
Polyisobutylene	23 wt.%	PHB	Solvent mixing and casting		Increase of ε (~2→~24%)		[16]
Propylene glycol	17 wt.%	PHBV (3% 3HV)	Melt blending and cast film extrusion		Reduction of EM and σ, no effect on ε	Increase of WVP and OP	[14]
Soybean oil	20 wt.%	PHBV (6% 3HV)	Solvent mixing and casting, press	Increase of T_g (−7→−3°C)	Increase of ε (~130→~75%)		[15]
Tributyrin	20 wt.%	PHB	Melt blending and moulding	Reduction of T_m and T_g, no significant effect on crystallinity	Reduction of EM, σ and ε (4.1→2.2 %)	Increase of WVP (0.6→4.3*10^{-11} g m^{-1} s^{-1} Pa^{-1} equal to 0.52→3.7 g 100 µm m^{-2} d^{-1} $mbar^{1}$)	[20]

(Continued)

TABLE 7.1 (CONTINUED)
Effect of Different Plasticisers on PHA-Based Films

Applied plasticiser	Amount of applied plasticiser	Base material	Preparation method	Effect on structural properties	Effect on mechanical properties	Effect on permeation properties	Ref.
Triethyl citrate	20 wt.%	PHBV (3% 3HV)	Melt blending and cast film extrusion	Reduction of T_m (173→164°C)	Reduction of EM and σ, increase of ε (0.8→2.3%)	Increase of WVP (0.21→1.96 g 100 μm m^{-2} d^{-1} $mbar^{-1}$) and OP (26.5→628 cm^3 (STP) 100 μm m^{-2} d^{-1} bar^{-1})	[14]
Triethyl citrate	20 wt.%	PHBV (6% 3HV)	Solvent mixing and casting, press	Reduction of T_g (−7→−30°C)	Increase of ε (~130→~235%)		[15]
Triethyl citrate	30 wt.%	PHB	Solvent mixing and casting	Reduction of crystallinity (69→65%)	Reduction of EM and σ, no effect on ε		[13]

EM: elastic modulus, σ: tensile strength, ε: elongation at break, WVP/ WVTR: water vapour permeability/transmission rate, OP: oxygen permeability has been little investigated, but some studies have found an increase in permeability if the additive is an effective plasticiser.

7.3.2 Blend Partners

Another way of improving certain properties is to process multi-component blend systems. This approach is convenient and economic and usually requires only minor adaptations to the processing. Blends are combinations of two or more polymers or biopolymers which are physically mixed, most often in a melt blending step. The blend partners can, but often do not, interact chemically. Therefore, miscibility and compatibility aspects are decisive for the stability of blends and their final properties. The miscibility of blends can be categorised into completely miscible, partially miscible and immiscible. These categories can be distinguished by a blend's morphology and T_g. Blends with a homogenous morphology and one single T_g are completely miscible, while blends with a partly homogenous morphology and a shift in the T_g are partially miscible. Blends with a heterogeneous morphology (e.g., inclusions, large interface area, minor interface reactions) and two different T_gs are immiscible [21]. Blends of biopolymer components are often not miscible or are only partially miscible.

Nevertheless, PHA-based blend systems can improve the processing and the final properties of the films. Examples of PHA-based blends and the effect of the blend partner are given in Table 7.2. Most PHA-based blends have poor miscibility, as demonstrated by two T_gs. Blending of PHA often leads to lower crystallinity, resulting in increased permeability and lower brittleness. Further effects of the incorporated blend partner are improved mechanical properties, such as lower elastic modulus and tensile strength and enhanced elongation. Again, the preparation method is a crucial factor affecting the final properties. Melt blending and film forming via extrusion lead to two thermal treatments which can, for example, reduce the polymer chain length. However, two thermal processing steps would probably be required for large-scale applications.

An effective blend partner affects several properties and shifts them to those of the blend partner. Blending highly crystalline and brittle PHBV with flexible polymers such as PBAT, TPU or EVA enhances the flexibility and also increases the permeability. The Young's modulus (elastic modulus) and water vapour permeability of a PHBV-based blend depend on the blend partner's concentration and show inverse proportionality (Figure 7.3). Due to the potential phase separation between the two polymers, there is a drastic change in properties at a certain concentration of the blend partner. At this concentration, the coherent PHBV network becomes disrupted, leading to a drastic decrease in crystallinity and Young's modulus and preferred permeation through the phase of the blend partner [23].

7.3.3 Compatibilisers

Blending is an effective way of tailoring the properties of PHA. However, multi-component PHA-based blends often have poor compatibility and miscibility. These properties can be improved by the addition of compatibilisers or crosslinkers which modify the interfacial tension between the blend partners by chemical interaction.

TABLE 7.2
Effect of Different Blend Partners on PHA-Based Films

Applied blend partner	Amount of applied blend partner	Base material	Preparation method	Effect on structural properties	Effect on mechanical properties	Effect on permeation properties	Ref.
Carboxylterminated butadiene acrylonitrile rubber	2 wt.%	PHB	Solvent mixing and casting	Two T_m, no effect on T_m, increase of crystallinity	Improved thermal stability		[22]
Ethylene vinyl acetate copolymer (EVA)	50 wt.%	PHBV (3% 3HV)	Melt blending and cast film extrusion	One T_m, reduction of T_m (172 → 164°C), reduction of crystallinity (67 → 61%)	Reduction of EM and σ, increase of ε (1.0 → 95%)	Increase of WVP (0.07 → 0.7 g 100 µm m^{-2} d^{-1} $mbar^{-1}$) and OP (7.9 → 142 cm^3 (STP) 100 µm m^{-2} d^{-1} bar^{-1})	[23]
Poly (butylene adipate-*co*-terephthalate) (PBAT)	30 wt.%	PHBV (3% 3HV)	Melt blending and blown film extrusion	Two T_m, reduction of $T_{m,PHBV}$ (176 → 171°C)	No significant effect on EM, increase of ε (~1.4 → ~11%)	Increase of WVP (~1.25 → ~1.6*10^{-11} g m m^{-2} s^{-1} Pa^{-1} equal to ~1.1 → ~1.4 g 100 µm m^{-2} d^{-1} $mbar^{-1}$)	[24]
PBAT	50 wt.%	PHBV (3% 3HV)	Melt blending and cast film extrusion	One T_m, slight reduction of T_m, reduction of crystallinity (67 → 29%)	Reduction of EM and σ, slight increase of ε	Increase of WVP (0.070.33 g 100 µm m^{-2} d^{-1} $mbar^{-1}$) and OP (7.9 → 48 cm^3 (STP) 100 µm m^{-2} d^{-1} bar^{-1})	[23]
PBAT	55 wt.%	PHBV	Injection moulding	Two T_m, reduction of $T_{m,PHBV}$ (166 → 157°C), two T_g, reduction of crystallinity (78 → 58%)	Reduction of specific EM (2.2 → 0.7 MPa kg^{-1} m^{-3}) and specific σ, increase of ε (2.7 → 6.9%)		[25]
PBAT+ PLA	30 wt.%	PHBV (3% 3HV)	Melt blending and cast film extrusion	One T_m, no effect on T_m, reduction of crystallinity (67 → 59%)	Reduction of EM and σ, no significant effect on ε	Increase of WVP and OP	[23]

(Continued)

TABLE 7.2 (CONTINUED)
Effect of Different Blend Partners on PHA-Based Films

Applied blend partner	Amount of applied blend partner	Base material	Preparation method	Effect on structural properties	Effect on mechanical properties	Effect on permeation properties	Ref.
Polybutylene succinate (PBS)	20 wt.%	PHBV (14% 3HV)	Solvent mixing and casting	One T_g, no effect on T_g, decrease of crystallisation rate			[26]
PBS	30 wt.%	PHBV (3% 3HV)	Melt blending and cast film extrusion	Two T_m, no effect on $T_{m,PHBV}$, no significant effect on crystallinity	Reduction of EM and σ, slight increase of ε	Increase of WVP and OP	[23]
Poly(ε-caprolactone) (PCL)	20 wt.%	PHBV (3% 3HV)	Melt blending and cast film extrusion	Two T_m, no effect on $T_{m,PHBV}$, no effect on crystallinity	Reduction of EM, no significant effect on σ, slight increase of ε	Increase of WVP and OP	[23]
PCL	50 wt.%	PHBV (12% 3HV)	Solvent mixing and casting	Two T_m, reduction of $T_{m,PHBV}$ (~148 → ~147°C)	Reduction of EM (1300 → 230 MPa) and σ (22 → 5 MPa), increase of ε (3 → 12%)		[27]
Poly epichlorhydrin (PECH)	60 wt.%	PHB	Solvent mixing and casting	Reduction of crystallinity	Reduction of σ, increase of ε		[28]
Polyglycidyl methacrylate (PGMA)	30 wt.%	PHB	Solvent mixing and casting	One T_m, slight reduction of T_m, increase of T_c, no effect on T_g			[29]
Polylactide (PLA)	10 wt.%	PHBV (8% 3HV)	Dry blending and flat film extrusion	Two T_m, slight reduction of $T_{m,PHBV}$ (171 → 168°C), one T_g	Reduction of EM (_4 → 3.4 GPa), increase of ε (1 → 26%)	Helium permeability lower than of pure PLA film	[30]

(Continued)

TABLE 7.2 (CONTINUED)
Effect of Different Blend Partners on PHA-Based Films

Applied blend partner	Amount of applied blend partner	Base material	Preparation method	Effect on structural properties	Effect on mechanical properties	Effect on permeation properties	Ref.
PLA	50 wt.%	PHBV	Melt blending and injection moulding	Two T_m, one T_g, reduction of crystallinity (51 → 18%)	Increase of tensile modulus, σ (~24 → ~40 MPa) and ε (~5.5 → ~10%)		[31]
PLA	50 wt.%	PHBV	Melt blending and compression moulding	Two T_m, no significant effect on $T_{m,PHBV}$, two T_g, reduction of crystallinity (47 → 26%)		Increase of WVP (0.23 → 1.42*10^{-11} g m^{-1} s^{-1} Pa^{-1} equal to 0.2 → 1.2 g 100 μm m^{-2} d^{-1} $mbar^{-1}$) and OP (~0.4 → ~1.9 cm^3 μm m^{-2} d^{-1} bar^{-1})	[32]
PLA	60 wt.%	PHBV (20% 3HV)	Solvent mixing and casting	Two T_g, one T_m, no effect on T_g and T_m, reduction of crystallinity	Increase of EM and σ, reduction of ε		[33]
PLA	75 wt.%	PHB	Melt blending and moulding	One T_m, no effect on T_m, reduction of crystallinity	Reduction of EM and σ, no effect on ε	Increase of OP (12 → 25 cm^3 mm m^{-2} d^{-1})	[34]
PLA	75 wt.%	PHBV (3% 3HV)	Melt blending and cast film extrusion		Increase of EM, σ (28 → 64 MPa) and ε (1.2 → 14%)	Increase of WVP (0.26 → 0.48 g 100 μm m^{-2} d^{-1} $mbar^{-1}$), OP (34 → 58 cm^3 (STP) 100 μm m^{-2} d^{-1} bar^{-1}) and NP, reduction of COP	[35]
PLA	80 wt.%	PHBV	Melt blending and injection moulding		Slight increase of EM, increase of impact strength, σ and ε (3.9 → 51%)		[36]

(Continued)

TABLE 7.2 (CONTINUED)
Effect of Different Blend Partners on PHA-Based Films

Applied blend partner	Amount of applied blend partner	Base material	Preparation method	Effect on structural properties	Effect on mechanical properties	Effect on permeation properties	Ref.
Poly(propylene carbonate) (PPC)	50 wt.%	PHBV (8 mol-% 3HV)	Melt blending, injection moulding and oriented flat film extrusion	One T_m, no significant effect on $T_{m,PHBV}$, two T_g, reduction of $T_{g,PHBV}$ (~1 → ~–4°C)	Reduction of EM (3951 → 2066 MPa) and stress at break (30 → 20 MPa), increase of ε (0.9 → 20%)	Increase of WVP (2.6 → 3.4*10^{-12} g m^{-1} s^{-1} Pa^{-1} equal to 0.22 → 0.29 g 100 μm m^{-2} d^{-1} mbar^{-1}) and OP (~0.26 → ~0.37 cm^3 μm m^{-2} d^{-1} bar^{-1})	[37]
Polyvinyl acetate (PVAc)	20 wt.%	PHBV (3% 3HV)	Melt blending and cast film extrusion	One T_m, slight reduction of T_m, reduction of crystallinity (67 → 55%)	Reduction of EM, slight increase of σ and ε	Increase of WVP and OP	[23]
Polyvinylpyrrolidone	2 wt.%	PHB	Solvent mixing and casting	Two T_m, no effect on T_m, increase of crystallinity	Improved thermal stability		[22]
Thermoplastic polyurethane elastomer (TPU)	50 wt.%	PHBV (3% 3HV)	Melt blending and cast film extrusion	One T_m, no effect on T_m, reduction of crystallinity (67 → 61%)	Reduction of EM and σ, increase of ε (1.0 → 607%)	Increase of WVP (0.07 → 0.44 g 100 μm m^{-2} d^{-1} mbar^{-1}) and OP (7.9 → 51 cm^3 (STP) 100 μm m^{-2} d^{-1} bar^{-1})	[23]
Thermoplastic starch (TPS)	20 wt.%	PHBV (3% 3HV)	Melt blending and cast film extrusion	One T_m, slight reduction of T_m, no effect on crystallinity	Reduction of EM and σ, no significant effect on ε	Increase of WVP and OP	[23]

EM: elastic modulus, σ: tensile strength, ε: elongation at break, WVP: water vapour permeability, OP: oxygen permeability, NP: nitrogen permeability, COP: carbon dioxide permeability

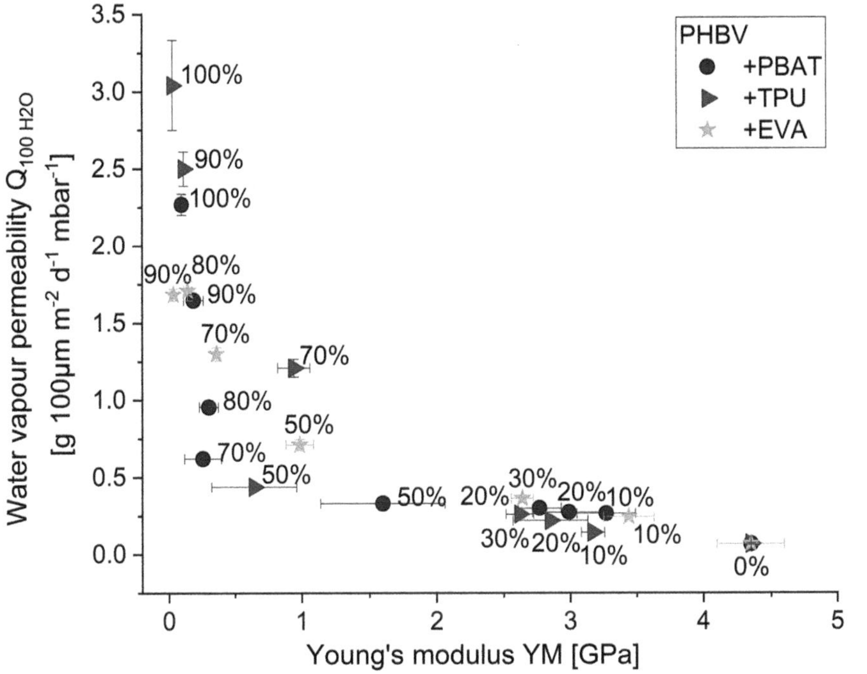

FIGURE 7.3 Water vapour permeability $Q_{100\ H2O}$ (23°C, 85 → 0% r.h.) versus Young's modulus of PHBV-based blends (number indicates the percentage of blend partner) based on [23].

While compatibilisers generally embed in the interface layer between the components and so lower the interfacial tension, crosslinkers induce additional, often covalent, bonds between the blend partners. This results in improved processing and enhanced performance of such compatibilised blends. Even though crosslinking can be induced by heat [38], irradiation [39] or enzymes [40], this chapter focuses on the effects of crosslinking by additives processed in a reactive blending step. The effects of different compatibilisers and crosslinkers on PHA-based blends are given in Table 7.3.

Compatibilisers mostly improve the miscibility of the PHA and the blend partner. This leads to lower crystallinity of the PHA phase and enhanced flexibility of the blends. However, the T_m is only rarely affected by these additives, so the long-term stability needs to be considered. As there have been few studies on compatibilised PHA-based blends, the efficiency of the interaction between the compatibiliser and the components of the blend cannot yet be generalised. The permeability of the blends is either not significantly affected or is increased as a result of the compatibilisation of the phases. However, as stated, there are still very few publications dealing with the permeation properties of compatibilised PHA-based blends.

7.3.4 Composites

Another approach for overcoming the limitations of an application of PHA – challenging processing, high crystallinity and brittleness, as well as low flexibility – is

TABLE 7.3
Effect of Different Compatibiliser on PHA-Based Films

Applied compatibiliser	Amount of applied compatibiliser	Base material	Preparation method	Effect on structural properties	Effect on mechanical properties	Effect on permeation properties	Ref.
Copper(II) acetyl acetonate ($Cu(acac)_2$)	0.44%	PLA+ PHBV (3% 3HV) (75:25)	Melt blending and cast film extrusion		Reduction of EM and σ, no significant effect on ε	No significant effect on WVP, OP and NP, increase of COP	[35]
Dicumyl peroxide (DCP)	0.5 wt.%	PHBV (12.7 mol-%)+ PBS (80:20)	Melt blending, compression and injection moulding	Two T_m, reduction of $T_{m,PHBV}$, increase of crystallinity and crystallisation rate			[41]
DCP	0.5 wt.%	PHBV (4 mol-% 3HV) + PCL (70:30)	Solvent mixing and casting	Two T_g and T_m	Reduction of EM and tensile stress at break		[42]
DCP	0.5 wt.%	PHBV (4.5 mol-%)+ PPC (50:50) + glycidyl methacrylate	Melt blending and casting	One T_m, reduction of crystallinity			[43]
DCP	0.1 wt.%	PLA+ PHBV (3% 3HV) (75:25)	Melt blending and cast film extrusion		Reduction of EM and ε, no significant effect on σ	Slight increase of WVP, OP, NP and COP	[35]
DCP	0.5 wt.%	PHBV (2 mol.% 3HV)+ PLA (70:30)	Melt blending and compression moulding	One T_g, two T_m, slight decrease of T_g and T_m, reduction of crystallinity (57 → 40%)	Increase of σ and impact toughness		[44]

(Continued)

TABLE 7.3 (CONTINUED)
Effect of Different Compatibiliser on PHA-Based Films

Applied compatibiliser	Amount of applied compatibiliser	Base material	Preparation method	Effect on structural properties	Effect on mechanical properties	Effect on permeation properties	Ref.
DCP	1.0 wt.%	PCL+ PHB (75:25)	Melt blending and compression moulding	Two T_m, no effect on T_m	Increase of σ (11.4 → 15.9 MPa) and ε (125 → 278%)		[45]
DCP+ vinyl benzene (VB)	0.26%, 0.13%	PLA+ PHBV (3% 3HV) (75:25)	Melt blending and cast film extrusion		Reduction of EM, σ and ε	Increase of WVP, OP, NP and COP	[35]
DCP+ hexane diol diacrylate (HDA)	0.1%, 0.08%	PLA+ PHBV (3% 3HV) (75:25)	Melt blending and cast film extrusion		Slight reduction of EM and σ, reduction of ε	Increase of WVP, OP, NP and COP	[35]
DCP+ diethylene glycol dimethacrylate (DEGDM) resp. DCP+ PEGDO	0.1%, 0.09% resp. 0.1%, 0.33%	PLA+ PHBV (3% 3HV) (75:25)	Melt blending and cast film extrusion		Reduction of EM, σ and ε	Increase of WVP, OP and COP, reduction of NP	[35]
Di-(2-tert-butyl-per oxyisopropyl)-b enzene (BIB)	0.5 wt.%	PCL+ PHB (75:25)	Melt blending and compression moulding	Two T_m, no effect on T_m	Increase of σ (11.4 → 16.0 MPa) and ε (125 → 305%)		[45]
Iron (III) acetyl acetonate (Fe(acac)$_3$)	0.37%	PLA+ PHBV (3% 3HV) (75:25)	Melt blending and cast film extrusion		Reduction of EM and σ (64 → 58 MPa), increase of ε (14.5 → 42%)	Increase of WVP and OP (slightly) and COP, slight reduction of NP	[35]
Lapol 108	7 wt.%	PLA+ PHB (75:25)		One T_g, reduction of T_g, no effect on T_m	Increase of ε		[46]
Maleic anhydride grafted PP	3 wt.%	PHBV+ PP (97.5:2.5)	Melt blending	Reduction of crystallisation	Increase of ε		[47]

(Continued)

TABLE 7.3 (CONTINUED)
Effect of Different Compatibiliser on PHA-Based Films

Applied compatibiliser	Amount of applied compatibiliser	Base material	Preparation method	Effect on structural properties	Effect on mechanical properties	Effect on permeation properties	Ref.
PEG 1.000		PHB+ cellulose acetate butyrate (60:40)	Melt blending	Reduction of T_g, T_m and crystallinity	Increase of σ (17.0 → 23.3 MPa) and ε (2.1 → 25.8%)		[48]
PEG 20.000	20 wt.%	PLA+ PHBV (1 mol % 3HV) (70:30)	Melt blending and compression moulding	No effect on T_m	Reduction of EM and σ (50 → 24 MPa), increase of impact strength and ε (2.1 → 237%)		[49]
Poly(ethylene glycol dioleat) (PEGDO)	2 wt.%	PLA+ PHBV (3% 3HV) (75:25)	Melt blending and cast film extrusion		Reduction of EM and σ (64 → 52 MPa), no significant effect on ε	No significant effect on WVP, OP, NP, increase of COP	[35]
Poly-(ethylene-block-polyethylene glycol) (PEG-PE) 1.400	2 wt.%	PLA+ PHBV (3% 3HV) (75:25)	Melt blending and cast film extrusion		Reduction of EM and σ (64 → 51 MPa), increase of ε (14.5 → 19%)	No significant effect on WVP, OP, NP, increase of COP	[35]
Polymeric methylene diphenyl diisocyanate	0.5 wt.%	PHBV+ PBAT+ (45:55)+ 20% dried distillers' Grains	Melt blending and injection moulding		Reduction of melt flow index, increase of EM, yield strength, impact strength and ε		[50]
Polymethyl methacrylate (PMMA)	2 wt.%	PLA+ PHBV (3% 3HV) (75:25)	Melt blending and cast film extrusion		Reduction of EM and σ (64 → 49 MPa), increase of ε (14.5 → 27%)	No significant effect on WVP, slight reduction of OP and NP, increase of COP	[35]

(Continued)

TABLE 7.3 (CONTINUED)
Effect of Different Compatibiliser on PHA-Based Films

Applied compatibiliser	Amount of applied compatibiliser	Base material	Preparation method	Effect on structural properties	Effect on mechanical properties	Effect on permeation properties	Ref.
PVAc	9.1 wt.%	PHB+ PLA (25:75)	Solvent mixing and casting	One T_m, reduction of T_m			[51]
Titanium butylate ($TiBu_4$)	0.94 %	PLA+ PHBV (3% 3HV) (75:25)	Melt blending and cast film extrusion		Reduction of EM and σ (64 → 51 MPa), increase of ε (14.5 → 29%)	Slight reduction of WVP and NP, no effect on OP, increase of COP	[35]
Zinc acetate	0.1 wt.%	PLA+ PHBV (1 mol-% 3HV) (70:30)	Melt blending and injection moulding	Two T_g, slight increase of $T_{g,PHBV}$, improved miscibility	No significant effect on EM, increase of ε (~8.6 → ~14.8%)		[52]
Zinc (II) acetate monohydrate (Zn acetate)	0.63 %	PLA+ PHBV (3% 3HV) (75:25)	Melt blending and cast film extrusion		Reduction of EM and σ (64 → 52 MPa), increase of ε (14.5 → 37%)	Slight reduction of WVP and NP, increase of OP and COP	[35]
Zinc (II) acetyl acetonate ($Zn(acac)_2$)	0.42 %	PLA+ PHBV (3% 3HV) (75:25)	Melt blending and cast film extrusion		Reduction of EM and σ (64 → 52 MPa), increase of ε (14.5 → 36%)	No significant effect on WVP, OP and NP, increase of COP	[35]
Zirconium (IV) acetyl acetonate ($Zr(acac)_4$)	0.41 %	PLA+ PHBV (3% 3HV) (75:25)	Melt blending and cast film extrusion		Reduction of EM, σ and ε	No significant effect on WVP, OP and NP, increase of COP	[35]

EM: elastic modulus, σ: tensile strength, ε: elongation at break, WVP: water vapour permeability, OP: oxygen permeability, NP: nitrogen permeability, COP: carbon dioxide permeability to incorporate fillers. Examples of inorganic fillers are minerals, clay and pigments such as titanium dioxide. Examples of organic fillers are bacterial cellulose, cellulose nanocrystals, lignin-based or hemicellulose-based substances and natural rubber. Fillers in PHA-based composites can reduce the usually high cost of PHA, enhance the crystallisation rate and improve the mechanical properties. While the thermal stability is reduced by lignin [53], it can be enhanced by clay [54] or cellulose nanocrystals [55]. The size of the added filler particles, the processing (preparation) method and their orientation in the final composite are decisive factors for their effect on the polymer. Particularly for layered silicates, the incorporation, level of dispersion and final structure in the composite are vital factors for their effect on the properties of the composite. The effects of different kinds of fillers on PHA-based composites are given in Table 7.4.

Fillers are mostly added in a single-digit percentage range and are sometimes incorporated in a PHA-based blend instead of pure PHA, due to the mentioned limitations. The effect of different fillers on the base material varies due to the differing interactions with the polymer matrix. Fillers affect the structural properties; they often act as a nucleating agent for PHA grades. The T_m is either not affected or is slightly reduced by fillers, but the crystallinity is mostly reduced. However, only small effects on the mechanical properties, such as reduced elastic modulus and tensile strength, have been observed. Moreover, the targeted increase in elongation at break is rarely realised (e.g. using organo-modified montmorillonite [57], carbon black [61] or lignin [65]). The effect on the permeability of PHA varies but is mostly increased. Importantly, the size of the filler is crucial: while cellulose fibres increase the permeability [62], cellulose nanocrystals reduce the permeability of a PHA-based composite (OP [64] and WVP [63]).

7.4 EFFECT OF DIFFERENT ADDITIVES ON THE PROPERTIES OF PHBV-BASED FILMS

The effect of an additive on PHA depends not only on the interaction, but also on various factors such as the M_w and monomer ratio of the PHA, the presence of other additives such as a nucleating agent and in particular, the conditions during processing (temperature, pressure, shear force, duration of treatment) which can lead to decomposition reactions. Therefore, general statements about the effectiveness of a specific additive are difficult. An approach for the comparative assessment of different kinds of additives is to consider publications describing the same PHA grade, preparation method and characterisation conditions: the effect of the low-molecular plasticiser triethyl citrate (TEC) [14], the blend partners PLA [35], PBAT and TPU [23] as well as the compatibiliser PMMA [35] on the mechanical and permeation properties of PHBV are compared in Figure 7.4 and Figure 7.5. The data show that an increase in flexibility correlates with an increase in permeability. Blends with PBAT or TPU show the strongest increase in the elongation at break. In combination with the relatively high tensile strength, these additives are very effective in improving the flexibility of PHBV. Films made from PHBV + PLA blends have lower elongation at break, but much higher strength than blends with PBAT or TPU. Even though the crystallinity is reduced by blending with PLA, the permeability increases

TABLE 7.4
Effect of Different Fillers on PHA-Based Films

Applied filler	Amount of applied filler	Base material	Preparation method	Effect on structural properties	Effect on mechanical properties	Effect on permeation properties	Ref.
Inorganic filler							
Hydroxyapatite	10 wt.%	PHB, PHBV (12.6 mol.%)	Mixing and compression moulding	Reduction of crystallinity	Increase of EM (with PHBV) and σ, no significant effect on ε		[56]
Organo-modified montmorillonite (Cloisite 10A)	3 wt.%	PHB, PHBV (5% 3HV)	Solvent mixing and casting	No significant effect on T_m and crystallinity	Increase of EM, σ and ε, effect stronger on PHBV	Reduction of WVP_{PHB} and WVP_{PHBV} (0.036 → 0.023 and 0.02 → 0.008 g mm m^{-2} d^{-1} mm Hg^{-1} equal to ~0.27 → ~0.17 and ~0.15 → ~0.12 g 100 μm m^{-2} d^{-1} $mbar^{-1}$)	[57]
Titanium dioxide (TiO_2)	4 wt.%	PHBV (4% 3HV)	Melt blending	Increase of T_m and crystallinity, TiO_2 acts as nucleating agent			[58]
Organic filler							
Babassu Filler	20 wt.%	PBAT+ PHB (50:50)	Melt blending and injection moulding		No significant effect on EM, reduction of σ and ε		[59]

(Continued)

TABLE 7.4 (CONTINUED)
Effect of Different Fillers on PHA-Based Films

Applied filler	Amount of applied filler	Base material	Preparation method	Effect on structural properties	Effect on mechanical properties	Effect on permeation properties	Ref.
Bacterial cellulose (BC)	2 wt.%	PHB+ Tributyrin (80:20)	Solvent mixing and casting	Slight reduction of T_m, T_g and crystallinity	No significant effect on EM, σ and ε	No significant effect on WVP	[60]
Carbon black	1 wt.%	PLA+ PHB+ citrate ester	Melt blending and blown film extrusion	Two T_m, no effect on T_m	Reduction of EM, no sign. effect on σ, increase of ε (122 → 233 %)		[61]
Cellulose fibres	10 wt.%	PHB+ Bis[(butoxyethoxy) ethyl] adipate	Melt blending	Slight reduction of T_m, reduction of crystallinity		Increase of WVP (by 75%)	[62]
Cellulose nanocrystals (CNC)	2 wt.%	PHB	Solvent mixing and casting	No significant effect on T_m, T_g and crystallinity	Increase of EM and σ, no significant effect on ε	Reduction of WVP	[63]
CNC	2 wt.%	PHB	Solvent mixing and casting	One T_m, no effect on T_m		Slight reduction of migration, reduction of OP (3.46 → 1.23*10^{-17} m^3 m m^{-2} s^{-1} Pa^{-1}) equal to 2.99 → 1.06*10^3 cm^3 100 μm m^{-2} d^{-1} bar^{-1})	[64]

(Continued)

TABLE 7.4 (CONTINUED)
Effect of Different Fillers on PHA-Based Films

Applied filler	Amount of applied filler	Base material	Preparation method	Effect on structural properties	Effect on mechanical properties	Effect on permeation properties	Ref.
CNC	4 wt.%	PHB+ Tributyrin (80:20)	Solvent mixing and casting	Slight reduction of T_m, T_g and crystallinity, acts as nucleating agent	No significant effect on EM, σ and ε	No significant effect on WVP	[60]
Lignin copolymers	33%	PHB	Solvent mixing, casting and electrospun nanofibers	No significant effect on T_m, T_g and crystallinity	Reduction of EM and σ, increase of ε (15 → 46%)		[65]
Natural rubber	10%	PHBV (5% 3HV)	Melt blending, injection and compression moulding	One T_m, no effect on T_m	Reduction of EM and σ, no significant effect on ε	Reduction of WVP (44 → 16 g m^{-2} d^{-1})	[66]
PBAT+ PLA+ filler	30 wt.%	PHBV (3% 3HV)	Melt blending and cast film extrusion	One T_m, slight reduction of T_m, reduction of crystallinity (67 → 51%)	Reduction of EM and σ, no significant effect on ε	Increase of OP and WVP	[23]

EM: elastic modulus, σ: tensile strength, ε: elongation at break, WVP: water vapour permeability, OP: oxygen permeability

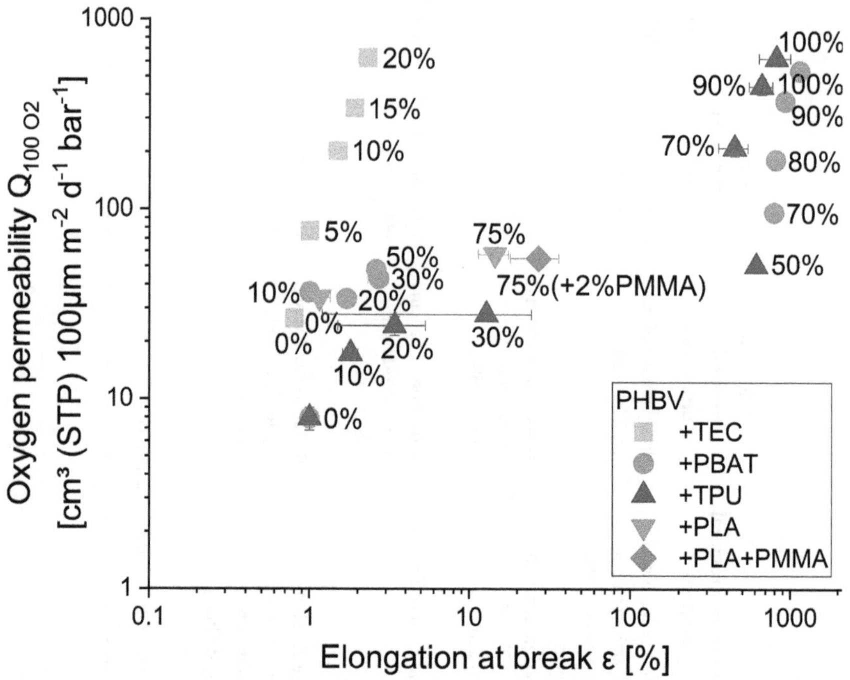

FIGURE 7.4 Oxygen permeability $Q_{100\ O2}$ (23°C, 50% r.h.) versus elongation at break of PHBV-based blends (number indicates the percentage of additive) based on literature (PHBV + TEC [14], PHBV + PLA and PHBV + PLA + PMMA [35], PHBV + PBAT and PHBV + TPU [23]).

only slightly. Generally, the elongation at break is enhanced drastically only when the blend partner is applied in high amounts. When using the application-oriented preparation method of melt blending, there is usually a maximum viable quantity of low molecular weight additives that can be used. Due to this limitation, the low molecular weight plasticiser TEC has a comparatively small effect on the mechanical properties of PHBV. The permeability is, however, markedly affected by TEC. The compatibilisation of a PHBV + PLA blend with PMMA affects the mechanical properties but had no significant effect on the permeability: the compatibilised films had lower strength and slightly increased elongation at break compared to the pure blend.

7.5 CONCLUSIONS AND OUTLOOK

This chapter has given an overview of the effects of different kinds of additives on PHA. PHA are a group of biobased and biodegradable polymers with interesting properties – in particular, their low permeability to oxygen and water vapour. However, there are still challenges to be overcome for broad application of these biopolymers. Examples are the slow crystallisation, high crystallinity and brittleness as well as low flexibility and thermal stability. Additives can be used to improve

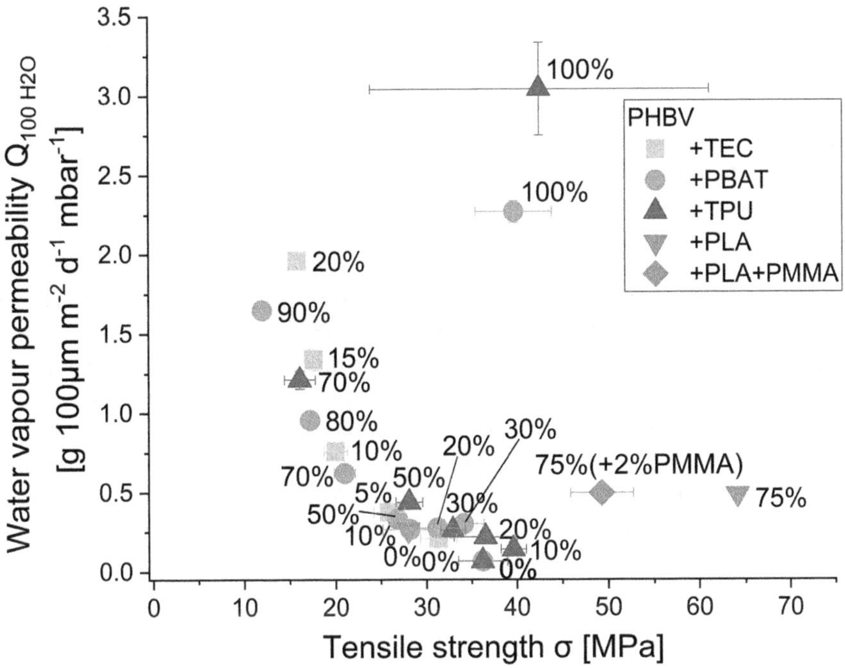

FIGURE 7.5 Water vapour permeability $Q_{100\ H2O}$ (23°C, 85 → 0% r.h.) versus tensile strength of PHBV-based blends (number indicates the percentage of additive) based on literature (PHBV + TEC [14], PHBV + PLA and PHBV + PLA + PMMA [35], PHBV + PBAT and PHBV + TPU [23]).

the processing, tailor certain properties and enable an industrial application. In this chapter, the additives are categorised as low molecular weight plasticisers, blend partners, compatibilisers and fillers.

Almost all references deal with PHB or PHBV grades because other grades are not readily (commercially) available. PHBV usually has a low 3HV ratio, despite the current research on producing grades with a higher 3HV content. The preparation method used in many studies was solvent mixing and casting which is understandable for a laboratory scale. For larger scales, thermal processing via melt blending and subsequent film formation by injection moulding, blown film extrusion or cast film extrusion is more realistic. However, the implementation of such a process usually involves at least two thermal treatments of the biopolymer, which can lead to a reduction of the M_w and modified final properties.

The incorporation of plasticiser leads to a lower T_m, T_g and crystallinity of PHA. Plasticised films show increased flexibility, reduced strength and enhanced permeability. However, the effect of plasticisers on PHA is limited and is usually insufficient for an application on a larger scale. PHA-based blends mostly have reduced crystallinity and strength and increased flexibility and permeability. These blends often have poor miscibility which can negatively affect the long-term stability. In order to improve the miscibility of PHA-based blends, compatibilisers can be incorporated.

The literature data show that compatibilisers reduce the crystallinity of the PHA phase and enhance the flexibility of the blend. The permeability of compatibilised blends is either maintained or increased. However, the efficiency of a compatibiliser needs to be evaluated for each case and thereby also the effect on the miscibility of the components and the long-term stability of the blend. Different types of fillers have different effects on PHA. They are mostly incorporated as nucleating agents to increase the crystallisation rate of PHA. This usually leads to reduced strength of the films. The flexibility and hence, permeability, of these composites is increased by some fillers, but not all.

The effect of additives on PHA cannot be generalised. The effect depends on the type and M_w of the additive, its concentration, the method of interaction and its miscibility with PHA, affecting, amongst other things, the long-term stability. In addition, the grade and purity of the PHA, the presence of nucleating agents, the processing conditions and the measuring conditions during characterisation are crucial. Many publications describe an improvement of the challenging processing and low flexibility of PHA by using additives. It remains unclear whether these improvements are sufficient and applicable for scale-up to commercial quantities.

REFERENCES

1. Holmes, P.A., Applications of phb - A microbially produced biodegradable thermoplastic. *Physics in Technology*, 1985. **16**(1): p. 32–36.
2. Chen, G.G.-Q., Polyhydroxyalkanoates. In: *Biodegradable Polymers for Industrial Applications*, R. Smith, Editor. 2005, Woodhead Publishing Limited and CRC Press LLC: Cambridge/Boca Raton. p. 32–56.
3. Holmes, P.A., Biologically produced (*R*)-3-hydroxy- alkanoate polymers and copolymers. In: *Developments in Crystalline Polymers*, D.C. Bassett, Editor. 1988, Springer Netherlands: Dordrecht. p. 1–65.
4. Pollet, E. and L. Avérous, Production, chemistry and properties of polyhydroxyalkanoates. In: *Biopolymers - New Materials for Sustainable Films and Coatings*, D. Plackett, Editor. 2011, John Wiley & Sons, Ltd: Chichester, UK. p. 65–86.
5. Barham, P.J., A. Keller, E.L. Otun, and P.A. Holmes, Crystallization and morphology of a bacterial thermoplastic - poly-3-hydroxybutyrate. *Journal of Materials Science*, 1984. **19**(9): p. 2781–2794. DOI: 10.1007/bf01026954.
6. El-Hadi, A., R. Schnabel, E. Straube, G. Muller, and S. Henning, Correlation between degree of crystallinity, morphology, glass temperature, mechanical properties and biodegradation of poly (3-hydroxyalkanoate) phas and their blends. *Polymer Testing*, 2002. **21**(6): p. 665–674. DOI: 10.1016/s0142-9418(01)00142-8.
7. Liu, W.J., H.L. Yang, Z. Wang, L.S. Dong, and J.J. Liu, Effect of nucleating agents on the crystallization of poly(3-hydroxybutyrate-co-3-hydroxyvalerate). *Journal of Applied Polymer Science*, 2002. **86**(9): p. 2145–2152. DOI: 10.1002/app.11023.
8. Furuhashi, Y., Y. Imamura, Y. Jikihara, and H. Yamane, Higher order structures and mechanical properties of bacterial homo poly(3-hydroxybutyrate) fibers prepared by cold-drawing and annealing processes. *Polymer*, 2004. **45**(16): p. 5703–5712. DOI: 10.1016/j.polymer.2004.05.069.
9. Jost, V., Packaging related properties of commercially available biopolymers – An overview of the *status quo. Express Polymer Letters*, 2018. **12**(5): p. 429–435. DOI: 10.3144/expresspolymlett.2018.36.

10. Bleisch, G., E. Herzau, H.-C. Langowski, J.-P. Majschak, J. Martin, B. Sadlowsky, H. Vogt, F. Weile, M. Weiß, and U. Weiß, *Lexikon verpackungstechnik*. Vol. 2. 2014, B. Behr's Verlag GmbH & Co. KG: Hamburg.
11. Wypych, G., *Handbook of Plasticizers*, 2nd edition. 2012, ChemTec Publishing: Toronto. p. 800.
12. Sears, J.K. and J.R. Darby, *The Technology of Plasticizers*. Spe Monographs. Vol. 1. 1982, John Wiley & Sons: New York.
13. Freier, T., C. Kunze, C. Nischan, S. Kramer, K. Sternberg, M. Saß, U.T. Hopt, and K.-P. Schmitz, *In vitro* and *in vivo* degradation studies for development of a biodegradable patch based on poly(3-hydroxybutyrate). *Biomaterials*, 2002. **23**(13): p. 2649–2657. DOI: 10.1016/S0142-9612(01)00405-7.
14. Jost, V. and H.C. Langowski, Effect of different plasticisers on the mechanical and barrier properties of extruded cast phbv films. *European Polymer Journal*, 2015. **68**: p. 302–312. DOI: 10.1016/j.eurpolymj.2015.04.012.
15. Choi, J.S. and W.H. Park, Effect of biodegradable plasticizers on thermal and mechanical properties of poly (3-hydroxybutyrate). *Polymer Testing*, 2004. **23**(4): p. 455–460. DOI: 10.1016/j.polymertesting.2003.09.005.
16. Savenkova, L., Z. Gercberga, V. Nikolaeva, A. Dzene, I. Bibers, and M. Kalnin, Mechanical properties and biodegradation characteristics of phb-based films. *Process Biochemistry*, 2000. **35**(6): p. 573–579. DOI: 10.1016/s0032-9592(99)00107-7.
17. Parra, D.F., J. Fusaro, F. Gaboardi, and D.S. Rosa, Influence of poly (ethylene glycol) on the thermal, mechanical, morphological, physical–chemical and biodegradation properties of poly (3-hydroxybutyrate). *Polymer Degradation and Stability*, 2006. **91**(9): p. 1954–1959. DOI: 10.1016/j.polymdegradstab.2006.02.008.
18. Catoni, S.E.M., K.N. Trindade, C.A.T. Gomes, A.L.S. Schneider, A.P.T. Pezzin, and V. Soldi, Influence of poly(ethylene grycol)-(peg) on the properties of influence of poly(3-hydroxybutyrate-co-3-hydroxyvalerate)-phbv. *Polimeros-Ciencia e Tecnologia*, 2013. **23**(3): p. 320–325. DOI: 10.4322/polimeros.2013.037.
19. Renard, E., M. Walls, P. Guérin, and V. Langlois, Hydrolytic degradation of blends of polyhydroxyalkanoates and functionalized polyhydroxyalkanoates. *Polymer Degradation and Stability*, 2004. **85**(2): p. 779–787. DOI: 10.1016/j.polymdegradstab.2003.11.019.
20. D'Amico, D.A., M.L.I. Montes, L.B. Manfredi, and V.P. Cyras, Fully bio-based and biodegradable polylactic acid/poly(3-hydroxybutirate) blends: Use of a common plasticizer as performance improvement strategy. *Polymer Testing*, 2016. **49**: p. 22–28. DOI: 10.1016/j.polymertesting.2015.11.004.
21. Muthuraj, R., M. Misra, and A.K. Mohanty, Biodegradable compatibilized polymer blends for packaging applications: A literature review. *Journal of Applied Polymer Science*, 2018. **135**(24). DOI: 10.1002/app.45726.
22. Hong, S.G., T.K. Gau, and S.C. Huang, Enhancement of the crystallization and thermal stability of polyhydroxybutyrate by polymeric additives. *Journal of Thermal Analysis and Calorimetry*, 2011. **103**(3): p. 967–975. DOI: 10.1007/s10973-010-1180-3.
23. Jost, V. and O. Miesbauer, Effect of different biopolymers and polymers on the mechanical and permeation properties of extruded PHBV cast films. *Journal of Applied Polymer Science*, 2018. **135**(15): p. 12. DOI: 10.1002/app.46153.
24. Cunha, M., B. Fernandes, J.A. Covas, A.A. Vicente, and L. Hilliou, Film blowing of phbv blends and PHBV-based multilayers for the production of biodegradable packages. *Journal of Applied Polymer Science*, 2016. **133**(2). DOI: 10.1002/app.42165.
25. Javadi, A., A.J. Kramschuster, S. Pilla, J. Lee, S. Gong, and L.-S. Turng, Processing and characterization of microcellular PHBV/PBAT blends. *Polymer Engineering and Science*, 2010. **50**(7): p. 1440–1448. DOI: 10.1002/pen.21661.

26. Qiu, Z.B., T. Ikehara, and T. Nishi, Miscibility and crystallization behaviour of biodegradable blends of two aliphatic polyesters. Poly(3-hydroxybutyrate-*co*-hydroxyvalerate) and poly(butylene succinate) blends. *Polymer*, 2003. **44**(24): p. 7519–7527. DOI: 10.1016/j.polymer.2003.09.029.
27. Del Gaudio, C., E. Ercolani, F. Nanni, and A. Bianco, Assessment of poly(ε-caprolactone)/poly(3-hydroxybutyrate-co-3-hydroxyvalerate) blends processed by solvent casting and electrospinning. *Materials Science and Engineering. Part A,-Structural Materials: Properties, Microstructure and Processing*, 2011. **528**(3): p. 1764–1772. DOI: 10.1016/j.msea.2010.11.012.
28. Tsukada, M., M. Romano, and A. Seves, Physical-properties of poly-D(-)(3-hydroxybutyrate)/poly(epichlorohydrin) blends. *Acta Polymerica*, 1992. **43**(6): p. 327–330. DOI: 10.1002/actp.1992.010430606.
29. Lee, M.S. and W.H. Park, Compatibility and thermal properties of poly(3-hydroxybutyrate)/poly(glycidyl methacrylate) blends. *Journal of Polymer Science Part A-Polymer Chemistry*, 2002. **40**(3): p. 351–358. DOI: 10.1002/pola.10128.
30. Boufarguine, M., A. Guinault, G. Miquelard-Garnier, and C. Sollogoub, Pla/phbv films with improved mechanical and gas barrier properties. *Macromolecular Materials and Engineering*, 2013. **298**(10): p. 1065–1073. DOI: 10.1002/mame.201200285.
31. Nanda, M.R., M. Misra, and A.K. Mohanty, The effects of process engineering on the performance of PLA and phbv blends. *Macromolecular Materials and Engineering*, 2011. **296**(8): p. 719–728. DOI: 10.1002/mame.201000417.
32. Zembouai, I., M. Kaci, S. Bruzaud, A. Benhamida, Y.-M. Corre, and Y. Grohens, A study of morphological, thermal, rheological and barrier properties of poly(3-hydroxyb utyrate-co-3-hydroxyvalerate)/ polylactide blends prepared by melt mixing. *Polymer Testing*, 2013. **32**(5): p. 842–851. DOI: 10.1016/j.polymertesting.2013.04.004.
33. Iannace, S., L. Ambrosio, S.J. Huang, and L. Nicolais, Poly(3-hydroxybutyrate)-co-(3-hydroxyvalerate)/poly-*l*-lactide blends: Thermal and mechanical properties. *Journal of Applied Polymer Science*, 1994. **54**(10): p. 1525–1535.
34. Arrieta, M.P., J. Lopez, A. Hernandez, and E. Rayon, Ternary pla-phb-limonene blends intended for biodegradable food packaging applications. *European Polymer Journal*, 2014. **50**: p. 255–270. DOI: 10.1016/j.eurpolymj.2013.11.009.
35. Jost, V. and R. Kopitzky, Blending of polyhydroxybutyrate-co-valerate with polylactic acid for packaging applications – Reflexions on miscibility and effects on the mechanical and barrier properties. *Chemical and Biochemical Engineering Quarterly*, 2015. **29**(2): p. 221–246. DOI: 10.15255/CABEQ.2014.2257.
36. Gerard, T., T. Budtova, A. Podshivalov, and S. Bronnikov, Polylactide/poly(hydroxyb utyrate-co-hydroxyvalerate) blends: Morphology and mechanical properties. *Express Polymer Letters*, 2014. **8**(8): p. 609–617. DOI: 10.3144/expresspolymlett.2014.64.
37. Corre, Y.M., S. Bruzaud, and Y. Grohens, Poly(3-hydroxybutyrate-co-3-hydroxyvaler ate) and poly(propylene carbonate) blends: An efficient method to finely adjust properties of functional materials. *Macromolecular Materials and Engineering*, 2013. **298**(11): p. 1176–1183. DOI: 10.1002/mame.201200345.
38. Hernandez-Munoz, P., R. Villalobos, and A. Chiralt, Effect of thermal treatments on functional properties of edible films made from wheat gluten fractions. *Food Hydrocolloids*, 2004. **18**(4): p. 647–654. DOI: 10.1016/j.foodhyd.2003.11.002.
39. Grondahl, L., A. Chandler-Temple, and M. Trau, Polymeric grafting of acrylic acid onto poly(3-hydroxybutyrate-co-3-hydroxyvalerate): Surface functionalization for tissue engineering applications. *Biomacromolecules*, 2005. **6**(4): p. 2197–2203. DOI: 10.1021/bm050127m.
40. Schmid, M., S. Sangerlaub, L. Wege, and A. Stabler, Properties of transglutaminase crosslinked whey protein isolate coatings and cast films. *Packaging Technology and Science*, 2014. **27**(10): p. 799–817. DOI: 10.1002/pts.2071.

41. Ma, P.M., X.X. Cai, W. Wang, F. Duan, D.J. Shi, and P.J. Lemstra, Crystallization behavior of partially crosslinked poly(beta-hydroxyalkonates)/poly(butylene succinate) blends. *Journal of Applied Polymer Science*, 2014. **131**(21). DOI: 10.1002/app.41020.
42. Avella, M., B. Immirzi, M. Malinconico, E. Martuscelli, and M.G. Volpe, Reactive blending methodologies for biopol. *Polymer International*, 1996. **39**(3): p. 191–204.
43. Li, J., C.R. Sun, and X.Q. Zhang, Preparation, thermal properties, and morphology of graft copolymers in reactive blends of phbv and PPC. *Polymer Composites*, 2012. **33**(10): p. 1737–1749. DOI: 10.1002/pc.22308.
44. Dong, W.F., P.M. Ma, S.F. Wang, M.Q. Chen, X.X. Cai, and Y. Zhang, Effect of partial crosslinking on morphology and properties of the poly(β-hydroxybutyrate)/poly(D,L-lactic acid) blends. *Polymer Degradation and Stability*, 2013. **98**(9): p. 1549–1555. DOI: 10.1016/j.polymdegradstab.2013.06.033.
45. Przybysz, M., M. Marc, M. Klein, M.R. Saeb, and K. Formel, Structural, mechanical and thermal behavior assessments of PCL/PHB blends reactively compatibilized with organic peroxides. *Polymer Testing*, 2018. **67**: p. 513–521. DOI: 10.1016/j.polymertesting.2018.03.014.
46. Abdelwahab, M.A., A. Flynn, B.S. Chiou, S. Imam, W. Orts, and E. Chiellini, Thermal, mechanical and morphological characterization of plasticized pla-phb blends. *Polymer Degradation and Stability*, 2012. **97**(9): p. 1822–1828. DOI: 10.1016/j.Polymdegradstab.2012.05.036.
47. Chen, L., J. Xu, and Z.Y. Qin, Crystallisation and mechanical properties of poly(3-hydroxybutyrate-co-hydroxyvalerate)/polypropylene blends. *Materials Research Innovations*, 2014. **18**(sup4): p. 848–853. DOI: 10.1179/1432891714z.000000000805.
48. Cheng, G.X., T.Z. Wang, Q. Zhao, X.L. Ma, and L.G. Zhang, Preparation of cellulose acetate butyrate and poly(ethylene glycol) copolymer to blend with poly(3-hydroxybutyrate). *Journal of Applied Polymer Science*, 2006. **100**(2): p. 1471–1478. DOI: 10.1002/app.23135.
49. Wang, S.A., P.M. Ma, R.Y. Wang, S.F. Wang, Y. Zhang, and Y.X. Zhang, Mechanical, thermal and degradation properties of poly(D,L-lactide)/poly(hydroxybutyrate-co-h ydroxyvalerate)/poly(ethylene glycol) blend. *Polymer Degradation and Stability*, 2008. **93**(7): p. 1364–1369. DOI: 10.1016/j.polymdegradstab.2008.03.026.
50. Zarrinbakhsh, N., A.K. Mohanty, and M. Misra, Improving the interfacial adhesion in a new renewable resource-based biocomposites from biofuel coproduct and biodegradable plastic. *Journal of Materials Science*, 2013. **48**(17): p. 6025–6038. DOI: 10.1007/s10853-013-7399-1.
51. El-Hadi, A.M., Effect of processing conditions on the development of morphological features of banded or nonbanded spherulites of poly(3-hydroxybutyrate) (PHB) and polylactic acid (PLLA) blends. *Polymer Engineering and Science*, 2011. **51**(11): p. 2191–2202. DOI: 10.1002/pen.21991.
52. Yang, J., H.J. Zhu, C.B. Zhang, Q.H. Jiang, Y. Zhao, P. Chen, and D.J. Wang, Transesterification induced mechanical properties enhancement of PLLA/PHBV bioalloy. *Polymer*, 2016. **83**: p. 230–238. DOI: 10.1016/j.polymer.2015.12.025.
53. Mousavioun, P., P.J. Halley, and W.O.S. Doherty, Thermophysical properties and rheology of PHB/lignin blends. *Industrial Crops and Products*, 2013. **50**: p. 270–275. DOI: 10.1016/j.indcrop.2013.07.026.
54. Erceg, M., T. Kovacic, and I. Klaric, Poly(3-hydroxybutyrate) nanocomposites: Isothermal degradation and kinetic analysis. *Thermochimica Acta*, 2009. **485**(1–2): p. 26–32. DOI: 10.1016/j.tca.2008.12.002.
55. Bhardwaj, U., P. Dhar, A. Kumar, and V. Katiyar, Polyhyroxyalkanoates, (PHA)-cellulose based nanobiocomposites for food packaging applications. In: *Food Additives and Packaging*, V. Komolprasert and P. Turowski, Editors. 2014. p. 275–314. Oxford University Press.

56. Bergmann, A. and A. Owen, Hydroxyapatite as a filler for biosynthetic PHB homopolymer and P(HB-HV) copolymers. *Polymer International*, 2003. **52**(7): p. 1145–1152. DOI: 10.1002/pi.1206.
57. Akin, O. and F. Tihminlioglu, Effects of organo-modified clay addition and temperature on the water vapor barrier properties of polyhydroxy butyrate homo and copolymer nanocomposite films for packaging applications. *Journal of Polymers and the Environment*, 2018. **26**(3): p. 1121–1132. DOI: 10.1007/s10924-017-1017-2.
58. Jaques, N.G., I.D. dos Santos Silva, M.d.C. Barbosa Neto, A. Ries, E.L. Canedo, and R.M. Ramos Wellen, Effect of heat cycling on melting and crystallization of PHB/TiO_2 compounds. *Polimeros-Ciencia e Tecnologia*, 2018. **28**(2): p. 161–168. DOI: 10.1590/0104-1428.12416.
59. Beber, V.C., S. de Barros, M.D. Banea, M. Brede, L.H. de Carvalho, R. Hoffmann, A.R.M. Costa, E.B. Bezerra, I.D.S. Silva, K. Haag, K. Koschek, and R.M.R. Wellen, Effect of babassu natural filler on pbat/phb biodegradable blends: An investigation of thermal, mechanical, and morphological behavior. *Materials*, 2018. **11**(5). DOI: 10.3390/ma11050820.
60. Seoane, I.T., P. Cerrutti, A. Vazquez, V.P. Cyras, and L.B. Manfredi, Ternary nanocomposites based on plasticized poly(3-hydroxybutyrate) and nanocellulose. *Polymer Bulletin*, 2019. **76**(2): p. 967–988. DOI: 10.1007/s00289-018-2421-z.
61. Mosnackova, K., M. Danko, A. Siskova, L.M. Falco, I. Janigova, S. Chmela, Z. Vanovcanova, L. Omanikova, I. Chodak, and J. Mosnacek, Complex study of the physical properties of a poly(lactic acid)/poly(3-hydroxybutyrate) blend and its carbon black composite during various outdoor and laboratory ageing conditions. *RSC Advances*, 2017. **7**(74): p. 47132–47142. DOI: 10.1039/c7ra08869h.
62. Tanase, E.E., M.E. Popa, M. Rapa, and O. Popa, Phb/cellulose fibers based materials, Physical, mechanical and barrier properties. In: *Conference Agriculture for Life, Life for Agriculture*, S.M. Cimpeanu, G. Fintineru, and S. Beciu, Editors. 2015, Elsevier Science Bv: Amsterdam. p. 608–615.
63. Seoane, I.T., E. Fortunati, D. Puglia, V.P. Cyras, and L.B. Manfredi, Development and characterization of bionanocomposites based on poly(3-hydroxybutyrate) and cellulose nanocrystals for packaging applications. *Polymer International*, 2016. **65**(9): p. 1046–1053. DOI: 10.1002/pi.5150.
64. Dhar, P., U. Bhardwaj, A. Kumar, and V. Katiyar, Poly, (3-hydroxybutyrate)/cellulose nanocrystal films for food packaging applications: Barrier and migration studies. *Polymer Engineering and Science*, 2015. **55**(10): p. 2388–2395. DOI: 10.1002/pen.24127.
65. Kai, D., H.M. Chong, L.P. Chow, L. Jiang, Q.Y. Lin, K.Y. Zhang, H.J. Zhang, Z. Zhang, and X.J. Loh, Strong and biocompatible lignin/poly (3-hydroxybutyrate) composite nanofibers. *Composites Science and Technology*, 2018. **158**: p. 26–33. DOI: 10.1016/j.compscitech.2018.01.046.
66. Modi, S.J., K. Cornish, K. Koelling, and Y. Vodovotz, Fabrication and improved performance of poly(3-hydroxybutyrate-co-3-hydroxyvalerate) for packaging by addition of high molecular weight natural rubber. *Journal of Applied Polymer Science*, 2016. **133**(37). DOI: 10.1002/app.43937.

Part III

Application

8 Competitive Advantage and Market Introduction of PHA Polymers and Potential Use of PHA Monomers

Konstantina Kourmentza, Vasiliki Kachrimanidou, Olga Psaki, Chrysanthi Pateraki, Dimitrios Ladakis, and Apostolos Koutinas

CONTENTS

8.1 INTRODUCTION

In 2018, the total production volume of bio-based building blocks and polymers reached 7.5 million metric tons [1], with global production capacity of bioplastics at 2.1 million metric tons [2]. This represents only 2%, and 1%, respectively, of the production volume of petrochemical polymers that has almost reached 360 million metric tons worldwide, with 62 million metric tons produced in Europe [3]. It is expected that production of bio-based building blocks and polymers will continue to grow with a compound annual growth rate (CAGR) of 4% until 2023, and their market share, compared to petrochemical polymers, will remain at around 2% [1, 4].

Bio-based polymers include non-biodegradable drop-in solutions, such as bio-based poly(ethylene terephthalate) (PET), poly(ethylene) (PE), poly(propylene) (PP), polyamides (PA), poly(trimethylene terephthalate) (PTT), polyurethanes (PUR), etc., as well as biodegradable ones, such as poly(butylene adipate-*co*-terephthalate) (PBAT), poly(butylene succinate) (PBS), poly(lactic acid) (PLA), polyhydroxy-alkanoates (PHA), starch blends, etc. [1]. Bio-based PE and PET are currently in the lead regarding their global production capacities, while bio-based PP and PEF poly(ethylene furanoate) are under development and will penetrate the market by 2023. PLA and PHA are leading the market of bio-based and biodegradable plastics, with their production capacities estimated to double and quadruple by 2023 compared to 2018 [4].

In 2018, the land needed to grow the feedstock that was subsequently transformed to 2.11 million metric tons of bio-based plastics was 0.81 million hectares, representing 0.016% of the global agricultural area [2]. Given the projected increase in the volume of bioplastics, it has been estimated that by 2022 around 0.020% of global arable land will be occupied with growing bio-plastics feedstock [2]. Another study shows that if global fossil plastics production was based on biomass as feedstock, feedstock demand would be around 5% of the total amount of biomass produced and harvested per year [5]. However, apart from biomass, alternative feedstocks such as food waste, waste by-products, and wastewaters, as well as agricultural residues can be used as feedstocks. Moreover, technological advancements in the context of a multistage biorefinery, i.e., for the co-production of biofuels, is expected to promote the sustainability of such a venture.

Regarding their benefits and valuable properties, bio-based plastics originate from renewable resources, therefore reducing the dependency on limited fossil resources. The environmental impacts of bio-based plastics and fossil-based plastics are in different impact categories. Compared to conventional plastics they have reduced greenhouse gas emissions (GHG), with several life cycle analyses showing significant CO_2 savings, and thus can contribute to meeting EU 2020 targets regarding the reduction of GHG emissions [2]. Moreover, the GHG emission reduction reached by bio-based plastics is significantly larger than that of biofuels [5]. In impact categories such as eutrophication and acidification, bio-based plastics seem to have a higher impact compared to fossil plastics. It is due to these differences in impact that no absolute rule can be given [6].

In general, biodegradable plastics should not be perceived as the solution to the problem of litter and "plastic soup". Biodegradability is a very useful characteristic for specific applications. However, this could stimulate consumers to dispose of a certified product in the environment. Therefore, a clear distinction should be made between certification and the most appropriate way to handle post-consumer plastic waste. In the following sections, the advantages of polyhydroxyalkanoates (PHA) will be discussed regarding their properties, certifications obtained and end-of-life options, and their current applications and potential ones will be presented.

8.2 POLYHYDROXYALKANOATES AND THEIR PROPERTIES

In recent years, plastic material has become an essential part of our society. Due to their mechanical and physical properties, low cost, and ease of manufacture, plastics are used in a wide range of products such as food packaging, fibrous building material, hygiene products, and so on [7]. However, the non-biodegradable nature of the synthetic plastics causes a serious problem due to their accumulation in the environment. In the past few years, research has been focused on the development of biodegradable plastics with the potential to replace petrochemical plastics [8]. Biodegradable polymers have similar properties to the conventional polymers such as poly(ethylene) (PE), poly(propylene) (PP), or poly(ethylene terephthalate) (PET) and can also be used as an alternative solution in food packaging [9]. Several bioplastics have already been used in food packaging and other applications, such as thermo-plasticized starch, poly(lactic acid) (PLA), bio-based poly(ethylene) (Bio-PE), poly(trimethylene terephthalate) (PTT), poly(butylene succinate) (PBS), poly(p-phenylene) (PPP), and microbial polyhydroxyalkanoates (PHA). PHA are the only biopolymers that microorganisms can synthesize from the utilization of renewable sources, with the polymerization taking place intracellularly by the microorganism [10, 11]. PHA are polyesters that are produced and accumulated intracellularly as a carbon and energy reserve, under stress conditions such as a nutrient limitation, while there is excess in carbon source [12]. PHA can be classified in groups depending on the number of carbon atoms in their monomeric unit. Thus, the mechanical properties of PHA depend on the chain length.

Scl-PHA consist of 3–5 carbon atoms and those polymers are stiff, brittle, and possess a high degree of crystallinity in the range of 60–80%. This category of polymers resembles the physicochemical properties of polypropylene (PP) and is considered a suitable alternative. It is worth noting that PP demand distribution within Europe in 2018 was up to 19.3%, with PP being used as food packaging, sweet and snack wrappers, hinged caps, microwave containers, pipes, automotive parts, bank notes, etc. [13]. On the other hand, *mcl*-PHA have 6–14 carbon atoms and their copolymers are flexible and elastic, with low crystallinity, low tensile strength, high elongation at break (300–450%), lower melting temperatures, and glass transition temperatures (T_g) below room temperature [14–16]. Their properties are similar to those of low-density poly(ethylene) (LDPE), which is used to produce reusable bags, trays and containers, agricultural film, food packaging film, etc., and represented 17.5% of the plastics demand distribution in EU for 2018 [13] (Table 8.1).

In Table 8.1, the physical and mechanical properties of polyhydroxyalkanoates, bio-based, and conventional polymers are summarized. As previously discussed, the properties of PHB are close to PP. The brittleness of PHB arises mainly due to the presence of large crystals in the form of spherulites [17]. However, the high tensile strength and Young's modulus (E) enable polymer modifications, since there is a large margin to increase deformability and toughness. As shown in Table 8.1, *mcl*-PHA and copolymers' properties are close to those of LDPE. However, they are not yet commercially available due to their higher production cost compared to *scl*-PHA. The melting temperatures (T_m) of PHA are within a reasonable range and their glass transition temperatures (T_g) lie somewhat below, but still relatively close to, room temperature. However, it is possible to lower the T_g, and thus make the materials ductile, i.e., by plasticizing them, blending them with miscible additives, etc. Varying 3-hydroxyvalerate (3HV) content in PHBV copolymers result in an increase of the impact strength, tensile strength, crystallinity, and also a decrease in T_m and T_g, but still the properties of PHBV copolymers are not significantly improved. However, copolymers of 3HB with *mcl*-PHA, i.e., with 3-hydroxyhexanoate (3HHx), can be prepared and provide improved properties since *mcl*-PHA form a separate crystalline lattice or not at all. These copolymers are usually characterized by lower T_g values compared to the PHB homopolymer [17,19].

PHA are also characterized by low water permeability, which is a very interesting property that suggests their possible applications for packaging. Food packaging plays an important role in the shelf-life of the food product and PHA can be used alone or in combination with other plastics for the development of packaging films via thermoforming [10]. Moreover, due to the degree of crystallinity and elasticity, PHA could form flexible foils acting as storage boxes and containers.

Key properties of materials used in the food packing applications are the CO_2, O_2 permeability/barrier, and the ability to resist moisture. The CO_2 barrier indicates the amount of CO_2 pervading a packing material per area unit and time. PHB is a good barrier material against CO_2 with similar values to PET and PVC [16]. The oxidative degradation can affect the physicochemical characteristics of food (color, flavor) and increase the possibility of microbial contamination. For that reason, a packaging film must have the ability to act as a barrier between the difference in oxygen partial pressures inside (0–2%) and outside the package (21%). Many researchers have reported that the O_2 permeability (PO_2) of PHA is much lower compared to conventional polymers such as PP, PE, and PS, close to the PO_2 of PET and PLA, but a lot higher compared to ethylene vinyl alcohol copolymer (EVOH). Moreover, the water vapor permeability of PHA is similar to EVOH, polyamide (PA), PET and PLA, but slightly higher than more nonpolar polymers such as PE and PP [16, 20, 21]. Moreover, according to previous studies, PHB has lower limonene permeability compared to PET, while in the case of PHBV polymers the limonene permeability was higher [21].

Perhaps the most important property of PHA is their ability to biodegrade through a wide variety of aerobic and anaerobic microorganisms present in the environment [22, 23]. However, the rate of degradation is affected not only by the physical and

TABLE 8.1
Physical and Mechanical Properties of Polyhydroxyalkanoates, Bio-Based, and Conventional Polymers [14, 17, 18]

Polymer	Density [g/cm³]	Melting Temperature, T_m [°C]	Glass transition temperature, T_g [°C]	Tensile strength [MPa]	E modulus [MPa]	Elongation at break [%]	Crystallinity [%]
			Conventional polymers				
PP	0.91	163	–5	33	150	400	56
LDPE	0.92	110	–120 to –40	10	200	600	50
PET		262		56	2200	7300	
PS		80–110	21	50	300	3–4	
			Polyhydroxyalkanoates				
PHB	1.25	175	4	43	3500	5	6
PHBV, 7% 3HV	1.25	153	5		900	15	51
PHBV, 20% 3HV		145	–1	20	800	50	
P(4HB)		130	–48	1000			
P(3HB-*co*-3HHx)		52	–4	20		850	
P(3HB-*co*-3HA), 6% *mcl*-HA		133	–	17	200	680	
P(3HO)		59, 61	–	6–10		300–450	
			Other bio-based polymers				
PLA	1.25	140–180	55	66	2000	4	0–40
PCL	1.11	65	–61		190	> 500	67
PBS	1.25	115	–32	34		560	34–45
PBAT	1.21	110–115	–30		52	> 500	20–35

chemical properties (crystallinity, composition, surface area, etc.), but it also depends on the environmental conditions (temperature, pH, moisture, etc.) [24–26]. PHA-degrading microorganisms, such as fungi and bacteria, excrete extracellular PHA depolymerases that hydrolyze PHA into monomers, and subsequently utilize them as nutritional components for their growth. Under anaerobic conditions, PHA degrades into water and carbon dioxide, while in anaerobic conditions PHA is converted to water and biogas (methane and carbon dioxide) [27]. Many researchers have isolated aerobic and anaerobic bacteria from soil, sludge, and seawater that are capable of degrading PHA polymers. Indicatively, P(3HB-*co*-3HV) was shown to completely degrade after an 8-week incubation period in aerobic sewage sludge [28], while PHB nanofibers can fully degrade in soil after 3 weeks [27]. Rapid degradation is favored for PHA films and nanofibers, due to the respective greater surface area and three-dimensional structure. Moreover, it has been shown that films prepared from PHB and treated by ultraviolet light (UV) had more cracks and accelerated degradation compared to PHB films without UV treatment [27].

8.3 CERTIFICATIONS AND LABELING OF PHA

Most consumers are not aware of the different types of plastics in the market, or sometimes they cannot easily distinguish different plastic products. Logos and labels linked to a certification system can be used to clearly communicate to the consumer whether a plastic is bio-based and/or compostable, and most importantly how to dispose it after use [5]. Several standards exist to define whether a material is compostable or biodegradable. Those international standards (EN 13432:2000, ISO 17088:2012, ASTM D6400-12, etc.) describe test schemes that have been used to evaluate and determine the compostability and biodegradability of materials such as PHA. In general, those standards include requirements to test parameters regarding the characterization of the material, its disintegration ability, its biodegradation into carbon dioxide, biomass, and water within 6 months, an assessment of heavy metal concentrations, and ecotoxicity tests [29]. PHA biodegradability has been examined under different environments and conditions, i.e., soil, water, marine, as well as industrial and home compost. PHA-producing industries have standardized their PHA products with certain certification organizations in order to verify the claims of biodegradability and compostability of each product and obtain the respective logo [30–33]. Standards and specifications have been developed by several authorities such as the American Society for Testing and Material (ASTM), the International Organization for Standardization (ISO), the British Standard Institution (BSI), etc. Table 8.2 shows the certifications and labels obtained for currently commercialized PHA resins.

PHA have been certified by the US Food and Drug Administration (FDA) for use that comes into contact with food. FDA approval means that specific PHA forms can be used in food packaging, caps, utensils, tubs, trays, and hot cup lips, as well as houseware, cosmetic, and medical products. This also means that these PHA grades can be used to store frozen food and can be used in microwaves and boiling water up to 212°F [37]. According to the regulations, materials and articles must be inert, in order to prevent the transport of their constituents into foods at levels that endanger human health, and also to avoid altering of food physicochemical composition [38, 39].

TABLE 8.2
Certifications regarding the PHA Bioplastic Resins Nodax™ PHA (Danimer Scientific) [34], Minerv™ PHA (Bio-On S.p.A.) [35], and Solon™ PHA (RWDC Industries) [30]

	Certifications		
Certification	**Nodax™ PHA**	**Minerv™ PHA**	**Solon™ PHA**
	Biodegradability		
• Anaerobic	ASTM D5511		
• Soil	ASTM D5988 SOIL OK biodegradable VINÇOTTE		SOIL OK biodegradable TÜV AUSTRIA SOIL S0720
• Freshwater	ASTM D5271 EN 29408 WATER OK biodegradable VINÇOTTE	WATER OK biodegradable VINÇOTTE	WATER OK biodegradable TÜV AUSTRIA WATER S0720
• Marine	ASTM D6691		MARINE OK biodegradable TÜV AUSTRIA MARINE S0720
	Compost		
• Industrial composting	ASTM D6400 EN 13432:2000 OK compost VINÇOTTE		OK compost TÜV AUSTRIA INDUSTRIAL S0720
• Home composting	EN 13432:2000 Lower temperature conditions HOME OK compost VINÇOTTE		HOME OK compost TÜV AUSTRIA HOME S0720
	Other		
• Bio-based	ASTM D6866 OK biobased VINÇOTTE	ASTM D6866 USDA CERTIFIED BIOBASED PRODUCT PRODUCT 98%	OK biobased TÜV AUSTRIA S0720
• Food contact	FDA [36]		
• Certificate of Material Excellence		Material ConneXion	

In the past, Mirel™ PHA F1005 and F1006 were the two food-contact injection molding grades commercialized by Telles (joint venture between Metabolix and Archer Daniels Midland) that were approved by FDA in 2010. In 2014, PHB with up to 25% 3-hydroxyvaleric acid, 3-hydroxyhexanoic acid, 3-hydroxyoctanoic acid, and/or 3-hydroxydecanoic acid was also approved for use in the manufacture of food contact materials, except for use in contact with infant formula and breast milk [36].

8.4 MARKET INTRODUCTION AND APPLICATIONS OF PHA

8.4.1 Global Bioplastics and PHA Market Data

Recent reports of the European Bioplastics' annual revision indicated an established rise in bioplastics manufacture, which is projected to expand by 25% up to 2023 [40]. The emerging need for environmentally benign products and the flourishing consumer perception of sustainability issues constitute the driving forces for the increase in bioplastics manufacture. On top of that, the deployment of circular economy and bio-economy concepts to mitigate the depletion of fossil resources and greenhouse gas emissions has encouraged bioplastics production. Within the bioplastics market, the share of polylactic acid (PLA) and polyhydroxyalkanoates (PHA) demonstrate the leading drivers in terms of growth. PHA have been undergoing research and development for quite a few decades (since the 1990s) and have been lately put into effect for the onset of industrial manufacturing, which is estimated to quadruple by 2023. The biodegradability and compostability, but also the mechanical properties of PHA have been widely recognized through various research studies [41].

More specifically, the worldwide PHA market is projected to increase from US$57 million in 2019 to US$98 million by 2024, characterized by a compound annual growth rate (CAGR) of 11.2% [42]. The wide range of applications that utilize conventional chemicals deriving from petroleum, combined with rigorous governmental policies and frameworks have imposed the industrialization of biodegradable plastics, including PHA. Likewise, the production volume of *scl*-PHA, and consequently their market share, constitute the quickest rate of increase between 2019 and 2024. The wide range of end uses for *scl*-PHA including packaging, the food and catering sector, biomedical applications, agriculture and horticulture, automotive, and electronics, can explain the expected increase in their production [42].

Currently, as illustrated in Table 8.3, industrial PHA production is occupied by several industrial facilities, viz., Matebolix Inc (USA), Shenzhen Ecomann Technology Co. Ltd (China), Tianjin GreenBio Materials Co. Ltd. (China), Danimer Scientific (USA), Bio-On Srl (Italy), Biomer (Germany), etc. [8, 43]. China is dominating global bioplastics production, whereas Europe accommodates one-fifth of the total bioplastics production, which is projected to show an increase of 27% by 2023 [42].

Nevertheless, it is unequivocal that the high cost of production for PHA is impeding large-scale manufacturing, regardless of the frameworks that recommend the introduction of biodegradable and bio-based plastics. Therefore, it is imperative to find solutions that will moderate the preventive cost of manufacture, deriving primarily from the cost of raw materials, such as high purity sugars, and downstream

TABLE 8.3
Pilot and Industrial Scale PHA Manufacturers Currently Active Worldwide [8]

Name of company	Product (Trademark)	Substrate	Biocatalyst	Production capacity
Biomatera, Canada	PHA resins (Biomatera)	Renewable raw materials	Non-pathogenic, non-transgenic bacteria isolated from soil	
Biomer, Germany	PHB pellets (Biomer®)	Sugar (sucrose)		
Bio-On Srl., Italy	PHB, PHBV spheres (minerv®-PHA)	Sugar beets	*Cupriavidus necator*	10000 t/a
BluePHA, China	Customized PHBVHHx, PHV, P3HP(3HB), P3HP4HB, P3HP, P4HB synthesis		Development of microbial strains via synthetic biology	
Danimer Scientific, USA	*mcl*-PHA (Nodax® PHA)	Cold pressed Canola oil		
Kaneka Corporation, Japan	PHB-PHHx (AONILEX®)	Plant oils		3500 t/a
Newlight Technologies LLC, USA	PHA resins (AirCarbon™)	Oxygen from air and carbon from captured methane emissions	Newlight's 9X biocatalyst	
PHB Industrial S.A., Brazil	PHB, PHBV (BIOCYCLE®)	Saccharose	*Alcaligenes* sp.	3000 t/a
PolyFerm, Canada	*mcl*-PHA (VersaMer™ PHA)	Sugars, vegetable oils	Naturally selected microorganisms	
Shenzhen Ecomann Biotechnology Co. Ltd, China	PHA pellets, resins, microbeads (AmBio®)	Sugar or glucose		5000 t/a

(Continued)

TABLE 8.3 (CONTINUED)
Pilot and Industrial Scale PHA Manufacturers Currently Active Worldwide [8]

Name of company	Product (Trademark)	Substrate	Biocatalyst	Production capacity
SIRIM Bioplastics Pilot Plant, Malaysia	Various types of PHA	Palm oil mill effluent (POME), crude palm kernel oil		2 000 t/a
TianAn Biologic Materials Co. Ltd, China	PHB, PHBV (ENMAT™)	Dextrose deriving from corn of cassava grown in China	*Ralstonia eutropha*	10 000 t/a, 50 000 t/a by 2020
Tianjin GreenBio Material Co., China	P(3,4HB) films, pellets/foam pellets (Sogreen®)	Sugar		10 000 t/a

PHB, P(3HB): poly(3-hydroxybutyrate); PHBV: poly(3-hydroxybutyrate-*co*-3-hydroxyvalerate);
PHBVHHx: poly(3-hydroxybutyrate-*co*-3-hydroxyvalerate-*co*-3-hydroxyhexanoate); PHV: poly(3-hydroxyvalerate); P3HP(3HB): poly(3-hydroxypropionate-*co*-3-hydroxybutyrate); P3HP4HB: poly(3-hydroxypropionate-*co*-4-hydroxybutyrate); P3HP: poly(3-hydroxypropionate); P4HB: poly(4-hydroxybutyrate); *mcl*-PHA: medium-chain-length PHA; P(3,4HB): poly(3-hydroxybutyrate-*co*-4-hydroxybutyrate)

processing. As an intracellular product, PHA extraction and purification represents around 60% of its production cost [44]. Therefore, the utilization of low cost and renewable resources, including municipal and food waste along with agricultural by-product streams as the onset material in fermentation processes exhibits a potential cost-effective solution. Likewise, enzymatic downstream processing to recover intracellular biopolymer has been proposed as an alternative to organic solvents [45].

On top of that, robust PHA producing strains should be developed by employing metabolic engineering or isolation of new candidate strains. In this case, high conversion yields, high productivity, tolerance to inhibitors, and consumption of a variety of carbon sources constitute the criteria to be met to attain the best PHA performing strains.

Moreover, within the development of biorefinery concepts to exploit renewable resources it is important to generate multiple high value-added products, which will find diversified end uses. Production of PHA with protein isolate and antioxidants [46] and pyrolysis products [47], succinate [48], L-arginine [49], carotenoids [50, 51] and various other products have been developed to reduce the cost of production. Moreover, the employment of microbial cell factories able to simultaneously produce intracellular PHA and extracellular high value-added products such as biosurfactants, in the form of rhamnolipids [52], bioemulsifiers/polysaccharides [53], or other secondary metabolites such as phenazines and pyrrolnitrins [54] can reduce the fermentation cost. Likewise, the production of targeted PHA that will entail specific applications, could enable the attainment of high market prices that will counterbalance the cost of manufacture, thus enhancing the profitability of the process. As such, the following section will elaborate on recent developments with respect to the applications of PHA as innovative materials with enhanced functional properties.

8.4.2 Applications of PHA

8.4.2.1 Packaging Applications

Around 40% of the plastics produced in the EU in 2018 were used for packaging purposes [55]. A study on the global flow of plastic packaging in 2013 showed that 14% of plastic packaging was collected for recycling and another 14% was sent to incineration and/or energy recovery. However, 72% of plastic packaging was not recovered, as 40% is landfilled and 32% leaks out of the collection system, with around 8 million metric tons of plastic packaging entering the oceans each year [56]. A large amount of petrol-based plastics for packaging purposes is used specifically for food and drink packaging. Common plastics used by the food and drinks sector include poly(ethylene terephthalate) (PET), low and high-density poly(ethylene) (LDPE, HDPE), poly(vinyl chloride) (PVC), poly(propylene) (PP) and poly(styrene) (PS).

Applications of PHA for packaging materials have been the primary sector implementing their utilization since the commercial production of P(3HB) with varying amounts of 3HV units under the name Biopol® by Imperial Chemical Industries to mitigate the use of conventional plastic. Utilization of PHA as packaging materials have to meet certain criteria, including the preservation of organoleptic characteristics (flavor, texture, color), protection from dust, humidity, dehydration, microbial contamination, and chemical modifications, stability during storage and environmental conditions, food grade quality, and allowing for modified atmosphere among others [57]. However, the brittle and rigid nature of P(3HB) and thermal stability often restrain packaging applications, particularly for film formulations. To overcome these mechanical properties and facilitate processing, efforts were targeted to synthesize copolymers of P(3HB-*co*-3HV) via the addition of precursors [58,59]. Likewise, the inclusion of medium-chain-length monomeric units enhanced the flexibility through a reduction in stiffness [19]. Moreover, blends of PHB with other materials have been also developed. For instance, blends of P(3HB) with poly(vinyl acetate) and poly(*cis*-1,4-propene) exhibited enhanced mechanical properties [60], P(3HB) and P(3HB-*co*-3HV), with PLA presenting better thermal stability, ductility, and toughness [61]. Another approach involves the formulation of multilayer structures, more specifically using P(3HB-*co*-3HV) and high barrier self-adhesive nanostructured interlayers of zein [20,62]. The same research group also developed hybrid electrospun P(3HB) fibers with crystalline bacterial cellulose nanowhiskers, suggesting improved barrier performance, aiming to be applied in food packaging [63].

Several food and drink industries are currently interested in incorporating biodegradable plastic products for their packaging needs. In January 2019, Nestlé Waters announced a global partnership with Danimer Scientific for the development of biodegradable water bottles made from Nodax™ PHA, in order to substitute PET. Nodax™ PHA is 100% bio-based, certified to be fully biodegradable in marine water, freshwater, and soil, as well as in home and industrial composting conditions by TUV Austria, and FDA approved for food contact [34]. PepsiCo, as an existing partner of Danimer Scientific, may also gain access to the PHA-based resins developed under this collaboration [64]. PepsiCo is also in collaboration with Danimer Scientific for the development of the next generation bio-based and compostable flexible packaging using Danimer 24365B and Danimer 01112 resins [65, 66]. Those resins are blends of PHA biopolymers and mineral filler and can be processed in blown film lines. PepsiCo and Danimer Scientific used these resins to develop a new industrial compostable snack bag that was awarded the annual Innovation in Bioplastics Award in 2018 and is currently being piloted in the U.S. and Chile [67]. Furthermore, Genpack®, a leading manufacturer of food quality packaging, partnered with Danimer Scientific in order to create a new line of biodegradable food containers branded as GenZero™ [68,69]. Moreover, the Italian PHA producing firm Bio-On S.p.A. in collaboration with Rivoira, a leading fruit producer and distributor in Europe, created a new company named Zeropack in 2018 with the aim of focusing on PHA-based food packaging for fruits and vegetables [35,70]. An overview of diverse areas of application for PHA is given in Figure 8.1.

8.4.2.2 Biomedical Applications

Apart from the biodegradability of PHA, another vital characteristic of PHA lies in their biocompatibility, rendering them non-harmful and non-toxic to soft tissues or blood of the host organism, if present at concentrations less than 20 mg/L [78,79]. Biocompatible materials do not release any substance (migration) that would

FIGURE 8.1 Examples of PHA applications in: a) compostable snack bags [67], b) sunscreen products [71, 72], c) disposable items [73], d) eco-sustainable toys [74], e) biodegradable bags [75], f) biodegradable plantable pots [76], and g) biodegradable escape panel for blue crab traps [77].

cause cytotoxicity and lead to inflammation through immune reactions and allergic responses.

PHA exhibit energy and storage components within microorganisms, plants, and animals, whereby 3HB is a compound of human blood, present at concentrations of 3–10 mg/100 ml. and *(R)*-3-hydroxybutanoic acid constitutes a product of cell metabolism, synthesized in the liver [41,78,79]. Likewise, the hydrolysis products of P(3HB) and P(4HB), and 3HB and 4HB respectively, are human metabolites indicated to be non-toxic [43]. Therefore, following the natural origin of PHA, they have been evaluated as drug delivery matrices. The biodegradation of PHA inside the tissues of the target end point demonstrates the option for controlled release of bioactive compounds. Biocompatibility of PHA has been studied via *in vitro* and *in vivo* experiments. For instance, P(3HB) Nebe *et al.* [80] employed osteosarcoma and epithelial cell lines to study structural modifications in the components of cell adhesion in cultures of P(3HB), indicating that cell adhesion was compromised in cultures of P(3HB), evidencing the biocompatibility of the polymer. *In vitro* cell cultures and experiments in laboratory animals, studying physiology, toxicology, medical, and histological factors, have indicated the biocompatibility of PHA in cells, tissues, and host organisms [41].

Recent reviews have extensively elaborated on the previous research conducted with respect to applications of PHA, distinguishing them primarily between biomedical and non-medical end uses [41,43]. Biomedical applications include heart valve tissue engineering to improve implications in cardiovascular diseases, bone and cartilage tissue engineering, nerve repair and regeneration, and drug delivery matrices, among others. For instance, tissue-engineered autologous aortic grafts using copolymers including PHA were prepared and evaluated as vascular substitutes. It is believed that PHA vascular tissue engineering will enable the tailored design of heart valves to replace conventional synthetic prosthetic valves, with enhanced durability. Recent progress on tissue engineering and stem cells are anticipated to provide solutions for patients with brain damage but also for the regeneration of injured or damaged nerve cells, through the utilization of PHA, e.g., P(3HB) and P(4HB), for nerve guide conduit formulation [41]. Significant research has been also performed to evaluate the exploitation of PHA as drug delivery carrier systems. Biocompatibility and biodegradability of PHA, which can be tailored using manifold approaches to include specific monomeric units, have enabled the administration of bioactive compounds (antibiotics, antitumor, and anticancer drugs) in specific sites of actions. Both *scl*-PHA and *mcl*-PHA have been successful for drug delivery that is correlated with their crystallinity, hydrophobicity, and melting behavior [81, 82]. It has been generally indicated that P(3HB), P(3HB-*co*-3HV), P(4HB), P(3HB-*co*-3HHx), and PHO have been implemented to generate sutures, repair devices and compartments, cardiovascular grafts, orthopedic pins, guided tissue repair, cartilage repair devices, etc. [83]. A broad spectrum of biomedical applications has been evidenced so far and the exquisite advantages of biodegradability and biocompatibility along with the potential to produced tailor-made PHA by fine tuning of the monomeric units has led to research to alleviate several health issues.

8.4.2.3 Agricultural Applications

At the moment, conventional plastics are used as greenhouse, mulch, and silage films, among others, and include mainly LLDPE (linear low-density poly(propylene)), LDPE (low density poly(propylene)), HDPE (high density poly(propylene)), EVA/EBA (ethylene vinyl acetate/ethylene butyl acrylate), and, to a lesser extent, PVC (poly(vinyl chloride)) and EVOH (ethylene vinyl alcohol). The global agricultural films market size is expected to reach around US$9.3 billion in 2021, with a respective volume of around 7.5 million metric tons [84]. Subsequently, pollution due to plastic film residues is a major issue as countermeasures are difficult to implement due to lack of policy, economic considerations, and difficulties in mechanical recycling [85].

PHA can replace the vast amounts of conventional plastics currently used in agriculture and horticulture due to their biodegradability, as they can decompose completely when buried into the soil. A recent study has shown that PHB nanofiber films can fully degrade within three weeks due to their three-dimensional structure and high surface area [27], which is faster than other PHB films. Applications include their use as biodegradable mulch films, nets, and grow bags, as well as controlled-release carriers to deliver pesticides [86].

Mulching is used for crop protection and as a mean to increase crop yield and improve crop quality. Mulch films act as protective layers on top of the soil in order to maintain good soil structure, suppress weed growth, conserve moisture, prevent contamination, and reduce fertilizer leaching. The amount of mulch films marketed in Europe reaches up to 80 metric kilotons per year, with only 5% being biodegradable [87]. Nodax™, a P(3HB-*co*-3HHx) copolymer produced by canola oil, as well as other PHA-based custom-made mulches are available from Danimer Scientific [34]. In addition, Mirel™ bioplastics are PHA-based resins produced by bacterial fermentation of sugars, which can form blown and cast mulch films as well as compost bags, retail bags, and packaging [33]. Both products have been certified regarding their bio-based carbon content (ASTM D6866), their compostability in municipal and industrial aerobic composting facilities (ASTM D6400), and their aerobic degradation in soil (ASTM D5988-96).

Plastic materials such as HDPE are also used in agriculture as nets, including shading nets for greenhouses and nets that protect crops from birds, insects, hail damage, and adverse weather conditions [88]. The benefits of using PHA in such applications include their direct disposal in soil, due to their effective decomposition, and their ability to be composted with other agricultural residues, as well as manure and food waste [88]. PHA, in the form of P(4HB) and ultra-high-molecular-weight P(3HB), as well as PLA/PHA blends are now considered as alternatives to conventional HDPE agricultural nets due to similar mechanical characteristics with HDPE, such as elongation at break and tensile strength [89–92].

Last, but not least, a considerable fraction of plastics that are used in agriculture are plastic fertilizer and grow bags. Grow bags, or seedling bags, are used to retain water, regulate temperature, enhance survival rates by isolating plants to grow individually and therefore avoid root interference, reposition young crops, and also protect crops from external factors. Conventional plastics used for such purposes include PE, and especially LDPE. Alternatively, SoilWrap® is a commercially available plantable pot, made of Mirel™ PHA and developed by the Ball Horticultural Company (US) [76, 93]. It acts as a protective cover while a plant is being shipped and also as informational packaging when the plant is on the shelf of the store. It can be planted right into the ground and break down in the soil or be composted. Those materials are non-photobiodegradable and they completely degrade within 3–6 months; to carbon dioxide and water under aerobic conditions, and to biogas (methane and carbon dioxide) and water when the biodegradation occurs under anaerobic conditions [26,27]. Other bio-based options for grow bags include the use of fiber fabrics originating from vegetable and agricultural residues, or green biocomposites by combining materials such as PHA, PLA, polycaprolactone, and natural fiber [94,95].

8.4.2.4 Marine Applications

According to the European Commission, it has been estimated that on European beaches plastics make up 80–85 % of marine litter by count. Single-use plastics and plastics used in fishing gear account for 49% and 27% of all marine litter by count respectively [96–98]. As plastic debris can travel over long distances, the issue of marine litter is universal, and several actions and laws worldwide were put together to address this issue.

Biodegradable polymers, such as PHA, provide the opportunity to replace the vast amounts of conventional plastics currently used in aquaculture/fisheries. The advantages of PHA in terms of biodegradability have also prompted fishery applications during the last years. Marine microbiota can break down PHA; thus, it can be ascertained that PHA are completely biodegraded in the aquatic environment, whereas the modification of the thickness layer can correspondingly manage the biodegradation rate [99]. In fact, PHA in aquatic environment exhibited a decomposition rate similar to cellulose, evidencing the potential for fish traps manufacture to alleviate the impact of the "ghost fishing" that causes a significant environmental burden [100]. Bilkovic *et al.* [101] designed biodegradable fishing panels for fish and/or crustaceans using PHA, to mitigate the losses in conventional pots that subsequently result in reductions in fish catch. Deployment of the PHA pots was successful in increasing the catch of blue crabs in a case study. In 2015, a PHA producing company announced the pilot-plant production of a freshwater fishing lure made of pure PHA in collaboration with a company producing fish traps [102]. Likewise, PHA were evaluated as the onset polymeric material to formulate biodegradable escape panels for crab, lobster, and fish traps, indicating that their application did not implicate the rate of fish catch [103]. Moreover, plastic shotgun wads are found on beaches, as they have become a common marine debris item. In an effort to provide a biodegradable and sustainable alternative, researchers have developed wads made of PHA that were shown to outperform some of the commercial ammunition and gave repeatable shot patterns along with acceptable velocities [104].

8.4.2.5 Disposable Items, Single-Use Plastics

More than 25 countries around the world either ban or tax single-use plastic items in order to reduce the vast amounts of plastic cutlery, cotton buds, straws and stirrers, food containers and plastic lids for hot drinks, as part of a sweeping law against plastic waste that despoils beaches and pollutes oceans [56]. To address today's plastic waste management challenge, several companies are working with biodegradable materials in order to substitute petrol-based single-use items with biodegradable ones.

PHA have been implemented in the manufacture of bottles (e.g., shampoo), bags and films via injection-molding, extrusion (blowing and casting), blowing, and thermoforming [105, 106]. More specifically, films produced with PHA have been applied for the formulation of shopping bags, containers and paper coatings, utensils, diapers, cosmetic containers and cups, and compostable bags and lids, among other everyday commodities [43].

RWDC industries launched its Solon brand last year, made of PHA biopolymers [107]. Solon™ was first introduced in Singapore as a drinking straw, aiming at launching biodegradable cutlery, cups, bags, coffee cups, and lids in the future [73, 108]. In a similar context, Danimer Scientific in a collaboration with WinCup, a manufacturer of foodservice packaging based in the U.S., launched a new line of straws and stirrers made from Nodax™ PHA, branded as phade™. Both products are certified to be fully biodegradable in marine water, freshwater, and soil, as well as in home and industrial composting conditions by TUV/Vincotte, and are available at commercially viable price [108, 109].

8.4.2.6 Other Applications

Several interesting applications of PHA produced by Bio-On S.p.A have been explored recently. Bio-On developed Minerv Bio Cosmetics PHA microspheres, to be used as biopolymeric ingredients for the cosmetics industry, and in 2018 was awarded the "Best Practices" Award by Frost & Sullivan for New Product Innovation in the Bio-based Ingredients for the Cosmetics Industry [110]. PHA microspheres can find a wide range of applications, from skin cream and make-up to hair care or hygiene products. They provide a sustainable and biodegradable alternative to synthetic polymers such as PET, PP, and PE used in cosmetics such as lipstick, nail polish, creams, shower gels, etc. Minerv Bio Cosmetics microspheres have a size ranging between 5 to 20 microns, allowing them to be used in the food, healthcare, and packaging sectors. They are characterized by extraordinary optical properties that help make cosmetics that exceed the mattifying effect of most products currently available in the market. Their high sphericity helps formulate smooth creams that are easier to apply and their high porosity helps control excess oil in cosmetic creams and easily absorb active substances. It is anticipated that in the near future, PHA microspheres could help with cancer treatment, acting as carriers for active substances. Moreover, they have a booster effect that drastically reduces the quantity of chemical UV filters needed to be used in the formulation of sun protection products [111]. As an example, in 2019, Bio-On, in collaboration with Unilever, launched a new line of ultra-green cosmetic ingredients for sun protection using Minerv Riviera PHB micro-powder, to formulate products under the brand My Kai [71]. PHB micro powders were used to significantly reduce the percentage of UV filters that are used for sun protection and also to boost water-resistance. Apart from providing an innovative and eco-friendly alternative for the consumers by incorporating PHA micro-powders, Unilever also pursues the goals stated in its Unilever Sustainable Living Plan business model: improve people's health and wellbeing by 2020 and halve the environmental impact of its products by 2030 [72, 112].

Bio-On has also developed a Minerv PHA polymeric liquid in order to replace the triacetin used in old- and new-generation cigarette filters. One cigarette contains around 50mg of triacetin that, apart from not being biodegradable, slows down the disintegration of cellulose acetates. It was shown that the replacement of triacetin with Minerv PHA resulted in blocking up to 60% of harmful substances to the human body (Reactive Oxygen Species), without changing the nicotine taste [113].

In addition, Bio-On has launched Fashion Development Materials in 2018, a new business unit aiming to identify processes, technologies, and patents in order to develop PHA-based fabrics, yarns, flexible surfaces, and films. Those biopolymers will aim to replace synthetic fibers used in clothes and textiles and contribute to the reduction of microplastics that are rinsed away by washing and end up in aquatic environments [114].

In the electronics sector, Eloxel, a new company created by Bio-On in 2018, aims to exploit the ability of PHB to produce and store electrical energy as a result of mechanical stress by taking advantage of its piezoelectric properties. New products developed may include new generation PHA-based batteries and fully green piezoelectric materials [115].

Other applications currently being pursued by Bio-On in collaboration with Italeri, a leading Italian toy manufacturer, is the production of Minerv PHA bioplastic toys, as those bioplastics possess the same thermo-mechanical properties as conventional plastics and have the benefit of being 100% eco-sustainable and biodegradable [74]. Moreover, Bio-On in a partnership with Kartell, a leading furniture design firm, worked together to produce an eco-sustainable line of furniture made from high quality PHA [116].

8.5 PHA MONOMERS AS BULK CHEMICALS

The development of bioprocesses for the production of bulk chemicals from renewable resources is challenging, especially the selection of bulk chemicals as suitable precursors for the development of sustainable processes. Poly(*R*)-3-hydroxyalkanoates (PHA) are ideal molecules for the industrial production of bio-based and biodegradable plastics, yet there is industrial potential for the production of *(R)*-3-hydroxyalkyl carboxylic acids for the production of versatile end products (i.e., novel polymers, fine chemicals, implant biomaterials, pheromones, medicines, and biofuels [117]. More than 150 different monomers of hydroxycarboxylic acids with different functional groups exist [118]. These consist of two functional groups, a carbonyl and a hydroxyl group, and a chiral center. Due to the chiral center that can easily be altered, hydroxyalkanoic acids (HAs) can be used as precursors or intermediates for the synthesis of versatile end products. Although chiral hydroxycarboxylic acids (HAs) have great market potential, only a few are commercially available, such as *(R)*-3-hydroxybutyric acid (R-3HB).

8.5.1 Production of HAs from PHA

PHA monomers can be produced via (i) chemical synthesis and chemical hydrolysis, (ii) enzymatic hydrolysis, and (iii) metabolic engineering of microorganisms.

8.5.1.1 Chemical Synthesis and Hydrolysis

HAs can be produced using different approaches for chemical synthesis. Organic synthesis may involve stereoselective functionalization through Sharpless' asymmetric epoxidation and hydroxylation, or through Brown's asymmetric allyboration [119]. These processes are expensive because a metal complex catalyst is involved and might leave impurities in the final product. The synthesis of enantiopure or enantiomerically enriched 3-hydroxycarboxylic acids involves enantioselective reduction using 3-keto acids [120]. This process often requires prochiral precursors (i.e., 3-keto esters) which might decrease the yield of the reaction and make the process more complex. Chemical hydrolysis has also been applied for the production of high purity 3-hydroxyoctanoic acid. Acidic methanolysis of PHA copolymers produced by *Pseudomonas putida* led to a high recovery yield [121]. Boric acid (10–20% mol) has been used for the selective esterification at high yields of a-hydroxycarboxylic acids in the presence of other carboxylic acids, in mild conditions [122].

8.5.1.2 Enzymatic Hydrolysis

Utilization of biochemical processes compared to chemical roots offers the advantage of selective reactions leading to higher enantiomeric excess [123]. PHA *in vitro* depolymerization is catalyzed by extracellular PHA depolymerases produced by many microorganisms isolated by different environments. PHA depolymerase by *A. faecalis* has been used in PHA films [121]. The production of 3-hydroxyalkanoates with different chain lengths (C4–C10) has been achieved by the degradation of PHA copolyesters with depolymerases produced by *W. eutropha* and *P. oleovorans.* The rate of enzymatic hydrolysis highly depends on the composition of the polyesters, decreasing with increased monomeric side chain length [124]. Another factor that has been proved to affect *in vitro* depolymerization is exposure to ultrasonication (37 kHz, 30% of power output, and 25°C). The highest degree of mlc-PHA depolymerization (around 50 wt.%) was observed at 120 min of sonication, whereas a shorter sonication time (90 min) resulted in around 30 wt.% depolymerization due to inefficient cell rapture, and a longer time (120 min) led to around 40 wt.% depolymerization due to denaturing of the depolymerase [125].

Lipases have also been used for *in vitro* degradation of polyesters to their respective monomers. Lipases from *P. aeruginosa* were used for the production of hydroxyhexanoate from polycaprolactone as substrate. Lipases can efficiently hydrolyze polyesters that consist of omega-hydroxyalkanoic acid (i.e., poly(6-hydroxyhexanoate) or poly(4-hydroxybutyrate)) rather than polyesters containing side chains in the polymer backbone (i.e., PHB) [126]. Different enzymes, nitrile hydratase and amidase from *Rhodococcus* sp. [127] and *Comamonas testosteroni* [128], have been used for the hydrolysis of 3-hydroxyalkanenitriles to 3HAs with relatively high conversion yields (63–83%).

In vivo depolymerization acts on intracellular PHA with hydrolytic enzymes that are produced by the same microorganisms that accumulate PHA. A successful example of (*R*)-3HB production from PHB producing *Alcaligenes latus* resulting in 96% recovery yield (g (*R*)-3HB/g PHB) was achieved at low fermentation pH (3–4) in order to inhibit reutilization of R-3HB by the microorganism [123]. *Pseudomonas putida* has been evaluated for the degradation of *mcl*-PHA at alkaline pH (9–11), resulting in the degradation of PHA into saturated monomers ((*R*)-3-hydroxyoctanoic acid) and unsaturated monomers (*(R)*-3-hydroxy-6-heptenoic acid, R-3-hydroxy-8-nonenoic acid, and 3-hydroxy-10-undecenoic acid) [129]. *Pseudomonas putida* has also been used for the production of eight enantiomerically pure (*R*)-3-hydroxycarboxylic acids ((*R*)-3-hydroxyoctanoic acid, (*R*)-3-hydroxyhexanoic acid, (*R*)-3-hydroxy-10-undecenoic acid, (*R*)-3-hydroxy-8-nonenoic acid, (*R*)-3-hydroxy-6-heptenoic acid, (*R*)-3hydroxyundecanoic acid, (*R*)-3-hydroxynonanoic acid, and (*R*)-3-hydroxyheptanoic acid) with a recovery yield of 78% and more than 95% purity in all (*R*)-3-hydroxycarboxylic acids [130]. Another case of *in vivo* depolymerization by *Pseudomonas putida* was observed in *mcl*-PHA, resulting in the liberation of monomers at a rate of 0.21 g/L/h with 98% recovery yield, at optimum temperature of 30°C and pH 9 [131]. Wood extract hydrolysate has also been used as renewable feedstock by *Burkholderia cepacia* for the production and *in vivo* hydrolysis of PHB, resulting in 14 g/L *R*-3HB in three days of fermentation [132].

8.5.1.3 Metabolic Engineering

PHA biosynthesis and depolymerization genes from *W. eutropha* have been successfully introduced to *E. coli* resulting in the production of dimers and monomers of 3HB via *in vivo* depolymerization of purified PHA [133]. Another case of a recombinant *E. coli* involved insertion of 3-ketothiolase (PhbA) and acetoacetyl-CoA reductase (PhbB) from *W. eutropha*, resulting in the extracellular production of 1 g/L 3-hydroxybutyric acid. Also, heterologous expression of *(R)*-3-hydroxydecanol-ACP:CoA transacylase (phaG) from *P. putida* resulted in the extracellular production of 0.6 g/L 3-hydroxydecanoic acid [134]. Simultaneous expression of 3-ketothiolase (phbA), acetoacetyl-CoA reductase (phbB), phosphor-transbutyrylase (ptb), and butyrate kinase (buk) in *E. coli* resulted in the production of 12 g/L of 3-hydroxybutyric acid in a fed-batch fermentation after 48 h [135]. Another engineering approach has been applied in *E. coli* for the production of R3HAs through *in vivo* depolymerization poly(3-HA). Initially, a PHA synthase (*PhaC2*) gene from *Pseudomonas* sp. and PHA depolymerase (*PhaZ*) gene from *P. aeruginosa* were simultaneously expressed in a recombinant *E. coli fadA* mutant, leading to 0.49 g/L production of R3HAs. Subsequently, a 3-ketoacyl-ACP reductase (*fabG*) gene was overexpressed leading to the production of 1 g/L of R3HAs [136]. Simultaneous extracellular production of 3HB and *mcl*-3HA (1:3 ratio of 3-hydroxyoctanoic acid and 3-hydroxydecanoic acid) was achieved by insertion of β-ketothiolase (*phbA*), acetoacetyl-CoA reductase (*phbB*), and 3-hydroxyacyl-ACP CoA transacylase (*phaG*) genes in *E. coli*, while the addition of acrylic acid in the fermentation medium resulted in six-fold increase of 3HB and *mcl*-3HA concentrations [137].

8.5.2 Applications of PHA Monomers

Novel tailor-made copolymers: Bacterial produced R-hydrocarboxylic acids will result in the production of polyesters with isotactic molecules creating a novel class of homo-polyesters. Polymerization of racemic starting materials of 3HB can occur in anhydrous solvents triggered by dicyclohexylcarbodiimide (DCC), p-toluenesulfonyl chloride (TosCl), or 2,4,6-triisopropylbenzenesulfonyl chloride (TPS), and at 25°C and triethyl amine at 0°C. Utilization of HCl or H_2SO_4 results in water release. Titanium isopropoxide catalyzes the polymerization at high pressure and temperature (140°C). Polymerization of 4HB produced by recombinant microorganisms was used as the precursor for the production of homopolymer poly(4-hydroxybutyrate) and copolymer poly(3-hydroxybutyrate-*co*-4-hydroxybutyrate) [138]. A PHA synthase by *Burkholderia contaminans* was expressed in a *Cupriavidus necator* mutant leading to biosynthesis of PHA polymers containing 4-hydroxybutyrate (4HB) and 5-hydroxyvalerate (5HV) monomers with varying content, low melting temperature and crystallinity, and lipase degradable for novel biomaterial applications [139, 140]

- *Medical applications*: *R*-3HB has been proven to have a wide variety of applications in medical treatment. Applications in hemorrhagic shock, burns, myocardial damage, cerebral hypoxia, anoxia, and ischemia are among the medical uses that have been applied so far. Other potential applications of 3HB are being investigated (increase in cardiac efficiency, prevention of brain damage) [117].

- *Antimicrobial agents*: RHAs act as antimicrobial or antiviral compounds in a similar way to detergents. RHAs are absorbed by the cell membrane and change its permeability. The length of the carbon chain affects the antimicrobial activity. *R*-3HO, *(R)*-3-hydroxy-8-nonenoic acid, and *(R)*-3-hydroxy-10-undecenoic acid have increased antimicrobial activity against *Listeria* and *Staphylococcus* species compared to nonhydroxylated compounds [130].
- *Chiral synthons*: RHAs can be used for chemical synthesis converted into aliphatic β-lactones via catalysis. So far *(R)*-3-hydroxyoctanoate has been used for the synthesis of α,β-unsaturated δ-lactone *(R)*-massoialactone through asymmetric hydrogenation and δ-lactone, 3,5-dihydroxydecanoic acid. Massoialactone is currently isolated from jasmine flowers and its production through biochemical roots would simplify the process. *(R)*-3-hydroxyoctanoate can be used for the production of gloeosporone, an autoinhibitor of spore germination, via asymmetric reduction of β-keto esters. *(R)*-3-hydroxyhept-6-enoate is a precursor for triquinane sesquiterpenes, components of essential oils found in plants and eremophilan carbolactones, valuable compounds for drug synthesis [141]. Catalysts for the dehydrative amide condensation of α-hydroxycarboxylic acids like alkylboronic acids have been used for large-scale synthesis of amides in a single step [142].

8.6 END-OF-LIFE OPTIONS

Plastics are affordable, durable, and versatile materials used in a wide variety of applications. Several benefits are associated with the production of plastic products related to society, in terms of economic activity, job creation, and ease of use. In many cases, plastics can help reduce energy consumption and furthermore greenhouse gas emissions, even in some packaging applications, when compared to other alternatives [143].

On the other hand, high quantities of plastic wastes contribute to negative environmental consequences [56]. The lengthy lifespan of plastics, due to their non-biodegradable nature, and the migration of complex blends of chemical substances, raise concerns regarding potential adverse effects on human health as well as the environment. In addition, lightweight plastics are easily carried away by the winds and sea currents, so giant masses of plastic waste debris have accumulated in the North Atlantic and Pacific Oceans which are very harmful to marine mammals, seabirds, and fish [143, 144].

Considering the above, the development of novel waste management techniques is of crucial importance. Significant improvement in this field is associated with an increase in plastics recycling, novel and innovative recycling technologies, and with the development of bio-based and biodegradable polymers (bioplastics), as alternatives to petrochemical ones.

Global plastic production reached 335 million metric tons in 2016, with around 60 million metric tons produced in Europe (EU28+NO/CH). On the other hand, in 2016 only 27.1 million metric tons of post-consumer plastic waste was collected through official schemes in the EU28+NO/CH in order to be further treated. The majority of the collected post-consumer plastic was incinerated (41.6%) for energy recovery, and the remaining was recycled (31.1%), and landfilled (27.3%) [145–147]. The evolution of the plastic waste treatment from 2006 to 2016 represented an increase in recycling by 79% (63% inside EU, 37% outside EU), and in energy recovery, by 61%, while landfilling was reduced by 43% [56]. It is worth mentioning that 62% of the total volume of collected plastic waste in 2016 originated from the plastic packaging sector, 40.8% of which was recycled, 38.8% incinerated, and 20.4% landfilled. By introducing biodegradable bio-based polymers into the market, the volume of plastic waste is expected to be reduced significantly, and provide novel and alternative options for their post-consumer management [148].

The major bottleneck of bioplastics production is their high production cost. For this reason, a combination of efficient waste management with appropriate End-of-Life (EoL) options is essential in order to adopt and promote the concept of the circular economy for those bio-based products. Current production lines follow the linear economic model, based on non-renewable fossil-based raw materials, utilizing significant amounts of non-renewable energy, while end products are usually disposed of after a short first-use cycle. Moreover, most plastics, and in particular plastics used for packaging, are only used once (single-use plastics), thus losing almost 95% of their value which is estimated to be US$80–120 billion annually [56]. On the other hand, the reduction of bioplastics' production costs, and their potential environmental impact by implementing the circular economy model, require defining proper EoL options for these products. In this section, different EoL options are discussed and their contribution to the circular economy concept is evaluated.

An overview of different End of Life options for PHA-based materials and their contribution to the process line of products is provided in Figure 8.2.

8.6.1 Reuse

The most effective and easy way to limit the amount of waste is the reusing of materials/products (including bio-based materials) [143,149]. Reusing materials/products leads to the reduction of raw material quantities used in the production phase and is considered the most economical way to use materials and the most effective one to preserve natural resources and protect the environment. By reusing materials, a reduction of manufacturing and supply chain carbon footprint is achieved, and waste management actions are limited, resulting in reducing the environmental impact [8,150]. PHA biopolymers are known for their biodegradability in different environmental conditions (e.g., industrial composting). However, PHA biopolymers can still be reused and therefore retain most of their value. Different options may include their enzymatic or chemical degradation to monomers, and subsequently the chemical synthesis of different PHA compositions; their use for gradually lower value applications compared to the original one; or their blending with other bioplastic recycled materials to formulate composites with different range of applications [146].

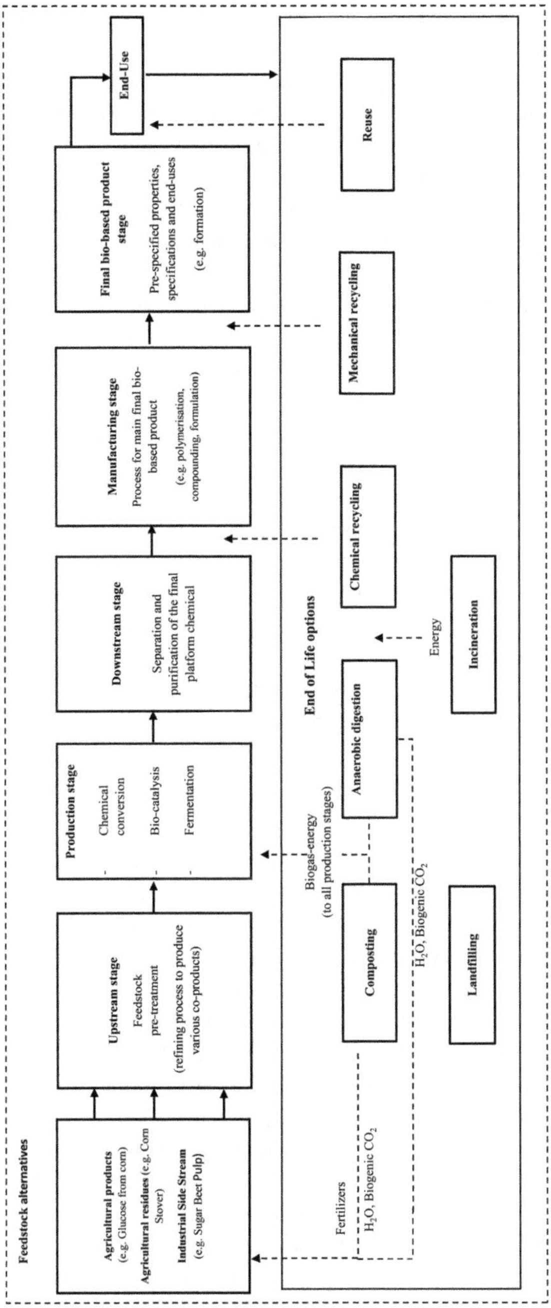

FIGURE 8.2 Different End-of-Life options and their contribution to the process line of the product.

8.6.2 Mechanical Recycling

Mechanical recycling is a method by which waste materials are recycled into secondary raw materials without changing the basic structure of the material. Mechanical recycling of plastics and bio-based plastics involves mainly size reduction (e.g., grinding), sorting, and reprocessing by conventional polymeric processing technologies [151]. The products of the recycling processing of plastic waste have the form of pellets, granules, or flakes, etc. and are called recyclates. The recyclates can be used directly (Figure 8.2), avoiding upstream, production, downstream, and manufacturing bioprocesses, and this may result in the lowest environmental impacts of recycling technology, if, for example, this is followed by chemical recycling and composting [152].

The non-biodegradable bio-based plastics waste (e.g., Bio-PE) can be easily managed through the mechanical recycling streams of the corresponding conventional plastics, as they are chemically identical [151]. On the other hand, the biodegradable and compostable (according to EN 13432) bio-based plastics waste (such as PHA), are designed for organic recycling (industrial composting), but may also be suitable for home composting or anaerobic digestion, depending on their characteristics. Possible contamination of conventional plastics recycling streams up to 10% by bio-based compostable plastics or mixed bio-based plastics waste is not expected to introduce any problems to the recycling processing or the quality of the produced recyclates, according to the relevant research works [151, 152]. The possibility of developing specific mechanical recycling streams and a recycling industry dedicated to bio-based products requires that the following pre-conditions are met: a) a growth of the bio-based plastics volumes to render mechanical recycling economically viable, and b) the development of new markets for bio-based plastics recyclates [153,154].

8.6.3 Chemical Recycling

The chemical recycling technique is used for breaking down plastic polymers into their constituent monomers. The recovered monomers can be used in refineries, biorefineries, or in fine-chemicals production. Several approaches have been proposed and applied for chemical recycling of different plastics or mixed plastics, including gasification, polymer cracking and conversion processes, and their use as reduction agents in blast furnaces [155,156]. Specifically, for bio-based products, research work has been performed for chemical recycling of PLA [157,158]. At the industrial level, NatureWorks LLC, the producer of the Ingeo™ PLA resin, describes chemical recycling (also known as feedstock recovery, or depolymerization) as a process for returning PLA-based products (or waste) back into the chemical monomer (lactic acid) [159]. NatureWorks LLC has already recycled more than 7.7 Mt of Ingeo™ PLA, obtained by breaking down PLA into lactic acid using hydrolysis. These monomers are then reused for the production of PLA resin. This approach is considered as a potential alternative EoL route for recovering this bio-based product. The monomers can be utilized in the manufacturing production stage of the whole production line (Figure 8.2) [146].

In the case of PHA, chemical recycling has been reported to occur by thermal degradation. Catalytic depolymerization of PHBV into vinyl monomers has been performed using alkali earth compound catalysts such as calcium oxide (CaO) and magnesium hydroxide ($Mg(OH)_2$) [160]. PHBV copolymers were selectively depolymerized into crotonic (CA) and 2-pentenoic acids (2-PA). Those two can be easily separated due to their different boiling points (180–185°C for CA and 199°C for 2-PA) and water solubilities (water-soluble CA and water-insoluble 2-PA). The monomers obtained by the chemical recycling of PHA can be subsequently used to produce polymers and copolymers, i.e., crotonic acid can be polymerized to poly(CA) and may also be copolymerized with acrylic acid to produce poly(crotonic acid-*co*-acrylic acid), P(CA-*co*-AA). This copolymer has high glass transmission temperature and excellent water solubility, and therefore can be used as an enzyme-stabilizing agent and hydrogel for bioabsorbents, and may find applications in wastewater treatment and agriculture [160, 161].

8.6.4 Composting and Anaerobic Digestion

Through anaerobic digestion and composting, agricultural, food bio-wastes, and bio-based products that cannot be recycled through mechanical or chemical methods can be utilized to produce fertilizers and biogas. PHA are degraded into water, carbon dioxide, and inorganic compounds through the aerobic process, and in the anaerobic process it produces water and biogas, a mixture of carbon dioxide and methane [26]. The produced biogas can be used in CHP plants for electricity and heat production or upgraded to bio-methane [146]. In addition, they emit less carbon dioxide than petrochemical polymers when they break down, and moreover, those emissions will be captured by crops/feedstock used to produce PHA, promoting carbon balance [162].

8.6.5 Energy Recovery – Incineration

According to European Bioplastics [146], energy recovery/incineration of bio-based waste products is considered the last EoL option, when the rest of the methods cannot be carried out. The bio-based part of incinerated materials releases the CO_2 amount that was sequestered by the plants, in this way closing the material carbon cycle. However, strong objections may be raised to this possibility as bio-based waste products may be used in higher value EoL routes, including organic recycling and anaerobic digestion. According to Zero Waste Europe [156], recycling makes much more sense than incineration, considering that recycling saves energy, is more profitable, and creates more jobs, while it is more flexible and dynamic. Furthermore, as bio-based products represent higher value products and potential secondary sources for new bio-based products, incineration should not be considered as an option among the alternative EoL routes for biodegradable bio-based products.

8.6.6 Landfilling

Landfilling is not recommended as an EoL route for biodegradable and/or bio-based products. In the case of non-biodegradable, non-recyclable bio-based materials, energy recovery is suggested [145].

8.7 CONCLUSIONS AND OUTLOOK

As discussed in this chapter, PHA production capacity is expected to quadruple within the next five years, as those polymers can replace conventional ones in various applications. Biodegradability is a great advantage. However, the end application should be wisely chosen, in order to preserve the value of those materials for as long as possible. *scl*-PHA are currently exploited for food packaging, agricultural, marine, cosmeceutical, and biomedical applications. Their stiff and brittle nature limits their applicability. However, research is currently focused on PHA blends with other bio-based and/or biodegradable polymers in order to overcome performance issues. Moreover, *mcl*-PHA are elastomers, but they are characterized by a high production cost and thus they may find use in high value targeted applications, such as medical implant biomaterials. The same applies for PHA monomers which may find several applications in biomedicine and pharmacy. Therefore, in the near future, PHA-based products are predicted to be used more regularly by consumers. As a result, efficient waste management strategies should be developed and information on how to dispose of such materials should be communicated by appropriate labeling. In addition, mechanical recycling with other bio-based and/or biodegradable plastics should be studied in order to understand whether the recyclates obtained can be used for specific applications. This will help avoid different recycling streams for each type of bio-based plastic, as their current volume is quite low, and it does not allow for an efficient recycling process.

REFERENCES

1. Chinthapalli, R., Skoczinski, P., Carus, M., Baltus, W., de Guzman, D., Kab, H., Raschka, A., Ravenstijn, J.: *Bio-Based Building Blocks and Polymers – Global Capacities, Production and Trends 2018–2023.* 16 (2019).
2. European Bioplastics: *Bioplastics Facts and Figures.* (2018).
3. PlasticsEurope: *Plastics – The Facts 2019: An Analysis of European Plastics Production, Demand and Waste Data.* (2019).
4. European Bioplastics, Nova-Institute: *Bioplastics Market Data 2018: Global Production Capacities of Bioplastics 2018-2023.* (2018).
5. van der Oever, M., Molenveld, K., van der Zee, M., Bos, H.: *Bio-Based and Biodegradable Plastics – Facts and Figures, Focus on Food Packaging in the Netherlands.* Wageningen Food Biobased Res. 65 (2017).
6. Koller, M., Sandholzer, D., Salerno, A., Braunegg, G., Narodoslawsky, M.: Biopolymer from industrial residues: Life cycle assessment of poly(hydroxyalkanoates) from whey. *Resour. Conserv. Recycl.* 73, 64–71 (2013). doi: 10.1016/j.resconrec.2013.01.017
7. Chanprateep, S.: Current trends in biodegradable polyhydroxyalkanoates. *J. Biosci. Bioeng.* 110, 621–632 (2010). doi: 10.1016/j.jbiosc.2010.07.014
8. Kourmentza, C., Plácido, J., Venetsaneas, N., Burniol-Figols, A., Varrone, C., Gavala, H.N., Reis, M.A.M.: Recent advances and challenges towards sustainable polyhydroxyalkanoate (PHA) production. *Bioengineering.* 4, 1–43 (2017). doi: 10.3390/bioengineering4020055
9. Kirwan, M.J., Strawbridge, J.M.: Food packaging technology. In: Coles, R., McDowell, D., and Kirwan, M.J. (eds.) *Packaging Technology and Science.* pp. 174–240. Blackwell Publishing, Oxford, UK (2004).
10. Koller, M.: Poly(hydroxyalkanoates) for food packaging: Application and attempts towards implementation. *Appl. Food Biotechnol.* 1(1), 3–15 (2014). doi: 10.22037/afb.v1i1.7127

11. van Crevel, R.: *Bio-Based Food Packaging in Sustainable Development.* Food and Agriculture Organization of the United Nations (FAO) (2016). http://www.fao.org/forestry/45849-023667e93ce5f79f4df3c74688c2067cc.pdf
12. Keshavarz, T., Roy, I.: Polyhydroxyalkanoates: Bioplastics with a green agenda. *Curr. Opin. Microbiol.* 13, 321–326 (2010). doi: 10.1016/j.mib.2010.02.006
13. PlasticsEurope: *Plastics – The Facts 2019, An Analysis of European Plastics Production, Demand and Waste Data.* (2019).
14. Anjum, A., Zuber, M., Zia, K.M., Noreen, A., Anjum, M.N., Tabasum, S.: Microbial production of polyhydroxyalkanoates (PHAs) and its copolymers: A review of recent advancements. *Int. J. Biol. Macromol.* 89, 161–174 (2016). doi: 10.1016/j.ijbiomac.2016.04.069
15. Bucci, D.Z., Tavares, L.B.B., Sell, I.: PHB packaging for the storage of food products. *Polym. Test.* 24, 564–571 (2005). doi: 10.1016/j.polymertesting.2005.02.008
16. Mensitieri, G., Di Maio, E., Buonocore, G.G., Nedi, I., Oliviero, M., Sansone, L., Iannace, S.: Processing and shelf life issues of selected food packaging materials and structures from renewable resources. *Trends Food Sci. Technol.* 22, 72–80 (2011). doi: 10.1016/j.tifs.2010.10.001
17. Chodak, I.: Polyhydroxyalkanoates: Origin, properties and applications. Monomers, Polym. *Compos. from Renew. Resour.* 451–477 (2008). doi: 10.1016/B978-0-08-0453 16-3.00022-3
18. Xu, J., Guo, B.H.: Poly(butylene succinate) and its copolymers: Research, development and industrialization. *Biotechnol. J.* 5, 1149–1163 (2010). doi: 10.1002/biot.201000136
19. Albuquerque, P.B.S., Malafaia, C.B.: Perspectives on the production, structural characteristics and potential applications of bioplastics derived from polyhydroxyalkanoates. *Int. J. Biol. Macromol.* 107, 615–625 (2018). doi: 10.1016/j.ijbiomac.2017.09.026
20. Fabra, M.J., Lopez-Rubio, A., Lagaron, J.M.: High barrier polyhydroxyalcanoate food packaging film by means of nanostructured electrospun interlayers of zein. *Food Hydrocoll.* 32, 106–114 (2013). doi: 10.1016/j.foodhyd.2012.12.007
21. Sanchez-Garcia, M.D., Gimenez, E., Lagaron, J.M.: Novel PET nanocomposites of interest in food packaging applications and comparative barrier performance with biopolyester nanocomposites. *J. Plast. Film Sheeting.* 23, 133–148 (2007). doi: 10.1177/8756087907083590
22. Kumar, S., Abe, H.: *Practical Guide to Microbial Polyhydroxyalkanoates.* Smithers Rapta, UK (2010).
23. Volova, T.G., Boyandin, A.N., Vasiliev, A.D., Karpov, V.A., Prudnikova, S. V., Mishukova, O. V., Boyarskikh, U.A., Filipenko, M.L., Rudnev, V.P., Bá Xuân, B., Vit Dũng, V., Gitelson, I.I.: Biodegradation of polyhydroxyalkanoates (PHAs) in tropical coastal waters and identification of PHA-degrading bacteria. *Polym. Degrad. Stab.* 95, 2350–2359 (2010). doi: 10.1016/j.polymdegradstab.2010.08.023
24. Akaraonye, E., Keshavarz, T., Roy, I.: Production of polyhydroxyalkanoates: The future green materials of choice. *J. Chem. Technol. Biotechnol.* 85, 732–743 (2010). doi: 10.1002/jctb.2392
25. Ren, Q., Roo, G. De, Ruth, K., Witholt, B., Zinn, M.: Simultaneous accumulation and degradation of polyhydroxyalkanoates: Futile cycle or clever regulation? *Biomacromolecules*, 10(4), 916–922 (2009).
26. Wang, S., Lydon, K.A., White, E.M., Grubbs, J.B., Lipp, E.K., Locklin, J., Jambeck, J.R.: Biodegradation of poly(3-hydroxybutyrate- co-3-hydroxyhexanoate) plastic under anaerobic sludge and aerobic seawater conditions: Gas evolution and microbial diversity. *Environ. Sci. Technol.* 52, 5700–5709 (2018). doi: 10.1021/acs.est.7b06688
27. Altaee, N., El-Hiti, G.A., Fahdil, A., Sudesh, K., Yousif, E.: Biodegradation of different formulations of polyhydroxybutyrate films in soil. *Springerplus.* 5(1), 1–12, (2016). doi: 10.1186/s40064-016-2480-2
28. Jendrossek, D.: Microbial degradation of polyesters. In: Babel, W. and Steinbüchel, A. (eds.) *Biopolyesters.* pp. 293–325. Springer Berlin Heidelberg, Berlin, Heidelberg (2001).

29. Harrison, J.P., Boardman, C., O'Callaghan, K., Delort, A.M., Song, J.: Biodegradability standards for carrier bags and plastic films in aquatic environments: A critical review. *R. Soc. Open Sci.* 5(5), 171792, (2018). doi: 10.1098/rsos.171792
30. RWDC Industries: *Solon is Certified*, https://www.rwdc-industries.com/technology
31. Danimer Scientific: *PHA Certifications*, https://danimerscientific.com/pha-the-future-of-biopolymers/pha-certifications/
32. Halonen, N.J., Palvölgyi, P.S., Bassani, A., Fiorentini, C., Nair, R., Spigno, G., Kordas, K.: Bio-based smart materials for food packaging and sensors – A review. *Frontiers in Materials*, 7, 82 (2020).
33. Mirel Plastics: *Mirel Bioplastics by Telles – Product Information*, http://www.mirelplastics.com/product-information/
34. Danimer Scientific: *A Family of Biopolymers*, https://danimerscientific.com/compostable-solutions/a-family-of-biopolymers/
35. BioOn Bioplastic: Minerv-PHA, http://www.bio-on.it/production.php
36. U.S. Food and Drug Administration: *Inventory of Effective Food Contact Substance (FCS) Notifications, FCN No. 1398.* (2014).
37. www.greenbiz.com: *FDA Approves Mirel Bioplastic for Food Packaging, Utensils*, https://www.greenbiz.com/news/2010/05/17/fda-approves-mirel-bioplastic-food-packaging-utensils
38. European Union: Commission regulation (EC) No 2023/2006 of 22 December 2006 on good manufacturing practice for materials and articles intended to come into contact with food (Text with EEA relevance). *Off. J. Eur. Union.* L 384/75, (2006).
39. European Union (EU): REGULATION (EC) No 1935/2004 OF THE EUROPEAN PARLIAMENT AND OF THE COUNCIL of 27 October 2004 on materials and articles intended to come into contact with food and repealing Directives 80/590/EEC and 89/109/EEC. (2004).
40. European Bioplastics: *Bioplastics Facts and Figures.* (2018).
41. Singh, A.K., Srivastava, J.K., Chandel, A.K., Sharma, L., Mallick, N., Singh, S.P.: Biomedical applications of microbially engineered polyhydroxyalkanoates: An insight into recent advances, bottlenecks, and solutions. *Appl. Microbiol. Biotechnol.* 103, 2007–2032 (2019). doi: 10.1007/s00253-018-09604-y
42. Research and Markets: *Polyhydroxyalkanoate (PHA) Market by Type (Short Chain Length, Medium Chain Length), Production Method (Sugar Fermentation, Vegetable Oil Fermentation, Methane Fermentation), Application, and Region – Global Forecast to 2024.* (2019).
43. Mathuriya, A.S., Yakhmi, J. V.: Polyhydroxyalkanoates: Biodegradable plastics and their applications. *Handb. Ecomater.* 4, 2873–2900 (2017). doi: 10.1007/978-3-319-68255-6_84
44. Pavan, F.A., Junqueira, T.L., Watanabe, M.D.B., Bonomi, A., Quines, L.K., Schmidell, W., de Aragao, G.M.F.: Economic analysis of polyhydroxybutyrate production by Cupriavidus necator using different routes for product recovery. *Biochem. Eng. J.* 146, 97–104 (2019). doi: 10.1016/j.bej.2019.03.009
45. Kachrimanidou, V., Kopsahelis, N., Vlysidis, A., Papanikolaou, S., Kookos, I.K., Monje Martínez, B., Escrig Rondán, M.C., Koutinas, A.A.: Downstream separation of poly(hydroxyalkanoates) using crude enzyme consortia produced via solid state fermentation integrated in a biorefinery concept. *Food Bioprod. Process.* 100, 323–334 (2016). doi: 10.1016/j.fbp.2016.08.002
46. Kachrimanidou, V., Kopsahelis, N., Alexandri, M., Strati, A., Gardeli, C., Papanikolaou, S., Komaitis, M., Kookos, I.K., Koutinas, A.A.: Integrated sunflower-based biorefinery for the production of antioxidants, protein isolate and poly(3-hydroxybutyrate). *Ind. Crops Prod.* 71, 106–113 (2015). doi: 10.1016/j.indcrop.2015.03.003

47. Zabaniotou, A., Kamaterou, P., Kachrimanidou, V., Vlysidis, A., Koutinas, A.: Taking a reflexive TRL3-4 approach to sustainable use of sunflower meal for the transition from a mono-process pathway to a cascade biorefinery in the context of circular bioeconomy. *J. Clean. Prod.* 172, 4119–4129 (2018). doi: 10.1016/j.jclepro.2017.01.151
48. Kang, Z., Du, L., Kang, J., Wang, Y., Wang, Q., Liang, Q., Qi, Q.: Production of succinate and polyhydroxyalkanoate from substrate mixture by metabolically engineered Escherichia coli. *Bioresour. Technol.* 102, 6600–6604 (2011). doi: 10.1016/j.biortech.2011.03.070
49. Xu, M., Qin, J., Rao, Z., You, H., Zhang, X., Yang, T., Wang, X., Xu, Z.: Effect of Polyhydroxybutyrate (PHB) storage on l-arginine production in recombinant *Corynebacterium crenatum* using coenzyme regulation. *Microb. Cell Fact.* 15, 1–12 (2016). doi: 10.1186/s12934-016-0414-x
50. Kumar, P., Jun, H.B., Kim, B.S.: Co-production of polyhydroxyalkanoates and carotenoids through bioconversion of glycerol by Paracoccus sp. strain LL1. *Int. J. Biol. Macromol.* 107, 2552–2558 (2018). doi: 10.1016/j.ijbiomac.2017.10.147
51. Obruca, S., Benesova, P., Kucera, D., Petrik, S., Marova, I.: Biotechnological conversion of spent coffee grounds into polyhydroxyalkanoates and carotenoids. *N. Biotechnol.* 32, 569–574 (2015). doi: 10.1016/j.nbt.2015.02.008
52. Kourmentza, C., Costa, J., Azevedo, Z., Servin, C., Grandfils, C., De Freitas, V., Reis, M.A.M.: Burkholderia thailandensis as a microbial cell factory for the bioconversion of used cooking oil to polyhydroxyalkanoates and rhamnolipids. *Bioresour. Technol.* 247, 829–837 (2018). doi: 10.1016/j.biortech.2017.09.138
53. Kourmentza, C., Araujo, D., Sevrin, C., Roma-Rodriques, C., Lia Ferreira, J., Freitas, F., Dionisio, M., Baptista, P. V., Fernandes, A.R., Grandfils, C., Reis, M.A.M.: Occurrence of non-toxic bioemulsifiers during polyhydroxyalkanoate production by *Pseudomonas* strains valorizing crude glycerol by-product. *Bioresour. Technol.* 281, 31–40 (2019). doi: 10.1016/j.biortech.2019.02.066
54. Sharma, P.K., Munir, R.I., Plouffe, J., Shah, N., de Kievit, T., Levin, D.B.: Polyhydroxyalkanoate (PHA) polymer accumulation and pha gene expression in phenazine (phz-) and pyrrolnitrin (prn-) defective mutants of *Pseudomonas chlororaphis* PA23. *Polymers (Basel).* 10(11), 1203 (2018). doi: 10.3390/polym10111203
55. PlasticsEurope: *Plastics – the Facts 2019.* (2019).
56. Ellen MacArthur Foundation: *The New Plastics Economy Rethinking the Future of Plastics.* (2016).
57. Israni, N., Shivakumar, S.: Polyhydroxyalkanoates in packaging. In: Kalia, V.C. (ed.) *Biotechnological Applications of Polyhydroxyalkanoates.* pp. 347–361 (2019). Springer Nature Singapore PTE Ltd., Singapore. ISBN 978-981-13-3758-1.
58. Kachrimanidou, V., Kopsahelis, N., Papanikolaou, S., Kookos, I.K., De Bruyn, M., Clark, J.H., Koutinas, A.A.: Sunflower-based biorefinery: Poly(3-hydroxybutyrate) and poly(3-hydroxybutyrate-co-3-hydroxyvalerate) production from crude glycerol, sunflower meal and levulinic acid. *Bioresour. Technol.* 172, 121–130 (2014). doi: 10.1016/j.biortech.2014.08.044
59. Gahlawat, G., Soni, S.K.: Valorization of waste glycerol for the production of poly (3-hydroxybutyrate) and poly (3-hydroxybutyrate-co-3-hydroxyvalerate) copolymer by Cupriavidus necator and extraction in a sustainable manner. *Bioresour. Technol.* 243, 492–501 (2017). doi: 10.1016/j.biortech.2017.06.139
60. Zhang, M., Thomas, N.L.: Blending polylactic acid with polyhydroxybutyrate: The effect on thermal, mechanical, and biodegradation properties. *Adv. Polym. Technol.* 30, 67–79 (2011).
61. Bonartsev, A.P., Boskhomodgiev, A.P., Iordanskii, A.L., Bonartseva, G.A., Rebrov, A. V, Makhina, T.K., Myshkina, V.L., Yakovlev, S.A., Filatova, E.A., Ivanov, E.A.,

Bagrov, D. V, Zaikov, G.E.: Hydrolytic degradation of poly(3-hydroxybutyrate), polylactide and their derivatives: Kinetics, crystallinity, and surface morphology. *Mol. Cryst. Liq. Cryst.* 556, 288–300 (2012). doi: 10.1080/15421406.2012.635982
62. Fabra, M.J., Lopez-Rubio, A., Lagaron, J.M.: Nanostructured interlayers of zein to improve the barrier properties of high barrier polyhydroxyalkanoates and other polyesters. *J. Food Eng.* 127, 1–9 (2014). doi: 10.1016/j.jfoodeng.2013.11.022
63. Fabra, M.J., López-Rubio, A., Ambrosio-Martín, J., Lagaron, J.M.: Improving the barrier properties of thermoplastic corn starch-based films containing bacterial cellulose nanowhiskers by means of PHA electrospun coatings of interest in food packaging. *Food Hydrocoll.* 61, 261–268 (2016). doi: 10.1016/j.foodhyd.2016.05.025
64. Nestlé: Press release: *Nestle and Danimer Scientific Partnering to Develop Biodegradable Water Bottle*, 2019. https://www.nestle.com/sites/default/files/asset-library/documents/media/press-release/2019-january/nestle-danimer-partnership-en.pdf
65. plasticstoday.com: *Danimer Scientific, PepsiCo Agreement Expands Biodegradable Resins for Flexible Packaging*, https://www.plasticstoday.com/packaging/danimer-scientific-pepsico-agreement-expands-biodegradable-resins-flexible-packaging/200936359056404
66. pepsico.com: *Danimer Scientific and PepsiCo to Collaborate on Biodegradable Resins*, https://www.pepsico.com/news/press-release/danimer-scientific-and-pepsico-to-collaborate-on-biodegradable-resins
67. plasticstoday.com: *Compostable Snacks Packaging Snags Bioplastic Award for Danimer Scientific, PepsiCo*, https://www.plasticstoday.com/packaging/compostable-snacks-packaging-snags-bioplastic-award-danimer-scientific-pepsico/83659095059493
68. genpack.com: *The Next Generation of Food Packaging Is Here*, https://www.genpak.com/genzero/#intro
69. prnewswire.com: *Danimer Scientific and Genpak Partner to Launch New Line of Biodegradable Food Packaging*, https://www.prnewswire.com/news-releases/danimer-scientific-and-genpak-partner-to-launch-new-line-of-biodegradable-food-packaging-300955587.html
70. bio-on, Rivoira: Press release: *Bio-on and Rivoira Present ZEROPACK, Bioplastic for Food Packaging of Fruits and Vegetables*, https://ml-eu.globenewswire.com/Resource/Download/78d894a8-d1f7-4ee8-a404-4aefc9665f73
71. my-kai.com: *MyKai Products*, http://www.my-kai.com/
72. bio-on.it: *Unilever and Bio-on Present My Kai the New Line of Ultra-Green Sun Creams*, http://www.bio-on.it/immagini/comunicati-finanziari/CS_76_MyKai_28_01_2019_ENG.pdf
73. bioplasticsmagazine.com: *Solon: The Solution to Single-Use Plastics*, https://www.bioplasticsmagazine.com/en/news/meldungen/20190617Solon--the-solution-to-single-use-plastics.php
74. bioplasticsnews.com: *Italians Switch Toys Production to Bioplastics with the Minerv Supertoys Program*, https://bioplasticsnews.com/2018/05/02/italians-toys-bioplastics-minerv-supertoys-program/
75. globalsources.com: *Products from Ecomann Biotechnology Co. Ltd, Biodegradable and Home Compostable Singlet Bags*, https://www.globalsources.com/si/AS/Ecomann-Biotechnology/6008850197990/pdtl/biodegradable-and-home-compostable-singlet-bags/1155813942.htm
76. Ball Horticultural Company: *Biodegradable, Printable and Plantable SoilWrap®*, https://hortcom.files.wordpress.com/2018/10/soilwrap-biopackaging.pdf
77. Virginia Institute of Marine Science, College of William & Mary: *FACT SHEET Polyhydroxyalkanoate (PHA) Biodegradable Escape Panel (Biopanel) for Crab, Lobster, and Fish Traps*, https://www.vims.edu/ccrm/_docs/marine_debris/biodegradablepanel_factsheet.pdf

78. Shrivastav, A., Kim, H.Y., Kim, Y.R.: Advances in the applications of polyhydroxyalkanoate nanoparticles for novel drug delivery system. *Biomed Res. Int.* 2013, (2013). doi: 10.1155/2013/581684
79. Ali, I., Jamil, N.: Polyhydroxyalkanoates: Current applications in the medical field. *Front. Biol. (Beijing).* 11, 19–27 (2016). doi: 10.1007/s11515-016-1389-z
80. Nebe, B., Forster, C., Pommerenke, H., Fulda, G., Behrend, D., Bernewski, U., Schmitz, K.P., Rychly, J.: Structural alterations of adhesion mediating components in cells cultured on poly-β-hydroxy butyric acid. *Biomaterials.* 22, 2425–2434 (2001). doi: 10.1016/S0142-9612(00)00430-0
81. Winnacker, M., Rieger, B.: Copolymers of polyhydroxyalkanoates and polyethylene glycols: Recent advancements with biological and medical significance. *Polym. Int.* 66, 497–503 (2017). doi: 10.1002/pi.5261
82. Fava, F., Totaro, G., Gavrilescu, M.: Material & energy recovery and sustainable development ECOMONDO 2014 18th International Trade Fair of Material & Energy Recovery and Sustainable Development. *Environ. Eng. Manag. J.* 14, 1475–1476 (2015). doi: 10.30638/eemj.2017.179
83. Zhang, J., Shishatskaya, E.I., Volova, T.G., da Silva, L.F., Chen, G.Q.: Polyhydroxyalkanoates (PHA) for therapeutic applications. *Mater. Sci. Eng. C.* 86, 144–150 (2018). doi: 10.1016/j.msec.2017.12.035
84. Research and Markets: $9.28 Billion Global Agricultural Films Market 2017-2021 with Focus on LLDPE, LDPE, EVA, Reclaim PE and Applications: *Greenhouse Films, Mulch Films, Silage Films.* (2017).
85. World Agriculture Net: *The Benefits and Challenge of Plastic Film Mulching in China*, http://www.world-agriculture.net/article/the-benefits-and-challenge-of-plastic-film-mulching-in-china
86. Grillo, R., Pereira, A. do E.S., de Melo, N.F.S., Porto, R.M., Feitosa, L.O., Tonello, P.S., Filho, N.L.D., Rosa, A.H., Lima, R., Fraceto, L.F.: Controlled release system for ametryn using polymer microspheres: Preparation, characterization and release kinetics in water. *J. Hazard. Mater.* 186, 1645–1651 (2011). doi: 10.1016/j.jhazmat.2010.12.044
87. European Bioplastics, EuropaBio: *Fertiliser Regulation: Biodegradable Mulch Films. 2016–2017* (2017).
88. Castellano, S., Scarascia Mugnozza, G., Russo, G., Briassoulis, D., Mistriotis, A., Hemming, S., Waaijenberg, D.: Plastic nets in agriculture: A general review of types and applications. *Appl. Eng. Agric.* 24, 799–808 (2008). doi: 10.13031/2013.25368
89. Iwata, T., Aoyagi, Y., Fujita, M., Yamane, H., Doi, Y., Suzuki, Y., Takeuchi, A., Uesugi, K.: Processing of a strong biodegradable poly[(*R*)-3-hydroxybutyrate] fiber and a new fiber structure revealed by micro-beam X-ray diffraction with synchrotron radiation. *Macromol. Rapid Commun.* 25, 1100–1104 (2004). doi: 10.1002/marc.200400110
90. Andrews, M.: *Mirel™ PHA Polymeric Modifiers & Additives*, https://www.slideshare.net/MetabolixInc/metabolix-mirel-pha-polymeric-modifiers-and-additives
91. Saito, Y., Doi, Y.: Microbial synthesis and properties of poly(3-hydroxybutyrate-co-4-hydroxybutyrate) in *Comamonas acidovorans. Int. J. Biol. Macromol.* 16, 99–104 (1994) https://doi.org/10.1016/0141-8130(94)90022-1
92. Williams, S., Said, R., Martin, D.: Poly-4-hydroxybutyrate (P4HB): A new generation of resorbable medical devices for tissue repair and regeneration, https://www.degruyter.com/view/j/bmte.2013.58.issue-5/bmt-2013-0009/bmt-2013-0009.xml, (2013).
93. inhabitat.com: *SoilWrap, Container You Can Plant*, https://inhabitat.com/revolutionary-soil-wrap-is-a-flower-container-you-can-plant/
94. Wei, L., McDonald, A.G.: A review on grafting of biofibers for biocomposites. *Materials (Basel).* 9(4), 303 (2016). doi: 10.3390/ma9040303

95. Shanks, R.A., Hodzic, A., Wong, S.: Thermoplastic biopolyester natural fiber composites. *J. Appl. Polym. Sci.* 91, 2114–2121 (2004). doi: 10.1002/app.13289
96. Jambeck, J.R., Ji, Q., Zhang, Y.-G., Liu, D., Grossnickle, D.M., Luo, Z.-X.: Plastic waste inputs from land into the ocean. *Science.* 347, 764–768 (2015). doi: 10.1126/science.1260879
97. European Union: DIRECTIVE (EU) 2019/904 OF THE EUROPEAN PARLIAMENT AND OF THE COUNCIL of 5 June 2019 on the reduction of the impact of certain plastic products on the environment. *Off. J. Eur. Union.* 1–19 (2019).
98. Bourguignon, D.: Briefing EU Legislation in Progress, Single-use plastics and fishing gear. *Eur. Parliam.* (2018).
99. Huntington, T.: Development of a best practice framework for the management of fishing gear. Part 1: Overview and current status. Glob. *Ghost Gear Initiat.* 52 (2017).
100. Higson, A.: *Market Perspective Bio-Based & Biodegradable Plastics in the UK.* NNFC The Bioeconomy Consultants, UK (2018).
101. Bilkovic, D.M., Havens, K.J., Stanhope, D.M., Angstadt, K.T.: Use of fully biodegradable panels to reduce derelict pot threats to marine fauna. *Conserv. Biol.* 26, 957–966 (2012). doi: 10.1111/j.1523-1739.2012.01939.x
102. BusinessWire: *MHG Debuts First Biodegradable Fishing Lures Produced by Rat-L-Traps,* https://www.businesswire.com/news/home/20150713006002/en/Biopolymer-Company-MHG-Exhibit-ICAST-Adventure-Products
103. Virginia Institute of Marine Science, College of William & Mary: *FACT SHEET Polyhydroxyalkanoate (PHA) Biodegradable Escape Panel (Biopanel) for Crab, Lobster, and Fish Traps.,* http://ccrm.vims.edu/marine_debris_removal/degradable_cull_panels/BiodegradablePanelFactSheet.pdf
104. Virginia Institute of Marine Science, College of William & Mary: *Biodegradable Shotgun Wads,* https://www.vims.edu/ccrm/research/marine_debris/solutions/wads/index.php
105. Clarinval, A.M., Halleux Crif, J.: Classification of biodegradable polymers. In: Smith, R. (ed.) *Biodegradable Polymers for Industrial Applications.* pp. 3–56. CRC Press, Boca Raton, Boston, New York, Washington DC (2005).
106. Crétois, R., Follain, N., Dargent, E., Soulestin, J., Bourbigot, S., Marais, S., Lebrun, L.: Microstructure and barrier properties of PHBV/organoclays bionanocomposites. *J. Memb. Sci.* 467, 56–66 (2014). doi: 10.1016/j.memsci.2014.05.015
107. RWDC Industries: *Products,* https://www.rwdc-industries.com/products
108. RWDC Industries: *History,* httpa://www.rwdc-industries.com/history
109. Danimer Scientific: *WinCup to Produce Biodegradable Straws Made from Danimer Scientific' s Nodax TM PHA Resin,* https://danimerscientific.com/2019/09/16/wincup-to-produce-biodegradable-straws-made-from-danimer-scientifics-nodax-pha-resin/
110. bioplasticsnews.com: *Frost & Sullivan Awards Bio-On for Best Cosmetic Innovation,* https://bioplasticsnews.com/2018/10/08/frost-sullivan-awards-bio-best-cosmetic-innovation/
111. bio-on.it: *Frost & Sullivan Gives Award for Best Innovation in the Cosmetics Sector to Bio-on,* https://bioplasticsnews.com/wp-content/uploads/2018/10/CS_65_Premio-Frost-Sullivan_UK-1.pdf
112. bio-on.it: *Frost & Sullivan Gives Award for Best Innovation in the Cosmetics Sector to Bio-On.*
113. ml-eu.globenewswire.com: *New Cigarette Filters Block up to 60 % of Harmful Substances Thanks to Bio-On Biopolymers,* https://ml-eu.globenewswire.com/Resource/Download/775fdfb3-f549-4869-b0e1-b0000f659caa
114. bioplasticsnews.com: *Bio-On Will Disrupt the Fashion Industry,* 09/11/2019.
115. ml-eu.globenewswire.com: *Organic Electronics: Bio-on Presents Eloxel, to Develop the Use of Bioplastics in the Electronics Sector at His Side Kartell, Strategic Partner,*

https://ml-eu.globenewswire.com/Resource/Download/21873d4d-156f-42a2-9274-0dd7f0c27f82
116. bioplasticsnews.com: *Bio-On and Kartell Launch Furniture Made from Revolutionary Bio-Material*, https://bioplasticsnews.com/2019/04/09/bio-on-and-kartell-launch-furniture-made-from-revolutionary-bio-material/
117. Chen, G.Q.: A microbial polyhydroxyalkanoates (PHA) based bio- and materials industry. *Chem. Soc. Rev.* 38, 2434–2446 (2009). doi: 10.1039/b812677c
118. Steinbüchel, A., Valentin, H.E.: Diversity of bacterial polyhydroxyalkanoic acids. *FEMS Microbiol. Lett.* 128, 210–228 (1995). doi: 10.1111/j.1574-6968.1995.tb07528.x
119. Brown, H.C., Ramachandran, P.V.: The boron approach to asymmetric synthesis. *Pure Appl. Chem.* 63(3), 307–316 (1991). doi: 10.1351/pac199163030307
120. Noyori, R., Kitamura, M., Ohkuma, T.: Toward efficient asymmetric hydrogenation: Architectural and functional engineering of chiral molecular catalysts. *Proc. Nat. Acad. Sci.* 101(15), 5356–5362 (2004). doi: 10.1073/pnas.0307928100
121. Lee, S.Y., Park, S.H., Lee, Y., Lee, S.H.: Production of chiral and other valuable compounds from microbial polyesters. *ChemInform.* (2003). doi: 10.1002/chin.200323288
122. Houston, T.A., Wilkinson, B.L., Blanchfield, J.T.: Boric acid catalyzed chemoselective esterification of α- hydroxycarboxylic acids. *Org. Lett.* 6(5), 679–681 (2004). doi: 10.1021/ol036123g
123. Lee, S.Y., Lee, Y., Wang, F.: Chiral compounds from bacterial polyesters: Sugars to plastics to fine chemicals. *Biotechnol. Bioeng.* (1999). doi: 10.1002/(SICI)1097-0290(19991105)65:3<363::AID-BIT15>3.0.CO;2-1
124. Kanesawa, Y., Tanahashi, N., Doi, Y., Saito, T.: Enzymatic degradation of microbial poly(3-hydroxyalkanoates). *Polym. Degrad. Stab.* 45, 179–185 (1994). doi: 10.1016/0141-3910(94)90135-X
125. Anis, S.N.S., Mohamad Annuar, M.S., Simarani, K.: *In vivo* and *in vitro* depolymerizations of intracellular medium-chain-length poly-3-hydroxyalkanoates produced by *Pseudomonas putida* Bet001. *Prep. Biochem. Biotechnol.* 47(8), 824–834 (2017). doi: 10.1080/10826068.2017.1342266
126. Jaeger, K.E., Steinbuchel, A., Jendrossek, D.: Substrate specificities of bacterial polyhydroxyalkanoate depolymerases and lipases: Bacterial lipases hydrolyze poly(omega-hydroxyalkanoates). *Appl. Environ. Microbiol.* 61(8), 3113–3118 (1995).
127. De Raadt, A., Klempier, N., Faber, K., Griengl, H.: Chemoselective enzymatic hydrolysis of aliphatic and alicyclic nitriles. *J. Chem. Soc. Perkin Trans. 1.*, 137–140 (1992). doi: 10.1039/p19920000137
128. Hann, E.C., Sigmund, A.E., Fager, S.K., Cooling, F.B., Gavagan, J.E., Ben-Bassat, A., Chauhan, S., Payne, M.S., Hennessey, S.M., DiCosimo, R.: Biocatalytic hydrolysis of 3-hydroxyalkanenitriles to 3-hydroxyalkanoic acids. *Adv. Synth. Catal.* 345(6–7), 775–782 (2003). doi: 10.1002/adsc.200303007
129. Ren, Q., Grubelnik, A., Hoerler, M., Ruth, K., Hartmann, R., Felber, H., Zinn, M.: Bacterial poly(hydroxyalkanoates) as a source of chiral hydroxyalkanoic acids. *Biomacromolecules.* 6(4), 2290–2298 (2005). doi: 10.1021/bm050187s
130. Ruth, K., Grubelnik, A., Hartmann, R., Egli, T., Zinn, M., Ren, Q.: Efficient production of (R)-3-hydroxycarboxylic acids by biotechnological conversion of polyhydroxyalkonoates and their purification. *Biomacromolecules.* 8(1), 279–286 (2007). doi: 10.1021/bm060585a
131. Anis, S.N.S., Annuar, M.S.M., Simarani, K.: Microbial biosynthesis and *in vivo* depolymerization of intracellular medium-chain-length poly-3-hydroxyalkanoates as potential route to platform chemicals. *Biotechnol. Appl. Biochem.* 65(6), 784–796 (2018). doi: 10.1002/bab.1666
132. Wang, Y., Liu, S.: Production of (*R*)-3-hydroxybutyric acid by *Burkholderia cepacia* from wood extract hydrolysates. *AMB Express.* 4(1) 28 (2014). doi: 10.1186/s13568-014-0028-9

133. Si, J.P., Sang, Y.L., Lee, Y.: Biosynthesis of (R)-3-hydroxyalkanoic acids by metabolically engineered *Escherichia coli*. In: *Applied Biochemistry and Biotechnology – Part A Enzyme Engineering and Biotechnology* 114(1–3), 373–379 (2004).
134. Wu, Q., Zheng, Z., Xi, J.Z., Gao, H., Chen, G.Q.: Production of hydroxyalkanoate monomers by microbial fermentation. *J. Chem. Eng. Japan.* 36(10), 1170–1173. (2003). doi: 10.1252/jcej.36.1170
135. Gao, H.J., Wu, Q., Chen, G.Q.: Enhanced production of D-(-)-3-hydroxybutyric acid by recombinant *Escherichia coli. FEMS Microbiol. Lett.* 213, 59–65 (2002). doi: 10.1016/S0378-1097(02)00788-7
136. Park, S.J., Lee, S.H., Oh, Y.H., Lee, S.Y.: Establishment of a biosynthesis pathway for (R)-3-hydroxyalkanoates in recombinant *Escherichia coli. Korean J. Chem. Eng.* 32(4), 702–706 (2015). doi: 10.1007/s11814-014-0240-y
137. Zhao, K., Tian, G., Zheng, Z., Chen, J.C., Chen, G.Q.: Production of D-(-)-3-hydroxyalkanoic acid by recombinant *Escherichia coli. FEMS Microbiol. Lett.* 218(1), 59–64 (2003). doi: 10.1016/S0378-1097(02)01108-4
138. Zhang, L., Shi, Z.Y., Wu, Q., Chen, G.Q.: Microbial production of 4-hydroxybutyrate, poly-4-hydroxybutyrate, and poly(3-hydroxybutyrate-co-4-hydroxybutyrate) by recombinant microorganisms. *Appl. Microbiol. Biotechnol.* 84(5), 909–916 (2009). doi: 10.1007/s00253-009-2023-7
139. Lakshmanan, M., Foong, C.P., Abe, H., Sudesh, K.: Biosynthesis and characterization of co and ter-polyesters of polyhydroxyalkanoates containing high monomeric fractions of 4-hydroxybutyrate and 5-hydroxyvalerate via a novel PHA synthase. *Polym. Degrad. Stab.* 163, 122–135 (2019). doi: 10.1016/j.polymdegradstab.2019.03.005
140. Al-Kaddo, K.B., Mohamad, F., Murugan, P., Tan, J.S., Sudesh, K., Samian, M.R.: Production of P(3HB-co-4HB) copolymer with high 4HB molar fraction by *Burkholderia contaminans* Kad1 PHA synthase. *Biochem. Eng. J.* 153, 107394 (2020). doi: 10.1016/j.bej.2019.107394
141. Ren, Q., Ruth, K., Thöny-Meyer, L., Zinn, M.: Enatiomerically pure hydroxycarboxylic acids: Current approaches and future perspectives. *Applied Microbiology and Biotechnology* 87(1), 41–52 (2010).
142. Yamashita, R., Sakakura, A., Ishihara, K.: Primary alkylboronic acids as highly active catalysts for the dehydrative amide condensation of α-hydroxycarboxylic acids. *Org. Lett.* 15(14), 3654–3657 (2013). doi: 10.1021/ol401537f
143. Shailendra, M., Mr, M., Lorcan, L.M., Bain, J., Débora Dias, M., Thibault, M., Johansson, L., Phil, M., Shields, D.M.L., Bowyer, M.C.: *European Commission (DG Environment) Plastic Waste in the Environment.* 171 (2011).
144. unenvironment.org: *Our Planet is Drowning in Plastic Pollution*, https://www.unenvironment.org/interactive/beat-plastic-pollution/
145. European Bioplastics: *Fact Sheet: Landfilling.* 0–3 (2015).
146. European Bioplastics: *Recycling and Recovery: End-of-Life Options for Bioplastics.* 1–2 (2017).
147. Singh, P., Sharma, V.P.: Integrated plastic waste management: Environmental and improved health approaches. *Procedia Environ. Sci.* 35, 692–700 (2016). doi: 10.1016/j.proenv.2016.07.068
148. Guzik, M.W., Kenny, S.T., Duane, G.F., Casey, E., Woods, T., Babu, R.P., Nikodinovic-Runic, J., Murray, M., O'Connor, K.E.: Conversion of post consumer polyethylene to the biodegradable polymer polyhydroxyalkanoate. *Appl. Microbiol. Biotechnol.* 98, 4223–4232 (2014). doi: 10.1007/s00253-013-5489-2
149. Takabatake, H., Satoh, H., Mino, T., Matsuo, T.: PHA (polyhydroxyalkanoate) production potential of activated sludge treating wastewater. *Water Sci. Technol.* 45, 119–126 (2002).

150. Yates, M.R., Barlow, C.Y.: Life cycle assessments of biodegradable, commercial biopolymers – A critical review. *Resour. Conserv. Recycl.* 78, 54–66 (2013). doi: 10.1016/j.resconrec.2013.06.010
151. Resch-Fauster, K., Klein, A., Blees, E., Feuchter, M.: Mechanical recyclability of technical biopolymers: Potential and limits. *Polym. Test.* 64, 287–295 (2017). doi: 10.1016/j.polymertesting.2017.10.017
152. European Bioplastics e.V.: Fact sheet: Mechanical recycling, July 2020. *Eur. Bioplastics* (2020). https://docs.european-bioplastics.org/publications/bp/EUBP_BP_recycling.pdf
153. European Bioplastics: *Waste Management and Recovery Options for Bioplastics*, https://www.european-bioplastics.org/bioplastics/waste-management/
154. Alaerts, L., Augustinus, M., Van Acker, K.: Impact of bio-based plastics on current recycling of plastics. *Sustain.* 10(5), 1487 (2018). doi: 10.3390/su10051487
155. Tukker, A., de Groot, H., Simons, L., Wiegersma, S.: Chemical Recycling of Plastics Waste (PVC and other resins). TNO-report STB-99-55, 1–132 (1999).
156. zerowasteeurope.eu: *El Dorado of Chemical Recycling – State of Play and Policy Challenges.* (2019).
157. Gironi, F., Frattari, S., Piemonte, V.: PLA chemical recycling process optimization: PLA solubilization in organic solvents. *J. Polym. Environ.* 24, 328–333 (2016). doi: 10.1007/s10924-016-0777-4
158. Clark, J., Farmer, T., Herrero-Davila, L., Moity, L., Arnaud, S., Sherwood, J., Short, G., Silva-Terra, G.: Opening bio-based markets via standards, labelling and procurement, Deliverable No 6.10: *Assessment of Chemical/Feedstock Recycling and Test Methods.* 44, 1–5 (2016).
159. www.natureworks.com: *NatureWorks – Chemical Recycling*, https://www.natureworksllc.com/What-is-Ingeo/Where-it-Goes/Chemical-Recycling
160. Ariffin, H., Nishida, H., Hassan, M.A., Shirai, Y.: Chemical recycling of polyhydroxyalkanoates as a method towards sustainable development. *Biotechnol. J.* 5, 484–492 (2010). doi: 10.1002/biot.200900293
161. Ariffin, H., Nishida, H., Shirai, Y., Hassan, M.A.: Determination of multiple thermal degradation mechanisms of poly(3-hydroxybutyrate). *Polym. Degrad. Stab.* 93, 1433–1439 (2008). doi: 10.1016/j.polymdegradstab.2008.05.020
162. Hermann, B.G., Debeer, L., De Wilde, B., Blok, K., Patel, M.K.: To compost or not to compost: Carbon and energy footprints of biodegradable materials' waste treatment. *Polym. Degrad. Stab.* 96, 1159–1171 (2011). doi: 10.1016/j.polymdegradstab.2010.12.026

9 Linking the Properties of Polyhydroxyalkanoates (PHA) to Current and Prospective Applications

Ana T. Rebocho, João R. Pereira, Cristiana A. V. Torres, Filomena Freitas, and Maria A. M. Reis

CONTENTS

9.1 INTRODUCTION

PHA are polyesters of hydroxyalkanoic acids synthesized by many Gram-positive and Gram-negative bacteria and several archaea that accumulate them intracellularly as carbon and energy storage compounds [1, 2]. Upon biosynthesis within bacterial cells, PHA form amorphous water-insoluble inclusions (see Figure 9.1), named carbonosomes [3], with an average size that ranges between 0.2 and 0.7 μm, depending

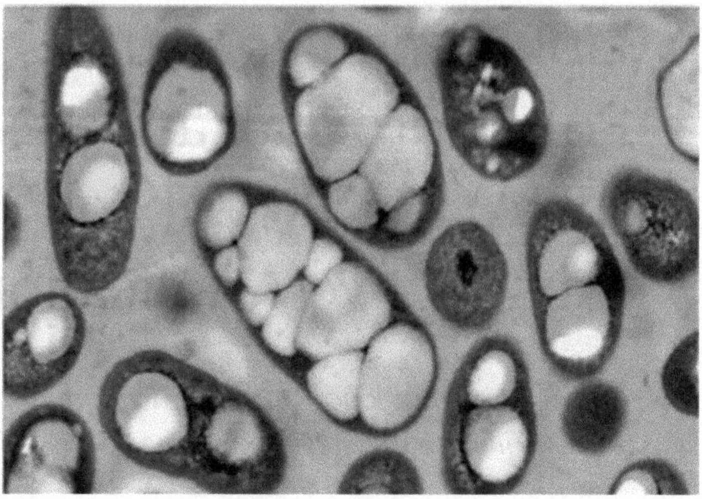

FIGURE 9.1 Morphology of PHA granules accumulated by *Cupriavidus necator* DSM 545 cells obtained by electron microscopy with magnification: 1/70 000; 48% of PHB in cell mass (Adapted from [9]).

on the producing species and the cultivation conditions. PHA granules surface contains a layer of proteins that stabilizes them and inhibit their coalescence and agglutination with surrounding granules [4–7]. During extraction, PHA granules are often subjected to physical and/or chemical treatments that alter their crystallinity, namely, that lead to their crystallization [4, 8].

PHA are water-insoluble polymers that present high degree of polymerization, a variable degree of crystallinity and are not cytotoxic. In addition, they exhibit optical activity, are more resistant to exposure to light, including ultraviolet radiation, than conventional plastics, possess antioxidant characteristics and are resistant to high temperatures [10–12]. PHA physicochemical and mechanical properties match those of many conventional petrochemical plastics, such as poly(propylene) (PP) and poly(ethylene) (PE), which cannot be naturally degraded in the environment [13]. In fact, one of the most relevant properties of PHA is their biodegradability, which renders them alternative ecological choices with advantages over synthetic polymers [10]. These characteristics, together with their biocompatibility, allowed PHA to be used in several applications, including biomedicine, pharmaceuticals, food, agriculture, as well as raw materials for enantiomeric pure chemicals production and in the paint industry [10, 13, 14].

Depending on the type of microbial producer and its cultivation conditions, namely, pH, temperature, cultivation mode (batch, fed-batch, continuous) and type and concentration of the carbon source, the synthesized PHA structures can vary in terms of monomer content, length and composition. Such variation in the polymers' macromolecules makes it possible to tailor PHA production towards a variety of potential applications [15, 16].

9.2 PHA PROPERTIES

9.2.1 Composition and Structure

PHA are predominantly linear thermoplastics, commonly composed of (*R*)-β-hydroxy fatty acids (see Figure 9.2), wherein the carboxyl group of a monomeric unit forms an ester bond with the hydroxyl group of the adjacent monomeric [1, 17].

The hydroxyalkanoic acid (HA) units in PHA macromolecules are all in the *R*(-) configuration due to the stereospecificity of biosynthetic enzymes, namely PHA polymerase [1, 19]. PHA chirality with an *R*-configuration confers the polymers' very crystalline structures with piezoelectric properties [20]. The piezoelectric effect is known to occur in numerous biopolymers, for instance, polysaccharides, proteins and deoxyribonucleotides, but is not often seen in plastics. Thus, this feature further enhances the interest in PHA. For example, this property of PHA is relevant for the use of these biopolymers in biomedical applications related to nerve repair, bone-filling augmentation material, or for ligament and tendon grafts [20].

Regarding the number of carbon atoms in the side-chain of the monomers, PHA can be classified in three major groups: short-chain-length (*scl*-PHA) whose monomers have 3–5 carbon atoms; medium-chain-length PHA (*mcl*-PHA) with 6–14 carbon atom monomers; and long-chain-length PHA (*lcl*-PHA) composed of monomers with chains longer than 14 carbon atoms [5, 20–22]. These monomers

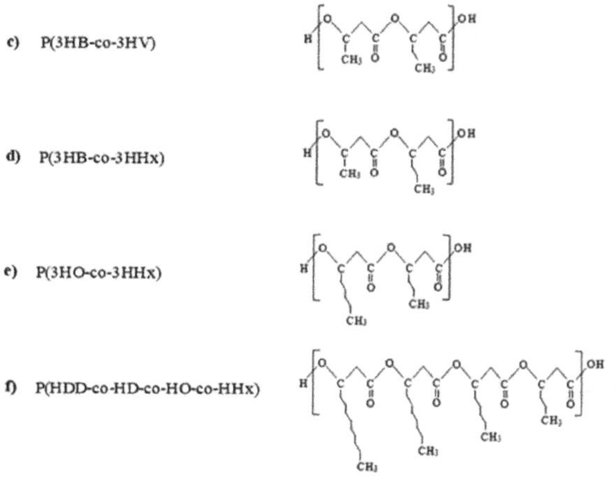

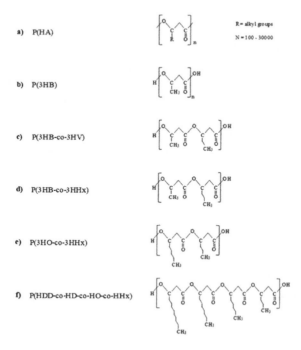

FIGURE 9.2 Chemical structures of PHA: general structure (a) and examples of homopolymers (b), copolymers (c, d, e) and terpolymers (f), with different monomer chain lengths (reproduced from [18]).

can have aliphatic saturated, unsaturated, straight or branched side-chains. Some microorganisms can synthesize PHA with aromatic, halogenic, pseudohalogenic or alkoxy groups [23]. All these possible HA monomers can be arranged into different molecular structures that include homopolyesters, copolyesters and terpolyesters (Figure 1.2), thus conferring the biopolymers' valuable distinct properties [18].

9.2.2 Thermal Properties

Concerning *scl*-PHA (e.g., the homopolymer poly(3-hydroxybutyrate), high glass transition (T_g) and melting temperatures (T_m) (−4–15°C and 170–180°C, respectively), as well as high crystallinity degree (X_c) values (around 70%), are observed [5, 16, 24, 25]. On the other hand, for *mcl*-PHA, lower T_g values, around −40°C, and lower crystallinity degrees (below about 40%) are usually attained, with a wider range of lower melting temperatures (30–60°C) (Table 9.1). Overall, T_m decreases as the side-chains become longer, the incorporation of other monomers in the *scl*-PHA homopolymer turning them more readily processable. As for T_g, the values also decrease by the incorporation of monomers with longer pendant groups, as observed for *mcl*-PHA [26]. Amongst *scl*-PHA, P(3HB) is the most common homopolymer. Its high T_m is close to the temperature where it decomposes thermally, which limits its processability [4]. By displaying a low T_g, *mcl*-PHA do not become brittle even at temperatures below freezing point, making them potentially interesting as rubber-like biological materials [27, 28].

9.2.3 Mechanical Properties

Depending on the branched length of the HA units or from the distance between the ester bonds in the polymer backbones, the mechanical properties of PHA vary from brittle to flexible and elastic. Usually, PHA with short pendant groups like *scl*-PHA (containing 3–5 carbon atoms) present properties of thermoplastics and are usually stiff crystalline materials, whereas PHA with longer pendant groups like *mcl*-PHA (polyesters of HAs containing 6–14 carbon atoms) feature properties of elastomers and latexes [4, 5, 26]. A comparison of the properties of PHA with conventional petrochemical-based plastics, such as poly(propylene) and poly(styrene), as well as other natural polymers, namely, PLA and polysaccharides, is shown in Table 9.2.

TABLE 9.1
Comparison of the Physical Properties of *scl*-PHA and *mcl*-PHA with Poly(propylene) (adapted from [19, 29])

Material Properties	*scl*-PHA	*mcl*-PHA	Poly(propylene) (PP)
Melting temperature (°C)	160–180	30–80	176
Glass transition temperature (°C)	−148 to −4	−14 to −150	−10
Crystallinity (%)	40–80	20–40	70
UV light resistance	Good	Good	Poor
Biodegradability	Good	Good	None

TABLE 9.2
Mechanical Properties of PHA and Other Polymers

Material	Tensile Strength [MPa]	Elongation at break [%]	Young Modulus [GPa]	Ref.
mcl*-PHA	6.50 ± 0.35	195 ± 46.5	5.27 ± 1.14 MPa	[30]
P(3HB)	43	5	3.5	[9]
P(3HB)	40	3–8	3.5–4	[16]
P(4HB)	50	1000	0.07	[25]
P(3HB-*co*-3HHx)	7.0 ± 0.5	400 ± 36	n.a.	[25]
P(3HB-*co*-3HV)	20–25	50	0.7–2.9	[25]
P(3HB-*co*-20 mol-% 3HV)	20	50	0.8	[7]
P(3HB-*co*-4HB)	26	444	n.a.	[25]
P(3HHx-co-3HO)	9	380	0.008	[25]
Poly(lactide acid)	50–62	5.2	0.3–0.4	[31]
Poly(butylene succinate)	25	175	0.2	[31]
Poly(ethylene)	22–29	298	0.1	[31]
Poly(propylene)	38	400	1.7	[7]
Isotactic poly(propylene)	32.3	375	1.2	[32]
Poly(styrene)	50	3–4	3.0–3.1	[16]
HDPE	17.9–33.1	12–700	0.4–1.0	[16]
LDPE	15.2–78.6	150–600	0.05–0.1	[16]
Poly(ethylene terephthalate)	56	7300	2.2	[16]
Nylon-6,6	83	60	2.8	[16]

mcl-PHA* composed of 3-hydroxyoctanoate and 3-hydroxydecanoate; P(3HB), poly(3-hydroxybutyrate); P(4HB), poly(4-hydroxybutyrate); P(3HB-*co*-3HHx), poly(3-hydroxybutyrate-*co*-3-hydroxyhexanoate); P(3HB-*co*-3HV), poly(3-hydroxybutyrate-*co*-3-hydroxyvalerate); P(3HB-*co*-4HB), poly(3-hydroxybutyrate-*co*-4-hydroxybutyrate); P(3HO), poly(3-hydroxyoctanoate); HDPE, high-density poly(ethylene); LDEP, low density poly(ethylene); n.a., data not available.

Scl-PHA are described as stiff and brittle biomaterials with a high degree of crystallinity (50–80%) [16]. This last characteristic supports *scl*-PHA being considered very similar to some thermoplastic petrochemical-based polymers, such as poly(propylene) (PP) and poly(ethylene) (PE). Due to its brittleness, P(3HB) is not very stress-resistant, showing poor mechanical properties, namely, Young's Modulus (~3.5 GPa) and elongation at break (3–5%), compared to petroleum-based materials (Table 9.2). For example, poly(propylene) presents a Young's Modulus of 1.7 GPa and an elongation at break of 400% [7, 16].

PHA copolymers in which *mcl* monomers are included in *scl*-PHA molecules, namely, P(3HB-*co*-3HV), P(3HB-*co*-3HHx) or P(3HB-*co*-4HB) (Table 9.2) are less stiff and brittle than P(3HB), thus improving their mechanical properties. This makes those biopolymers competitive with conventional plastics such as PP, poly(styrene), poly(ethylene terephthalate) and high-density poly(ethylene) (HDPE) [4, 7, 16]. Upon the formation of the copolymer, the properties of the material change, namely, their crystallinity, melting temperature, stiffness and toughness. Consequently, there is an impact on the biopolymer's mechanical properties, such as a decrease in Young's Modulus and an increase of impact strength, producing more desirable properties for commercial applications. The copolymer P(3HB-*co*-3HV) is one of the most well-known, and presents this set of characteristics when compared to P(3HB) [16]. A range of properties that have a balance of stiffness and toughness, and softness and toughness, is found in copolymers with the incorporation of 3HV, 4HB or 3HO monomers [4, 25].

Mcl-PHA are commonly composed of monomers such as 3-hydroxyhexanoate (3HHx), 3-hydroxyoctanoate (3HO), 3-hydroxydecanoate (3HD), 3-hydroxydodecanoate (3HDd) and 3-hydroxytetradecanoate (3HTd). Duo to the presence of monomers, such as 3HO, 3HD and 3HDd, *mcl*-PHA present mechanical properties with improved elastic and flexibility features in contrast to those of *scl*-PHA. *Mcl*-PHA display high elongation at break (above 100%) and low tensile strength (up to 10 MPa) (Table 9.2) [27, 30]. The characteristics of *mcl*-PHA resemble those of elastomers, latexes and resins. Therefore, materials based on *mcl*-PHA are considered suitable candidates for a variety of applications, such as rubbers, adhesives and glues [27, 28].

In comparison with other well-known biodegradable or biobased polymers, such as polylactic acid, polybutylene succinate or poly(ethylene), PHA display a much wider diversity in their thermal and mechanical properties [33]. Only poly(lactic acid) presents tensile strength and deformation at break values similar to those of P(3HB), while polybutylene succinate and poly(ethylene) present more elastic features that are similar to *mcl*-PHA and other copolymers, namely P(3HB-*co*-3HHx), P(3HB-*co*-4HB) and P(3HHx-*co*-3HO) (Table 9.2) [25, 31].

9.2.4 Biodegradability

PHA have received widespread attention due to their inherent biodegradability [19]. The key advantage of PHA over petroleum-based plastics is that they can be degraded in either aerobic or anaerobic environments through thermal degradation or enzymatic hydrolysis [4]. The rate of biodegradation of PHA in natural environments, namely, soil, seawater, lake water and sludge, is influenced by several factors, such as the microbial population (bacteria, algae and fungi) in a specified environment, temperature, moisture level, pH and nutrients. Hence, microbial populations, present in such environments, are capable of degrading PHA using PHA depolymerases, which are intra- or extracellular enzymes that hydrolyze PHA into water-soluble oligomers and monomers, or through PHA hydrolysis [4, 19, 34, 35]. However, the activity of these enzymes varies depending on the monomeric composition, crystallinity, additives and the surface area of the polymer, thus affecting their degradation rate [19, 34]. The final products obtained from PHA degradation in an aerobic environment are water and carbon dioxide, while in anaerobic conditions, methane is obtained. Subsequently, microorganisms use the resulting units of the polymer degradation as nutrients for biomass accumulation [19, 35, 36]. *In vivo* and *in vitro*, PHA degradation is characterized for the polymers' surface erosion and resulting weight loss, molecular weight decrease, increase in the degree of crystallinity and loss of mechanical properties [4]. In mammals, the hydrolysis of the polymer, in the medical context, happens gradually with PHA being degraded enzymatically, resulting in the innocuous 3-hydroxybutyric acid (3HB), a known component of blood plasma, which is non-toxic in nature [4, 19].

9.2.5 Biocompatibility

A key factor that distinguishes a biomaterial is its ability to be in contact with human body tissues without causing any negative response from the organism. The positive biological response to PHA polymers *in vivo* represents an important property of these biomaterials if a medical application is being considered, such as tissue engineering or drug delivery. The biocompatibility of PHA can be affected by several factors, for instance, shape, surface porosity, surface hydrophilicity, surface energy, material chemistry and degradation of the polymer [34]. So far, polymers, namely, P(3HB) and the copolymers P(3HB-*co*-3HV), P(4HB), P(3HB-*co*-4HB), P(3HB-*co*-3HHx) and P(3HHx-*co*-3HO) have been demonstrated to be compatible in various host systems upon animal testing and *in vivo* tests for tissue response [14]. In 2007, the US Food and Drug Administration (FDA) approved the use of P(4HB) for surgical sutures in clinical applications [34].

Most studies reported in the literature were based on the use of industrial grade PHA rather than medical ones, allowing for the presence of pyrogenic contaminants that were copurified alongside PHA, for example, the outer membrane lipopolysaccharide (LPS) endotoxin. The presence of LPS can induce a strong immunogenic adverse response, as opposed to the monomeric composition of the PHA. Therefore, it is fundamental that several conditions must be fulfilled upon the recovery process of the polymer from bacteria cells to guarantee its purity and suitability for use in biomedicine [6, 14, 23].

9.3 CURRENT PHA APPLICATIONS

9.3.1 Packaging, Molding and Coatings

The food sector represents the main customer of the packaging in processing industry [11]. Most of the materials used in this sector have short service times, so the majority of them are discarded in landfills. The use of biodegradable plastics can serve as a response to this issue. PHA fits this purpose due to their biodegradability in different environments. They can be processed for use in many applications, including packaging, molded goods, paper coatings, nonwoven fabrics, adhesives, films and performance additives [11, 37].

PHA films display a distinct hydrophobicity whilst exhibiting a high water vapor barrier, as well as a barrier against gases such as carbon dioxide, and a low oxygen transmission rate, which is required to restrict microbial growth and the oxidative spoilage of unsaturated fatty acids. These characteristic barriers make PHA interesting as raw materials for producing bottles for liquid foods and carbonated soft drinks [38]. So far, PHB, PHV, P(HB-*co*-HV), P(HB-*co*-HO) and P(HB-*co*-HD) have been tested. They present good tensile strength, printability, flavor and odor barrier properties, grease and oil resistance, temperature stability and are easy to dye [39]. For example, the FDA-approved "Metabolix PHA", a blend of PHB and poly(3-hydroxyoctanoate), P(3HO), is produced by Metabolix in the US for the production of food additives and making packages to maintain the performance of non-degradable plastics [39].

The use of PHB has shown some limitations, such as high production costs, but mostly due to the brittleness and low thermal stability in its melted state, which narrows the processing window for this polymer. Copolymerisation of PHB with other monomers, such as 3-hydroxyvalerate or 3-hydroxyhexanoate, can partially avoid these limitations [37]. Another way of tailoring the properties of PHA is by blending with other biopolymers, namely, starch, PLA, polycaprolactone and polyvinyl alcohol.

There have been a few studies on PHA as coatings for paper which could be suitable for food packaging. For example, preparation of PHB and PHBV granules as paper-sizing agents, where they act as coatings on paper or cardboard to reduce the moisture absorption and the water vapor permeability of these materials [37]. Using biopolymers for the coating of paper provides a way to decrease the high hydrophilicity of the paper. Recently, an innovative alternative to current paper-plastic two-layer latex composite sheets that are commonly used in retail was proposed, based on the use of *mcl*-PHA as bioresins, since they display a sticky characteristic, as a thin layer on paper sheets. The uses of *mcl*-PHA include biodegradable cheese coatings and fast food service ware, which take advantage of superior properties inherent to the PHA [38].

Regarding other applications, the production costs are still the major drawback concerning the introduction of PHA in the foam packaging market. Present industrial applications of PHA foams have yet to be established due to the thermal instability of PHA. The thermal degradation of the polymer makes foaming difficult, due to the low viscosity after the addition of a foaming agent, which results in the cell collapse. The literature reports a mixture of PHB, PVOH and starch with optimum viscosity was produced, and azodicarbonamide was used as a foaming agent, which presented a faster rate of biodegradation than other bulk materials with similar composition [37].

A novel field has been under development in the food packaging industry, which is the incorporation of nanoparticles in the PHA matrix [38]. Nanoparticles hold the potential to improve materials feature such as gas permeability and certain thermal and mechanical properties, but also act as an antimicrobial agent. Thus, copper or silver nanoparticles have been incorporated into PHA matrixes to enhance the functionality of polymer films, especially in food packaging applications, either in the form of coatings or wrappings [38]. Moreover, the enhancement of material properties through the development of nanocomposites based on PHA and nanofillers are increasing for food packaging. The nanofillers allow the modification of morphology, crystallization behavior, thermal stability, mechanical and barrier properties and the biodegradation rate, all of which are relevant from the food packaging perspective [37]. The first attempt to produce an *scl*-PHA paper composite started with the production of *scl*-PHA sheets that were further compressed on Kraft paper sheets. The biodegradation of P(4HB) coated Kraft paper has been shown to be successful [38].

9.3.2 Medical Applications

Nowadays, PHA are being used to develop various products for different medical purposes. For instance, in the cardiovascular system, PHA can have an important role in the development of pericardial patches, vascular grafts or heart valves; in wound management, these biopolymers can be used as sutures, dressing materials or as scaffolds for soft tissue repair; in orthopedics, the use of PHA for guided bone regeneration, cartilage repair and bone fixation has been reported; several other applications for PHA have also been reported, such as nerve repair, meshes, adhesives, 3D-scaffolds and other medical devices [26, 40].

9.3.2.1 Cardiovascular System

In the cardiovascular system, PHA have found various applications, including their use in sutures, stents, synthetic grafts or scaffolds. The use of mechanical valves and synthetic grafts for the treatment of heart valve diseases, although commonly practiced, often results in the formation of clots that could break loose and be carried by the bloodstream and plug other vessels (thrombosis). In this case, PHA by having inherent biocompatibility and good structural properties are great candidates for the development of heart valves that can provide both cellular support and differentiation (Figure 9.3). More specifically, *mcl*-PHA due to their higher flexibility have shown to be a promising polyester scaffold to be used in the fabrication of heart valves [34].

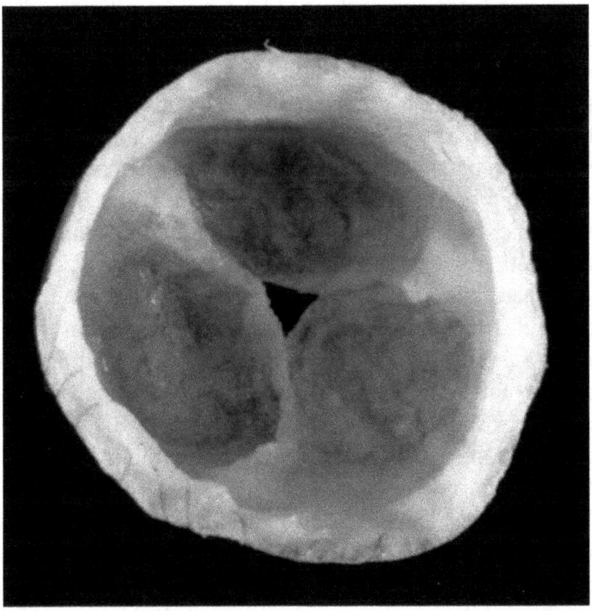

FIGURE 9.3 Tissue engineered heart valve construct derived from a PHA-based scaffold (reproduced, with permission, from [26]).

Another example where PHA could have a significant impact is in the treatment of coronary artery disease. Usually, this procedure uses a metal stent coated with a petrochemical plastic to serve as tubular support or to help in the healing process and/or releasing any existing obstructions inside the blood vessel. Recently, the use of different biodegradable materials to serve as artery stents to help the blood vessels' healing process has been studied. However, many biodegradable polymers failed to provide the mechanical support needed for this type of application, and their use could cause inflammatory reactions [41]. PHA, on the other hand, offer the required support for the blood vessels without damaging them, and since their degradation results in non-toxic products, their use also reduces the risk of inflammatory responses [42]. In this case, it is necessary to develop a porous polymer with some structural resistance that is also flexible enough to be introduced inside a blood vessel. In line with this, Puppi *et al.* [43] used P(3HB-*co*-3HHx) copolymer (which can provide the structure, due to the 3HB monomers, and flexibility, due to the 3HHx monomers, needed for this application) with a computer-aided wet-spinning (CAWS) strategy to successfully develop a porous, biodegradable and biocompatible coronary artery stent [43]. These are just two examples of the huge potential of PHA for the development of medical devices for treatment of diseases in the cardiovascular system.

9.3.2.2 Wound Management

An ideal topical dressing material for wound treatment should fulfill the following requirements: maintain high humidity and water balance at the wound–dressing interface, permit the exchange of gases, provide thermal insulation, prevent secondary infection having a rapid and prolonged bactericidal activity, adhere satisfactorily enough to maintain good wound–dressing contact, absorb wound exudate and associated toxic compounds and it should not provoke adverse reactions through prolonged tissue contact (e.g., toxic or allergic reactions) [44, 45]. A wide range of materials such as films, foams, hydrogels, hydrocolloids and hydrofibers have been investigated for wound management and many products, mostly based on synthetic polymers, are available on the market (e.g., Cutifilm, Curagel, Lyofoam, Algisite) [46, 47]. PHA polymers, due to their biodegradability, biocompatibility and structural properties, together with some inherent biological activity, can outperform the synthetic dressing materials currently available [40, 42]. PHA ranging from rigid and brittle polymers to flexible and elastomeric polyesters can form films with tunable properties that can serve as dressing materials for wound healing systems [48–51]. These films can be obtained by solvent casting, melting, compression or electrospinning processes. Electrospun membranes can be used as 3D-scaffolds for cellular support and proliferation, which could have an important role in the treatment of skin injuries and tissue regeneration [52].

Recently, the use of *mcl*-PHA has been studied as potential wound-dressing materials due to the elastic and even glue-like behavior exhibited by some of these polymers. One example is the use of *mcl*-PHA to produce adhesives for wound closure or films to serve as wound dressing materials (Figure 9.4). Moreover, it has also been shown that these types of polyesters are more suitable candidates than *scl*-PHA, as, due to their lower melting temperatures, higher elasticity and flexibility, *mcl*-PHA

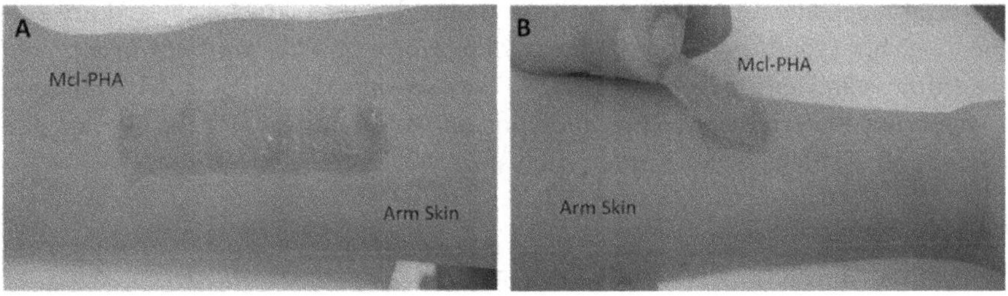

FIGURE 9.4 *In vivo* peeling test in human arm skin. A: *mcl*-PHA applied adhered to the arm skin; B: peeling the *mcl*-PHA film (reproduced, with permission, from [51].

can melt, adhere and adapt well to the body shape without causing any major trauma to the skin; also, upon peeling, there was no harm to the skin and there was no evidence of residue of *mcl*-PHA left on the skin. These are very important features that revealed that PHA as raw polymers are potentially interesting for the development of novel bio-based adhesives (e.g., bandages) [51].

9.3.2.3 Orthopedic

Bone is a complex cellular structure with dense, porous and calcified connective tissue. An ideal material for bone tissue regeneration should provide an appropriate scaffolding for cell attachment and maintenance of cellular functions. The treatments for bone loss usually use bone grafts (autografts, allografts and xenografts); however, the use of synthetic materials in these applications generally leads to nerve damage, pain, infection or immune responses [53]. As biodegradable and biocompatible materials with enhanced biological activity, PHA appear to be promising candidates for the development of bone regeneration systems. In orthopedy, PHA can be used as scaffolds for cartilage repair, bone graft substitutes and spinal cages. Various classes of PHA have been found to be suitable for bone tissue engineering; more specifically, P(3HB) polymers (due to their inherent stiffness and rigidity) were revealed to have the mechanical properties suitable for this type of applications. P(3HB) were also shown to have piezoelectric properties, which initiate bone growth and healing. In addition, *in vivo* degradation of P(3HB) results in 3-hydroxybutyrate, which is a natural component present in the bloodstream that has a promising effect in promoting cell viability and proliferation (in osteoblasts, fibroblasts and glial cells) [4, 40]. However, the incorporation of 3HV or 3HHx monomers in P(3HB) polymers can provide flexibility, which is an important factor for bone and cartilage repair. Cartilage is a flexible tissue that covers the joint surfaces and allows the bones to slide over each other, reducing friction and preventing bone damage. The current treatments for cartilage defects are usually related to surgical procedures or growth factors administration, which represent a risk of surgery complications and immune responses, respectively. One of the alternative approaches for treatment of cartilage defects is seeding chondrocytes in three-dimensional polymeric scaffolds to produce articular cartilage tissue; many different biodegradable polymers (including PHA) have already been studied in this application [54]. Furthermore, it has been reported that porous P(3HB-*co*-3HHx) scaffolds (produced using the CAWS strategy) have enhanced mechanical properties with good chondrocytes attachment and are suitable to anchor type II collagen filaments, outperforming PCL scaffolds in compressive stiffness and tensile deformation [41]. These results showed how PHA can serve as supportive materials for cell proliferation and differentiation for the development of bone and cartilage regeneration systems.

9.3.2.4 Nerve Repair

As mentioned before, PHA have low inflammatory response, permeability for nutrient penetration and minimal cytotoxicity, which are the main requirements for an ideal supportive material for nerve regeneration [40]. Moreover, 3-hydroxybutyrate, which is the final product from degradation of P(3HB), can promote differentiation of neuronal stem cells and neural progenitor cells. Due to these properties, PHA have been studied for their potential use as active materials in nerve repair applications [34]. More specifically, PHA have been studied for use as nerve guidance conduits, instead of synthetic tubes, in the treatment of damaged peripheral nerves. The main disadvantages of using synthetic conduits are their risk of trigger immune reactions, the induction of scar tissue, the difficult application in surgery and their degradation usually releases products that are harmful to the regeneration process. Also, the use of nonbiodegradable materials generally needs a follow-up surgery for the removal of the material used in the treatment procedure [54]. PHA, on the other hand, as biocompatible materials, can easily overcome these problems; also, since they are biodegradable polymers, their use as nerve guidance conduits prevents nerve compression, as they gradually degrade throughout the healing process, avoiding any follow-up surgery for material removal [55]. Moreover, P(3HB) polymers and copolymers were shown to have enhanced cell growth and cellular differentiation when compared with other PLA and PCL polyesters; more specifically, the P(3HB-*co*-3HO) copolymer (with 25% of 3HO monomer) was revealed to have better support and mechanical properties for nerve guidance applications [41, 56]. Additionally, these P(3HB) polyesters can also be used as epineural sutures to heal peripheral nerve tissue, or as scaffolds for mechanical support of Schwann cells, promoting their growth and differentiation to treat spinal cord injuries [34]. These studies exhibit how PHA can be useful and active materials in nerve regeneration.

9.3.3 Pharmaceutical Applications

For a successful and efficient management of any disease, a good pharmaceutical approach is needed. Therefore, it is truly important to have a good drug delivery system that ensures the proper dosage of the pharmaceutical substance for the required time, at a specific site and without affecting any other tissue or organ. However, the major drawbacks in current drug delivery systems are associated with the toxicity of the compounds used in the development of pharmaceutical carriers, the efficiency of the active substances, and the burst delivery of these agents throughout the targeted and non-targeted tissues [57]. For these reasons, polymeric materials, such as PHA, are gaining special attention due to the inherent biocompatibility, biological activity, biodegradability and tunable structured properties. In the pharmaceutical field,

PHA are known to be involved in the production of tablets, in targeted drug delivery systems (as microparticulate carriers) and in the development of new drugs with prolonged activity (since PHA are stereoregular compounds, they can serve as chiral precursors for chemical synthesis of active substances) [4, 26]. The use of PHA as microcarriers for targeted drug delivery of anesthetics, hormones, antibiotics, anti-inflammatory and anticancer agents has been reported. More specifically, *in vitro* studies showed that P(3HB) is a potential carrier for rifampicin and rubomycin, an antibiotic and an anticancer agent, respectively. On the other hand, P(3HB-*co*-3HV) was revealed to be a promising candidate for the delivery of gentamycin and sulperazone antibiotics. Moreover, P(3HB-*co*-3HO) was also studied for targeted drug delivery of folic acid and doxorubicin (an anticancer drug) [58]. In most of these studies, PHA had greater performance in terms of biocompatibility, biodegradability and controlled release of the pharmaceutical agent over long periods of time when compared to other biodegradable polyesters, such as PGA, PLA or PLGA. PHA also showed the presence of fewer toxic products from the polymerization process and a reduced inflammatory response to the human body (since their degradation results in hydroxy acids which are natural compounds found in the bloodstream, e.g. 3-hydroxybutyric acid) [59].

9.3.4 Prospective Applications

As raw materials, PHA have already found a wide range of applications in different significant areas, including packaging, medicine and pharmaceutics. These potential applications are intrinsically related to the great thermomechanical features found in PHA polymers [31, 60, 61]. Moreover, the tuneability of these biopolymers makes them suitable for the development of new advanced smart products with improved physical and chemical properties [41, 60].

The functionalization of PHA with other materials (biopolymers, bioactive substances, metal nanoparticles and other synthetic compounds) could be one of the next steps towards the production of smart polymeric matrices. Bionanocomposites produced from PHA could be a key concept for the development of new reactive biopolymers and current scientific investigation is not that far behind this futuristic view of PHA polymers. The number of reports concerning the use of PHA matrices with metallic nanoparticles has been increasing over the past decade. It has been shown that *scl*-PHA P(3HB) can be used for the development of bionanocomposites containing silver, copper or zinc nanoparticles, whose antimicrobial activity led to the proposal of such materials as suitable for food packaging [62]. The use of P(3HB) and iron oxide nanoparticles to target cancer cells has also been reported [63]. The copolymer P(3HB-*co*-3HV) and iron oxide nanoparticles have also been found to be usable for magnetic controlled release drug delivery systems [64]. According to Zhao *et al.* [65], polymer-based nanocomposites made of PHB and titanium dioxide nanoparticles could be used for degradation of methylene blue pollutants. This is only a small example of the use of bionanocomposites for different noteworthy applications. However, there is still more to learn about these smart biopolymers' properties and possible applications – for instance, there are only a few reports about their application in cancer therapy, in electro-stimulated drug release, in hyperthermia therapy or in cellular labeling for magnetic resonance imaging (MRI) [66, 67] – not forgetting that, currently, there are many studies being published regarding this issue.

PHA are also widely studied for their great structural properties. The use of PHA blends with other PHA and other synthetic or natural polymers have already been revealed to improve the mechanical properties of the regular PHA polymers [37, 41]. In addition, the production of scaffolds is used in almost all medical applications that requires tissue engineering. The production of these scaffolds using PHA blends, conductive metals, ceramics or synthetic constructs can be useful for wound management, cardiac tissue engineering and nerve or bone repair. Also, thermal heating and solvent casting methods used to manufacture the scaffolds should be replaced by techniques, such as sol-gel method and 3D printing, which do not involve heating the polymer nor the use of hazardous solvents that are not environmentally friendly [54].

9.4 CONCLUSIONS AND OUTLOOK

PHA's structural diversity, adaptable properties, controllable degradation and biocompatibility are very interesting features that render these biopolymers suitable for use in several areas of application. Over the past few years PHA have been at the forefront of scientific investigation for their potential use in medical applications, due to their inherent biocompatibility, biological activity and biodegradability.

As natural thermoplastics, PHA also have unique thermal and mechanical properties that currently have a wide range of applications, from packaging and coating to tissue repair and drug carriers. Moreover, these properties are deeply integrated with the PHA's monomeric composition, which are intrinsically related to the bacterial strain, carbon source, fermentation strategy and downstream processing used in these biopolymers' production processes. Furthermore, by changing one of these steps, PHA properties can be tailored to satisfy all the requirements needed for almost any specific application.

PHA have already demonstrated their advantages over synthetic petrochemical materials for use in packaging, tissue engineering, drug delivery and nerve or cartilage repair. The development of PHA bionanocomposites and blends could be a major step towards the fabrication of new advanced smart biomaterials with improved properties. For these reasons, it is crucial to surpass the main limitation of PHA: that is, the costs associated with production and extraction of these biopolymers. To solve this issue, suitable cheap carbon sources (or industrial waste) and recombinant bacterial species are, currently, being studied for their potential use in the production of PHA. Although the currently economic feasible area of PHA application is mainly in the biomedical area, reducing the costs associated with the production of PHA will surely unlock the viability of their use in a wider range of areas.

ACKNOWLEDGMENTS

This work was supported by the Unidade de Ciências Biomoleculares Aplicadas (UCIBIO), which is financed by national funds from FCT/MEC (UID/Multi/04378/2013) and co-financed by the ERDF under the PT2020 Partnership Agreement (POCI-01-0145-FEDER-007728).

REFERENCES

1. Reddy CSK, Ghai R, Kalia V, Kalia VC. Polyhydroxyalkanoates: An overview. *Bioresour Technol*, 2003; 87(2): 137–146.
2. Keshavarz T, Roy I. Polyhydroxyalkanoates: Bioplastics with a green agenda. *Curr Opin Microbiol*, 2010; 13(3): 321–326.
3. Jendrossek D, Pfeiffer D. New insights in the formation of polyhydroxyalkanoate granules (carbonosomes) and novel functions of poly(3-hydroxybutyrate). *Environ Microbiol*, 2014; 16(8): 2357–2373.
4. Muhammadi S, Afzal M, Hameed S, Hameed S. Bacterial polyhydroxyalkanoates-eco-friendly next generation plastic: Production, biocompatibility, biodegradation, physical properties and applications. *Green Chem Lett Rev*, 2015; 8(3–4): 56–77.
5. Koller M, Niebelschütz H, Braunegg G. Strategies for recovery and purification of poly [(R)-3-hydroxyalkanoates](PHA) biopolyesters from surrounding biomass. *Eng Life Sci*, 2013; 13(6): 549–562.
6. Kunasundari B, Sudesh K. Isolation and recovery of microbial polyhydroxyalkanoates. *Exp Polym Lett*, 2011; 5(7): 620–634.
7. Sudesh K, Abe H, Doi Y. Synthesis, structure and properties of polyhydroxyalkanoates: Biological polyesters. *Prog Polym Sci*, 2000; 25(10): 1503–1555.
8. Martino L, Cruz MV, Scoma A, *et al.* Recovery of amorphous polyhydroxybutyrate granules from *Cupriavidus necator* cells grown on used cooking oil. *Int J Biol Macromol*, 2014; 71: 117–123.
9. Koller M, Gasser I, Schmid F, Berg G. Linking ecology with economy: Insights into polyhydroxyalkanoate-producing microorganisms. *Eng Life Sci*, 2011; 11(3): 222–237.
10. Volova TG, Zhila NO, Shishatskaya EI, *et al.* The physicochemical properties of polyhydroxyalkanoates with different chemical structures. *Polym Sci A*, 2013; 55(7): 427–437.
11. Bugnicourt E, Cinelli P, Lazzeri A, Alvarez V. Polyhydroxyalkanoate (PHA): Review of synthesis, characteristics, processing and potential applications in packaging. *Express Polym Lett*, 2014; 8(11): 791–808.
12. Prados E, Maicas S. Bacterial production of hydroxyalkanoates (PHA). *Univers J Microbiol Res*, 2016; 4(1): 23–30.
13. Madkour MH, Heinrich D, Alghamdi MA, *et al.* PHA recovery from biomass. *Biomacromolecules*, 2013; 14(9): 2963–2972.
14. Valappil SP, Misra SK, Boccaccini AR, Roy I. Biomedical applications of polyhydroxyalkanoates, an overview of animal testing and *in vivo* responses. *Expert Rev Devic*, 2006; 3(6): 853–868.
15. Verlinden RA, Hill DJ, Kenward MA, *et al.* Bacterial synthesis of biodegradable polyhydroxyalkanoates. *J Appl Microbiol*, 2007; 102(6): 1437–1449.
16. Anjum A, Zuber M, Zia KM, *et al.* Microbial production of polyhydroxyalkanoates (PHAs) and its copolymers: A review of recent advancements. *Int J Biol Macromol*, 2016; 89: 161–174.
17. Costa SS, Miranda AL, Morais MG, *et al.* Microalgae as source of polyhydroxyalkanoates (PHAs) — A review. *Int J Biol Macromol*, 2019; 131: 536–547.
18. Freitas F, Alves VD, Coelhoso I, Reis MAM. Production and food applications of microbial biopolymers. In: *Engineering Aspects of Food Biotechnology. Part I: Use of Biotechnology in the Development of Food Processes and Products*, Teixeira José A and António A Vicente, Editors, 2013. CRC Press/Taylor & Francis Group: Boca Raton, FL; 61–88.
19. Akaraonye E, Keshavarz T, Roy I. Production of polyhydroxyalkanoates: The future green materials of choice. *J Chem Technol Biotechnol*, 2010; 85(6): 732–743.
20. Roy I, Visakh PM, Eds. *Polyhydroxyalkanoate (PHA) Based Blends, Composites and Nanocomposites*. Royal Society of Chemistry: Cambridge, UK; 30, 2015; 18–46.

21. Ashby RD, Solaiman DK. Poly(hydroxyalkanoate) biosynthesis from crude Alaskan pollock (*Theragra chalcogramma*) oil. *J Polym Environ*, 2008; 16(4): 221–229.
22. Wang S, Chen W, Xiang H, *et al.* Modification and potential application of short-chain-length polyhydroxyalkanoate (SCL-PHA). *Polymers*, 2016; 8(8): 273.
23. Koller M, Salerno A, Dias M, *et al.* Modern biotechnological polymer synthesis: A review. *Food Technol Biotechnol*, 2010; 48(3): 255–269.
24. Misra SK, Valappil SP, Roy I, Boccaccini AR. Polyhydroxyalkanoate (PHA)/inorganic phase composites for tissue engineering applications. *Biomacromolecules*, 2006; 7(8): 2249–2258.
25. Nigmatullin R, Thomas P, Lukasiewicz B, *et al.* Polyhydroxyalkanoates, a family of natural polymers, and their applications in drug delivery. *J Chem Technol Biotechnol*, 2015; 90(7): 1209–1221.
26. Williams SF, Martin DP. Applications of PHAs in medicine and pharmacy. *Biopolymers*, 2002; 4: 91–127.
27. Muhr A, Rechberger EM, Salerno A, *et al.* Biodegradable latexes from animal-derived waste: Biosynthesis and characterization of *mcl*-PHA accumulated by *Ps. citronellolis*. *React Funct Polym*, 2013; 73(10): 1391–1398.
28. Muhr A, Rechberger EM, Salerno A, *et al.* Novel description of mcl-PHA biosynthesis by *Pseudomonas chlororaphis* from animal-derived waste. *J Biotechnol*, 2013; 165(1): 45–51.
29. Zinn M, Hany R. Tailored material properties of polyhydroxyalkanoates through biosynthesis and chemical modification. *Adv Eng Mater*, 2005; 7(5): 408–411.
30. Pappalardo F, Fragalà M, Mineo PG, *et al.* Production of filmable medium-chain-length polyhydroxyalkanoates produced from glycerol by *Pseudomonas mediterranea*. *Int J Biol Macromol*, 2014; 65: 89–96.
31. Chen GQ. A microbial polyhydroxyalkanoates (PHA) based bio-and materials industry. *RSC*, 2009; 38(8): 2434–2446.
32. Villaluenga JPG, Khayet M, López-Manchado MA, *et al.* Gas transport properties of polypropylene/clay composite membranes. *Eur Polym J*, 2007; 43(4): 1132–1143.
33. Meng DC, Chen GQ. Synthetic biology of polyhydroxyalkanoates (PHA). In: *Synthetic Biology – Metabolic Engineering. Advances in Biochemical Engineering/Biotechnology*, vol. 162, Zhao H and AP Zeng, Editors, 2017. Springer: Berlin, Heidelberg; 147–174.
34. Grigore ME, Grigorescu RM, Iancu L, *et al.* Methods of synthesis, properties and biomedical applications of polyhydroxyalkanoates: A review. *J Biomat Sci Polym E*, 2019; 30(9): 695–712.
35. Chanprateep S. Current trends in biodegradable polyhydroxyalkanoates. *J Biosci Bioeng*, 2010; 110(6): 621–632.
36. Numata K, Abe H, Iwata T. Biodegradability of poly(hydroxyalkanoate) materials. *Materials*, 2009; 2(3): 1104–1126.
37. Plackett D, Siró I. Polyhydroxyalkanoates (PHAs) for food packaging. In: *Multifunctional and Nanoreinforced Polymers for Food Packaging*, Lagaron JM, Editor, 2011. Woodhead Publishing Ltd: Cambridge, UK; 498–526.
38. Rai V, Kumar P. Polyhydroxyalkanoates (PHAs) – Biodegradable polymers for "green" food packaging materials. In: *Recent Advances in Biotechnology*, Rai V, Editor, 2017. Shree Publishers & Distributors: New Delhi; 149–165.
39. Mangaraj S, Yadav A, Bal LM, *et al.* Application of biodegradable polymers in food packaging industry: A comprehensive review. *J Packag Technol Res*, 2019; 3(1): 77–96.
40. Bonartsev AP, Bonartseva GA, Reshetov IV, *et al.* Application of polyhydroxyalkanoates in medicine and the biological activity of natural poly(3-hydroxybutyrate). *Acta Nat*, 2019; 11(2): 4–16.
41. Koller M. Biodegradable and Biocompatible polyhydroxyalkanoates (PHA): Auspicious microbial macromolecules for pharmaceutical and therapeutic applications. *Molecules*, 2018; 23(2): 1–20.

42. Souza L, Shivakumar S. Polyhydroxyalkanoates (PHA) Applications in wound treatment and as precursors for oral drugs. In: *Biotechnological Applications of Polyhydroxyalkanoates*, Kalia V, Editor, 2019. Springer: Singapore; 227–270.
43. Puppi D, Pirosa A, Lupi G, *et al.* Design and fabrication of novel polymeric biodegradable stents for small caliber blood vessels by computer-aided wet-spinning. *Biomed Mater*, 2017; 12(3): 1–15.
44. Kennedy JF, Knill CJ, Thorley M. Natural polymers for healing wounds. In: *Recent Advances in Environmentally Compatible Polymers. Cellucon '99. Proceedings*, Kennedy JF, Phillips GO, Williams PA, Editors, 2001. Cambridge, UK; 97–104.
45. Lipsky BA, Hoey C. Topical antimicrobial therapy for treating chronic wounds. *Clin Infect Dis*, 2009; 49(10): 1542–1549.
46. Han G, Ceilley R. Chronic wound healing: A review of current management and treatments. *Adv Ther*, 2017; 34(3): 599–610.
47. Grey JE, Enoch S, Harding KG. ABC of wound healing pressure ulcers. *BMJ*, 2006; 332(7539): 472–475.
48. Cruz MV, Freitas F, Paiva A, *et al.* Valorization of fatty acids-containing wastes and by products into short- and medium-chain length polyhydroxyalkanoates. *New Biotechnol*, 2016; 33(1): 206–215.
49. Laycock B, Halley P, Pratt S, *et al.* The chemomechanical properties of microbial polyhydroxyalkanoates. *Prog Polym Sci*, 2014; 39(2): 397–442.
50. Rai R, Keshavarz T, Roether JA, *et al.* Medium chain length polyhydroxyalkanoates, promising new biomedical materials for the future. *Mat Sci Eng R*, 2011; 12: 2126–2136.
51. Pereira JR, Araujo D, Marques AC, *et al.* Demonstration of the adhesive properties of the medium-chain-length polyhydroxyalkanoate produced by *Pseudomonas chlororaphis* subsp. *aurantiaca* from glycerol. *Int J Biol Macromol*, 2019; 122: 1144–1151.
52. Shishatskaya EI, Nikolaeva ED, Vinogradova ON, Volova TG. Experimental wound dressings of degradable PHA for skin defect repair. *J Mater Sci Mater Med*, 2016; 27(11): 1–16.
53. Sopyan I, Mel M, Ramesh S, Khalid KA. Porous hydroxyapatite for artificial bone applications. *Sci Technol Adv Mat*, 2007; 8(1–2): 116–123.
54. Lizarraga-Valderrama LR, Panchal B, Thomas C, *et al.* Biomedical applications of polyhydroxyalkanoates. In: *Biomaterials from Nature for Advanced Devices and Therapies*, 1st edition, Neves NM, Reis RL, Editors, 2016. John Wiley & Sons: New Jersey; 339–383.
55. Babu P, Behl A, Chakravarty B, *et al.* Entubulation techniques in peripheral nerve repair. *J Neurotrauma*, 2008; 5(1): 15–20.
56. Basnett P, Ching KY, Stolz M, *et al.* Novel poly(3-hydroxyoctanoate)/poly(3-hydroxyb utyrate) blends for medical applications. *React Funct Polym*, 2013; 73(10): 1340–1348.
57. Bhatia SK, Wadhwa P, Bhatia RK, *et al.* Strategy for biosynthesis of polyhydroxyalkonates polymers/copolymers and their application in drug delivery. In: *Biotechnological Applications of Polyhydroxyalkanoates*, Kalia VC, Editor, 2019. Springer: Singapore; 13–34.
58. Balogun-Agbaje OA, Odeniyi OA, Odeniyi MA. Applications of polyhydroxyalkanoates in drug delivery. *Fabad J Pharm Sci*, 2019; 44(2): 147–158.
59. Rodríguez-Contreras A, Rupérez E, Marqués-Calvo MS. PHAs as matrices for drug delivery. In: *Bio Med Mater Eng*, Holban A-M, Grumezescu AM, Editors, 2019. Elsevier: Amsterdam, Netherlands; 183–213.
60. Berezina N, Martelli SM. Polyhydroxyalkanoates: Structure, properties and sources. In: *Polyhydroxyalkanoates--Plastic Materials of the 21st Century: Production, Properties, Applications*, Volova TG, Editor, 2004. Nova Publishers: New York.
61. Chen GQ. Introduction of bacterial plastics PHA, PLA, PBS PE, PTT and PPP. In: *Plastics from Bacteria*, Chen GQ, Editor, Steinbüchel A, Series Editior, 2010. Springer: Berlin, Heidelberg; 1–16.

62. Castro-Mayorga JL, Freitas F, Reis MAM, *et al.* Biosynthesis of silver nanoparticles and polyhydroxybutyrate nanocomposites of interest in antimicrobial applications. *Int J Biol Macromol*, 2018; 108: 426–435.
63. Erdal E, Kavaz D, Sam M, *et al.* Preparation and characterization of magnetically responsive bacterial polyester based nanospheres for cancer therapy. *J Biomed Nanotech*, 2012; 8(5): 800–808.
64. Oka C, Ushimaru K, Horiishi N, *et al.* Core–shell composite particles composed of biodegradable polymer particles and magnetic iron oxide nanoparticles for targeted drug delivery. *J Magn Magn Mater*, 2015; 381: 278–284.
65. Zhao X, Lv L, Pan B, *et al.* Polymer-supported nanocomposites for environmental application: A review. *Chem Eng J*, 2011; 170(2–3): 381–394.
66. Laurent S, Forge D, Port M, *et al.* Magnetic iron oxide nanoparticles: Synthesis, stabilization, vectorization, physicochemical characterizations, and biological applications. *Chem Rev*, 2008; 108(6): 2064–2110.
67. Shi Z, Gao X, Ullah MW, *et al.* Electroconductive natural polymer-based hydrogels. *Biomaterials*, 2016; 111: 40–54.

10 Hydrogen-Oxidizing Producers of Polyhydroxyalkanoates
Synthesis, Properties, and Applications

Tatiana G. Volova, Ekaterina I. Shishatskaya, Natalia O. Zhila, and Evgeniy G. Kiselev

CONTENTS

10.1 INTRODUCTION

The diversity of the forms of living matter and the new knowledge in physics and chemistry of living systems have provided the basis for constructing biological systems of different complexity and arrangements in order to synthesize a wide range of macromolecules.

Microorganisms are a source for producing various foodstuffs as well as medical and technical products. Development of new biotechnologies should be based on knowing how physical and chemical environmental conditions influence the synthesis of intracellular macromolecules. Processes of microbial synthesis can be of two kinds: processes related to cell growth and biomass formation, which are positively correlated to cell proliferation rate, and processes occurring and speeding up as cell growth rates decrease [1]. These processes can be optimized in various ways, depending on how closely the growth rates of the producer correspond to the rates of synthesis of different macromolecules or how much they differ from each other. To achieve the best results in accumulation of cell biomass in the culture and synthesis of the primary products of metabolism, it is enough to optimize the conditions of nutrition of the cells and to arrange conditions for the balanced growth of the culture. Cells usually begin accumulating storage compounds (polyphosphates, polysaccharides, lipids, etc.) under unbalanced growth conditions, when the medium is depleted of a nutrient, and when growth and synthesis of the major (nitrogen-containing) components is limited. It is more difficult to optimize synthesis of storage compounds, as one should first gain insight into the principles of formation of macromolecules and develop special approaches for every particular case. Nutrient deficiency slows down cell growth, causing considerable changes in the chemical composition of cells, chiefly the ratio of the primary to storage macromolecules. It is critical to know the principles underlying these changes for targeted synthesis of macromolecules and fabrication of tailored products.

Lipid storage compounds produced by prokaryotes – polyhydroxyalkanoates (PHA) – are valuable products of biotechnology. These polymers have a wide range of useful properties, and thus, they are promising materials for various applications [2–8]. Except for a few species, most PHA producers synthesize PHA under unbalanced growth conditions, when intracellular production of the major (nitrogen-containing) compounds is limited by deficiency in one of the nutrients.

PHA-producing strains play a key role in microbial biotechnologies. PHA producers include various taxa of organisms, and their list keeps growing.

Among diverse PHA producers, *Cupriavidus* hydrogen-oxidizing bacteria occupy a special position, as they are capable of synthesizing PHA with different chemical compositions under autotrophic and heterotrophic conditions on various carbon substrates (CO_2, sugars, organic acids, alcohols, etc.) at high yields (up to 80–90 wt.%). These microorganisms have a specific constructive metabolism: the ratio between intracellular nitrogen-containing compounds and storage PHA changes during batch culture. Even when cells are grown on a complete nutrient medium, by the middle of the linear growth phase, intracellular protein synthesis slows down, and they begin to synthesize and accumulate considerable amounts of the storage compound, poly(3-hydroxybutyrate) [9,10]. The systematic position and the name of *Cupriavidus* have been changed several times: *Hydrogenomonas*, *Alcaligenes*, *Wautersia*, *Ralstonia* [11].

Hydrogen-oxidizing bacteria began to attract practical interest in the early 1970s as a potential regenerative component in closed biotechnological life support systems. Biosynthesis of hydrogen-oxidizing bacteria in combination with electrolysis was proposed as a solution to the main issues of human life support during space flights: oxygen generation and carbon dioxide consumption, water treatment, and production of the protein part of the human diet. As hydrogen-oxidizing bacteria grow much faster than other chemotrophs, they have attracted the attention of researchers as a potential source of protein. Their metabolism and growth were extensively studied in the US, the Federal Republic of Germany, Japan, and the USSR (Russian Federation since 1991). The late 1980s and the early 1990s saw an upsurge of interest in hydrogen-oxidizing bacteria as very promising producers of polyhydroxyalkanoates (polyesters of alkanoic acids) – polymers similar to polypropylene, but degradable in the natural environment.

At the Institute of Biophysics SB RAS, studies related to hydrogen-oxidizing bacteria were first undertaken in the late-1960s–early-1970s, thanks to cooperation with

Georgy Zavarzin – an outstanding scientist and a Full Member of the Academy of Sciences. Zavarzin donated his strains, including *Alcaligenes eutrophus* Z-1, a superproducer of protein, to our team. The strain was later registered in the All-Union (since 1991, Russian) National Collection of Industrial Microorganisms (VKPM) as *Cupriavidus necator* B-3358. The properties of this strain are similar to those of the strain *R. eutropha* H-16 of Prof. Hans Günter Schlegel, which has been studied by numerous research teams all over the world.

The research team that has formed at the Institute of Biophysics SB RAS has unique experience in hydrogen-based biosynthesis. The team has covered the distance from bacterial cultures in small-size laboratory reactors to pilot production plants. The major research directions that have been pursued by the team are "hydrogen-based biosynthesis for human life support systems"; "hydrogen-oxidizing bacteria as a new protein source"; "hydrogen-based biosynthesis and degradable bioplastics" [7, 8, 12, 13].

As the original task was to produce cell biomass of hydrogen-oxidizing bacteria with the highest possible protein content, it was necessary to determine the conditions that would minimize intracellular poly(3-hydroxybutyrate) (P(3HB), aka PHB) synthesis. Prof. H. Schlegel produced mutant bacterial strains with defective pathways of polymer synthesis [14]. The approach used by our team was to employ parametric control of continuous fermentation of bacterial cells. First, we needed to determine the factors inducing polymer synthesis in bacterial cells. The non-optimal pH and temperature of the medium, hydrogen deficiency, and CO_2 or O_2 limitation or inhibition of *Al. eutrophus* Z-1 growth were not found to induce accumulation of the polymer, and its concentration reached just a few percent. The only factor that induced P(3HB) accumulation in the cells of the study strain was a change in the flow rate of the mineral elements fed to the culture medium. In autotrophic chemostat culture on a single carbon substrate (CO_2), with cell growth limited by the low flow rates of the mineral elements fed to the medium ($D = 0.1\ h^{-1}$), intracellular polymer content increased. The highest P(3HB) content (about 40%) was obtained under nitrogen deficiency, 20–25% under sulfur and phosphorus deficiency, and the lowest – (about 15%) PHA content – under potassium or magnesium deficiency. An advantage of the chemostat is the possibility of polymer synthesis for indefinitely long time periods and with no strict sterility measures; its disadvantages are low cell concentration in the culture and low intracellular polymer content (no higher than 40%). Moreover, the process involves the use of the continuously operating unit for the separation of the low-density culture medium, which consumes a lot of power. These data were mainly published in the Russian journals *Mikrobiologiya* and *Prikladnaya biokhimiya i mikrobilogiya*. They were summarized in a 2009 book by Volova [8].

Only in the fed-batch culture could we manage to obtain polymer concentrations of 70–80%, using glycerol, fructose, and autotrophic substrate (CO_2+O_2+H_2) [16]. That was the first study in our research on various aspects of PHA biotechnology.

The key issues to be resolved in order to develop and implement effective processes of PHA biosynthesis are related to the following aspects of synthesis of these valuable macromolecules:

First, all PHA-producing prokaryotes synthesize these macromolecules under specific conditions of unbalanced growth, as an endogenous energy and carbon storage. The growth of PHA producers is limited by different elements: nitrogen, phosphates, oxygen, or other substrate components. Thus, it is important to find and evaluate new PHA producers and determine the conditions that would maximize polymer yields.

Second, PHA include polymers with various compositions, whose physicochemical properties (molecular weight, melting point, thermal degradation temperature, crystallinity, and degradation rates in biological environments) differ significantly, depending on the polymer chemical composition. A necessary condition for developing the approaches to the synthesis of novel PHA with tailored properties is to find microorganisms and conditions for producing polymers with different chemical structures.

Third, the scale of production and use of PHA is largely related to their cost, which is mainly determined by the type and cost of the growth substrate used. As the C-substrate can account for up to 40% of the costs of PHA production, the investigations aimed at determining the strategy of commercial production of PHA are currently focused on a search for new substrates. Thus, an important prerequisite for PHA production is to find the strains capable of PHA synthesis from the available feedstocks, including industrial waste.

Extensive experimental studies provided the basis for developing and performing effective synthesis of polymers with different chemical compositions from various substrates using wild-type strains of hydrogen-oxidizing bacteria, and constructing and testing high-performance products for diverse applications: biomedicine, experimental pharmacology, agriculture, municipal engineering, etc. [13, 15].

Our team carried out pioneering basic research on the cell cycle of PHA, including their molecular structure and activities of the key enzymes of polymer synthesis and intracellular degradation. Natural prokaryotes were used for the first time to synthesize PHA copolymers composed of short- and medium-chain-length monomers, and those results were supported by studies performed by our colleagues in other countries several years later.

10.2 PHA SYNTHESIS FROM AUTOTROPHIC AND HETEROTROPHIC SUBSTRATES

Potential raw materials for PHA synthesis are various substrates with different oxidation states of carbon, energy content, and cost, including individual compounds (carbon dioxide and hydrogen, sugars, alcohols, organic acids) and by-products of alcohol and sugar industries, chemical processing of plant raw materials, and production of olive, soybean, and palm oils [16–18]. PHA can be synthesized from both individual carbon compounds and various industrial wastes, making them economically feasible materials for different applications. Studies have been performed to investigate potential C_1 carbon sources for PHA production such as methane, methanol, and CO_2 [19] and the respective bacterial taxa capable of PHA production from these substrates (*Methylocystis*, *Methylobactrerium*, and *Cupriavidus*).

The ability of hydrogen-oxidizing bacteria to synthesize PHA under autotrophic conditions, without organic media, with CO_2 as a source of constructive metabolism and H_2 as a source of energy metabolism, makes them good candidates for commercial production of PHA [5, 6, 19–22]. The efficiency of P(3HB) production from hydrogen as an energy substrate is very high; the yield coefficient of polymer production from hydrogen reaches 1.0. The use of the poorly soluble and explosive substrate is the major technological challenge for the implementation of this process.

There are, however, rather few published studies reporting PHA synthesis from feedstocks containing hydrogen. Among the research teams that have been investigating CO_2 and H_2 as growth substrates for PHA biosynthesis are the Institute of Biophysics SB RAS and Kyushu University (Japan). Researchers of the Institute of Biophysics SB RAS tested different modes of cultivation of hydrogen-oxidizing bacteria on the gaseous substrate, such as autotrophic synthesis of protein or poly(3-hydroxybutyrate) and PHA copolymers using electrolytic hydrogen and synthesis gas [5, 22–25]. The research team headed by Prof. A. Ishizaki (Japan) investigated production of poly(3-hydroxybutyrate) in *Alcaligenes eutrophus* culture in fermenters with different mass exchange parameters [21, 26, 27]. The Japanese team scaled up P(3HB) synthesis in fermenters with high mass exchange parameters, in the culture of *Alcaligenes eutrophus*, and achieved cell biomass and polymer content of 91.3 g/L and 61.9%, respectively [21, 26].

Because of the specific metabolism of PHA, which are accumulated by bacterial cells whose growth is limited, and intracellular degradation and utilization of these storage macromolecules, simultaneous production of high yields of polymers and high cell concentrations presents a problem. Therefore, studies were performed to determine culture conditions that would enable achieving both high intracellular PHA content and high concentrations of bacterial cells under limited cultivation conditions. Due to their specific growth physiology and constructive metabolism, in the middle of the linear growth phase, hydrogen-oxidizing bacteria stop synthesizing protein and begin accumulating PHA, even on a complete nutrient medium [9, 15]. Based on this fact, we investigated the cell growth and polymer yields in the autotrophic batch culture of *C. eutrophus* B-10646 with different inflowing amounts of nitrogen in a 10 L fermenter of the original design, with a working volume of 3 L (see Figure 10.1). In the experiment with the amount of the nitrogen supplied to the culture corresponding to the physiological requirement of the cells (120 mg/g cells), for 70 h of cultivation, the intracellular P(3HB) content had reached 55% and cell concentration 30 g/L (Option I). The second cultivation mode consisted of two stages: in the first, the cells were grown on a complete nutrient medium with a continuous inflow of nitrogen in the amounts corresponding to the physiological requirement of cells (120 mg/g cells); in the second, the cells were grown on a nitrogen-free medium. Under these conditions, the polymer yield reached 80%, but the duration of the process was increased to 80–85 h (Option II). In the third mode of batch cultivation, cells were grown under limited nitrogen supply in the first stage and without nitrogen at all in the second. The best results were obtained with a nitrogen supply amounting to 50% of that required by cells in the first stage (60 mg/g cells) (Option III). By the end of the experiment (70 h), intracellular polymer content had reached 85% and cell concentration 48 g/L.

Consumption of the gaseous substrate by the culture accumulating P(3HB) varied significantly throughout the process. In the first stage, when the culture was growing and accumulating the polymer, specific rates of consumption of gas mixture

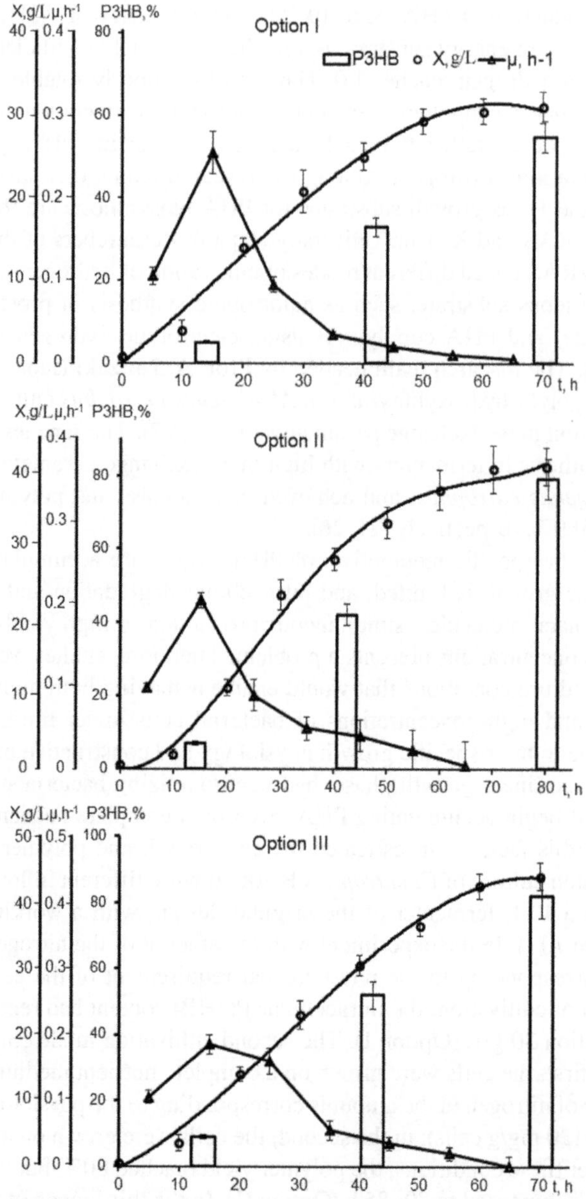

FIGURE 10.1 Parameters of *C. eutrophus* B-10646 autotrophic batch culture: a – a single-stage process on complete medium; b – a two-stage process: Stage I on complete medium and Stage II on nitrogen-free medium; c – a two-stage process: Stage I (I) at 50% nitrogen supply and Stage II on nitrogen-free medium; X – cell concentration of the culture (g/L), μ – cell specific growth rate, h^{-1}; P(3HB) – intracellular polymer concentration (% of dry matter).

components by the culture were 0.035–0.05 mol./(g·h) for hydrogen, 0.17–0.18 mol./(g·h) for oxygen, and 0.11–0.14 mol./(g·h) for carbon dioxide. The average yield coefficients of the culture were 3.0 g/mol. on H_2, 1.2 g/mol. on O_2, and 1.7 g/mol. on CO_2. In the second stage, when the cells were actively accumulating the polymer, the rates of consumption of the carbon substrate (CO_2), the energy substrate (H_2), and oxygen were significantly lower, due to a dramatic decrease in the cell growth rate and an almost complete cessation of cell growth at the end of the stage. Yield coefficients of the culture were 2.3; 0.8; 1.3 g/mol., respectively. The average yield coefficients of the culture for the whole process were 2.6 g/mol. on H_2; 1.0 g/mol. on O_2; and 1.5 g/mol. on CO_2. Thus, the two-stage batch cultivation with cells grown under nitrogen deficiency in the first stage and without nitrogen in the second resulted in polymer yields and cell concentrations reaching 85 and 50 g/L. As preparation of the inoculum for the production fermenter also takes some time (30–35 h), the full cycle of autotrophic batch cultivation of bacterial cells lasts about 100–110 h.

Based on these results, a more productive strain, *C. eutrophus* B-10646, was used later to synthesize PHA in two-stage autotrophic chemostat batch cultivation (see Figure 10.2). In the first stage, cells were grown under flow chemostat conditions, and the amount of nitrogen supplied to the medium was 100 mg/g cells. The resulting inoculum had cell concentration and intracellular polymer content (4±1) g/L and (8±2)%, respectively. Then, the flow was stopped, and nitrogen supply was limited to 60 mg/g cells. After 35 h, nitrogen supply was stopped. In that stage, intracellular polymer content was (47±2)% and cell concentration (15±1) g/L. In the second

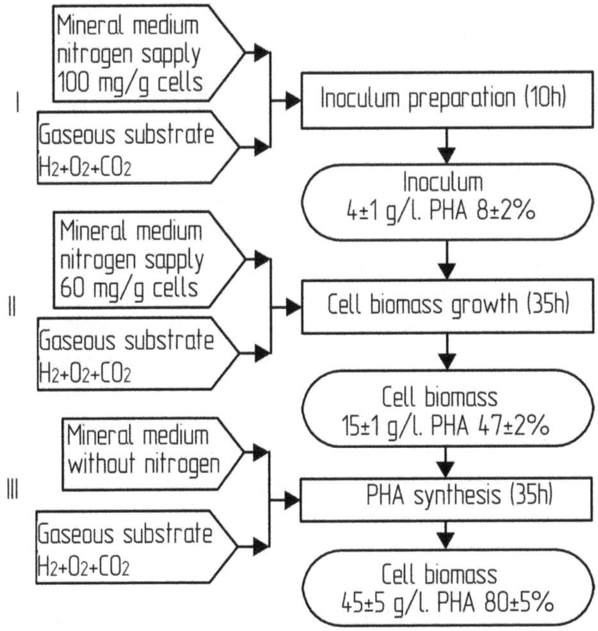

FIGURE 10.2 A flow-chart of chemostat autotrophic batch culture of *C. eutrophus* B-10646.

stage, cultivation was conducted in nitrogen-free medium for 35 h. By the end of the experiment (70 h), intracellular polymer content reached 80±5% and cell concentration 45±5 g/L.

Recently, PHA synthesis from a gaseous substrate has attracted increasing attention from researchers. The previously known and new organisms utilizing CO_2 and H_2 to synthesize biomass and PHA have been described in the literature.

The fact that some of the hydrogen-oxidizing bacterial species are tolerant of CO suggests the possibility of cultivating them using such feedstocks as industrial hydrogen (hydrogen-containing products of natural gas and coal conversion, wood waste, hydrolysis lignin, etc.) [22, 24, 28–32]. The first bacterial strain that was described as CO resistant was *Alcaligenes eutrophus* Z1 – a strain maintained in Zavarzin's collection at the Institute of Microbiology RAS (Moscow) [15, 33]. The hydrogenase of this microorganism is not inhibited by CO [34]. The first *R. eutropha* B5786 was studied for CO resistance, and the ability to grow and synthesize PHA in the presence of CO. In *R. eutropha* B5786 cells, carbon monoxide enhances hydrogenase activity and cytochrome synthesis and strengthens cell membranes; a compensatory increase occurs in expenditure of the energy substrate for biosynthesis [24]. The defense response of cells includes higher levels of saturation and cyclization of fatty acids of membrane lipids, causing increased rigidity and decreased permeability of cell membranes [24]. Moreover, the study of PHA synthesis in the presence of CO showed that CO did not inhibit either the key enzymes involved in the PHA intracellular cycle or the polymer synthesis reactions [24]. Based on those results, for the first time in practical biotechnology, PHA synthesis was conducted on model gas mixtures in the presence of CO, and on syngas produced by gasification of brown coals from the Kansk-Achinsk coalmines and hydrolysis lignin. That research proved the feasibility of achieving intracellular PHA concentrations of 80–85% from hydrogenous products of processing of natural carbonaceous materials (natural gas, low-grade coals, hydrolysis lignin, etc.); the processes developed by our team have been protected by patents [35,36].

In a relatively recent study, Tanaka *et al.* reported CO tolerance of the hydrogen-oxidizing bacteria *R. eutropha* ATCC7697 and *Alcaligenes latus* ATCC29712 and their ability to synthesize PHA in the presence of CO; the authors isolated and studied a new microorganism, *Ideonella* sp. O-1, capable of synthesizing P(3HB) in high yields in the presence of CO [30]. Do *et al.* described *Rhodospirillum rubrum* culture growth and P(3HB-*co*-3HV) synthesis on model gas mixtures containing CO, and on the syngas produced by gasification of corn waste [31]. A study by Volova *et al.* in 2015 [25] reported the ability of another autotrophic hydrogen-oxidizing organism – the *Seliberia carboxydohydrogena* Z-1062 aerobic bacterium – to synthesize PHA in the presence of CO. Poly(3-hydroxybutyrate) yields were investigated in experiments with limiting concentrations of mineral nutrients in batch culture of *S. carboxydohydrogena* Z-1062 grown on gas mixtures consisting of CO_2, O_2, H_2, and CO. CO concentrations of 10, 20, and 30% v/v did not affect synthesis of the polymer, whose content after 56-h cultivation under limiting concentrations of nitrogen and sulfur was 52.6–62.8% of biomass weight at a productivity of 0.13–0.22 g/(L·h). The inhibitory effect of CO on cell concentration was revealed at CO concentration of 30% v/v. That also caused a decrease in substrate (H_2 and O_2) use efficiency.

Thus, this carboxydobacterium can be regarded as a potential producer of PHA from industrial hydrogenous sources.

Another advantage of hydrogen-oxidizing microorganisms as PHA producers is that using the heterotrophic substrate instead of the gaseous one is a way to substantially simplify the technical/technological parameters of the fermentation system and obtain higher cell concentrations and polymer yields in a shorter time period. Factors affecting PHA accumulation were studied using a two-stage batch cultivation mode in mineral medium containing fructose, with limited nitrogen supply in Stage 1 and without nitrogen in Stage 2, which had been previously developed in autotrophic mode. By varying the starting cell concentration in the inoculum and intracellular polymer content, we determined the conditions necessary to achieve the highest possible cell concentration and polymer content in a shorter time period. By using concentrated inoculum (5–7 g/L or higher), with physiologically active cells containing no more than 20% polymer, we minimized (or even eliminated) the lag phase and achieved the highest growth rates and polymer synthesis rates. Inoculum was prepared in continuous culture, then bacterial suspension was concentrated by centrifugation under aseptic conditions. In Stage 1 of the two-stage cultivation, when nitrogen supply was limited to 50% of the strain's physiological requirement, the highest cell concentration was reached, with intracellular polymer content of 45–50%. In Stage 2, in the nitrogen-free medium, polymer content reached its highest values (80–90%).

Various substrates (acetate, sugars, fatty acids, plant oils, glycerol) were tested to achieve effective PHA synthesis by hydrogen-oxidizing bacteria under heterotrophic conditions. Hydrogen-oxidizing bacteria, which use a truncated version of the Entner–Doudoroff pathway to assimilate sugars, utilize only fructose, but a number of strains readily generate mutants capable of utilizing glucose [10].

Using the collection of productive strains (the fast-growing *Ralstonia eutropha* B5786, glucose-assimilating mutant *Ralstonia eutropha* B8562, and *Cupriavidus eutrophus* B10646, which is tolerant to precursor substrates needed for synthesis of PHA copolymers), we achieved productive synthesis of P(3HB) from sugars, organic acids, and glycerol. Glycerol has been recently regarded as a promising substrate for large-scale production of PHA, whose manufacture scale has been growing together with the increased production of biodiesel as a renewable energy source – an alternative to oil [37]. The most effective processes are shown in Figure 10.3.

The most productive and stable process was obtained in a medium with glucose and glycerol, which produced a wide range of physiological effects (from 3 to 30 g/L) as opposed to fructose, where bacterial growth was inhibited at a concentration in the culture medium exceeding 15 g/L, and the process became unstable.

Thus, the cost-effectiveness of PHA can be enhanced by using new strains and carbon sources and by improving manufacturing processes, including large-scale ones. Scaling up laboratory biotechnologies to pilot production (PP) is a necessary step towards commercial production. The pilot production facility, established at the Siberian Federal University (Krasnoyarsk, Russia) (see Figure 10.4), contains units for media and inoculum preparation, a unit for fermentation, and a unit for polymer extraction and purification. The PP fermentation unit includes a steam generator

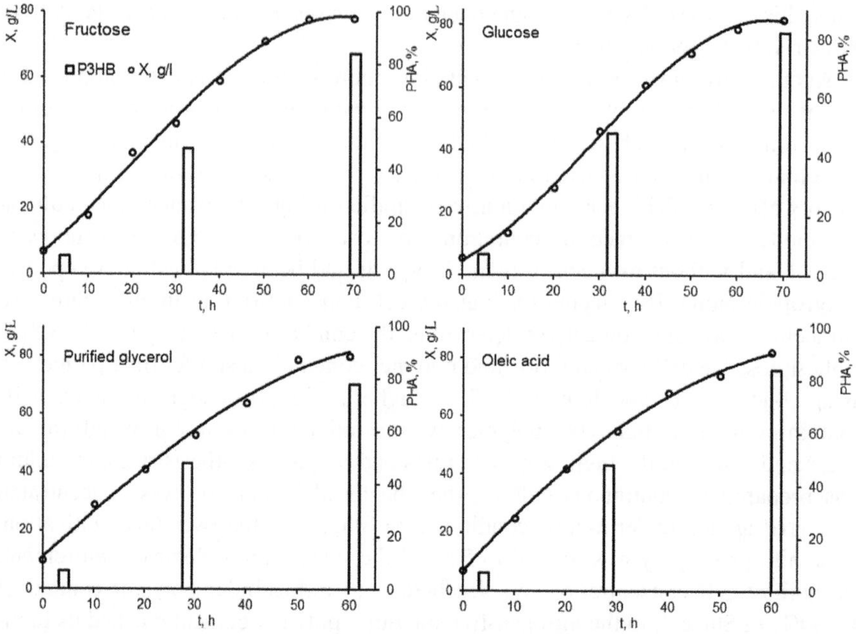

FIGURE 10.3 Dynamics of accumulation of cell biomass and P(3HB) in a 30 L fermenter used to cultivate various wild-type strains of hydrogen-oxidizing bacteria from different C-substrates: a – fructose, b – glucose; c – purified glycerol; d – oleic acid.

FIGURE 10.4 A photo of the fermentation line of the PHA pilot production.

(Biotron, South Korea) for sterilizing fermenters and connecting lines, a compressor (Remeza, Belarus) for air supply, a 30 L seed culture fermenter (Bioengineering AG, Switzerland), a 150 L production fermenter (Bioengineering AG, Switzerland), an ultrafiltration unit (Vladisart, Russia) to concentrate the culture, and a unit for freeze drying of the condensed bacterial suspension (LP10R ILSHIN C, South Korea).

The synthesis of P(3HB) was scaled up to pilot production in a 150 L fermenter on sugars (fructose and glucose) and purified and crude glycerol in a culture of the *C. eutrophus* B-10646 wild-type strain. Over 60 h of cultivation, a cell concentration of 150–160 g/L was obtained on purified glycerol and glucose; cultivation on fructose and crude glycerol resulted in a cell concentration of 130±10 g/L. Polymer content and yield coefficients for the biomass were similar on all substrates (80–85 wt.% and 0.29–0.33 kg biomass/kg carbon substrate, respectively) (see Table 10.1).

Varying C-substrates significantly affect the cost of the polymer. Costs of different carbon substrates used in the present study for polymer production are compared in Table 10.2.

Carbon substrates (sugars or glycerol) are consumed at comparable rates, about 3 kg/kg polymer. Fructose has the highest cost – $5.6/kg polymer. The replacement of fructose by purified glycerol decreases the cost of the carbon substrate by 43%, and by crude glycerol, by 57%. Glucose is less expensive than fructose, and if glucose is replaced by glycerol, the cost of the carbon substrate decreases by 14% and 33.3%, depending on whether purified or crude glycerol is used. However, the substrate cost is still too high relative to the polyethylene price.

TABLE 10.1
Summary of Fed-Batch Cultures in a 150 L Fermenter with Different Substrates at the End of Cultivation: Biomass Productivity (Px) and Yield Coefficient of the Polymer (Y_p)

Substrate	Biomass, g/L	Polymer, wt.%	P_x, g/(L·h)	Y_p, g PHA/ g substrate
Glucose	150 ± 10	85 ± 5	2.4 ± 0.1	0.28 ± 0.02
Fructose	130 ± 10	80 ± 5	2.2 ± 0.1	0.25 ± 0.02
Purified glycerol	160 ± 10	85 ± 5	2.5 ± 0.2	0.29 ± 0.01
Crude glycerol	130 ± 10	85 ± 5	2.3 ± 0.2	0.27 ± 0.02

TABLE 10.2
Costs of Carbon Substrates for Production of 1 kg P(3HB)

Carbon substrate	Substrate cost, $/kg	Substrate consumption [kg/kg P(3HB)]	Substrate unit cost [$/kg P(3HB)]
Fructose, EU	1.50	3.70	5.6
Glucose, China	1.00	3.57	3.6
Purified glycerol, Dutch glycerol refinery, Netherlands	0.89	3.45	3.1
Crude glycerol, Prisma Comercial Exportadora de Oleoquímicos LTDA, Brazil	0.66	3.70	2.4

These results allowed us to obtain the initial data and scale up the process in order to design a larger-scale polymer production facility. By using the collection of natural strains of hydrogen-oxidizing bacteria with different organotrophic potential, we managed to synthesize PHA from different substrates (mixtures of hydrogen and carbon dioxide, sugars, glycerol, organic acids synthesis gas obtained by gasification of hydrolysis lignin, brown coals, natural gas conversion, etc.).

10.3 SYNTHESIS OF PHA OF DIFFERENT COMPOSITION

The best studied and most common PHA, poly(3-hydroxybutyrate), however, has high crystallinity (75±5%), and the products based on this material have rather low mechanical strength and are prone to "physical aging" [38]. Synthesis of PHA copolymers, which are more readily processable materials, is a very complicated technological task. The culture medium has to be supplemented with precursor substrates, which are toxic to bacterial cells and thus, reduce both cell concentration and polymer yield [39–41]. PHA copolymers are usually produced by recombinant strains or in mixed cultures.

Bacterial culture synthesizing PHA copolymers is a very complex system controlled by various factors. First, for PHA synthesis to occur, a sufficient amount of carbon substrate should be present in the culture medium; at the same time, substrate concentration should be maintained within the physiological range for each specific strain, to avoid both substrate deficiency and its inhibitory effect. Second, to promote PHA accumulation, one of the substrates of constructive metabolism (for the study strain, this is nitrogen) must limit cell growth. Third, the culture medium must contain precursor substrate concentrations that would be sufficient to enable formation of the target monomer units but would not profoundly inhibit the cell culture. Synthesis of PHA containing medium-chain-length monomers presents an even greater challenge. Precursor substrates (hexanoate, etc.) are metabolized in the fatty acid cycle, breaking down into shorter carbon chains, which prevents accumulation of high concentrations of medium-chain-length monomers in the polymer. Therefore, in order to facilitate incorporation of 3-hexanoate into the P(3HB-*co*-3HHx) copolymers, reactions of fatty acid β-oxidation should be blocked by, e.g. sodium acrylate, which, however, may inhibit cell growth. Thus, it was necessary to study the relationship between specific growth rate of the cells and sodium acrylate concentration and estimate the influence of this substrate on the yield of P(3HB-*co*-3HHx) copolymers and their 3-hydroxyhexanoate fraction and on the total production of bacterial biomass.

PHA synthase – one of the key enzymes of PHA synthesis – catalyzes formation of ester bonds during polymerization of monomers [42]. Based on the notion of substrate specificity of synthases, all known types of PHA have been divided into three groups: short-chain-length (*scl*-PHA), medium-chain-length (*mcl*-PHA), and long-chain-length (*lcl*-PHA) PHA. *Scl*-PHA are composed of monomers consisting of three to five carbon atoms (C3–C5), *mcl*-PHA of C6–C14, and *lcl*-PHA – more than C14 [43]. To provide the evidence for the ability of hydrogen-oxidizing bacteria to synthesize medium-chain-length polymers, we investigated PHA synthase of the study strain. The PHA synthase gene of *R. eutropha* B5786 was cloned and characterized, and molecular structure of the enzyme was compared with PHA synthases of several strains accumulating *scl-co-mcl* PHA [44]. Homology of the PHA synthase of *R. eutropha* B5786 to the PHA synthase of *R. eutropha* H16 was 99%. Homology of the *R. eutropha* B5786 synthase to the synthases of some strains producing *scl-co-mcl* PHA was between 26% and 41%. Thus, no direct relationship was found between molecular organization of PHA synthases and their functions, namely, their ability to synthesize PHA with certain structures. Wild-type strains of other taxa, which have PHA synthases with broadened substrate specificity, similar to *R. eutropha* B5786 and *R. eutropha* H16 synthase, were isolated and described as well [45–48].

Based on the authors' knowledge that 1) rates of intracellular synthesis and accumulation of 3-hydroxybutyrate differ from the rates of synthesis of monomers with a longer C-chain; 2) precursor substrates inducing formation of PHA copolymers are toxic to bacteria and, thus, their concentrations must be no more than the limits determined for each strain; 3) the precursor substrate should be fed to the culture medium when PHA synthesis rate is the highest (10–15 h of fermentation) and concentration of the major carbon substrate in the culture medium is the lowest; 4) the fraction of the second monomer incorporated into the 3-hydroxybutyrate chain reaches its maximum in a certain time period after the precursor is fed to the medium (8 to 15 h); and 5) the monomer ratio in the PHA is not a constant value because of polymer endogenous metabolism; and hence, the fraction of a second, third, etc. monomers other than 3HB decreases over time while the 3HB fraction increases, we developed two modes of precursor substrate feeding. The first mode was periodic feeding of the precursor substrate to the culture medium, within the safe limits determined for each substrate. The second mode was continuous feeding of the precursor substrate to the medium. By using the continuous mode, we achieved the following results: a) as the diluted valerate solution was fed in small doses (tenth and hundredth parts of mg/h/L culture medium), the substrate was not toxic to cells; b) the steady and continuous feeding of the valerate resulted in the stable valerate fraction in the copolymer and the stable monomer ratio; c) the length of the fermentation period was increased (in previous experiments, because of the instability of the monomer ratio in the copolymer, cultivation had to be stopped 10–15 h after the addition of the valerate, and that produced a negative effect on cell concentrations and copolymer yields).

The wild-type and mutant strains with broadened organotrophic potential, tolerant to various carbon substrates, were used to synthesize bi-, ter-, and quaterpolymers, including structurally new polymers. The detailed investigation of the polymers showed a relationship between their structure and physicochemical properties. The results of the research conducted for the past few years are summarized in Table 10.3. In the early phase of the research, we faced a serious challenge of constructing copolymers with major fractions of monomer units other than 3-hydroxybutyrate. However, later, after the strains were adapted to the precursor substrates and the mode of feeding the substrates to the medium was optimized, we managed to achieve considerably better results.

By adding valerate or γ-butyrolactone periodically or continuously to the cell culture synthesizing PHA, we produced PHA copolymers containing major molar fractions of two monomers (3HB-*co*-4HB, 3HB-*co*-3HV) that contained up to

TABLE 10.3
Composition and Properties of PHA Synthesized by Wild-Type Strains of Hydrogen-Oxidizing Bacteria from Different Substrates

Strain/substrate and precursor	Polymer [%]	PHA composition, mol.%						M_w [kDa]	D	C_x, %	T_{melt}, °C	T_{degr}, °C	T_g, °C	T_c, °C	Ref.
		3HB	4HB	3HV	3HHx	3H4MV	DEG								
R. eutropha B5786/CO_2:H_2:O_2	65–69	99.2	–	t.a.	–	–	–	1220	1.6	76	180	280	n.d.	n.d.	[49]
R. eutropha B5786/fructose	80–90	100	–	–	–	–	–	1300	1.6	76	n.d.	n.d.	n.d.	n.d.	[50]
C. eutrophus B–10646/ CO_2:H_2:O_2	85	99.1	–	t.a.	–	–	–	922	2.5	76	179	295	n.d.	n.d.	[51]
C. eutrophus B–10646/glucose	85	100	–	–	–	–	–	740	2.5	72	178	296	n.dt.	95	[52]
R. eutropha B5786/syngas	75	100	–	–	–	–	–	432	n.d.	65	168	200	n.dt.	112	[6]
R. eutropha B5786/syngas	75	100	–	–	–	–	–	720	n.d.	70	167	196	n.dt.	110	[6]
C. eutrophus B–10646/ fructose	85	100	–	–	–	–	–	690	2.6	74	176	293	n.dt.	94	[52]
C. eutrophus B–10646/ purified glycerol	82	100	–	–	–	–	–	370	3.5	52	174	294	2.8	97	[52]
C. eutrophus B10646/crude glycerol	80	100	–	–	–	–	–	304	3.5	52	172	295	2.7	99	[53]
Copolymers															
C. eutrophus B–10646/ CO_2:H_2:O_2:valerate	79	35.6	–	64.3	t.a.	–	–	1111	3.5	53	153	249	n.d.	n.d.	[51]
C. eutrophus B–10646/ CO_2:H_2:O_2:γ–butyrolactone	70	48.2	51.3	t.a.	t.a.	–	–	837	4.0	38	169	291	n.d.	n.d.	[51]
C. eutrophus B–10646/ CO_2:H_2:O_2: hexanoate	71	79.3	–	t.a.	20.0	–	–	421	3.0	42	164	256	n.d.	n.d.	[51]

(Continued)

TABLE 10.3 (CONTINUED)
Composition and Properties of PHA Synthesized by Wild-Type Strains of Hydrogen-Oxidizing Bacteria from Different Substrates

Strain/substrate and precursor	Polymer [%]	PHA composition, mol.% 3HB	4HB	3HV	3HHx	3H4MV	DEG	M_w [kDa]	D	C_x, %	T_{melt}, °C	T_{degr}, °C	T_g, °C	T_c, °C	Ref.
C. eutrophus B–10646/ fructose: γ–butyrolactone	70	25.0	75	–	–	–	–	700	4.7	30	158	295	n.dt.	88	[54]
C. eutrophus B–10646/ purified glycerol: valerate	85	62.0	–	38.0	–	–	–	330	2.9	46	172 157	280	–0.9	67 65	[52]
C. eutrophus B-10646/purified glycerol:ε-caprolactone	69	90.2	–	9.8	–	–	–	290	2.6	46	162 151	263	-0.4	77 74	[53]
C. eutrophus B-10646/purified glycerol:γ-butyrolactone	79.2	–	20.8	–	–	–	–	315	2.9	43	174 161	281	–0.6	73 71	[52]
C. eutrophus B-10646/purified glycerol: propionate	70	64.3	28.5	–	–	–	–	273	2.7	45	154 166	291	1.1	89 87	[53]
C. eutrophus B-10646/ fructose:diethylene glycol	80–85	97.0	–	–	–	–	3.0	72 4205	2.6 1.2	58	162 172	276	n.d.	n.d.	[55]
Terpolymers															
C. eutrophus B-10646/ fructose: 4-methylvalerate	20–70	80.7	–	8.0	–	11.3	–	691	4.2	41	144 163	297	0.61	68	[56]
R. eutropha B5786/ CO_2:H_2:O_2: hexanoate	49	79.6	–	1.5	18.0	–	–	n.d.	n.d.	53	155	253	n.d.	n.d.	[40]
C. eutrophus B-10646/ CO_2:H_2:O_2:valerate: hexanoate	70	67.3	–	7.6	25.1	–	–	970	3.5	40	164	263	n.d.	n.d.	[51]

(Continued)

TABLE 10.3 (CONTINUED)
Composition and Properties of PHA Synthesized by Wild-Type Strains of Hydrogen-Oxidizing Bacteria from Different Substrates

Strain/substrate and precursor	Polymer [%]	PHA composition, mol.%						M_w [kDa]	D	C_x, %	T_{melt}, °C	T_{degr}, °C	T_g, °C	T_c, °C	Ref.
		3HB	4HB	3HV	3HHx	3H4MV	DEG								
C. eutrophus B-10646/ CO_2:H_2:O_2: propionate: γ-butyrolactone	65	56.2	23.6	20.2	–	–	–	645	3.7	22	171	280	n.d.	n.d.	[51]
C. eutrophus B-10646/ glucose: propionate: γ-butyrolactone	70	26.2	60.4	13.4	–	–	–	507	3.4	17	158	274	n.d.	n.d.	[57]
C. eutrophus B-10646/ glucose:valerate: hexanoate	63	84.6	–	1.8	13.6	–	–	924	4.1	63	172	270	n.d.	n.d.	[57]
C. eutrophus B-10646/purified glycerol: valerate:γ-butyrolactone	75	58.5	19.4	22.1	–	–	–	310	2.7	36	169	283	-4.7	68	[52]
C. eutrophus B-10646/ fructose:valerate: γ-butyrolactone	70	27.4	55.6	17.0	–	–	–	540	3.4	26	166	296	n.dt.	25	[54]
C. eutrophus B-10646/ glucose:valerate: hexanoate	n.d.	66.4	–	23.4	10.2	–	–	540	3.5	60	176	271	1.3	52	[58]
C. eutrophus B-10646/ glucose:4-methylvalerate	n.d.	68.5	–	24.9	–	6.6	–	592	3.5	49	142 158	295	1.1	65	[58]

(*Continued*)

TABLE 10.3 (CONTINUED)
Composition and Properties of PHA Synthesized by Wild-Type Strains of Hydrogen-Oxidizing Bacteria from Different Substrates

Strain/substrate and precursor	Polymer [%]	PHA composition, mol.%						M_w [kDa]	D	C_x, %	T_{melt}, °C	T_{degr}, °C	T_g, °C	T_c, °C	Ref.
		3HB	4HB	3HV	3HHx	3H4MV	DEG								
					Quaterpolymers										
C. eutrophus B-10646/ glucose:valerate: hexanoate:γ-butyrolactone	n.d.	85.2	5.3	7.4	2.1	–	–	562	4.7	36	167	285	-4.7	75	[58]
C. eutrophus B-10646/ glucose:valerate: hexanoate:γ-butyrolactone	n.d.	63.5	12.3	19.4	4.8	–	–	437	6.1	30	168	286	-8.6	84	[59]
C. eutrophus B-10646/ fructose:valerate: hexanoate:γ-butyrolactone	n.d.	74.7	10.1	12.7	2.5	–	–	580	4.5	44	165	290	-3.1	70	[54]

t.a. – trace amounts, n.d. – not determined, n.dt. – not detected

64 and 75 mol.% 3HV and 4HB, respectively (see Figure 10.5, Table 10.3). The first data showing the ability of the wild-type strain *R. eutropha* B5786 to synthesize P(3HB-*co*-3HHx) copolymers with 3HHx reaching 10–16 mol.% when grown in autotrophic culture on CO_2 substrate were reported by Volova *et al.* [60]. The study comparing two strains – *R. eutropha* B5786 and *R. eutropha* H16 – showed that the H16 strain was also capable of synthesizing P(3HB-*co*-3HHx) [40]. The molar fraction of 3HHx in P(3HB-*co*-3HHx) reached 50 mol.% in experiments with the *Cupriavidus eutrophus* B10646 strain grown heterotrophically on fructose, in the medium supplemented with sodium hexanoate and sodium acrylate (the latter blocking reactions of the fatty acid oxidation cycle and preventing the carbon chain of hexanoic acid from shortening) [61]. Green *et al.* [62] also showed that the wild-type *R. eutropha* H16 was capable of synthesizing *scl-co-mcl* PHA containing 3HHx. In the copolymers synthesized by the cells grown on octanoic acid as a carbon source, in the medium supplemented with different concentrations of sodium acrylate, the molar fraction of 3HHx reached 5.7 mol.%. The less studied P(3HB)/DEG diblock copolymers containing diethylene glycol (DEG) from 0.2 to 3.0 mol.% were synthesized as well (Table 10.3).

PHA terpolymers were synthesized by cultivating the cells in the same modes, but with two precursor substrates fed to the medium (see Figure 10.6 and Table 10.3). Thus, PHA terpolymers synthesized in autotrophic and heterotrophic modes consisted of major fractions of three monomer units (3HB/4HB/3HV, 3HB/3HV/3HHx)

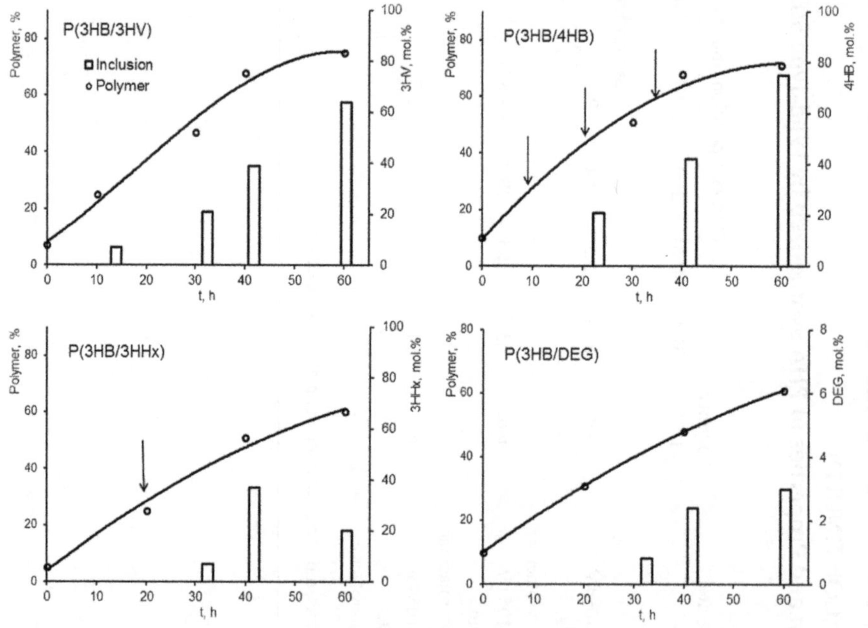

FIGURE 10.5 Synthesis of PHA copolymers (with 3HV, 4HB, 3HHx, DEG) in fed-batch culture of wild-type strains of hydrogen-oxidizing bacteria. Arrows show additions of the precursors to the culture medium (sodium valerate, γ-butyrolactone, sodium hexanoate and sodium acrylate, diethylene glycol).

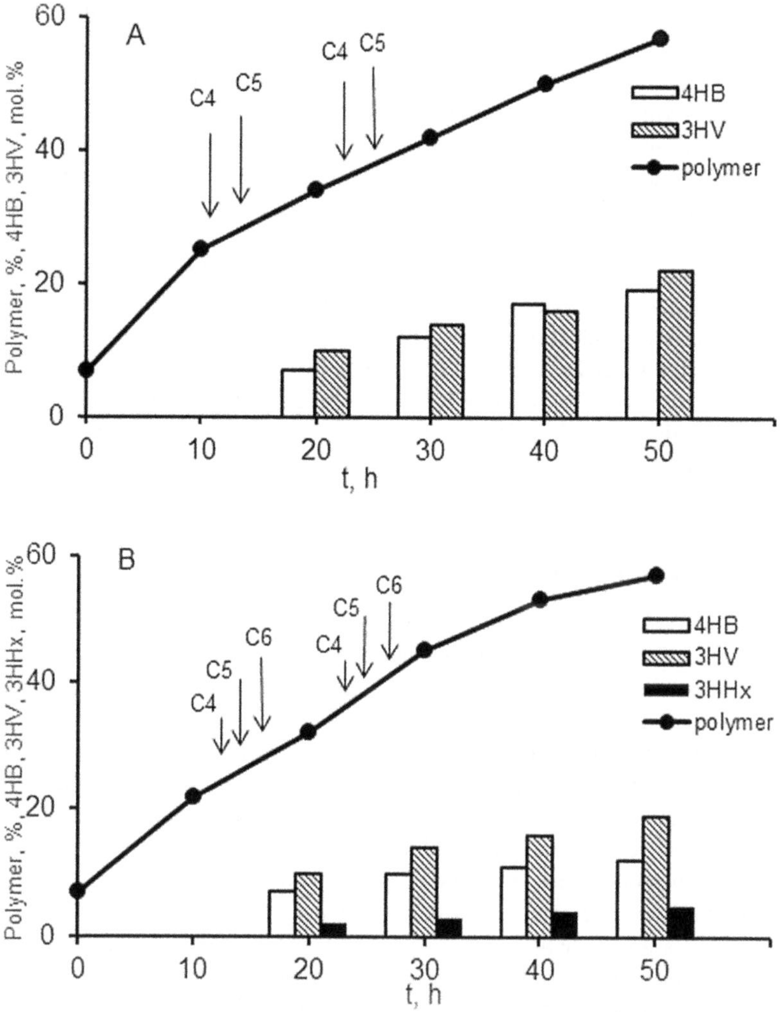

FIGURE 10.6 Synthesis of P(3HB/3HV/4HB) terpolymer (A) and P(3HB/3HV/4HB/3HHx) quaterpolymer (B) in batch culture of the wild-type hydrogen-oxidizing bacterium. Arrows show additions of the precursors to the culture medium (C5 – sodium valerate, C4 – γ-butyrolactone, C6 – sodium hexanoate + sodium acrylate).

and contained 1.5–26, 5–60, and 2–25 mol.% 3HV, 4HB, and 3HHx monomer units, respectively [51,57].

Finally, following the previously developed procedure, *Cupriavidus eutrophus* B10646, the strain tolerant to various precursor substrates, was used to synthesize a series of polyhydroxyalkanoate (PHA) quaterpolymers composed of the short-chain-length 3-hydroxybutyrate (3HB), 4-hydroxybutyrate (4HB), and 3-hydroxyvalerate (3HV) and the medium-chain-length 3-hydroxyhexanoate (3HHx) [59] (Figure 10.6, Table 10.3). The molar fraction of 3HB in the quaterpolymers varied between 63.5

and 93.1 mol.%, 3HV between 1.1 and 24.6 mol.%, 4HB between 2.4 and 15.6 mol.%, and 3HHx between 0.4 and 4.8 mol.%. PHA terpolymers containing 3-hydroxy-4-methylvalerate from 2.7 to 11.3 mol.% [P(3HB/3HV/3H4MV)] were synthesized using 4-methylvaleric acid as a precursor substrate [56].

It has been known that among *Cupriavidus* (*Ralstonia*) species, there are strains capable of synthesizing both poly(3-hydroxybutyrate) and PHA copolymers containing short-chain-length monomers (3-hydroxybutyrate and 3-hydroxyvalerate) and medium-chain-length monomers (3-hydroxyhexanoate and 3-hydroxyoctanoate) [38, 40, 62–68]. However, synthesis of PHA copolymers by *Cupriavidus* species has only been achieved under heterotrophic conditions, with sugars or organic acids, palm oil, biodiesel production wastes, and other feedstocks used as the main growth C-substrate. In a study by Volova *et al.* in 2013 [51], a strain of hydrogen-oxidizing bacterium, *Cupriavidus eutrophus* B-10646, was cultivated under autotrophic conditions; the cells synthesized PHA copolymers with a different chemical structure with CO_2 as the main carbon substrate as effectively as from heterotrophic substrates [54, 57, 59].

10.4 PHA PROPERTIES

Considerable research efforts in biotechnology are directed toward finding polymer-producing strains and determining growth conditions in order to produce polymers with different chemical compositions. The main goal is to create polymer materials with tailored properties. Properties of PHA are determined by their structure, chiefly the structure of the lateral groups in the polymer chain and the distance between ester groups in a molecule. The database collected by researchers of the Siberian Federal University and the Institute of Biophysics SB RAS shows that the properties of PHA vary considerably depending on the type of monomers in the polymer chain and their proportions. However, despite the obvious significance of studies on PHA with different chemical compositions, the available evidence is far from being exhaustive. Moreover, the literature data on the effect of chemical composition of the PHA on its molecular weight, degree of crystallinity, and temperature parameters vary greatly, even in descriptions of the studies using PHA with similar chemical compositions [38]. The discrepancies between the data reported by different authors at different times are likely to be associated with dissimilarities in PHA specimens used: they were synthesized by different producers on different media, the amounts of residual impurities in the samples were not equal, the polymers were processed by various techniques, and, finally, the authors employed diverse methods. The ambiguity and insufficiency of the available data on the effect of the chemical composition of a PHA on its basic properties prompted the authors of this study to investigate this aspect.

PHA composed of different types of monomer units and containing dissimilar proportions thereof were recovered from bacterial cells, purified following the same procedure, and examined using the available equipment. Thus, an accurate relationship between the composition of the PHA and its properties was obtained (Table 10.3).

The properties of the homogenous PHA differed from those of copolymers and were determined by the type of carbon substrate. PHA specimens produced under autotrophic cultivation of bacterial cells showed the highest weight average molecular weight (1 000 kDa or higher) and the lowest polydispersity. The most important parameter, the degree of crystallinity, was the highest in the P(3HB) specimens synthesized in autotrophic culture. P(3HB) specimens synthesized by cells grown on organic substrates, especially on glycerol, exhibited reduced M_w and C_x values.

In almost all types of PHA copolymers, polydispersity was considerably increased, and the amorphous regions prevailed over the crystalline ones. However, some of the specimens containing 3-hydroxyvalerate had the M_w close to the M_w of the specimens produced on CO_2. It is important that in almost all types of PHA copolymers, although their molecular weight and degree of crystallinity decreased, the difference between the melting point and the thermal degradation temperature remained unchanged, i.e. the polymers were thermally stable. Copolymers composed of 3HB and 3HV or 4HB and synthesized from glycerol as the main carbon substrate showed altered thermal behavior: two peaks were registered in the melting region. That may indicate the presence of mixtures of different fractions of the copolymers in the samples, which are separated during melting. Diblock copolymers containing 3HB, DEG, and 3-hydroxy-4-methylvalerate showed similar thermal behavior.

The most significant changes were observed in the degree of crystallinity of PHA bi-, ter, and quaterpolymers. The extent to which C_x decreased varied considerably, depending on the type and concentration of monomers other than 3HB. For instance, incorporation of 3-hydroxyvalrate monomers into the carbon chain of 3-hydroxybutyrate decreased the C_x of the copolymer regardless of 3HV concentration, but it did not drop below 50%. In the copolymers containing 4HB or 3HHx, the C_x decreased more noticeably, to 30–40%.

More complex PHA – terpolymers and quaterpolymers – had lower M_w and higher polydispersity, but their temperature parameters did not change considerably. Their degree of crystallinity decreased substantially (below 40–50%) at lower concentrations of monomer units other than 3-hydroxybutyrate, in contrast to PHA copolymers.

Thus, productive strains of hydrogen-oxidizing bacteria cultivated both under autotrophic and heterotrophic conditions in the medium supplemented with appropriate dosages of precursor substrates were able to synthesize PHA with different chemical compositions and tailored properties, including degree of crystallinity, molecular weight, and temperature characteristics.

10.5 PHA APPLICATIONS

At the present time, the use of PHA is limited by their rather high cost, but their application areas are continuously expanding. The growing importance of environmental protection, on the one hand, and the possibility of reducing the cost of biopolymers by increasing the production efficiency, on the other, make polyhydroxyalkanoates promising materials for the 21st century.

There are two possible ways to broaden PHA applications. One way is to develop large-scale PHA production, i.e. to increase PHA outputs and reduce their cost by manufacturing inexpensive items such as packaging materials, everyday articles, films and pots for agriculture, etc. The other way is to establish small-scale facilities for the production of high-cost specialized items. PHA show the greatest potential in medicine and pharmacology. The mild immune response to PHA implants and the sufficient duration of PHA degradation in biological media make these polymers attractive candidates for use as drug carriers in controlled-release drug delivery systems, implants and grafts for tissue and organ regeneration, and materials for tissue engineering and designing of bioartificial organs.

The Institute of Biophysics SB RAS and Krasnoyarsk State University (Siberian Federal University since 2007) were the first in Russia to start biomedical studies of PHA and experimental products made from them. Between 2000 and 2006, IBP SB RAS and SFU, in cooperation with the V.I. Shumakov Center of Transplantology and Artificial Organs (Moscow, Russia), conducted studies of films and monofilament sutures made from poly(3-hydroxybutyrate) (P(3HB), aka. PHB) and P(3HB-*co*-3HV) copolymers (PHBV) and found that they neither produced cytotoxic effects when contacting with the cultured cells nor caused negative tissue response when implanted *in vivo* for long periods of time. High purity polymer specimens were found to be suitable for contact with blood [69–71]. The results of these studies were summarized in the first Russian book on PHA: *Polyhydroxyalkanoates – Biodegradable Polymers for Medicine* [5, 72].

Since then, the scope of its PHA studies has been considerably widened. New types of PHA were synthesized and used to construct experimental films, barrier membranes, granules, filling materials, ultrafine fibers produced by electrospinning, solid and porous 3D implants for bone tissue defect repair, tubular biliary stents, mesh implants modified by PHA coating, microparticles for drug delivery, etc. A large number of those studies were performed between 2009 and 2014, within the framework of the project "Biotechnologies of novel biomaterials" in accordance with Resolution No. 220 of the Government of the Russian Federation of April 9, 2010, "On measures designed to attract leading scientists to the Russian institutions of higher learning". The project was headed by one of the leading experts in biotechnology of biomaterials and bioengineering, Prof. Anthony John Sinskey (MIT, US).

Films prepared from PHA that differed in their chemical structure, containing different monomers in dissimilar proportions – 3-hydroxybutyrate, 3-hydroxyvalerate, 3-hydroxyhexanoate, and 4-hydroxybutyrate – were tested in cell cultures and implanted to laboratory animals for long time periods. They were found to be highly biocompatible and effective as scaffolds for cell cultures and as wound dressings for skin defect repair [73–77]. An advanced method – electrospinning from polymer solutions [78] – was used to produce nonwoven membranes composed of variously oriented ultrafine fibers from polymers with different chemical compositions. The membranes differed in their structure and physical/mechanical properties and were effective as scaffolds for cell growth and differentiation [79].

P(3HB-*co*-4HB) membranes prepared by electrospinning and composed of ultrafine fibers facilitated cell attachment and proliferation of NIH 3T3 fibroblasts *in vitro*. Those membranes were successfully tested as scaffolds for cell cultures and dressings for burn wounds [80]. Another type of hybrid biotechnological wound dressing was constructed using two biomaterials: P(3HB-*co*-4HB) and bacterial cellulose (BC). The *in vitro* study showed that the most effective scaffolds for growing fibroblasts were composite BC/P(3HB/4HB) films loaded with actovegin. Two types of the experimental biotechnological wound dressings – BC/P(3HB/4HB)/actovegin and BC/P(3HB/4HB)/fibroblasts – were tested *in vivo*, on laboratory animals with model third-degree skin burns. Wound planimetry, histological examination, and biochemical and molecular methods of detecting factors of angiogenesis, inflammation, type I collagen, and keratin 10 and 14 were used to monitor wound healing. Experimental wound dressings promoted healing more effectively than VoskoPran – a commercial wound dressing [81].

Poly(3-hydroxybutyrate) and composites of the polymer with ceramics (hydroxyapatite, wollastonite) were used to prepare osteoplastic filling material and 3D implants. Experiments were performed on laboratory animals, and ectopic bone formation assay and models of segmental osteotomy proved that the polymer had osteoconductive and osteoinductive properties. Experiments with animals with chronic osteomyelitis showed that powdered P(3HB) and P(3HB)/tienam used as filling materials were suitable for filling bone cavities infected with *Staphylococcus aureus*. Filling of infected defects of long bones, which had been subjected to surgical debridement, with hydrophobic material resulted in quicker subsidence of infection, regeneration of bone defects, and recovery of the supportability of the affected limb than in the control group of animals (with defects filled with bone allograft). Biodegradable 3D implants and P(3HB)-based filling materials showed adequate osteoplastic properties and degraded *in vivo* at a slow rate, enabling normal reparative osteogenesis [82–85]. P(3HB) 3D implants shaped as flat plates and subjected to laser cutting to make them more porous were seeded with the primary osteoblast culture differentiated from adipose tissue MSCs. The experimental grafts had adequate osteoplastic properties, were slowly degraded *in vivo*, and completely closed the model defects of cranial flat bones in laboratory animals over 120 d [86].

Innovative research was conducted to provide the data on *in vivo* PHA degradation. PHA monofilament sutures, 3D constructs, films, and microparticles were implanted in laboratory animals for long periods of time (3–6 months, to 12 months). PHA devices were implanted intramuscularly, subcutaneously, into the bone tissue, and into internal organs (microparticles produced from the polymer synthesized on ^{13}C-labeled carbon substrate were administered intravenously). The studies showed that PHA biodegradation occurred via humoral and cellular pathways, involving mononuclear macrophages and foreign-body giant cells, and was determined by the chemical composition of the polymer, the technique employed to produce the implant, the shape of the implant, and the site of implantation. That was a slow controllable process [87, 88].

PHA solutions, emulsions, and powders were used to construct polymer drug carriers shaped as films, pressed pellets, and microparticles. Their biocompatibility and release kinetics were studied *in vitro* and *in vivo*. The drugs were released slowly, and the rate of release did not increase sharply in the early phase of the experiment, suggesting that PHA are promising materials for constructing controlled drug delivery systems. Microcarriers loaded with various drugs were constructed, and their biocompatibility and the kinetics of release of the active ingredient from the degrading matrix were studied *in vivo*. Polymer microparticles were successfully administered intramuscularly, subcutaneously, and intravenously [89–92]. Sustained-release controlled microsystems for delivering cytostatic drugs were constructed and administered locally (to the tumor site), significantly inhibiting tumor development [93, 94]. Transdermal systems containing non-steroidal anti-inflammatory drugs (NSAIDs) were constructed and tested in treatment of burns [95].

A new line of research is enhancing the functional properties of the surface of polymer scaffolds and implants. In cooperation with the L.V. Kirensky Institute of Physics SB RAS and Tomsk Polytechnic University, we studied the effects of physicochemical treatment (laser cutting, plasma treatment) on the structure and properties of polymer surfaces and produced scaffolds with enhanced adhesive properties, facilitating attachment of cells cultured *in vitro* [96–99].

In cooperation with the Voino-Yasenetsky Krasnoyarsk State Medical University, we started limited clinical trials of experimental polymer products (modified meshes for hernia repair, absorbable monofilament sutures, biliary stents, bone replacement implants, and filling material). In clinical trials, hernias of the anterior abdominal wall were treated surgically, using PHA-coated mesh implants, resulting in better clinical outcomes after hernia repair. Degradable P(3HB) monofilament was successfully used to suture musculo-fascial wounds and create various intestinal anastomoses such as entero-entero-, gastro-entero-, and cholecysto-entero-anastomoses. A method of treating obstructive jaundice with varying degrees of extrahepatic biliary obstruction using fully resorbable P(3HB-*co*-3HV) biliary stents was proposed and proven to be clinically effective. 3D implants used to treat patients with chronic osteomyelitis were found to be clinically effective. Hybrid wound dressings – elastic nonwoven ultrafine membranes of P(3HB-*co*-4HB) and bacterial cellulose – were effective in treatment and epithelization of infected wounds. They served as matrices for new tissue formation and performed a number of important functions: provided a barrier against secondary infections, controlled fluid loss, and, at the same time, enabled wound aeration, thus speeding up the healing of the wound. Results of experiments and clinical trials of PHA devices performed by researchers at SFU, IBP SB RAS, and the General Surgery Department at the Voino-Yasenetsky KSMU were summarized in a book by Volova *et al.* in 2017 [8].

The brief review of the results obtained to date suggests that PHA show good potential as materials for biomedical applications. In this sphere, the high functional properties of PHA and the ability to produce much-needed materials, high performance implants, and devices necessary for surgical reconstruction, and to improve clinical outcomes and patients' quality of life outweigh the cost of PHA production.

At the present time, commercial use of PHA as materials for manufacturing degradable packaging and containers may be possible only if the cost of their production is reduced. This is one of the main issues in PHA biotechnology. However, as the trend towards using degradable, including PHA-based, packaging becomes universal, more research effort goes into understanding the biodegradation behavior of these polymers in the complex and ever-changing environment, in ecosystems with different structures.

Results obtained in laboratory experiments cannot be used to construct prognostic models and predict PHA behavior and degradation in diverse and changeable natural ecosystems. This can only be achieved in integrated studies, which should answer the following key questions: how does the microbial community composition in a given environment influence the process of PHA degradation and what microorganisms are the most effective PHA degraders under given conditions? How do the chemical composition of a PHA, the process used to prepare PHA-based devices, and the shape and size of the devices influence the PHA degradation rate? How do the macro- and microstructure of PHA and their properties (crystallinity, molecular weight, polydispersity) change during degradation? Do the physicochemical conditions of the environment (temperature, pH, oxygen availability, salinity, etc.) considerably affect this process? How will the process of PHA degradation be affected by the weather and climate of different regions? Analysis of the available literature shows that rather few authors reported integrated studies of various aspects of PHA degradation, which is a very complex process.

PHA degradation behavior was studied in different environments: Siberian soils under broadleaved and coniferous trees [100], tropical soils (in the environs of Hanoi and Nha Trang) [101], seawater (the South China Sea) [102], a brackish lake (Lake Shira) [103], and freshwater recreational water bodies in Siberia [104]. Those studies showed that degradation occurred at different rates depending on the polymer composition, shape of the specimen (film or 3D construct), climate and weather conditions, and microbial community composition. The time over which the polymer loses 50% of its mass may vary between 68.5 and 270 days in Siberian soils, between 16 and 380 days in tropical soils of Vietnam, between 73 and 324 days in the brackish lake (Shira), between 127 and 220 days in the seawater of the South China Sea, and between 17 and 65.9 days in freshwater lakes. Research showed that PHA biodegradation is influenced by the chemical structure of the polymer, its geometry and the technique used to process it, climate and weather, the type of the natural ecosystem, and its microbial component in particular, as the factor determining the mechanism of PHA biodegradation: preferential attack of the amorphous regions of the polymer or equal degradation of both crystalline and amorphous phases. PHA-degrading microorganisms that dominate microbial populations in some soil and aquatic ecosystems have been isolated and identified. Results of this extensive research can be used as the basis for predicting the ability of a natural ecosystem to rid itself of PHA – a promising packaging material.

The data on PHA degradation behavior in soil provided the basis for another socially and environmentally significant application of these polymers: designing slow-release and targeted formulations for crop protection. We designed a pioneering series of pesticides loaded into a degradable polymer matrix and studied the effectiveness of their use. This research has considerable importance, since the wide use of products of chemical synthesis based on nonrenewable resources has led to an excessive and progressive increase in the amounts of unrecycled waste, causing environmental concerns, and posing a global environmental problem. A possible solution is the wider use of biotechnological tools and methods, which, on the one hand, protect useful biota and increase agricultural production and, on the other, decrease toxic impact on different ecosystems and the entire biosphere. An important scientific task is to provide a fundamental basis for construction and agricultural use of new-generation agrochemicals in order to decrease the risk from uncontrolled spreading and accumulation of chemical products of the technosphere in the biosphere.

This is the goal of the project "Agropreparations of the new generation: a strategy of construction and realization" supported in accordance with the Resolution of the Government of the Russian Federation. The leading scientist in this project is Prof. Sabu Thomas – the Director of the International and Inter University Centre for Nanoscience and Nanotechnology, Mahatma Gandhi University, Kottayam, Kerala, India. Prof. Thomas is a world-renowned scientist and expert in engineering of polymers and polymer nanocomposites. PHA were blended with herbicides and fungicides, slow-release formulations were produced, and their properties were studied [105–108]. To increase the cost-effectiveness of the formulations, available natural materials (clay, peat, wood flour) were used as fillers in the polymer matrix. The kinetics of release of the active ingredients were studied during the degradation of the formulations in soil [109–111]. The effectiveness of fungicidal and herbicidal preparations, differing in biodegradation rates and the kinetics of the release of active substances, has been proven in crop plant systems infected with fusarium and weeds [112,113]. The experimental formulations provide targeted and controlled delivery of agrochemicals to plants. Their use is intended to reduce the rates of application and the risk of uncontrolled spread of xenobiotics in the biosphere. The results were published in a book by Volova *et al.* in 2019 [114].

10.6 CONCLUSIONS AND OUTLOOK

The results of the research suggest the following conclusions:

- Biotechnological processes make it possible to produce new-generation materials that do not accumulate in the biosphere;
- Hydrogen-oxidizing bacteria are a promising PHA-producer for commercial synthesis;
- PHA can be produced from various raw materials, including industrial waste;
- PHA are suitable for the construction of high-performance medical devices to improve the treatment and the quality of life of patients;
- PHA are an expedient candidate for the replacement of synthetic non-degradable plastics as material for degradable packaging and eco-friendly products for agriculture.

ACKNOWLEDGMENTS

This study was financially supported by the projects: "Biotechnologies of novel biomaterials" (Agreement No. 11.G34.31.0013) 2010–2014, "Agropreparations of the new generation: a strategy of construction and realization" (No. 074-02-2018-328) 2018–2020 in accordance with Resolution No. 220 of the Government of the Russian Federation of April 09, 2010, "On measures designed to attract leading scientists to the Russian institutions of higher learning" and by the State assignment of the Ministry of Science and Higher Education of the Russian Federation No. FSRZ-2020-0006.

REFERENCES

1. Pirt SJ. *Principles of Microbe and Cell Cultivation*. New York: John Wiley & Sons 1975.
2. Lütke-Eversloh T, Steinbüchel A. Novel precursor substrates for polythioesters (PTE) and limits of PTE biosynthesis in *Ralstonia eutropha*. *FEMS Microbiol Lett* 2003; 221(2): 191–196.
3. Chen GQ. Industrial production of PHA. In: Chen GQ, Ed., Steinbüchel A, Series Ed., *Plastics from Bacteria*. Berlin, Heidelberg: Springer 2010: 121–132.
4. Koller M. Chemical and biochemical engineering approaches in manufacturing polyhydroxyalkanoate (PHA) biopolyesters of tailored structure with focus on the diversity of building blocks. *Chem Biochem Eng Q* 2018; 32(4): 413–438.
5. Volova TG. *Polyhydroxyalkanoates - Plastic Materials of the 21st Century: Production, Properties, Application*. New York: Nova Science Pub Inc 2004; 282 p.
6. Volova TG. *Hydrogen-Based Biosynthesis*. New York: Nova Science Pub Inc 2009; 287 p.
7. Volova TG, Shishatskaya EI, Sinskey AJ. *Degradable Polymers: Production, Properties, Applications*. Ney: York. Nova Science Pub Inc 2013; 380 p.
8. Volova TG, Vinnik YS, Shishatskaya EI, *et al. Natural-Based Polymers for Biomedical Applications*. Toronto: Appl Acad Press 2017; 440 p.
9. Schlegel HG, Gottschalk G, Von Bartha R. Formation and utilization of poly-β-hydroxybutyric acid by Knallgas bacteria (*Hydrogenomonas*). *Nature* 1961; 191: 463–465.
10. Zavarzin GA. *Litothrophic Microorganisms. (Litotrophnye Mikroorganizmy)*. Moscow: Nauka 1972, 330 p (in Russian).
11. Vandamme P, Coenye T. Taxonomy of the genus *Cupriavidus*: A tale of lost and found. *Int J Syst Evol Microbiol* 2004; 54(6): 2285–2289.
12. *Hydrogen Based Protein Production (Proizvodstvo Belka Na Vodorode)*. Gitel'zon II, Ed., Novosibirsk: Nauka 1980 (in Russian).
13. Volova TG, Kalacheva GS. *Polyhydroksybutyrate – Thermoplastic Biodegradable Polymer (Polioksibutirat – Termoplastichnyy Biodegradiruemyy Polimer)*. Krasnoyarsk: Institute of Biophysics SB AS USSR 1990; 23 p (in Russian).
14. Steinbüchel A, Schlegel HG. Excretion of pyruvate by mutants of *Alcaligenes eutrophus*, which are impaired in the accumulation of poly(β-hydroxybutyric acid) (PHB), under conditions permitting synthesis of PHB. *Appl Microbiol Biotechnol* 1989; 31(2): 168–175.
15. Zavarzin GA. *Hydrogen Bacteria and Carboxydobacteria (Vodorodnye I Karboksidobakterii)*. Moscow: Nauka 1978; 204 p (In Russian).
16. Du C, Sabirova J, Soetaert W, *et al.* Polyhydroxyalkanoates production from low-cost sustainable raw materials. *Curr Chem Biol* 2012; 6(1): 14–25.
17. Możejko-Ciesielska J, Kiewisz R. Bacterial polyhydroxyalkanoates: Still fabulous? *Microbiol Res* 2016; 192: 271–282.
18. Sabbagh F, Muhamad II. Production of poly-hydroxyalkanoate as secondary metabolite with main focus on sustainable energy. *Renew Sust Energ Rev* 2017; 72: 95–104.
19. Khosravi-Darani K, Mokhtari ZB, Amai T, Tanaka K. Microbial production of poly(hydroxybutyrate) from C_1 carbon sources. *Appl Microbiol Biotechnol* 2013; 97(4): 1407–1424.
20. Volova TG, Konstantinova VM, Guseinov OA. The formation of thermoplastic degradable polyhydroxyalkanoates by chemolithotrophic bacteria (Obrazovanie termoplastichnih razrushaemih polioksialkanoatov hemolitotrofnimi bakteriyami). *Biotechnology (Biotehnologiya)* 1992; 5: 81–83.

21. Ishizaki A, Tanaka K, Taga N. Microbial production of poly-d-3-hydroxybutyrate from CO_2. *Appl Microbiol Biotechnol* 2001; 57(1–2): 6–12.
22. Volova TG, Voinov NA. Biosynthesis of biodegradable polymers polyhydroxyalkanoates (PHAs) on hydrogen of different origin: Kinetic aspects. In: Zaikov GE, Jimenez A, Eds. *Quantitative Level of Chemical Reactions.* New York: Nova Science Pub Inc 2003: 173–184.
23. Volova TG, Terskov IA, Sidko FYa. *Hydrogen-Based Microbiological Synthesis.* (*Mikrobiolgicheskiy sintez na vodorode*). Moscow: Nauka 1985; 148 p.
24. Volova TG, Kalacheva GS, Altukhova OV. Autotrophic synthesis of polyhydroxyalkanoates by the bacteria *Ralstonia eutropha* in the presence of carbon monoxide. *Appl Microbiol Biotechnol* 2002; 58(5): 675–678.
25. Volova TG, Zhila NO, Shishatskaya EI. Synthesis of poly(3-hydroxybutyrate) by the autotrophic CO-oxidizing bacterium *Seliberia carboxydohydrogena* Z-1062. *J Ind Microbiol Biotechnol* 2015; 42(10): 1377–1387.
26. Tanaka K, Ishizaki A, Kanamuru T, Kawano T. Production of poly(d-3-hydroxybutyrate) Rom CO_2, H_2, and O_2 by high cell density autotrophic cultivation of *Alcaligenes eutrophus. Biotechnol Bioeng* 1995; 45(3): 268–275.
27. Sugimoto T, Tsuge T, Tanaka K, Ishizaki A. Control of acetic acid concentration by pH-stat continuous substrate feeding in heterotrophic culture phase of two-stage cultivation of *Alcaligenes eutrophus* for production of P(3HB) from CO_2, H_2, and O_2 under non-explosive conditions. *Biotechnol Bioeng* 1999; 62(6): 625–631.
28. Volova TG, Voinov NA. Study of a *Ralstonia eutropha* culture producing polyhydroxyalkanoates on products of coal processing. *Appl Biochem Microbiol* 2004; 40(3): 249–252.
29. Volova TG, Guseinov OA, Kalacheva GS, *et al.* Effect of carbon monoxide on metabolism and ultrastructure of carboxydobacteria. *World Appl Microbiol Biotechnol* 1993; 9(2): 160–163.
30. Tanaka K, Miyawaki K, Yamaguchi A, *et al.* Cell growth and P(3HB) accumulation from CO_2 of a carbon monoxide-tolerant hydrogen-oxidizing bacterium, *Ideonella* sp. O-1. *Appl Microbiol Biotechnol* 2011; 92(6): 1161–1169.
31. Do YS, Smeenk J, Broer KM, *et al.* Growth of *Rhodospirillum rubrum* on synthesis gas: Conversion of CO to H_2 and poly-β-hydroxyalkanoate. *Biotechnol Bioeng* 2007; 97(2): 279–286.
32. Przybylski D, Rohwerder T, Dilßner C, *et al.* Exploiting mixtures of H_2, CO_2, and O_2 or improved production of methacrylate precursor 2-hydroxyisobutyric acid by engineered *Cupriavidus necator* strains. *Appl Microbiol Biotechnol* 2015; 99(5): 2131–2145.
33. Kesler. (Volova) TG, Stasishina GE, Kalacheva GS. (On the possibility of culturing hydrogen bacteria in the presence of CO) (O vozmozhnosti kultivirovaniya vodorodnykh bakteriy v prisutstvii CO). *Microbiology. (Mikrobiologiya)* 1978; 47(1): 17–20 (in Russian).
34. Gruzinsky IV, Gogotov IN, Bechina EM, *et al.* Dehydrogenase activity of hydrogen-oxidizing bacteria *Alcaligenes eutrophus* (Aktivnost degidrogenazy vodorodokislyayushchikh bakteriy *Alcaligenes eutrophus*). *Microbiology. (Mikrobiologiya)* 1977; 46(4): 625–630 (in Russian).
35. Volova TG, Kalacheva GS, Gitelson II, *et al.* A technique of producing a polymer - β-hydroxybutyric acid (Sposob polucheniya polimera—β-oksimaslyanoi kisloty). RF Patent 2003; No. 2207375 (In Russian).
36. Volova TG, Shishatskaya EI. Bacterial strain VKPM B-10646 - A producer of polyhydroxyalkanoates and a method of their production (Shtamm bakterii VKPM B-10646 - produtsent poligidroksialkanoatov I sposob ikh polucheniya). RF Patent: 2012; 2439143 (In Russian).

37. Fernández-Dacosta C, Posada JA, Kleerebezem R, *et al.* Microbial community-based polyhydroxyalkanoates (PHAs) production from wastewater: Techno-economic analysis and ex-ante environmental assessment. *Bioresour Technol* 2015; 185: 368–377.
38. Laycock B, Peter H, Pratt S, *et al.* The chemomechanical properties of microbial polyhydroxyalkanoates. *Prog Polym Sci* 2013; 38(34): 536–583.
39. Zhao W, Chen GQ. Production and characterization of terpolyester poly(3-hydroxybutyrate-*co*-3-hydroxyvalerate-*co*-3-hydroxyhexanoate) by recombinant *Aeromonas hydrophila* 4AK4 harboring genes *pha*AB. *Process Biochem* 2007; 42(9): 1342–1347.
40. Volova TG, Kalacheva GS, Steinbüchel A. Biosynthesis of multi-component polyhydroxyalkanoates by the bacterium *Wautersia eutropha. Macromol Symp* 2008; 269(1): 1–7.
41. Bhubalan K, Rathi DN, Abe H, *et al.* Improved synthesis of P(3HB–*co*–3HV–*co*–3HHx) terpolymers by mutant *Cupriavidus necator* using the PHA synthase gene of *Chromobacterium* sp. USM2 with high affinity towards 3HV. *Polym Degrad Stab* 2010; 95(8): 1436–1442.
42. Rehm BHA. Polyesters synthases: Natural catalysts for plastics. *J Biochem* 2003; 376(1): 15–33.
43. Steinbüchel A, Valentin HE. Diversity of bacterial polyhydroxyalkanoic acids. *FEMS Microbiol Lett* 1995; 128(3): 219–228.
44. Kozhevnikov IV, Volova TG, Hai T, *et al.* Cloning and molecular organization of the polyhydroxyalkanoic acid synthase gene (*phaC*) of *Ralstonia eutropha* strain B5786. *Appl Biochem Microbiol* 2010; 46(2): 140–147.
45. Pieper U, Steinbüchel A. Identification, cloning and sequence analysis of the poly(3-hydroxyalkanoic acid) synthase gene of the Gram-positive bacterium *Rhodococcus ruber. FEMS Microbiol Letts* 1992; 96(1): 73–80.
46. Fukui T, Doi Y. Cloning and analysis of the poly(3-hydroxybutyrate-co-3-hydroxyhexanoate) biosynthesis genes of *Aeromonas caviae. J Bacteriol* 1997; 179(15): 4821–4830.
47. Clemente T, Shah D, Tran M, *et al.* Sequence of PHA synthase gene from two strains *of Rhodospirillum rubrum* and *in vivo* substrate specificity of four PHA synthases across two heterologous expression systems. *Appl Microbiol Biotechnol* 2000; 53(4): 420–429.
48. Zhang S, Kolvek S, Goodwin S, Lenz RW. Poly(hydroxyalkanoic acid) biosynthesis in Ectothiorhodospira shaposhnikovii: Characterization and reactivity of a type III PHA synthase. *Biomacromolecules* 2004; 5(1): 40–48.
49. Zhila NO, Nikolaeva ED, Syrvacheva DA, *et al.* Microbial synthesis and characterization of poly (3-hydroxybutyrate-*co*-4-hydroxybutyrate) copolymers. *J Sib Fed Univ Biol* 2011; 4(2): 158–171.
50. Volova TG, Zhila NO, Kalacheva GS, *et al.* Synthesis of 3-hydroxybutyrate-*co*-4-hydroxybutyrate copolymers by hydrogen-oxidizing bacteria. *Appl Biochem Microbiol* 2011; 47(5): 494–499.
51. Volova T, Kiselev E, Shishatskaya E, *et al.* Cell growth and PHA accumulation from CO_2 and H_2 of a hydrogen-oxidizing bacterium, *Cupriavidus eutrophus* B-10646. *Bioresour Technol* 2013; 146: 215–222.
52. Volova T, Kiselev E, Zhila N, Shishatskaya E. Synthesis of polyhydroxyalkanoates by hydrogen-oxidizing bacteria in a pilot production process. *Biomacromolecules* 2019; 20(9): 3261–3270.
53. Volova TG, Demidenko A, Kiselev E, *et al.* Polyhydroxyalkanoate synthesis based on glycerol and implementation of the process under conditions of pilot production. *Appl Microbiol Biotechnol* 2019; 103(1): 225–237.
54. Zhila NO, Shishatskaya EI. Properties of PHA bi-, ter-, and quarter-polymers containing 4-hydroxybutyrate monomer units. *Int J Biol Macromol* 2018; 111: 1019–1026.

55. Volova T, Zhila N, Kiselev E, Shishatskaya E. A study of synthesis and properties of poly-3-hydroxybutyrate/diethylene glycol copolymers. *Biotechnol Prog* 2016; 32(4): 1017–1028.
56. Volova T, Menshikova O, Zhila N, *et al.* Biosynthesis and properties of P(3HB-*co*-3HV-*co*-3H4MV) produced by using the wild-type strain *Cupriavidus eutrophus* B-10646. *J Chem Technol Biotechnol* 2019; 94(1): 195–203.
57. Volova TG, Kiselev EG, Vinogradova ON, *et al.* A glucose-utilizing strain, *Cupriavidus eutrophus* B-10646: Growth kinetics, characterization and synthesis of multicomponent PHAs. *PLOS ONE* 2014; 9(2): e87551.
58. Volova TG, Vinogradova ON, Zhila NO, *et al.* Physicochemical properties of multicomponent polyhydroxyalkanoates: Novel aspects. *Polym Sci A* 2017; 59(1): 98–106.
59. Volova TG, Vinogradova ON, Zhila NO, *et al.* Properties of a novel quaterpolymer P(3HB/4HB/3HV/3HHx). *Polymer* 2016; 101: 67–74.
60. Volova TG, Belyaeva OG, Gitelzon II, *et al.* Obtain and research of microbial heteropolymeric polyhydroxyalkanoates. *Dok akad nauk* 1996; 347(2): 256–258.
61. Volova TG, Syrvacheva DA, Zhila NO, Sukovatiy AG. Synthesis of P(3HB-*co*-3HHx) copolymers containing high molar fraction of 3-hydroxyhexanoate monomer by *Cupriavidus eutrophus* B10646. *J Chem Technol Biotechnol* 2016; 91(2): 416–425.
62. Green PR, Kemper J, Schechtman L, *et al.* Formation of short chain length/medium length polyhydroxyalkanoate copolymers by fatty acid β-oxidation inhibited *Ralstonia eutropha. Biomacromolecules* 2002; 3(1): 208–213.
63. Chanprateep S, Kulpreecha S. Production and characterization of biodegradable terpolymer poly(3-hydroxybutyrate-*co*-3-hydroxyvalerate-*co*-4-hydroxybutyrate) by *Alcaligenes* sp. A-04J. *J Biosci Bioeng* 2006; 101(1): 51–56.
64. Dai Y, Yuan Z, Jack K, Keller J. Production of targeted poly(3-hydroxyalkanoates) copolymers by glycogen accumulating organisms using acetate as sole carbon source. *J Biotechnol* 2007; 129(3): 489–497.
65. Chanprateep S. Current trends in biodegradable polyhydroxyalkanoates. *J Biosci Bioeng* 2010; 110(6): 621–632.
66. López-Cuellar MR, Alba-Flores J, Gracida Rodríguez JN, Pérez-Guevara F. Production of polyhydroxyalkanoates (PHAs) with canola oil as carbon source. *Int J Biol Macromol* 2011; 48(1): 74–80.
67. Aziz NA, Sipaut CS, Abdullah AAA. Improvement of the production of poly(3-hydroxybutyrate-*co*-3-hydroxyvalerate-*co*-4-hydroxybutyrate) terpolyester by manipulating the culture condition. *J Chem Technol Biotechnol* 2012; 87(11): 1607–1614.
68. Cavalheiro JMBT, Raposo RS, de Almeida MCMD, *et al.* Effect of cultivation parameters on the production of poly(3-hydroxybutyrate-*co*-4-hydroxybutyrate) and poly(3-hydroxybutyrate-4-hydroxybutyrate-3-hydroxyvalerate) by *Cupriavidus necator* using waste glycerol. *Bioresour Technol* 2012; 111: 391–397.
69. Sevastianov VI, Perova NV, Dovzhik IA, *et al.* Sanitary-chemical, toxicological and hemocompatible properties of polyoxyalkanoates - Biodegradable bacterial polymers (Sanitarno-khimicheskie, toksilogicheskie I gemosovmestimye svoystva poligidroksialkanoatov – biorazrushaemykh polimerov). *J Adv Mater (Perspectivnye Materialy)* 2001; 5: 47–55 (in Russian).
70. Shishatskaya EI, Eremeev AV, Gitel'zon II, *et al.* Cytotoxicity of polyhydroxyalkanoates in animal cell cultures. *Dok akad nauk* 2000; 374: 539–542.
71. Volova TG, Shishatskaya EI, Sevastianov VI, *et al.* Results of biomedical investigations of PHB and PHB/PHV fibers. *Biochem Eng J* 2003; 16(2): 125–133.
72. Volova TG, Sevastianov VI, Shishatskaya EI. *Polyhydroxyalkanoates (PHA) – Biodegradable Polymers for Medicine (Polioksialkanoaty (PHA) – Biorazrushaemye Polimery Dlya Meditsiny).* Novosibirsk: SB RAS 2003; 330 p (in Russian).

73. Shishatskaya EI, Volova TG. A comparative investigation of biodegradable polyhydroxyalkanoate films as matrices for *in vitro* cell cultures. *J Mater Sci Mater Med* 2004; 15(8): 915–923.
74. Nikolaeva ED, Shishatskaya EI, Mochalov KE, *et al.* Comparative investigation of polyhydroxyalkanoate scaffolds with various chemical compositions. *Cell Transplant Tissue Eng* 2011; 6(4): 54–63.
75. Shishatskaya EI, Volova TG, Goreva AV, *et al.* An in vivo study of 2D PHA matrixes of different chemical compositions: Tissue reactions and biodegradations. *Mater Sci Technol* 2014; 30(5): 549–557.
76. Borovkova NV, Evseev AK, Andreev YuV, *et al.* Wound dressings made of biodegradable natural polymers polyhydroxyalkanoates (PHAs): Production and properties. *J Sib Fed Univ Biol* 2016; 9(1): 88–97 (in Russian).
77. Borovkova NV, Evseev AK, Makarov MS, *et al.* Study of biocompatible films and nonwoven membranes made of copolymer of 3-hydroxybutyric acid and 4-hydroxybutyric acid in vitro. *J Sib Fed Univ Biol* 2016; 9(1): 43–52 (in Russian).
78. Volova TG, Shishatskaya EI, Gordeev SA. Characterization of ultrathin fibers obtained by electrostatic molding of thermoplastic polyester [poly(hydroxybutyrate/hydroxyvalerate)] (Kharakteristika ultratonkikh volokon, poluchennykh elektrostaticheskim formovaniem termoplastichnogo poliephira [poli(gidroksibutirat/gidroksivalerat)]). *J Adv Mater (Perspectivnye Materialy)* 2006; 3: 25–29 (in Russian).
79. Volova T, Goncharov D, Sukovatyi A, *et al.* Electrospinning of polyhydroxyalkanoate fibrous scaffolds: Effects on electrospinning parameters on structure and properties. *J Biomater Sci Polym Ed* 2014; 25(4): 370–393.
80. Shishatskaya EI, Nikolaeva ED, Vinogradova ON, Volova TG. Experimental wound dressings of degradable PHA for skin defect repair. *J Mater Sci Mater Med* 2016; 27(11): 165.
81. Volova TG, Shumilova AA, Nikolaeva ED, *et al.* Biotechnological wound dressings based on bacterial cellulose and degradable copolymer P(3HB/4HB). *Int J Biol Macromol* 2019; 131: 230–240.
82. Shishatskaya EI. Biocompatible and functional properties of the polyhydroxybutyrate/hydroxyapatite hybrid composite (Biosovmestimye I funktsionalnye svoystva gibridnogo Komposita poligidroksibutirat/gidroksiapatit). *Russ J Transplantol Artif Organs (Vestn Transplantol Iskusstvennyh Organ)* 2006; 8(3): 34–38 (in Russian).
83. Shishatskaya EI, Chlusov IA, Volova TG. A hybrid PHA-hydroxyapatite composite for biomedical application: Production and investigation. *J Biomater Sci Polym Ed* 2006; 17(5): 481–498.
84. Shishatskaya EI, Kamendov IV, Starosvetskiy SI, *et al.* Study of the osteoplastic properties of matrices made from resorbable hydroxybutyric acid polyester. *Cell Transplantology and Tissue Engineering (Kletochnaya transplantalogiya and tkanevaya inzheneriya)* 2008; 3(4): 41–47.
85. Shishatskaya EI, Kamendov IV, Starosvetsky SI, *et al.* An *in vivo* study of osteoplastic properties of resorbable poly-3-hydroxybutyrate in models of segmental osteotomy and chronic osteomyelitis. *Artif Cells Nanomed Biotechnol* 2014; 42(5): 344–355.
86. Shumilova AA, Myltygashev MP, Kirichenko AK, *et al.* Porous 3D implants of degradable poly-3-hydroxybutyrate used to enhance regeneration of rat cranial defect. *J Biomed Mater Res A* 2017; 105(2): 566–577.
87. Shishatskaya EI, Volova TG, Efremov SN, *et al.* Tissue response to the implantation of biodegradable polyhydroxyalkanoate sutures. *J Mater Sci Mater Med* 2004; 15(6): 719–728.
88. Shishatskaya EI, Volova TG, Gordeev SA, Puzyr AP. Degradation of P(3HB) and P(3HB–*co*–3HV) in biological media. *J Biomater Sci Polym Edn* 2005. V; 16(5): 643–657.

89. Shishatskaya EI. Biodegradation of PHA *in vivo*. *J Sib Fed Univ Biol* 2016; 9(1): 21–32 (in Russian).
90. Shishatskaya EI, Voinova ON, Goreva AV, *et al.* Biocompatibility of polyhydroxybutyrate microspheres: *In vitro* and in vivo evaluation. *J Mater Sci Mater Med* 2008; 19(6): 2493–2502.
91. Shishatskaya EI, Goreva AV, Kalacheva GS, Volova T. Biocompatibility and resorption of intravenously administered polymer microparticles in tissue of internal organs of laboratory animals. *J Biomater Sci Polym Ed* 2011; 22(16): 2185–2203.
92. Goreva AV, Shishatskaya EI, Volova TG, Sinskey AJ. Characterization of polymeric microparticles based on resorbable polyesters of oxyalkanoic acids as a platform for deposition and delivery of drugs. *Polym Sci A* 2012; 54(2): 94–105.
93. Goreva AV, Shishatskaya EI, Kuzmina AM, *et al.* Microparticles prepared from biodegradable polyhydroxyalkanoates as matrix for encapsulation of cytostatic drug. *J Mater Sci Mater Med* 2013; 24(8): 1905–1915.
94. Shishatskaya EI, Goreva AV, Kuzmina AM. Study of the efficiency of doxorubicin deposited in microparticles from resorbable bioplastotane™ on laboratory animals with Ehrlich's solid carcinoma. *Bull Exp Biol Med* 2013; 154(6): 773–777.
95. Murueva AV, Shershneva AM, Sishatskaya EI, Volova TG. The use of polymeric microcarriers loaded with anti-inflammatory substances in the therapy of experimental skin wounds. *Bull Exp Biol Med* 2014; 157(5): 597–602.
96. Volova TG, Tarasevich AA, Golubev AI, *et al.* Laser processing of polymer constructs from poly(3-hydroxybutyrate). *J Biomater Sci Polym Ed* 2015; 26(16): 1210–1228.
97. Syromotina DS, Surmenev RA, Surmeneva MA, *et al.* Oxygen and ammonia plasma treatment of poly(3-hydroxybutyrate) films for controlled surface zeta potential and improved cell compatibility. *Mater Lett* 2016; 163: 277–280.
98. Syromotina DS, Surmenev RA, Surmeneva MA, *et al.* Surface wettability and energy effects on the biological performance of poly-3-hydroxybutyrate films treated with RF plasma. *Mater Sci Eng C* 2016; 62: 450–457.
99. Surmenev P, Syromotina D, Syromotina DS, *et al.* Low-temperature argon and ammonia plasma treatment of poly-3-hydroxybutyrate films: Surface topography and chemistry changes affect fibroblast cells *in vitro*. *Europ Polym J* 2019; 112: 137–145.
100. Boyandin AN, Volova TG, Prudnikova SV, *et al.* Biodegradation of polyhydroxyalkanoates by soil microbial communities of different structures and detection of PHA degrading microorganisms. *Appl Biochem Microbiol* 2012; 48(1): 28–36.
101. Boyandin AN, Prudnikova SV, Karpov VA, *et al.* Microbial degradation of polyhydroxyalkanoates in tropical soils. *Int Biodeterior Biodegrad* 2013; 83: 77–84.
102. Volova TG, Boyandin AN, Vasiliev AD, *et al.* Biodegradation of polyhydroxyalkanoates (PHAs) in tropical coastal waters and identification of PHA-degrading bacteria. *Polym Degrad Stab* 2010; 95(12): 2350–2359.
103. Zhila NO, Prudnikova SV, Zadereev ES, *et al.* Degradation of polyhydroxyalkanoate films in Brackish Lake Shira. *J Sib Fed Univ Biol* 2012; 5(2): 210–215 (in Russian).
104. Volova TG, Gladyshev MI, Trusova MY, Zhila NO. Degradation of polyhydroxyalkanoates in eutrophic reservoir. *Polym Degrad Stab* 2007; 92(4): 580–586.
105. Volova TG, Zhila NO, Vinogradova ON, *et al.* Constructing herbicide metribuzin sustained-release formulations based on the natural polymer poly-3-hydroxybutyrate as a degradable matrix. *J Environ Sci Health B* 2016; 51(2): 113–125.
106. Volova T, Zhila N, Vinogradova O, *et al.* Characterization of biodegradable poly-3-hydroxybutyrate films and pellets loaded with the fungicide tebuconazole. *Environ Sci Pollut Res* 2016; 23(6): 5243–5254.
107. Boyandin AN, Zhila NO, Kiselev EG, Volova TG. Constructing slow-release formulations of metribuzin based on degradable poly(3-hydroxybutyrate). *J Agric Food Chem* 2016; 64(28): 5625–5632.

108. Volova T, Prudnikova S, Boyandin A, *et al.* Constructing slow-release fungicide formulations based on poly(3-hydroxybutyrate) and natural materials as a degradable matrix. *J Agricult Food Chem* 2019; 67(33): 9220–9231.
109. Volova T, Zhila N, Kiselev E, *et al.* Poly(3-hydroxybutyrate)/metribuzin formulations: Characterization, controlled release properties, herbicidal activity, and effect on soil microorganisms. *Environ Sci Pollut Res* 2016; 23(23): 23936–23950.
110. Volova TG, Prudnikova SV, Zhila NO, *et al.* Efficacy of tebuconazole embedded in biodegradable poly-3-hydroxybutyrate to inhibit the development of *Fusarium moniliforme* in soil microecosystems. *Pest Manag Sci* 2017; 73(5): 925–935.
111. Shershneva AM, Murueva AV, Zhila NO, Volova TG. Antifungal activity of P3HB microparticles containing tebuconazole. *J Env Sci Health B* 2019; 54(3): 196–204.
112. Zhila N, Murueva A, Shershneva A, *et al.* Herbicidal activity of slow-release herbicide formulations in wheat stands infested by weeds. *J Environ Sci Health B* 2017; 52(10): 729–735.
113. Volova TG, Prudnikova SV, Zhila NO. Fungicidal activity of slow-release P(3HB)/TEB formulations in wheat plant communities infected by *Fusarium moniliforme. Environ Sci Pollut Res* 2018; 25(1): 552–561.
114. Volova TG, Shishatskaya EI, Zhila NO, *et al. New Generation Formulations of Agrochemicals: Current Trends and Future Priorities.* Toronto: Appl Acad Press 2019; 268 p.

11 Polyhydroxyalkanoates, Their Processing and Biomedical Applications

Emmanuel Asare, David A. Gregory, Annabelle Fricker, Elena Marcello, Alexandra Paxinou, Caroline S. Taylor, John W. Haycock, and Ipsita Roy

CONTENTS

11.1 INTRODUCTION

Innovative biomedical materials that are non-immunogenic, bioresorbable and easily processable for the production of a wide range of biomedical products are highly sought after. Commonly used biomaterials include polylactic acid (PLA), poly(DL) lactic acid (PDLLA), poly(L-lactic acid) (PLLA), poly(lactic acid-*co*-glycolic acid) (PLGA) and polycaprolactone (PCL). These polymers, however, are not suitable for all biomedical applications, as their material properties and degradation rates are not easily tunable. The aforementioned undergo bulk degradation and lead to the release of acidic degradation products that can induce an inflammatory response.

Polyhydroxyalkanoates (PHA) are indeed a very promising group of natural biomaterials, exhibiting structural diversity, bioresorbability, biocompatibility and surface degradation properties that distinguish them as excellent candidates for various biomedical applications. These include tissue engineering applications, implants and controlled drug delivery. PHA are intracellular biopolymers accumulated by many Gram-positive and Gram-negative bacteria and can be produced by bacterial fermentation under nutrient limiting conditions [1]. Classification of PHA is based on the number of carbon atoms present in their monomeric units, which has an influence on their mechanical and thermal properties. Short-chain-length (*scl-*) PHA have 3 to 5 carbon atoms in their monomer units, whereas medium-chain-length (*mcl-*) PHA have 6 to 13 carbon atoms in their monomeric units [1]. Comparatively, *mcl*-PHA are more elastomeric, with lower melting and glass transition temperatures. In addition to this, they also exhibit a much lower level of crystallinity. These features make *mcl*-PHA-based scaffolds suitable for soft tissue engineering applications. On the other hand, *scl*-PHA, like P(3HB), are more brittle and stiff, with high melting temperatures, lower elongation at break values and high crystallinity. Hence, *scl*-PHA are more suitable for hard tissue engineering applications, such as bone tissue engineering. Other applications of PHA include stents, cardiac patches, drug carriers for drug delivery systems, artificial skin and heart valves [2].

The degradation of PHA occurs via surface erosion [3], and by altering their monomer composition, the rate of degradation can be tailored to match the needs of the specific applications. In addition to this, the degradation products are natural metabolites, such as 3-hydroxybutyrate-CoA, or 3-hydroxyacyl-CoAs, making PHA highly biocompatible polymers. In a study conducted by Wu *et al.* [4] to examine the biodegradation of P(3HB-*co*-4HB), they determined that lipases, esterases and other hydrolytic enzymes are responsible for their degradation into hydroxy-acids. In particular, the conformation of 4HB renders it more prone to *in vivo* enzymatic degradation [4].

Several *in vitro* and *in vivo* studies indicate that PHA allow nutrient supply to cells, thereby supporting their growth and maintenance of tissue [5]. PHA also have the ability to provide immuno-isolation for cells, a great advantage in tissue engineering applications [6–8].

The group led by Prof. Roy, based at the University of Sheffield, have shown that P(3HB), P(3HO), P(3HO-*co*-3HD), P(3HO-*co*-3HD-*co*-3HDD) and their blends exhibit excellent cytocompatibility with several cell lines such as HaCaT keratinocytes, C2C12 myoblasts, NG108-15 neuronal cells, BRIN-BD11 pancreatic cells, and MG63 osteoblast like cells [9–13]. Peng *et al.* [14], have shown that members of the PHA family, such as P(3HB), P(3HB-*co*-3HV), P(3HB-*co*-4HB), P(3HB-*co*-3HHx) and P(3HB-*co*-3HV-*co*-3HHx) can support cell proliferation or tissue regeneration without tumor induction [14]. Both *scl*-PHA and *mcl*-PHA have good biocompatibility with epithelial cells [15], adrenocortical cells [16], rabbit aorta smooth muscle cells and smooth muscle cells [17]. All these studies suggest that PHA are highly biocompatible with a vast number of host tissues and cells [18].

PHA received FDA approval in 2007 for their use as biodegradable sutures [19]. A potential pathway for the development of PHA-based tissue engineered constructs for medical applications is illustrated below (Figure 11.1). The first step involves the production of the PHA using microbial fermentation, followed by their extraction and purification. Next the polymer is characterized in order to determine the chemical structure and physico-chemical and thermal properties. Upon the establishment of these properties, appropriate fabrication techniques are explored for the construction of 2D/3D scaffolds of appropriate shape, porosity and dimensions. The scaffolds are then tested *in vitro* for cytocompatibility using suitable cell type. Finally, *in vivo* testing enables the confirmation of their biocompatibility and biodegradability. Knowledge acquired from these studies, along with computer-aided techniques and mathematical modeling allows the production of functional prototypes that can closely match native organs and tissues, ready for clinical trials.

11.2 PROCESSING OF PHA FOR MEDICAL APPLICATIONS

As described above, focus on PHA has heightened in recent years in view of their demonstrated ability to overcome many biomedical challenges. In this section, different processing methods that enable the production of versatile scaffold materials and implants using PHA are highlighted.

11.2.1 Solvent Casting

One of the simplest methods of creating PHA films is solvent casting. This involves the dissolution of the PHA in a solvent and then casting it into e.g., a petri dish, leaving the solvent to volatilize. With this method it is possible to create polymer blends simply by altering the ratios of the polymers dissolved in the solvent. By altering the blend ratios, it is then possible to alter the mechanical properties of the films, which can also affect other properties such as cell viability. Basnett *et al.* [1] give an example of this by blending Poly(3-hydroxyoctanoate), P(3HO) and Poly(3-hydroxybutyrate), P(3HB) in varying ratios to produce films with different properties (Figure 11.2). Blending the two polymers resulted in a control of degradation rate depending on the polymer ratio, where P(3HB) exhibited a higher degradation rate as compared to P(3HO).

11.2.2 Melt Molding

Melt molding is another simple method that allows the control of scaffold pore size, as well as giving the ability to produce different shapes, dependent on the mold design (Figure 11.3). Here, the PHA is mixed in a powder form with e.g., a well-defined monodisperse crystalline NaCl powder to produce an evenly mixed powder. This is then heated under pressure above the polymer T_g, fusing the polymer particles together. The fused construct is then removed from the mold and immersed into e.g., water to dissolve the salt crystals and a well-defined porous scaffold network remains, where the porosity can be tuned, depending on the crystal size and structure of the salt added. An example of this method was described by Baek *et al.* [21],

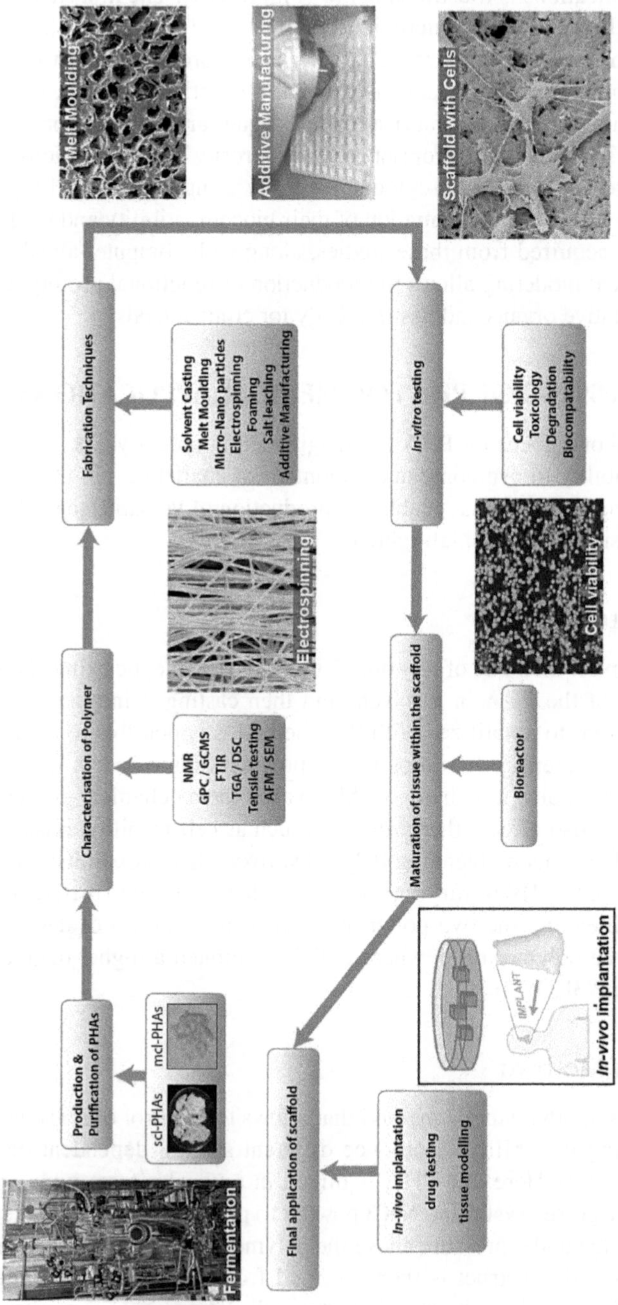

FIGURE 11.1 A schematic of the potential application pathway of the use of PHA in tissue engineering, with adaptations from [20–23].

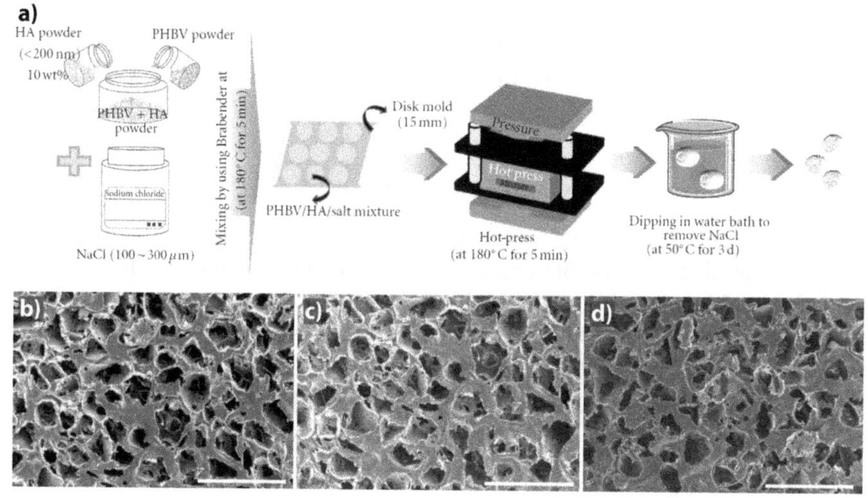

FIGURE 11.2 SEM images of the P(3HO)/P(3HB) blend films revealing a change in their surface topography with changing ratios: (a) neat P(3HO) film, (b) P(3HO)/P(3HB) 20:80, (c) P(3HO)/P(3HB) 50:50 and (d) neat P(3HB) film. Adapted with permission from [1].

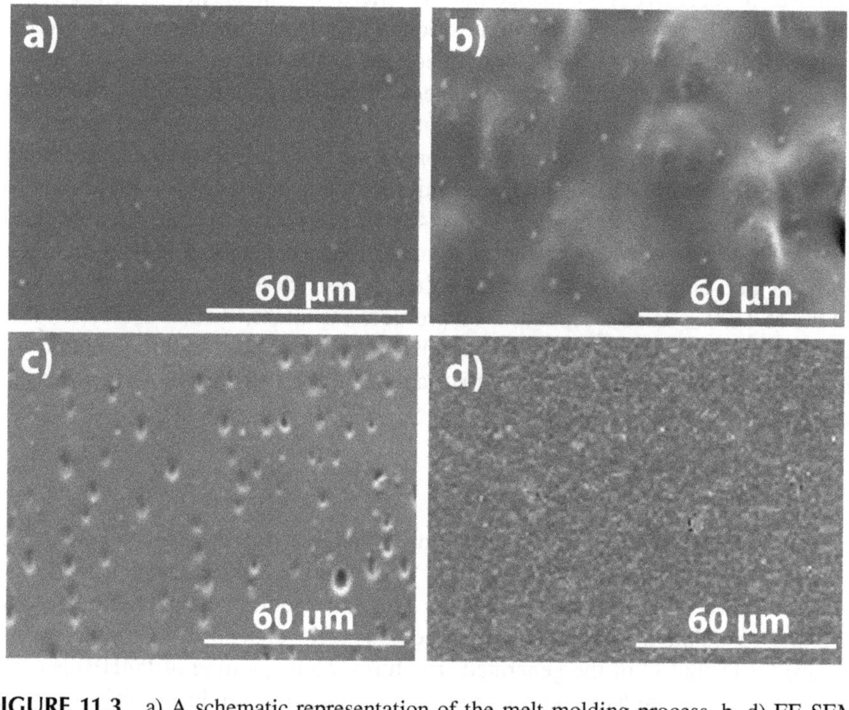

FIGURE 11.3 a) A schematic representation of the melt molding process, b–d) FE-SEM images of porous PHBV (a), PHBV/HA (b), and PHBV/HA/Col (c) scaffolds. Scale bar 1 mm. Adapted with permission from [21].

where salt crystals ranging from 100–300 μm in diameter were mixed with PHBV/hydroxyapatite powder (9:1 w/w) and heated to 180°C. After allowing the salt to leach from the produced scaffold the porous network exhibited pore sizes ranging from a few microns to ~400 μm.

11.2.3 Injection Molding and Film Extrusion

Injection molding and film extrusion are processes that are frequently used in the industry to produce 2D and 3D objects. Injection molding is where a heated polymer is injected into a mold to produce a specific shape for rapid production. Blends can be used to create more porous structures, for example, with the addition of blowing agents. Film extrusion is carried out by heating and ejecting the heated polymer through a specially shaped die, dependent on the final application [24]. This process has been used to create trademark products such as TephaFLEX® by Tepha Inc [25].

11.2.4 Production of Micro- and Nanoparticles

The production of PHA-based colloids is of particular interest when pursuing drug delivery applications [26]. The idea here is to deliver a drug or compound to a specific target at a high dosage, without the side-effects generally associated with a high drug dosage. Drugs are loaded into the micro- or nanoparticles and are gradually released during particle decomposition. The release can also be triggered by environmental conditions, such as temperature or pH changes. There are several methods of producing microspheres; these include the solid-in-oil-water (s/o/w) technique (also known as a Pickering emulsion), which was employed to produce Poly(3-hydroxybutyrate) microspheres for the delivery of gentamicin by Francis *et al.* [27] (Figure 11.4). The produced microsphere sizes ranged from 1.54 μm to 2.00 μm and a drug loading efficiency of 48% was reported. Another double emulsion protocol employed for the fabrication of folate-conjugated P(3HB) nanoparticles for anti-cancer drug delivery resulted in particle sizes ranging from 45 ± 0.4 nm to 1100 ± 2.2 nm dependent on protocol variations, such as sonication time and PVA content addition. Here, the emulsion was stirred until all the organic solvent had evaporated, followed by centrifugation and re-suspension of the pellet into an aqueous phase [28].

11.2.5 Fiber Spinning

There are various methods of fiber spinning which include wet-, dry- and melt-spinning. Here the polymer is either heated or dissolved in a solvent and extruded via a narrow nozzle e.g., a blunt syringe needle. This extrusion can either be into air (dry) or into a coagulation bath (wet). This process is often enhanced with motorization to have CAD control of the generated structures. Melt-spinning of P(4HB) is commercially used to produce sutures or processed further to produce surgical meshes [25, 29].

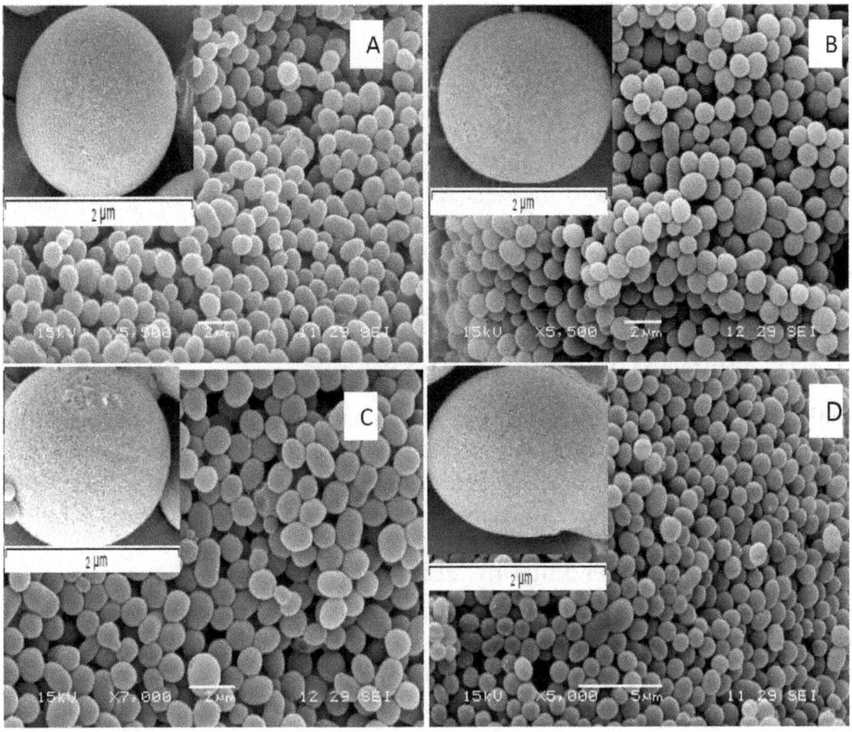

FIGURE 11.4 SEM images of P(3HB) microspheres prepared under different conditions: A) P(3HB) 1 g/L, 0.5% PVA, stirring rate 800 rpm, B) P(3HB) 3 g/L, 0.5% PVA stirring rate 300 rpm, C) P(3HB) 1 g/L, 1.0% PVA stirring rate 300 rpm, D) P(3HB) 3 g/L, 1.0% PVA, stirring rate 800 rpm, reproduced with permission from [27].

11.2.6 Electrospinning

Electrospinning is currently one of the most frequently employed techniques to process polymers in the lab and on an industrial scale. In this technique, a high voltage (ranges from 20–45 kV) is applied across a capillary bearing the polymer solution, essentially pulling it towards the collector electrode. During this process the solvent contained in the polymer solution evaporates and the polymer solidifies into thin fibers that can range from µm to several nm in diameter, depending on the acceleration voltage and polymer/solvent mixture. These micro- and nanostructure systems can be used for several biomedical applications, such as drug release and filtration membranes. Huerta-Angeles *et al.* [22], for example, produced a hybrid copolymer P(3HB)-g-HA (hyaluronic acid) which was soluble in water, allowing a green electrospinning process (Figure 11.5). Variation of the M_w of polymer allowed for tuning of the fibers, where the optimized process resulted in diameter ranges from 100–150 nm, with low polydispersity. In order to produce fibers that were water-insoluble, the copolymer also contained photo cross-linkable moieties that were activated post-electrospinning.

FIGURE 11.5 SEM images of electrospun nanofibers obtained from solutions of A) HA, B) P(3HB)-g-HA (GD = 8.3%) and C) P(3HB)-g-HA (GD = 12.8%). Scale bar: 1 μm. Adapted with permission from [22].

11.2.7 Additive Manufacturing – 3D Printing

Additive manufacturing, which is also known as 3D printing, is a manufacturing approach by which a 3D model is generated in a CAD (computer-aided design) program, and this model is then sliced in a slicer program. The material is then deposited or annealed in a layer-by-layer approach gradually creating the desired 3D shape.

11.2.7.1 Selective Laser Sintering (SLS)

Selective laser sintering (SLS) is an additive manufacturing technique where a laser is scanned across a power bed and used to fuse together plastic powder particles. Once a layer is completed, the bed is lowered and a new layer of powder is added and scanned with the laser again, thus eventually producing a 3D structure. SLS printing appears to have been the first printing technique used for the 3D printing of PHA structures (Figure 11.6) [30, 31]. As this process uses a powder, the fabricated structures tend to have a particulate nature, depending on how well the particles have been fused together via the laser. The pore size within the resulting scaffold depends on the size of the powder as well as the degree of fusion between the powder particles.

11.2.7.2 Fused Deposition Modeling (FDM)

Fused deposition modeling (FDM), also known as 3D extrusion printing, is one of the most well-known types of currently available 3D printing techniques. Extensive

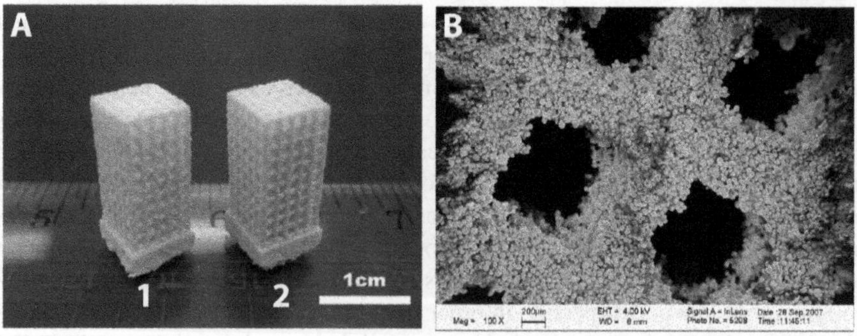

FIGURE 11.6 (A) SLS printed samples of PHBV (1) and Ca-P/PHBV (2), overview picture, (B) SEM close-up of PHBV sample. Adapted with permission from [31].

research has gone into the printing of various polymers such as PLA; however, at present, little has been done with regard to PHA as printing materials. This technology has many similarities to heated fiber spinning, as well as the injection and film extrusion methods described above. There are two major types of extrusion systems, one which uses a filament that passes through a heated nozzle (Figure 11.7a) and the other where pellets or powders are put into a heated syringe and melted, and driven out through a nozzle via a plunger (Figure 11.7c). In addition to these there is also a variant where the heated polymer is fed through a screw (Figure 11.7b) [20]. In recent years, commercial companies have produced filaments of PHA blends with well characterized synthetic polymers, which have been used for research purposes [32]. However, the details of how these filaments are made and their composition are not in the public domain. Very little has currently been published on the production of neat PHA filaments, and the preliminary work has indicated that the filaments are very brittle and that the polymer degraded during the process. Wu *et al.* have reported the production of several types of PHA-based composite material filaments; however, little research has gone into the printing of these filaments [33, 34].

The syringe extruder therefore appears to be more desirable in the case of thermoplastic printing of PHA. Here, the pellet or powder form of the polymers are loaded into a metal syringe-like extruder and then extruded via a piston either mechanically or with a pressure system. However, in this system, it is desirable that the polymer is melted as close to the nozzle as possible to avoid extended elevated temperatures of the PHA material before being extruded, in order to avoid degradation. Successful printing of PHA scaffolds has been pioneered by Prof. Roy's group as shown in Figure 11.8.

11.2.7.3 Computer-Aided Wet-Spinning (CAWS)

Computer-aided wet-spinning is a combination of 3D printing CAD design capability together with traditional wet-spinning. Here the nozzle is controlled via x-and-y motors, together with the syringe pump to enable control of the positioning of the fibers into the coagulation bath [35] (Figure 11.9a). This process allows for much more well-defined structures than traditional wet-spinning and results in the

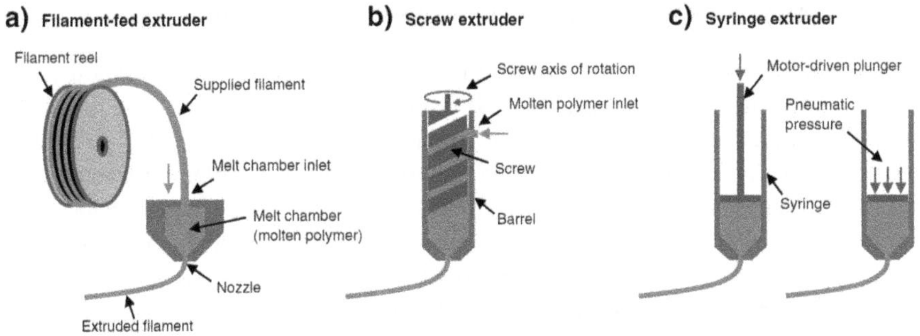

FIGURE 11.7 Three different types of extruders are illustrated schematically. a) Filament-fed extruder, b) Screw extruder, c) Syringe extruders with either a mechanically driven plunger or a pneumatic pressure plunger. Reproduced with permission from Gleadall *et al.* [20].

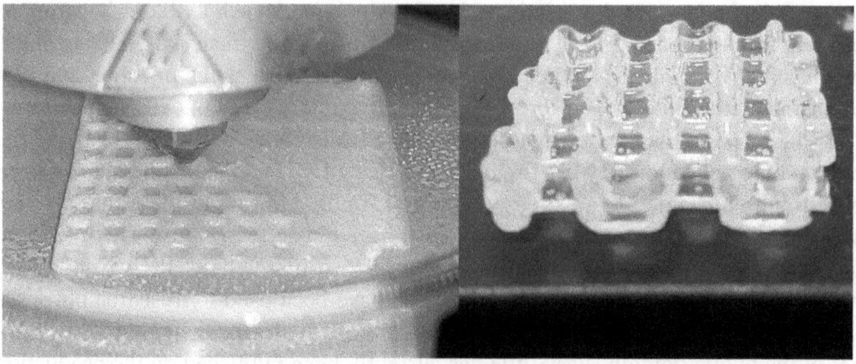

FIGURE 11.8 Melt extruded P(3HB) (left) and P(3HO-*co*-3HD) (right). 3D printed by Gregory and Marcello *et al.* in Prof. Roy's research group.

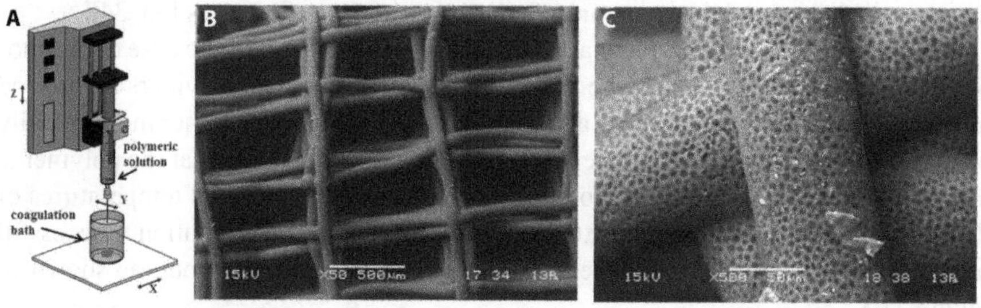

FIGURE 11.9 A) A schematic of CAWS, B) SEM image of top down view of printed lattice of P(3HB-*co*-3HHx), C) High magnification SEM showing surface morphology. Adapted with permission from [36].

production of fine scaffold designs. Puppi *et al.* [36] fabricated P(3HB-*co*-3HHx) scaffolds with critical shape and size regimes for a New Zealand rabbit radius model to investigate optimal bone regeneration conditions.

11.3 APPLICATIONS OF PHA IN NERVE TISSUE ENGINEERING

Despite rapidly advancing surgical methods, full recovery from peripheral nerve injuries (PNI) has yet to be achieved. PNI is a worldwide problem causing long-term disabilities and reduced quality of life in patients. In the USA alone, 20 million people suffer every year from PNIs due to either medical disorders or trauma injuries. These injuries are conventionally treated by means of tensionless end-to-end suturing, allografts and autografts. However, this 'gold standard' has known disadvantages. The harvesting of donor nerves from other sites in the patient requires a second surgery, potentially resulting in depletion of donor nerves themselves or loss of their function. In order to address these limitations, nerve guidance conduits (NGCs), tubular polymeric structures, have been developed. In their application, the proximal and distal stumps of the severed/injured nerve are sutured/glued to each end of the NGC, bridging the nerve defect gap [37]. NGCs can bridge nerve defect gaps of 30–40 mm [38]. The first generation of commercial hollow designs used as nerve conduits were made in 1982 using silicone, a non-resorbable and brittle material which often required a post-surgery procedure for their removal [39]. PGA was the first approved biodegradable material for fabrication of NGCs in 1999, known commercially as a Neurotube®. The latest version available in the market for nerve regeneration, since 2015, is called Reaxon Plus® and is constructed using chitosan [19]. Despite extensive studies and FDA approval, there are issues associated with the current FDA-approved NGCs [38]. Major disadvantages include increased immune reactivity, lack of sufficient biocompatibility and formation of scar tissue, as well as the poor mechanical and degradation features [19].

PHA are attractive natural polymer candidates for nerve tissue engineering applications due to their high biocompatibility and unique properties. They have a range of mechanical properties and degradation rates that allow tailoring of the properties of the NGCs. Hence both *scl*-PHA and *mcl*-PHA have been explored in the development of NGCs, either as the outer tube material, or as aligned internal structures, both as neat PHA and as blends [4].

P(3HB) was the first type of PHA to be used for nerve tissue engineering in 1999. Hazari *et al.* used P(3HB) to bridge a 1 cm nerve defect of sciatic nerves of a rat [40]. Using P(3HB) as a nerve guidance conduit, a 4 cm long nerve defect gap was successfully bridged and the incorporation of Schwann cells in such scaffolds exhibited strong enhancement in the regeneration of the area [41]. Enhancement to these P(3HB) tubular structures was achieved with the addition of fibers in the inner part of the nerve conduit [42] and growth factors such as glial growth factor (GGF) [41]. Besides P(3HB), P(3HB-*co*-3HHx), P(3HB-*co*-4HB) and P(3HB-*co*-3HV), copolyesters which are more elastomeric than P(3HB) have been investigated for their possible use in nerve tissue engineering and have shown good biocompatibility [43, 120]. Xu *et al.* cultured neural stem cells onto P(3HB), P(3HB-*co*-3HV), P(3HB-*co*-3HHx) and P(3HB-*co*-4HB) films and nanofibers and found that PHA nanofibers increased neural stem cell (NSC) adhesion, compared to PHA films, and that P(3HB-*co*-3HHx) nanofiber scaffolds promoted better NSC differentiation compared to the other material fibers [120]. Besides electrospun mats, P(3HB-*co*-3HHx) has also been investigated as the outer tube material for peripheral nerve repair. Bian *et al.* [43] fabricated porous P(3HB-*co*-3HHx) conduits by the dipping-leaching method, using NaCl crystals as porogens. P(3HB-*co*-3HHx) conduits, with non-uniform porosity, were comparable to that of the autograft controls [43]. Blending P(3HB-*co*-3HHx) with other polymers and bioactive molecules, such as graphene, has also been investigated for nerve tissue engineering applications [119].

The use of *mcl*-PHA in nerve tissue engineering has gained attention in recent years due to these biopolymers exhibiting advantageous mechanical properties for soft tissue engineering applications [5]. Hazer *et al.* manufactured P(3HO) tubes using solvent casting and implanted them into 10 mm sciatic nerve defects in Wistar male rats. The study reported favorable mechanical properties and a good tissue response, but the results were not comparable to those of an autograft [44]. P(3HO-*co*-3HD) and P(3HO-*co*-3HD-*co*-3HDD), copolymers of P(3HO), are also of interest in peripheral nerve repair and are currently being investigated.

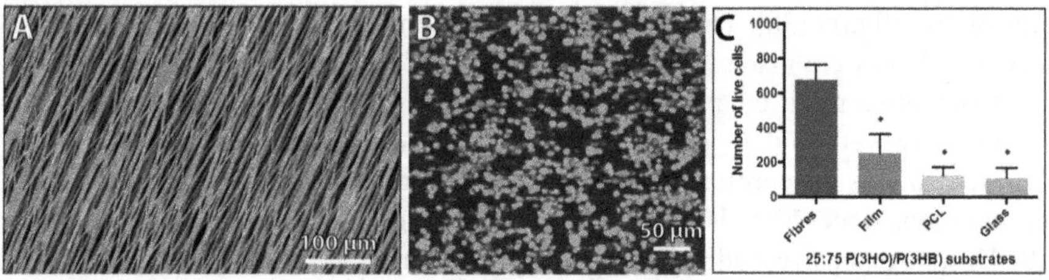

FIGURE 11.10 (A) P(3HO)/P(3HB) 25:75 blend electrospun fibers, (B) NG108-15 cells on an aligned electrospun mat, (C) Comparison of NG108-15 cell proliferation on a range of substrates. Adapted from [23].

The exploitation of blends of PHA have led to the fabrication of novel biodegradable polymers. Lizarraga-Valderrama *et al.* developed P(3HO)/P(3HB) blends with ratios of 75:25, 50:50 and 25:75. Neuronal proliferation was observed in all blends, but the neurite outgrowth was superior in the cases of 25:75 P(3HO)/P(3HB) and P(3HB) compared to other blends [45]. The blend of 25:75 P(3HO)/P(3HB) was fabricated into aligned electrospun fibers, with varying diameters, for use as an internal guidance scaffold for the regenerating nerve. Although all fiber diameters supported NG108-15 neuronal cell adhesion and differentiation, large fibers of 13 μm significantly increased neuronal cell adhesion [23] (Figure 11.10).

To further improve these scaffolds, new designs, surface functionalization, incorporation of electroconductive molecules, growth factors and support cells have all been explored for nerve tissue engineering applications [46, 47]. Utilization of bioresorbable, biocompatible biopolymers with tailored parameters for use as nerve conduits will help improve the motor and function recovery and shorten the rehabilitation time of nerve injuries. For such purposes, PHA demonstrate themselves as perfect candidates for next generation nerve guidance conduits.

11.4 BONE TISSUE ENGINEERING

Hard tissue engineering is one of the most investigated areas for the biomedical applications of PHA. The characteristics of scaffolds suited for bone regeneration are extensively described in the literature [48–51]. Osteo-conductivity is considered one of the central roles of the scaffold, which is described as the ability of bone to grow on the surface of the material [52, 53]. Moreover, the construct should degrade in the human body in a controlled manner to counterbalance the progressive regrowth of the tissue, and its degradation products should be well-tolerated without inducing an inflammatory response. The material needs to have adequate mechanical characteristics to equal those of the surrounding site of implantation. Finally, the scaffold should possess a porous structure with interconnected pores able to favor the flow of nutrients and waste and allow cell growth.

Considering these parameters, several approaches have been investigated for the production of scaffolds for bone regeneration using PHA.

Scl-PHA are considered the best candidates for the regeneration of bone owing to their high crystallinity, Young's modulus and ultimate tensile strength values, comparable to native bone [53]. P(3HB) and P(3HB-*co*-3HV) have been mostly investigated for their use in bone regeneration. However, recently *mcl*-PHA have also been investigated for non-load bearing applications [54]. As PHA are not inherently osteoconductive, most of the research conducted has focused on the development of composite materials using hydroxyapatite (HA), the inorganic component of bone, due to its excellent biocompatibility and bioactivity [55]. Alternatively, the use of bioactive glasses (BG) and carbon nanotubes (MWCNTs) as fillers have also been explored (Figure 11.11). BG are degradable silicate-based glasses that are able to create a strong bond with native bone through the formation of a carbonate layer on their surface upon contact with biological fluids [56, 57]. MWCNT act as a scaffold strengthening filler and also introduces electrical conductivity, usable for non-invasive sensing.

Two main techniques have been employed for the development of the constructs, salt leaching and electrospinning (described in Section 11.2.6). The former is characterized by mixing the polymer with a porogen (e.g., salt or sugar) in suitable solvent (e.g., chloroform for *scl*-PHA). After casting the suspension in an appropriate mold, the scaffold is dried and then placed in water to allow the release of the porogen. The size and the amount of porosity are strictly controlled by the porogen dimensions and the porogen/polymer ratio [59]. In all the studies conducted to develop porous PHA scaffolds, crystal dimensions ranging from 200–300 μm were found to be optimal for the development of interconnected pores of similar dimensions [60–66]. A high ratio of porogen over polymer (80–90:20–10, porogen:polymer) was also necessary to achieve high porosity (average values >75%) [54, 60, 61, 65]. Composite scaffolds can be developed using solvent casting through the incorporation of hydroxyapatite/bioactive glass (HA/BG) in the polymer/porogen mixture. The introduction of the filler within the porous scaffold resulted in a positive effect on the mechanical properties of the constructs, increasing compressive strength and modulus compared to neat P(3HB) and P(3HB-*co*-3HV) scaffolds, obtaining values more similar to the mechanical properties of natural bone [67–70]. The dimension of the filler was found to be a crucial parameter for the reinforcing effect on the polymeric matrix. A few studies have, in fact, demonstrated the ability of nano-HA or BG to create a better interface with the polymeric matrix, due to a higher dispersion compared to

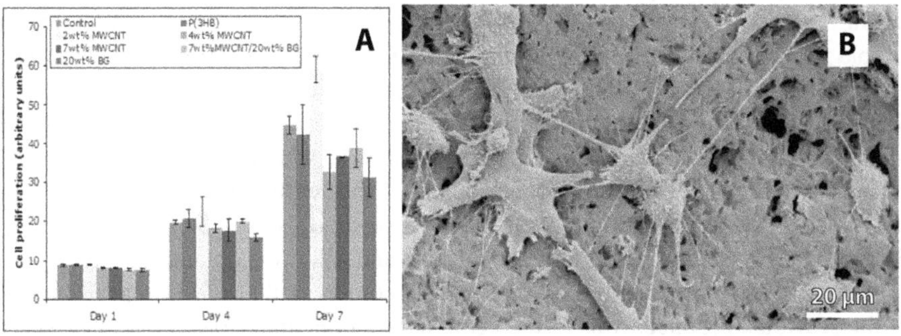

FIGURE 11.11 (A) Comparative study of MG63 proliferation on P(3HB) scaffolds with BG and MWCNT, (B) MG-63 proliferation on P(3HB)-based scaffold adapted from [58].

micro-size particles, which have higher tendency to form agglomerates [64, 72, 73]. The developed composite materials exhibited better *in vitro* compatibility compared to neat materials [65, 66]. In a recent study [71], a different approach was implemented to develop porous composite scaffolds based on P(3HB) and HA, through a thermally induced phase separation technique. Structures with a pore size lower than <50 μm were obtained through phase separation by dissolving P(3HB) in dioxane at a high temperature (100°C), followed by cooling at –20°C and subsequent removal of the solvent. Moreover, composite scaffolds were obtained by *in situ* formation of HA in the polymeric reaction mixture or a simple mixing of HA in the polymer solution during the heating phase leading to non-porous scaffolds [71].

The electrospinning technique described in Section 11.2.6 has also been investigated for the production of PHA-based micro- or nano-fibrous mats for bone regeneration. Such a method is based on the application of a high electric potential to a polymer solution to generate a charge imbalance that leads to the generation and deposition of fibers on a collector [68]. Micro-fibrous matrices with an average fiber diameter between 2–4 μm were obtained using P(3HB), P(3HB-*co*-3HV) and 50/50 blends of these materials [70]. Composite fibers have also been produced by mixing HA or BGs in the polymer solution at a concentration of up to 15 wt.% and electrospinning such a mixture [37, 67, 72]. A novel approach was evaluated by Ramier *et al.* for the incorporation of HA in fibrous matrices by simultaneous electrospinning of a P(3HB) solution and electrospraying of an HA dispersion. In this way, the composite scaffolds produced showed the presence of HA on the surface of the fibers, compared to HA being entrapped within the fibers in the case of electrospinning of a mixture of P(3HB)/HA [37]. Better biological properties were induced by the presence of HA on the surface, but such structures exhibited lower mechanical properties compared to both neat P(3HB) and HA-entrapped P(3HB) mats [37].

Recently, 3D printing technology (described in Section 11.2.7) with PHA has been investigated for the development of bone substitutes to obtain structures with controlled and repetitive geometries. Among the various techniques in additive manufacturing, laser and extrusion-based technologies have been employed to develop PHA-based 3D structures. P(3HB) was successfully printed using Selective Laser Sintering (SLS), which is based on the induction of sintering of particles through the application of a laser beam on layers of powder materials deposited on top of each other [30]. To improve the bioactivity of the obtained 3D P(3HB) scaffolds, Saska *et al.* coated the surface of the material with osteogenic growth factors by simple physical absorption [69]. In another study, Duan *et al.* improved the osteoconductivity of P(3HB-*co*-HV) scaffolds by loading it with calcium phosphates. In this study, the oil-in-water emulsion technique was used to produce microspheres of P(3HB-*co*-3HV) loaded with calcium phosphate, which were then used as the starting powder material for the development of the scaffold using SLS technology [31]. Solution-based 3D printing was also investigated using P(3HB-*co*-3HHx) and BG [74]. In this case, a mixture of the polymer and filler was obtained using organic solvents which was then extruded by applying compressed air. *In vivo* performance of the scaffolds produced was tested using a rat model and showed that the materials were able to stimulate bone repair 8 weeks after implantation [74]. Finally, Yang *et al.* [75] produced composite scaffolds by melt 3D printing of P(3HB-*co*-3HHx), followed by immersion in a solution containing BG to obtain surface functionalization. This material showed enhanced cell proliferation and ALP activity of human mesenchymal stem cells compared to non-loaded P(3HB-*co*-3HHx) [75].

11.5 CARTILAGE TISSUE ENGINEERING

Limited research has been conducted on the use of PHA as scaffolds for cartilage repair. For the regeneration of cartilage tissue, the scaffold needs to be biocompatible, biodegradable and possess appropriate mechanical features. Due to the intrinsic viscoelastic nature of this tissue conferring elasticity and flexibility, the use of blends of *mcl*- and *scl*-PHA was investigated [76–78]. A blend ratio of 1:2 P(3HB): P(3HB-*co*-3HHx) was found to favor better proliferation of chondrocytes isolated from rabbit articular cartilage, compared to that of P(3HB) or blends with a higher ratio of P(3HB) (i.e., 2:1) [79, 80]. This behavior was attributed to the mechanical properties of the material, characterized by a more flexible property, with a lower tensile strength and a higher elongation at break [78]. Such constructs were tested *in vivo* in rabbit articular cartilage defect model, and shown to be able to induce full thickness cartilage repair [66]. Blending of P(3HO) and P(3HB) was also investigated to produce scaffolds using electrospinning. Scaffolds with P(3HB)/P(3HO) ratio of 1:0.25 showed randomly oriented fibers with a diameter close to that of the collagen fibrous network, typical of native cartilage. The material was cultured for three weeks with human articular chondrocytes and showed the deposition of cartilage-like tissue [80, 81]. P(3HB-*co*-3HV) has also been investigated for the development of scaffolds for cartilage repair through the salt leaching technique and the materials showed the capability of inducing early cartilage formation in *in vivo* studies [82–84]. In addition, the combination of P(3HB) with natural polymers (i.e., chitosan and cellulose) was also studied using a range of techniques. Akarayonye *et al.* [85] investigated the use of P(3HB) and micro-fibrillated bacterial cellulose to develop composite scaffolds using a combination of compression molding and salt leaching technique. Electrospinning and lyophilization were employed to blend the polyester with chitosan. All the materials showed good *in vitro* compatibility as they favored the attachment and proliferation of a chondrogenic cell line [85, 86]. Finally, P(3HB-*co*-4HB) was also used for the fabrication of fibrous scaffolds using electrospinning, and such constructs induced the formation of new cartilage-like tissue in *in vivo* tests, showing the potential of this material for cartilage repair [87].

11.6 DRUG DELIVERY APPLICATION

Drug administration by conventional methods is often unable to deliver appropriate concentrations and at the desired location required for the drug to be efficacious. Hence, a higher dosage is often used in order to be able to elicit a therapeutic response which in turn increases the risks of unintended therapeutic side-effects in healthy tissue [88, 89].

Alternatively, there has been an emergence of innovative controlled drug delivery systems in recent times. Basically, a localized controlled drug delivery system involves the encapsulation of a drug or active ingredient in a suitable matrix, that allows a controlled release of the drug to a targeted site of action within the body [89]. Polymeric microspheres/nanospheres have been widely used as microencapsulation vehicles to deliver drugs to cells or tissues of interest. The ideal microencapsulation polymer allows desirable concentrations of drugs to be administered, acts as a protective agent to prevent drugs or active ingredients from degradation within the extracellular milieu and provides predictable drug release kinetics [13, 88, 89]. Another advantage of microsphere/nanosphere encapsulation is the capability of administering multiple drugs within a single injection, since each microsphere/nanosphere-encapsulated drug is separated from the others, thereby avoiding drug compatibility issues encountered with conventional methods [11, 88]. Moreover, the use of targeted drug delivery systems can prevent healthy cells from exposure to drugs and reduce risks of cytotoxicity. An insight into the physico-chemical and biocompatibility properties of the polymer is a prerequisite for the success of controlled drug delivery [13, 27]. Mechanisms for release of the drug from an encapsulating carrier include diffusion and degradation. In the diffusion mechanism, the concentration gradient results in the release of the drug, whereas degradation of the polymer is normally due to hydrolysis which can either occur by bulk or surface erosion.

As previously mentioned, biocompatibility, biodegradability and the ability to tailor their chemical composition are the key characteristics of the versatile PHA family of polymers that make them highly suitable for drug delivery applications [27]. Both *scl*- and *mcl*-PHA have been explored for drug delivery [13, 27]. A study conducted by Gogolewski *et al.* [91] compared polylactides and P(3HB), P(3HB-*co*-3HV) in terms of their degradation profiles. A degradation rate of 50% and 1.6% was observed for the polylactides and PHA respectively, over a period of 6 months. Thus, PHA exhibit a slow degradation rate which has attracted much attention for their exploitation in the design of long-term drug delivery formulations [13, 91]. Also, the ability to tailor their physical properties to allow active factor release rates suited for a particular application, is another attractive feature [100]. Akhtar and Pouton [88] argued that biosynthesized PHA, which inherently lack chemical catalysts and initiators, were better suited to producing safer therapeutic products for drug delivery rather than their chemically synthesized counterparts [88, 89].

Francis *et al.* developed a multifunctional P(3HB)/45S5 Bioglass® composite system for bone tissue engineering, including P(3HB) microspheres with encapsulated gentamicin, coated onto 45S5 Bioglass® scaffolds (Figure 11.12). The composite system exhibited increased compressive strength, bioactivity and surface nanotopography compared to the neat Bioglass® scaffold. Under *in vitro* conditions, the gentamicin release was found to be bimodal, with an initial burst release followed by a diffusion mediated sustained release [90]. Francis *et al.* also successfully encapsulated three model proteins, Bovine Serum Albumin (BSA), Lucentis® and Ribonuclease A (RNase A), within P(3HB) microspheres in order to understand delivery of protein drugs, especially in the intraocular environment, in the case of Lucentis® [11]. Controlled release kinetics of the proteins, maintaining the structural integrity of the proteins was observed, further confirming the potential of PHA to deliver proteins in a controlled fashion [11].

Another study demonstrated the suitability of PHA-based patches in drug delivery with the required adhesiveness and drug permeation for optimal performance

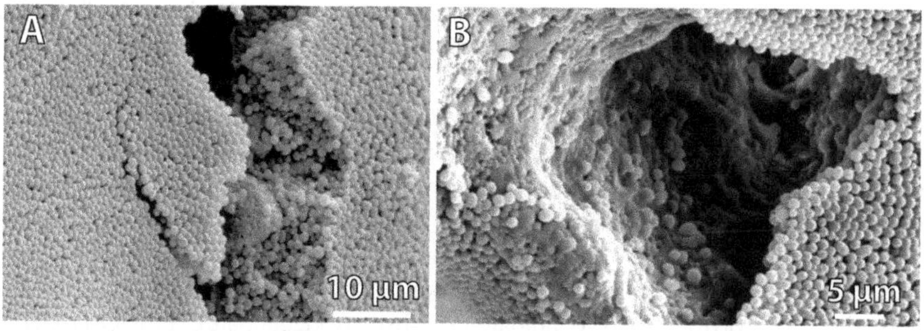

FIGURE 11.12 P(3HB) microsphere with encapsulated gentamycin coated on a Bioglass® scaffold. Adapted from [11, 26].

[95]. Further, PHA rods made from a polymer matrix of P(3HB-*co*-4HB) and P(3HB-*co*-3HV), were cast using drug/polymer pastes, and loaded with different antibiotics, namely sulbactam-cefoperazone, sulbactam-ampicillin and gentamicin. Rods with high drug loading were assessed *in vitro* and a burst release phase followed by a slower protracted release of up to 2 months was observed. The rods were further implanted into rabbit tibias containing *S. aureus* infected metal implants. The PHA rod-antibiotic delivery system resulted in an effective treatment of the infection within a 6-week period, as the antibiotics were released from the rods in a controlled manner [94, 96–98].

The concept of targeted drug delivery can potentially eradicate or substantially reduce the unacceptable effects of chemotherapy for cancer and side-effects of other therapeutic agents [94]. PHA have also been investigated for their prospects in targeted drug delivery. In their research, Yao *et al.* [92] developed a receptor-mediated drug delivery system where PhaP, a PHA granule binding protein, was fused with polypeptide ligands on PHA nanoparticles, encapsulating mostly hydrophobic drugs. *In vitro* studies proved that hepatocellular carcinoma cells, BEL7402, and macrophages preferentially absorbed the drugs upon recognition and binding to their respective receptors [92]. Eldridge *et al.* observed efficient targeted delivery of *Staphylococcal* enterotoxin B formalinized vaccine to the gut-associated lymphoid tissues using P(3HB) microspheres [93].

11.7 CARDIAC TISSUE ENGINEERING

Cardiovascular diseases (CVD) are the largest cause of death globally. Some of the main types include myocardial infarction, coronary artery disease, heart valve diseases, and congenital heart diseases. The human heart has poor capacity for regeneration, and while current treatments do well to mitigate symptoms of these conditions, they do not treat the underlying problems. For example, when someone suffers a myocardial infarction, an area of the heart muscle is subjected to ischemia and the cells in that area die, leaving scar tissue. This scar tissue does not contract, as the necrosed cells cannot conduct the electrical impulse that coordinates heart muscle contraction; therefore, this significantly reduces the functionality of the heart and its blood pumping efficiency. This decrease in functionality results in many sufferers of myocardial infarction dying within a month and many more within the first year [99]. With the current lack of donor hearts for transplantation and current treatments not regenerating the damaged area, there is a need for regenerative medical treatments that can enable the cardiovascular system to regain its function.

The use of biomaterials for cardiac repair requires the materials to have certain mechanical properties. The heart is continuously contracting and relaxing in order to pump blood efficiently around the body, and therefore the heart muscle moves with significant continuous force. *mcl*-PHA are highly flexible, elastomeric polymers with low crystallinity and low glass transition temperatures (T_g) and include polymers such as Poly(3-hydroxyoctanoate-*co*-3-hydroxydecanoate), P(3HO-*co*-3HD) and Poly(3-hydroxyoctanoate), P(3HO) [18, 100–103]. Their flexible properties make these polymers suitable for tissue engineering of organs such as the heart.

PHA have been found to have excellent biocompatibility, as previously discussed, and due to their good mechanical properties they have been and are being investigated for their suitability for many cardiac tissue engineering applications, in particular cardiac repair patches for particular use in post-myocardial infarction sufferers, as well as cardiovascular applications including valve replacements and vascular stents [104–106].

PHA, in particular P(3HB), have previously been researched for their potential use to prevent adhesions after cardiac surgery [107, 108] and also for the augmentation of arteries [109]. Recently there has been focused research into the generation of cardiac patches for the repair of the myocardium in patients who have suffered from a myocardial infarction. The myocardium of a human heart has been reported to have a similar Young's modulus to that of *mcl*-PHA – between 0.02 and 0.05 MPa [110]. Combined with their ability to be processed into complex 3D structures, this makes *mcl*-PHA highly suitable for use as myocardial patches.

P(3HO) has been investigated as a left ventricular cardiac patch for myocardial infarct repair [102] (Figure 11.13). Bagdadi *et al.* analyzed the mechanical properties of P(3HO) at body temperature (37°C) and measured a Young's modulus of 1.5 +/– 0.4 MPa. This value decreased to 0.41 +/– 0.03 MPa with an increased porosity of the patch, putting it in the range of that of the adult human myocardium [102]. This polymer is also highly elastomeric at body temperature, with this study reporting an elongation at break of 447 +/– 5% for the porous patch, enabling it to cope with the regular beating of the heart and thus the continuous contraction and relaxation of the heart muscle. Another essential factor to consider when using PHA for this application is whether cells can adhere to them and subsequently survive and proliferate. This is essential when creating a cardiac repair patch, as healthy cells must be introduced into the area in order to regenerate the tissue that has undergone necrosis. In this study [103], C2C12 cells were cultured on P(3HO) scaffolds and found to not only adhere to the polymer scaffold but also to survive and proliferate (Figure 11.13). The researchers also demonstrated that the incorporation of the cell attachment enhancer RGD-peptide, as well as the vascular endothelial growth factor (VEGF) resulted in better cell adhesion and increased cell proliferation, ideal for a patch that needs to integrate with the host tissue in order to repair an area of damaged tissue.

PCL is another polymer that when blended with PHA has been shown to decrease hydrophobicity and subsequently increase cell adhesion to the polymer. A study [111]

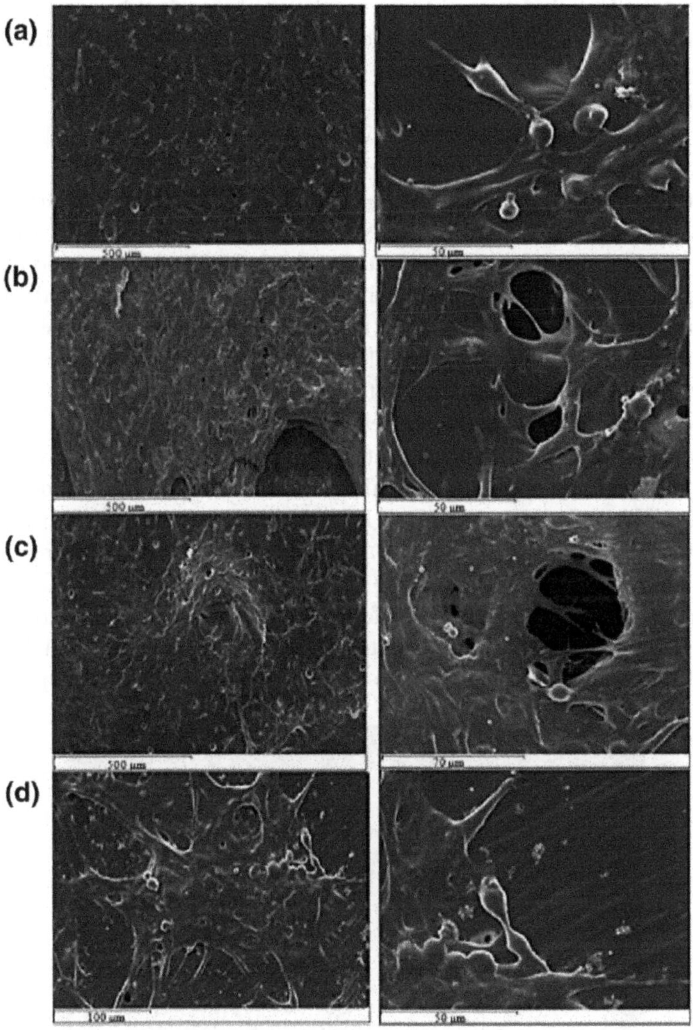

FIGURE 11.13 Scanning electron microscopy (SEM) images of C2C12 cells on a cardiac patch made using P(3HO). (a) non-porous cardiac patch, (b) porous cardiac patch, (c) non-porous patch with additional 750 nm electrospun fibers, (d) porous patch with additional 750 nm electrospun fibers. Reproduced with permission from [103].

which blended *mcl*-PHA with 5% PCL to form a porous film and seeded it with cardiac progenitor cells, containing cardiac stem cells, endothelial cells and fibroblasts, found that cell adhesion was enhanced. When injected into the mouse myocardium, the density of cells on the film was found to decrease, and histological analysis indicated cardiac progenitor cell retention on the polymer scaffold and in the surrounding myocardium.

Another, more well-established cardiac application for PHA is in valve grafts. The current standard treatment to repair valve defects is by replacing them with a mechanical valve; however, these are not very biocompatible and can lead to thrombus formation without anti-coagulation medication [112]. Also, this treatment is not able to change shape and therefore is unable to grow with the patient in the case of a child requiring a valve replacement.

Many studies have been conducted which show the suitability of PHA for this application, including those using P(4HB) and P(3HB-*co*-3HHx) [113]. These PHA have good thermoplastic properties, allowing them to be formed into structures that constitute cardiac valves. Further study showed that porous P(3HO) can also be formed into valve structures and subsequently seeded with vascular endothelial cells [114] (Figure 11.14). This study investigated the polymer when implanted into a lamb pulmonary artery and showed evidence that it was able to grow with the lamb as the valve increased in lumen diameter and length, and it also became more elastomeric over time. The P(3HO) was also shown to decrease in molecular weight over time *in vivo*, suggesting that it was gradually being replaced by viable tissue.

Another option for valve replacements is to use decellularized valves from either human (homograft) or animal (xenograft) donors. These can be coated using PHA such as P(3HB) or P(3HB-*co*-4HB) [115], which research has shown can reduce platelet activation compared to an uncoated xenograft *in vitro*, and therefore reduce the formation of blood clots resulting in thrombosis. Another PHA, P(3HB-*co*-3HHx), has also been investigated and found to again reduce the formation of blood clots in comparison to its uncoated control. A study using P(3HB-*co*-3HHx)-coated valves [116] in sheep also gave evidence that they maintained valvular structure well

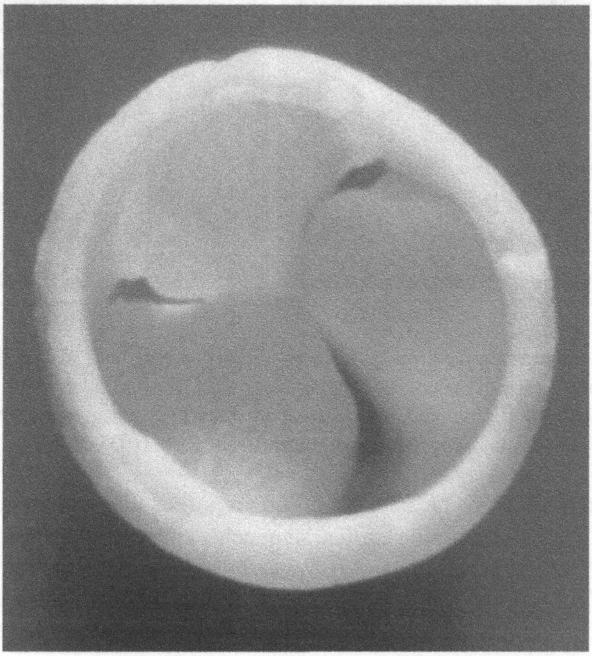

FIGURE 11.14 A tricuspid heart valve made using porous P(4HB). Reproduced with permission from [113].

and had a confluent endothelial monolayer covering the surface, as well as reducing calcification which is seen in non-coated grafts.

Finally, stents are a routine treatment for opening coronary arteries, allowing blood to continue to flow to the heart muscle after restriction. Metallic stents are the standard; however, these can lead to restenosis and require further treatment. The use of a biocompatible and even biodegradable material can overcome this issue, and therefore PHA have been investigated as a potential stent material. In addition, these PHA stents can be made drug-eluting so that they can release anti-proliferative drugs which can help prevent restenosis [117]. A large study conducted within a FP7 project ReBioStent [118] set out to develop a drug-eluting, biocompatible, biodegradable and mechanically suitable stent. Researchers used different blends of P(3HB) and other *mcl*-PHA to find the material with the most suitable mechanical properties. These PHA were molded into tubular structures and then a stent design was created by laser cutting, resulting in a design similar to the metallic stents so that they can be expanded and used to maintain an increased lumen diameter once in place (Figure 11.15). These PHA stents could then be sprayed with the anti-proliferative drugs.

Hence, PHA have been used in many cardiovascular applications and have a huge potential in this medical field. This is of great importance, as cardiovascular disease is the leading cause of death worldwide and the need for effective and long-lasting treatments is vital. The properties of PHA, including mechanical characteristics, processability, biocompatibility and biodegradability, make them a promising material for future cardiovascular applications.

11.8 CONCLUSIONS AND OUTLOOK

PHA have thus been established as natural biopolymers with the potential to make a great impact in biomedical applications including tissue engineering, drug delivery and implant production. Their processability, structural diversity, controlled surface degradation, high biocompatibility (due to their degradation products being natural metabolites), tuneability of material properties and degradation rates make them preferred candidate materials for several medical interventions. Further continuous and effective research is required to unleash the full potential of PHA to the benefit

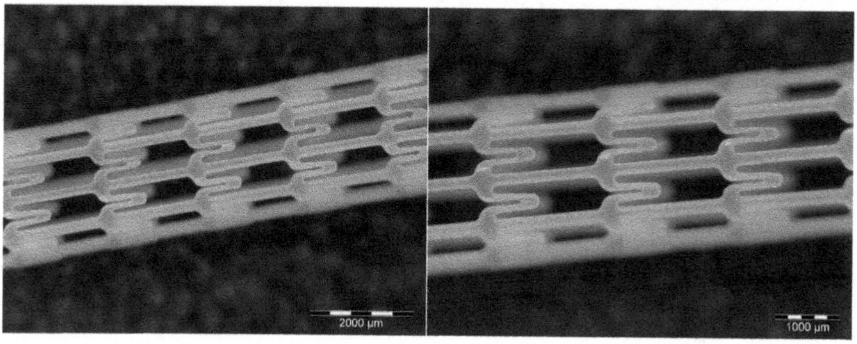

FIGURE 11.15 PHA-based stents developed in ReBioStent. Picture courtesy Dr Iban Quintana, Tekniker, Spain (https://rebiostent.eu).

of society. Currently only one type of PHA, P(4HB), has received FDA approval for application as sutures. The biomedical material properties reviewed herein for PHA and the range of potential applications currently under investigation would support wider FDA approval for PHA-based biomedical implant devices for these applications. In conclusion, PHA are reviewed in this context as an emerging family of biodegradable medical polymers, with tremendous potential to revolutionize the world of biomedical implants.

REFERENCES

1. P. Basnett *et al.*, "Novel poly(3-hydroxyoctanoate)/poly(3-hydroxybutyrate) blends for medical applications," *Reactive and Functional Polymers*, vol. 73, no. 10, pp. 1340–1348, 2013. doi: 10.1016/j.reactfunctpolym.2013.03.019.
2. Y. Deng *et al.*, "Poly(hydroxybutyrate-co-hydroxyhexanoate) promoted production of extracellular matrix of articular cartilage chondrocytes *in vitro*," *Biomaterials*, vol. 24, no. 23, pp. 4273–4281, 2003. doi: 10.1016/s0142-9612(03)00367-3.
3. F. Masood, T. Yasin, and A. Hameed, "Polyhydroxyalkanoates – What are the uses? Current challenges and perspectives," *Critical Reviews in Biotechnology*, vol. 35, pp. 1–8, 2014. doi: 10.3109/07388551.2014.913548.
4. Q. Wu, Y. Wang, and G. Q. Chen, "Medical application of microbial biopolyesters polyhydroxyalkanoates," *Artificial Cells Blood Substitutes and Biotechnology*, vol. 37, no. 1, pp. 1–12, 2009, Art no. Pii 907476591. doi: 10.1080/10731190802664429.
5. S. Ray and V. C. Kalia, "Biomedical applications of polyhydroxyalkanoates," *Indian Journal of Microbiology*, vol. 57, no. 3, pp. 261–269, 2017. doi: 10.1007/s12088-017-0651-7.
6. A. Rodriguez-Contreras, "Recent advances in the use of polyhydroyalkanoates in biomedicine," *Bioengineering*, vol. 6, p. 82, 2019. doi: 10.3390/bioengineering6030082.
7. M. Koller, "Biodegradable and biocompatible polyhydroxy-alkanoates (PHA): Auspicious microbial macromolecules for pharmaceutical and therapeutic applications," *Molecules*, vol. 23, no. 2, 2018, Art no. Unsp 362. doi: 10.3390/molecules23020362.
8. G. F. Bunster, "Polyhydroxyalkanoates: Production and use," *Encyclopedia of Biomedical Polymers and Polymeric Biomaterials*, vol. 11, pp. 6412–6421, 2016. doi: 10.1081/E-EBPP-120049915.
9. L. R. L. Valderrama, R. Nigmatullin, C. Taylor, J. W. Haycock, F. Claeyssens, and I. Roy, "Nerve tissue engineering using blends of polyhydroxyalkanoates for peripheral nerve regeneration," *Tissue Engineering Part A*, vol. 21, pp. S325–S325, 2015. [Online]. Available: <Go to ISI>://WOS:000360205202364.
10. P. Dubey, "Development of cardiac patches using medium chain length polyhydroxyalkanoates for cardiac tissue engineering," PhD, School of Life Sciences, University of Westminster, WestminsterResearch, London, 2017.
11. Lydia Francis, "Biosynthesis of polyhydroxyalkanoates and their medical applications," PhD, School of Life Sciences, University of Westminster, WestminsterResearch, 2011.
12. Moyinoluwa Odugbemi, "Biopolymers for bioartificial pancreas," PhD, School of Life Sciences, University of Westminster, WestminsterResearch, London, 2018.
13. Barbara Lukasiewicz, "Biosynthesis of polyhydroxyalkanoates, their novel blends and composites for biomedical applications," PhD, School of Life Sciences, University of Westminster, WestminsterResearch, London, 2014.
14. S. W. Peng *et al.*, "An assessment of the risks of carcinogenicity associated with polyhydroxyalkanoates through an analysis of DNA aneuploid and telomerase activity," *Biomaterials*, vol. 32, no. 10, pp. 2546–2555, 2011. doi: 10.1016/j.biomaterials.2010.12.051.

15. T. Volova, *Polyhydroxyalkanoates--Plastic Materials of the 21st Century: Production, Properties, Applications.* Nova Publishers, New York, 2004.
16. L. B. Wu, H. Zhang, J. C. Zhang, and J. D. Ding, "Fabrication of three-dimensional porous scaffolds of complicated shape for tissue engineering. I. Compression molding based on flexible-rigid combined mold," *Tissue Engineering*, vol. 11, no. 7–8, pp. 1105–1114, 2005. doi: 10.1089/ten.2005.11.1105.
17. X. H. Qu, Q. Wu, and G. Q. Chen, "*In vitro* study on hemocompatibility and cytocompatibility of poly(3-hydroxybutyrate-co-3-hydroxyhexanoate)," *Journal of Biomaterials Science-Polymer Edition*, vol. 17, no. 10, pp. 1107–1121, 2006. doi: 10.1163/15685620 6778530704.
18. Q. Liu, G. Luo, X. R. Zhou, and G.-Q. Chen, "Biosynthesis of poly(3-hydroxydecanoate) and 3-hydroxydodecanoate dominating polyhydroxyalkanoates by β-oxidation pathway inhibited Pseudomonas putida," *Metabolic Engineering*, vol. 13, no. 1, pp. 11–17, 2011. doi: 10.1016/j.ymben.2010.10.004.
19. S. Kehoe, X. F. Zhang, and D. Boyd, "FDA approved guidance conduits and wraps for peripheral nerve injury: A review of materials and efficacy," *Injury-International Journal of the Care of the Injured*, vol. 43, no. 5, pp. 553–572, 2012. doi: 10.1016/j.injury. 2010.12.030.
20. A. Gleadall, D. Visscher, J. Yang, D. Thomas, and J. Segal, "Review of additive manufactured tissue engineering scaffolds: Relationship between geometry and performance," *Burns Trauma*, vol. 6, p. 19, 2018. doi: 10.1186/s41038-018-0121-4.
21. J.-Y. Baek *et al.*, "Fabrication and characterization of collagen-immobilized porous PHBV/HA nanocomposite scaffolds for bone tissue engineering," *Journal of Nanomaterials*, vol. 2012, pp. 1–11, 2012. doi: 10.1155/2012/171804.
22. G. Huerta-Angeles *et al.*, "Aligned nanofibres made of poly(3-hydroxybutyrate) grafted to hyaluronan for potential healthcare applications," *Journal of Materials Science: Materials in Medicine*, vol. 29, no. 3, p. 32, 2018. doi: 10.1007/s10856-018-6045-5.
23. L. R. Lizarraga-Valderrama, C. S. Taylor, F. C. Aeyssens, J. W. Haycock, J. C. Knowles, and I. Roy, "Unidirectional neuronal cell growth and differentiation on aligned polyhydroxyalkanoate blend microfibres with varying diameters," *Journal of Tissue Engineering and Regenerative Medicine*, vol. 13, no. 9, pp. 1581–1594, 2019. doi: 10.1002/ term.2911.
24. V. Raeisdasteh Hokmabad, S. Davaran, A. Ramazani, and R. Salehi, "Design and fabrication of porous biodegradable scaffolds: A strategy for tissue engineering," *Journal of Biomaterials Science, Polymer Edition*, vol. 28, no. 16, pp. 1797–1825, 2017. doi: 10.1080/09205063.2017.1354674.
25. "Tepha medical devices." https://www.tepha.com/ (accessed 12 December, 2019).
26. D. Meng *et al.*, "Tetracycline-encapsulated P(3HB) microsphere-coated 45S5 bioglass((R))-based scaffolds for bone tissue engineering," *Journal of Materials Science: Materials in Medicine*, vol. 24, no. 12, pp. 2809–17, 2013. doi: 10.1007/s10856-013-5012-4.
27. L. Francis, D. Meng, J. Knowles, T. Keshavarz, A. R. Boccaccini, and I. Roy, "Controlled delivery of gentamicin using poly(3-hydroxybutyrate) microspheres," *International Journal of Molecular Sciences*, vol. 12, no. 7, pp. 4294–314, 2011. doi: 10.3390/ijms12074294.
28. A. Althuri, J. Mathew, R. Sindhu, R. Banerjee, A. Pandey, and P. Binod, "Microbial synthesis of poly-3-hydroxybutyrate and its application as targeted drug delivery vehicle," *Bioresource Technology* vol. 145, pp. 290–6, 2013. doi: 10.1016/j.biortech.2013.01.106.
29. "Galateas surgical." https://www.galateasurgical.com/ (accessed 12 December, 2019).
30. T. F. Pereira, M. F. Oliveira, I. A. Maia, J. V. L. Silva, M. F. Costa, and R. M. S. M. Thiré, "3D printing of poly(3-hydroxybutyrate) porous structures using selective laser sintering," *Macromolecular Symposia*, vol. 319, no. 1, pp. 64–73, 2012. doi: 10.1002/masy.201100237.

31. B. Duan, W. L. Cheung, and M. Wang, "Optimized fabrication of Ca-P/PHBV nanocomposite scaffolds via selective laser sintering for bone tissue engineering," *Biofabrication*, vol. 3, no. 1, p. 015001, 2011. doi: 10.1088/1758-5082/3/1/015001.
32. J. Rydz *et al.*, "3D-printed polyester-based prototypes for cosmetic applications-future directions at the forensic engineering of advanced polymeric materials," *Materials (Basel)*, vol. 12, no. 6, 2019. doi: 10.3390/ma12060994.
33. C. S. Wu and H. T. Liao, "Interface design of environmentally friendly carbon nanotube-filled polyester composites: Fabrication, characterisation, functionality and application," *Express Polymer Letters*, vol. 11, no. 3, pp. 187–198, 2017. doi: 10.3144/expresspolymlett.2017.20.
34. C.-S. Wu, H.-T. Liao, and Y.-X. Cai, "Characterisation, biodegradability and application of palm fibre-reinforced polyhydroxyalkanoate composites," *Polymer Degradation and Stability*, vol. 140, pp. 55–63, 2017. doi: 10.1016/j.polymdegradstab.2017.04.016.
35. D. Puppi *et al.*, "Additive manufacturing of wet-spun polymeric scaffolds for bone tissue engineering," *Biomedical Microdevices*, vol. 14, no. 6, pp. 1115–27, 2012. doi: 10.1007/s10544-012-9677-0.
36. D. Puppi, A. Pirosa, A. Morelli, and F. Chiellini, "Design, fabrication and characterization of tailored poly[(R)-3-hydroxybutyrate-co-(R)-3-hydroxyexanoate] scaffolds by computer-aided wet-spinning," *Rapid Prototyping Journal*, vol. 24, no. 1, pp. 1–8, 2018. doi: 10.1108/rpj-03-2016-0037.
37. J. Ramier *et al.*, "Biocomposite scaffolds based on electrospun poly(3-hydroxybutyrate) nanofibers and electrosprayed hydroxyapatite nanoparticles for bone tissue engineering applications," *Materials Science & Engineering C-Materials for Biological Applications*, vol. 38, pp. 161–169, 2014. doi: 10.1016/j.msec.2014.01.046.
38. C. J. Pateman *et al.*, "Nerve guides manufactured from photocurable polymers to aid peripheral nerve repair," *Biomaterials*, vol. 49, pp. 77–89, 2015. doi: 10.1016/j.biomaterials.2015.01.055.
39. G. Lundborg, R. H. Gelberman, F. M. Longo, H. C. Powell, and S. Varon, "In vivo regeneration of cut nerves encased in silicone tubes: Growth across a six-millimeter gap," *Journal of Neuropathology & Experimental Neurology*, vol. 41, no. 4, pp. 412–422, 1982. doi: 10.1097/00005072-198207000-00004.
40. A. Hazari, M. Wiberg, G. Johansson-Ruden, C. Green, and G. Terenghi, "A resorbable nerve conduit as an alternative to nerve autograft in nerve gap repair," *British Journal of Plastic Surgery*, vol. 52, no. 8, pp. 653–657, 1999. doi: 10.1054/bjps.1999.3184.
41. R. C. Young, M. Wiberg, and G. Terenghi, "Poly-3-hydroxybutyrate (PHB): A resorbable conduit for long-gap repair in peripheral nerves," *British Journal of Plastic Surgery*, vol. 55, no. 3, pp. 235–240, 2002. doi: 10.1054/bjps.2002.3798.
42. P. N. Mohanna, R. C. Young, M. Wiberg, and G. Terenghi, "A composite poly-hydroxybutyrate-glial growth factor conduit for long nerve gap repairs," *Journal of Anatomy*, vol. 203, no. 6, pp. 553–565, 2003. doi: 10.1046/j.1469-7580.2003.00243.x.
43. Y. Z. Bian, Y. Wang, G. Aibaidoula, G. Q. Chen, and Q. Wu, "Evaluation of poly(3-hyd roxybutyrate-co-3-hydroxyhexanoate) conduits for peripheral nerve regeneration," *Biomaterials*, vol. 30, no. 2, pp. 217–225, 2009. doi: 10.1016/j.biomaterials.2008.09.036.
44. D. B. Hazer *et al.*, "*In vivo* application of poly-3-hydroxyoctanoate as peripheral nerve graft," (in Eng), *Journal of Zhejiang University Science B*, vol. 14, no. 11, pp. 993–1003, 2013. doi: 10.1631/jzus.B1300016.
45. L. R. Lizarraga-Valderrama *et al.*, "Nerve tissue engineering using blends of poly(3-hydroxyalkanoates) for peripheral nerve regeneration," *Engineering in Life Sciences*, vol. 15, no. 6, pp. 612–621, 2015. doi: 10.1002/elsc.201400151.
46. W. Daly, L. Yao, D. Zeugolis, A. Windebank, and A. Pandit, "A biomaterials approach to peripheral nerve regeneration: Bridging the peripheral nerve gap and enhancing functional recovery," *Journal of the Royal Society Interface*, vol. 9, no. 67, pp. 202–221, 2012. doi: 10.1098/rsif.2011.0438.

47. G. C. W. Ruiter, M. J. A. Malessy, M. J. Yaszemski, A. J. Windebank, and R. J. Spinner, "Designing ideal conduits for peripheral nerve repair," *Neurosurgical Focus*, vol. 26, no. 2, pp. 1–20, 2009, Art no. E5. doi: 10.3171/foc.2009.26.2.E5.
48. P. Chocholata, V. Kulda, and V. Babuska, "Fabrication of scaffolds for bone-tissue regeneration," *Materials*, vol. 12, no. 4, p. 568, 2019, Art no. 568. doi: 10.3390/ma12040568.
49. D. W. Hutmacher, "Scaffolds in tissue engineering bone and cartilage," *Biomaterials*, vol. 21, no. 24, pp. 2529–2543, 2000. doi: 10.1016/s0142-9612(00)00121-6.
50. L. Polo-Corrales, M. Latorre-Esteves, and J. E. Ramirez-Vick, "Scaffold design for bone regeneration," *Journal of Nanoscience and Nanotechnology*, vol. 14, no. 1, pp. 15–56, 2014. doi: 10.1166/jnn.2014.9127.
51. G. Turnbull *et al.*, "3D bioactive composite scaffolds for bone tissue engineering," *Bioactive Materials*, vol. 3, no. 3, pp. 278–314, 2018. doi: 10.1016/j.bioactmat.2017.10.001.
52. T. Albrektsson and C. Johansson, "Osteoinduction, osteoconduction and osseointegration," *European Spine Journal*, vol. 10 Supplement 2, pp. S96–S101, 2001. doi: 10.1007/s005860100282.
53. V. Karageorgiou and D. Kaplan, "Porosity of 3D biomaterial scaffolds and osteogenesis," *Biomaterials*, vol. 26, no. 27, pp. 5474–5491, 2005. doi: 10.1016/j.biomaterials.2005.02.002.
54. N. F. Ansari, M. S. M. Annuar, and B. P. Murphy, "A porous medium-chain-length poly(3-hydroxyalkanoates)/hydroxyapatite composite as scaffold for bone tissue engineering," *Engineering in Life Sciences*, vol. 17, no. 4, pp. 420–429, 2017. doi: 10.1002/elsc.201600084.
55. H. Yoshikawa and A. Myoui, "Bone tissue engineering with porous hydroxyapatite ceramics," (in Eng), *Journal of Artificial Organs*, vol. 8, no. 3, pp. 131–6, 2005. doi: 10.1007/s10047-005-0292-1.
56. A. A. El-Rashidy, J. A. Roether, L. Harhaus, U. Kneser, and A. R. Boccaccini, "Regenerating bone with bioactive glass scaffolds: A review of *in vivo* studies in bone defect models," *Acta Biomaterialia*, vol. 62, pp. 1–28, 2017. doi: 10.1016/j.actbio.2017.08.030.
57. J. R. Jones, "Reprint of: Review of bioactive glass: From Hench to hybrids," (in Eng), *Acta Biomaterialia*, vol. 23 Suppl, pp. S53–82, 2015. doi: 10.1016/j.actbio.2015.07.019.
58. S. K. Misra, Philip, S. E., Chrzanowski, W., Nazhat, S. N., Roy, I., and Knowles, J. C., "Incorporation of vitamin E in poly(3hydroxybutyrate)/bioglass composite films: Effect on surface properties and cell attachment," *Journal of the Royal Society Interface*, vol. 6, no. 33, pp. 401–409, 2009. doi: doi:10.1098/rsif.2008.0278.
59. A. Prasad, M. R. Sankar, and V. Katiyar, "State of art on solvent casting particulate leaching method for orthopedic scaffolds fabrication," *Materials Today-Proceedings*, vol. 4, no. 2, pp. 898–907, 2017. [Online]. Available: <Go to ISI>://WOS:000410686100101.
60. H. Hajiali, S. Karbasi, M. Hosseinalipour, and H. R. Rezaie, "Preparation of a novel biodegradable nanocomposite scaffold based on poly (3-hydroxybutyrate)/bioglass nanoparticles for bone tissue engineering," *Journal of Materials Science-Materials in Medicine*, vol. 21, no. 7, pp. 2125–2132, 2010. doi: 10.1007/s10856-010-4075-8.
61. A. N. Hayati, S. M. Hosseinalipour, H. R. Rezaie, and M. A. Shokrgozar, "Characterization of poly(3-hydroxybutyrate)/nano-hydroxyapatite composite scaffolds fabricated without the use of organic solvents for bone tissue engineering applications," *Materials Science & Engineering C-Materials for Biological Applications*, vol. 32, no. 3, pp. 416–422, 2012. doi: 10.1016/j.msec.2011.11.013.
62. G. T. Kose, H. Kenar, N. Hasirci, and V. Hasirci, "Macroporous poly(3-hydroxybutyra te-co-3-hydroxyvalerate) matrices for bone tissue engineering," *Biomaterials*, vol. 24, no. 11, pp. 1949–1958, 2003. doi: 10.1016/s0142-9612(02)00613-0.
63. H. P. Lu, Y. Liu, J. Guo, H. L. Wu, J. X. Wang, and G. Wu, "Biomaterials with antibacterial and osteoinductive properties to repair infected bone defects," *International Journal of Molecular Sciences*, vol. 17, no. 3, 2016, Art no. 334. doi: 10.3390/ijms17030334.

64. B. Rai, W. Noohom, P. H. Kithva, L. Grondahl, and M. Trau, "Bionanohydroxyapatite/poly(3-hydroxybutyrate-co-3-hydroxyvalerate) composites with improved particle dispersion and superior mechanical properties," *Chemistry of Materials*, vol. 20, no. 8, pp. 2802–2808, 2008. doi: 10.1021/cm703045u.
65. A. Saadat, A. Behnamghader, S. Karbasi, D. Abedi, M. Soleimani, and A. Shafiee, "Comparison of acellular and cellular bioactivity of poly 3-hydroxybutyrate/hydroxyapatite nanocomposite and poly 3-hydroxybutyrate scaffolds," *Biotechnology and Bioprocess Engineering*, vol. 18, no. 3, pp. 587–593, 2013. doi: 10.1007/s12257-012-0744-4.
66. Y. Wang, Y. Z. Bian, Q. Wu, and G. Q. Chen, "Evaluation of three-dimensional scaffolds prepared from poly(3-hydroxybutyrate-co-3-hydroxyhexanoate) for growth of allogeneic chondrocytes for cartilage repair in rabbits," *Biomaterials*, vol. 29, no. 19, pp. 2858–2868, 2008. doi: 10.1016/j.biomaterials.2008.03.021.
67. R. Iron, M. Mehdikhani, E. Naghashzargar, S. Karbasi, and D. Semnani, "Effects of nanobioactive glass on structural, mechanical and bioactivity properties of Poly (3-hydroxybutyrate) electrospun scaffold for bone tissue engineering applications," *Materials Technology*, vol. 34, no. 9, pp. 540–548, 2019. doi: 10.1080/10667857.2019.1591728.
68. I. Jun, H. S. Han, J. R. Edwards, and H. Jeon, "Electrospun fibrous scaffolds for tissue engineering: Viewpoints on architecture and fabrication," *International Journal of Molecular Sciences*, vol. 19, no. 3, 2018, Art no. 745. doi: 10.3390/ijms19030745.
69. S. Saska *et al.*, "Three-dimensional printing and *in vitro* evaluation of poly(3-hydroxybutyrate) scaffolds functionalized with osteogenic growth peptide for tissue engineering," *Materials Science & Engineering C-Materials for Biological Applications*, vol. 89, pp. 265–273, 2018. doi: 10.1016/j.msec.2018.04.016.
70. K. Sombatmankhong, N. Sanchavanakit, P. Pavasant, and P. Supaphol, "Bone scaffolds from electrospun fiber mats of poly (3-hydroxybutyrate), poly(3-hydroxybutyrate-co -3-hydroxyvalerate) and their blend," *Polymer*, vol. 48, no. 5, pp. 1419–1427, 2007. doi: 10.1016/j.polymer.2007.01.014.
71. M. Degli Esposti, F. Chiellini, F. Bondioli, D. Morselli, and P. Fabbri, "Highly porous PHB-based bioactive scaffolds for bone tissue engineering by in situ synthesis of hydroxyapatite," *Materials Science and Engineering C*, vol. 100, pp. 286–296, 2019. doi: 10.1016/j.msec.2019.03.014.
72. M. Kouhi, M. P. Prabhakaran, M. Shamanian, M. Fathi, M. Morshed, and S. Ramakrishna, "Electrospun PHBV nanofibers containing HA and bredigite nanoparticles: Fabrication, characterization and evaluation of mechanical properties and bioactivity," *Composites Science and Technology*, vol. 121, pp. 115–122, 2015. doi: 10.1016/j.compscitech.2015.11.006.
73. S. K. Misra *et al.*, "Fabrication and characterization of biodegradable poly(3-hydroxybutyrate) composite containing bioglass," *Biomacromolecules*, vol. 8, no. 7, pp. 2112–2119, 2007. doi: 10.1021/bm0701954.
74. S. C. Zhao *et al.*, "Three dimensionally printed mesoporous bioactive glass and poly(3-hydroxybutyrate-co-3-hydroxyhexanoate) composite scaffolds for bone regeneration," *Journal of Materials Chemistry B*, vol. 2, no. 36, pp. 6106–6118, 2014. doi: 10.1039/c4tb00838c.
75. S. B. Yang *et al.*, "Mesoporous bioactive glass doped-poly (3-hydroxybutyrate-co-3-hydroxyhexanoate) composite scaffolds with 3-dimensionally hierarchical pore networks for bone regeneration," *Colloids and Surfaces B-Biointerfaces*, vol. 116, pp. 72–80, 2014. doi: 10.1016/j.colsurfb.2013.12.052.
76. Z. Cao, C. Dou, and S. W. Dong, "Scaffolding biomaterials for cartilage regeneration," *Journal of Nanomaterials*, vol. 2014, pp. 1–8, 2014, Art no. 489128. doi: 10.1155/2014/489128.

77. Y. Liu, G. D. Zhou, and Y. L. Cao, "Recent progress in cartilage tissue engineering-our experience and future directions," *Engineering*, vol. 3, no. 1, pp. 28–35, 2017. doi: 10.1016/j.Eng.2017.01.010.
78. K. Zhao, Y. Deng, J. C. Chen, and G. Q. Chen, "Polyhydroxyalkanoate (PHA) scaffolds with good mechanical properties and biocompatibility," *Biomaterials*, vol. 24, no. 6, pp. 1041–1045, 2003, Art no. Pii s0142-9612(02)00426-x. doi: 10.1016/s0142-9612(02)00426-x.
79. Z. Zheng, Y. Deng, X. S. Lin, L. X. Zhang, and G. Q. Chen, "Induced production of rabbit articular cartilage-derived chondrocyte collagen II on polyhydroxyalkanoate blends," *Journal of Biomaterials Science-Polymer Edition*, vol. 14, no. 7, pp. 615–624, 2003. doi: 10.1163/156856203322274888.
80. P. Basnett *et al.*, "Novel poly(3-hydroxyoctanoate)/poly(3-hydroxybutyrate) blends for medical applications," *Reactive & Functional Polymers*, vol. 73, no. 10, pp. 1340–1348, 2013. doi: 10.1016/j.reactfunctpolym.2013.03.019.
81. K. Y. Ching *et al.*, "Nanofibrous poly(3-hydroxybutyrate)/poly(3-hydroxyoctanoa te) scaffolds provide a functional microenvironment for cartilage repair," *Journal of BiomaterialsApplications*, vol. 31, no. 1, pp. 77–91, 2016. doi: 10.1177/0885328216639749.
82. G. T. Kose *et al.*, "Tissue engineered cartilage on collagen and PHBV matrices," *Biomaterials*, vol. 26, no. 25, pp. 5187–5197, 2005. doi: 10.1016/j.biomaterials.2005.01.037.
83. K. Xue, X. D. Zhang, Z. X. Gao, W. Y. Xia, L. Qi, and K. Liu, "Cartilage progenitor cells combined with PHBV in cartilage tissue engineering," *Journal of Translational Medicine*, vol. 17, p. 104, 2019, Art no. 104. doi: 10.1186/s12967-019-1855-x.
84. J. Wu, K. Xue, H. Y. Li, J. Y. Sun, and K. Liu, "Improvement of PHBV scaffolds with bioglass for cartilage tissue engineering," *Plos One*, vol. 8, no. 8, p. e71563, 2013, Art no. e71563. doi: 10.1371/journal.pone.0071563.
85. E. Akaraonye *et al.*, "Composite scaffolds for cartilage tissue engineering based on natural polymers of bacterial origin, thermoplastic poly(3-hydroxybutyrate) and micro-fibrillated bacterial cellulose," *Polymer International*, vol. 65, no. 7, pp. 780–791, 2016. doi: 10.1002/pi.5103.
86. M. Giretova *et al.*, "Polyhydroxybutyrate/chitosan 3D scaffolds promote *in vitro* and *in vivo* chondrogenesis," *Applied Biochemistry and Biotechnology*, vol. 189, no. 2, pp. 556–575, 2019. doi: 10.1007/s12010-019-03021-1.
87. G. Li *et al.*, "Poly(3-hydroxybutyrate-co-4-hydroxybutyrate) based electrospun 3D scaffolds for delivery of autogeneic chondrocytes and adipose-derived stem cells: Evaluation of cartilage defects in rabbit," *Journal of Biomedical Nanotechnology*, vol. 11, no. 1, pp. 105–116, 2015. doi: 10.1166/jbn.2015.2053.
88. C. W. Pouton and S. Akhtar, "Biosynthetic polyhydroxyalkanoates and their potential in drug delivery," *Advanced Drug Delivery Reviews*, vol. 18, no. 2, pp. 133–162, 1996. doi: 10.1016/0169-409x(95)00092-l.
89. P. Basnett, "Biosynthesis of polyhydroxyalkanoates, their novel blends and composites for biomedical applications," PhD Experimental, Life Sciences, University of Westminster, London, UK, 2014. [Online]. Available: http://www.westminster.ac.uk/research/westminsterresearch.
90. L. Francis, D. C. Meng, J. Knowles, T. Keshavarz, A. R. Boccaccini, and I. Roy, "Controlled delivery of gentamicin using poly(3-hydroxybutyrate) microspheres," *International Journal of Molecular Sciences*, vol. 12, no. 7, pp. 4294–4314, 2011. doi: 10.3390/ijms12074294.
91. S. Gogolewski, M. Jovanovic, S. M. Perren, J. G. Dillon, and M. K. Hughes, "Tissue response and *in vivo* degradation of selected polyhydroxyacids: Polylactides (PLA), poly(3-hydroxybutyrate) (PHB), and poly(3-hydroxybutyrate-co-3-hydroxyvalerate)

(PHB/VA)," *Journal of Biomedical Materials Research*, vol. 27, no. 9, pp. 1135–48, 1993. doi: 10.1002/jbm.820270904.
92. Y. C. Yao *et al.*, "A specific drug targeting system based on polyhydroxyalkanoate granule binding protein PhaP fused with targeted cell ligands," *Biomaterials*, vol. 29, no. 36, pp. 4823–4830, 2008. doi: 10.1016/j.biomaterials.2008.09.008.
93. J. H. Eldridge, J. A. Meulbrook, J. K. Staas, R. M. Gilley, and T. R. Tice, "Controlled vaccine release in the gut-associated lymphoid tissues, I: Orally administered biodegradable microspheres target the Payer's patches," *Journal of Controlled Release*, vol. 11, pp. 205–214, 1990. doi: 10.1016/0168-3659(90)90133-E.
94. R. Nigmatullin, P. Thomas, B. Lukasiewicz, H. Puthussery, and I. Roy, "Polyhydroxyalkanoates, a family of natural polymers, and their applications in drug delivery," *Journal of Chemical Technology and Biotechnology*, vol. 90, no. 7, pp. 1209–1221, 2015. doi: 10.1002/jctb.4685.
95. Z. X. Wang *et al.*, "Mechanism of enhancement effect of dendrimer on transdermal drug permeation through polyhydroxyalkanoate matrix," *Journal of Bioscience and Bioengineering*, vol. 96, no. 6, pp. 537–540, 2003. doi: 10.1016/s1389-1723(04)70146-2.
96. I. Gursel, F. Korkusuz, F. Turesin, N. G. Alaeddinoglu, and V. Hasirci, "*In vivo* application of biodegradable controlled antibiotic release systems for the treatment of implant-related osteomyelitis," *Biomaterials*, vol. 22, no. 1, pp. 73–80, 2001. doi: 10.1016/s0142-9612(00)00170-8.
97. I. Gursel, F. Yagmurlu, F. Korkusuz, and V. Hasirci, "*In vitro* antibiotic release from poly(3-hydroxybutyrate-co-3-hydroxyvalerate) rods," *Journal of Microencapsulation*, vol. 19, no. 2, pp. 153–164, 2002. doi: 10.1080/02652040110065413.
98. F. Turesin, I. Gursel, and V. Hasirci, "Biodegradable polyhydroxyalkanoate implants for osteomyelitis therapy: *In vitro* antibiotic release," *Journal of Biomaterials Science-Polymer Edition*, vol. 12, no. 2, pp. 195–207, 2001. doi: 10.1163/156856201750180924.
99. K. Thygesen *et al.*, "Universal definition of myocardial infarction," *Circulation*, vol. 116, no. 22, pp. 2634–2653, 2007. doi: 10.1161/circulationaha.107.187397.
100. P. Basnett, B. Lukasiewicz, E. Marcello, H. K. Gura, J. C. Knowles, and I. Roy, "Production of a novel medium chain length poly(3-hydroxyalkanoate) using unprocessed biodiesel waste and its evaluation as a tissue engineering scaffold," *Microbial Biotechnology*, vol. 10, no. 6, pp. 1384–1399, 2017. doi: 10.1111/1751-7915.12782.
101. R. Rai *et al.*, "The homopolymer poly(3-hydroxyoctanoate) as a matrix material for soft tissue engineering," *Journal of Applied Polymer Science*, vol. 122, no. 6, pp. 3606–3617, 2011. doi: 10.1002/app.34772.
102. R. Rai *et al.*, "Poly-3-hydroxyoctanoate P(3HO), a medium chain length polyhydroxyalkanoate homopolymer from *Pseudomonas mendocina*," *Biomacromolecules*, vol. 12, no. 6, pp. 2126–2136, 2011. doi: 10.1021/bm2001999.
103. A. V. Bagdadi *et al.*, "Poly(3-hydroxyoctanoate), a promising new material for cardiac tissue engineering," *Journal of Tissue Engineering and Regenerative Medicine*, vol. 12, no. 1, pp. E495–E512, 2018. doi: 10.1002/term.2318.
104. C. M. Bunger *et al.*, "A biodegradable stent based on poly(L-lactide) and poly(4-hydroxybutyrate) application: Preliminary for peripheral vascular experience in the pig," *Journal of Endovascular Therapy*, vol. 14, no. 5, pp. 725–733, 2007. doi: 10.1583/1545-1550(2007)14[725:Absbop]2.0.Co;2.
105. N. Grabow *et al.*, "Development of a sirolimus-eluting poly (L-lactide)/poly(4-hydroxybutyrate) absorbable stent for peripheral vascular intervention," *Biomedical Engineering-Biomedizinische Technik*, vol. 58, no. 5, pp. 429–437, 2013. doi: 10.1515/bmt-2012-0050.
106. S. Kischkel *et al.*, "Biodegradable polymeric stents for vascular application in a porcine carotid artery model," *Gefasschirurgie*, vol. 21, pp. S30–S35, 2016. doi: 10.1007/s00772-015-0011-z.

107. O. Duvernoy, T. Malm, J. Ramstrom, and S. Bowald, "A biodegradable patch used as a pericardial substitute after cardiac surgery-6 month and 24 month evaluation with CT," *Thoracic and Cardiovascular Surgeon*, vol. 43, no. 5, pp. 271–274, 1995. doi: 10.1055/s-2007-1013226.
108. T. Malm, S. Bowald, A. Bylock, T. Saldeen, and C. Busch, "Regeneration of pericardial tissue on absorbable polymer patches implanted into the pericardial sac - an immunohistochemical, ultrastructural and biochemical study in the sheep," *Scandinavian Journal of Thoracic and Cardiovascular Surgery*, vol. 26, no. 1, pp. 15–21, 1992. doi: 10.3109/14017439209099048.
109. T. Malm, S. Bowald, A. Bylock, C. Busch, and T. Saldeen, "Enlargement of the right-ventricular outflow tract and the pulmonary-artery with a new biodegradable patch in transannular position," *European Surgical Research*, vol. 26, no. 5, pp. 298–308, 1994. doi: 10.1159/000129349.
110. S. Ramadan, N. Paul, and H. E. Naguib, "Standardized static and dynamic evaluation of myocardial tissue properties," *Biomedical Materials*, vol. 12, no. 2, 2017, Art no. 025013. doi: 10.1088/1748-605X/aa57a5.
111. C. Constantinides *et al.*, "*In vivo* tracking and (1)H/(19)F magnetic resonance imaging of biodegradable polyhydroxyalkanoate/polycaprolactone blend scaffolds seeded with labeled cardiac stem cells," (in Eng), *ACS Applied Materials & Interfaces*, vol. 10, no. 30, pp. 25056–25068, 2018. doi: 10.1021/acsami.8b06096.
112. G. D. Dangas *et al.*, "Prosthetic heart valve thrombosis," *Journal of the American College of Cardiology*, vol. 68, no. 24, pp. 2670–2689, 2016. doi: 10.1016/j.jacc.2016.09.958
113. R. Sodian *et al.*, "Evaluation of biodegradable, three-dimensional matrices for tissue engineering of heart valves," *Asaio Journal*, vol. 46, no. 1, pp. 107–110, 2000. doi: 10.1097/00002480-200001000-00025.
114. R. Sodian *et al.*, "Early *in vivo* experience with tissue-engineered trileaflet heart valves," (in Eng), *Circulation*, vol. 102, no. 19 Suppl 3, pp. Iii22–9, 2000. doi: 10.1161/01.cir.102.suppl_3.iii-22.
115. C. Stamm *et al.*, "Biomatrix/polymer composite material for heart valve tissue engineering," *Annals of Thoracic Surgery*, vol. 78, no. 6, pp. 2084–2092, 2004. doi: 10.1016/j.athoracsur.2004.03.106.
116. S. Wu, Y. L. Liu, B. Cui, X. H. Qu, and G. Q. Chen, "Study on decellularized porcine aortic valve/poly (3-hydroxybutyrate-co-3-hydroxyhexanoate) hybrid heart valve in sheep model," *Artificial Organs*, vol. 31, no. 9, pp. 689–697, 2007. doi: 10.1111/j.1525-1594.2007.00442.x.
117. M. G. Mennuni, P. A. Pagnotta, and G. G. Stefanini, "Coronary stents: The impact of technological advances on clinical outcomes," *Annals of Biomedical Engineering*, vol. 44, no. 2, pp. 488–496, 2016. doi: 10.1007/s10439-015-1399-z.
118. https://rebiostent.eu (accessed 06/01/2020, 2020).
119. Q. Liu *et al.*, "Synthesis of an electrospun PHA/RGO/Au scaffold for peripheral nerve regeneration: An in vitro study," *Applied Nanoscience*, vol. 10, no. 3, pp. 687–694, 2020. doi:10.1007/s13204-019-01130-1
120. X. Y. Xu, X. T. Li, S. W. Peng *et al.*, "The behaviour of neural stem cells on polyhydroxyalkanoate nanofiber scaffolds," *J. Biomaterials*, vol. 31, no. 14, pp. 3967–3975, 2010. doi:10.1016/j.biomaterials.2010.01.132

12 Polyhydroxyalkanoates (PHA)-Based Materials in Food Packaging Applications

State of the Art and Future Perspectives

Lucía Pérez Amaro, David Barsi, Tommaso Guazzini, Vassilka Ivanova Ilieva, Arianna Domenichelli, Ilaria Chicca, and Emo Chiellini

CONTENTS

12.1 BRIEF INTRODUCTION TO PHAS' STRUCTURAL FEATURES AND PRODUCTION

12.1.1 Statistics and Standards on Packaging Applications

The packaging industry is, undoubtedly, the principal sector that consumes and generates the most plastic in the world per year. The amount of plastic used in the production of packaging, as well as the amount of plastic generated world-wide as waste from that activity, was estimated to be around 140–145 million tons in the year 2018.

Among the application sectors, packaging is the one whose mean product lifetime (from production to disposal) is the lowest (0.5 years) if compared, for example, with building and construction (35 years) or textiles (5 years).

It is also important to take a look at the plastic waste generation by country or geographical area and to find the correlation between the generation and the ability of such countries to manage such waste quantities. We found that PR China is the number one country that generates the major amount of plastic in the world (59.8 million tons per year), followed by the United States of America with 37.8 million tons per year, Germany with 14.5 million tons, and Brazil with 11.9 million tons. Even though India generates "only" 4.5 million tons of plastic waste, this country represents the second-largest country in the world in terms of population.

It has to be considered that more than 59% of the world population, estimated at 7.6 billion in 2018, is located in only 12 countries (Bangladesh, Brazil, Canada, PR China, Ethiopia, Germany, India, Indonesia, Nigeria, Pakistan, the Russian Federation, and the United States of America), and 50% of these countries share the major percentage of plastic waste that is inadequately managed (Bangladesh, Pakistan, India, Indonesia, Nigeria, China). These six countries represent 48% of the world population (3.7 billion) that to date share the total plastic waste that is inadequately managed. Inadequately disposed waste is not appropriately managed and often is subjected to disposal in dumps or open, uncontrolled landfills, where it is not safely enclosed.

So, we are facing a concomitance of factors that severely influence all areas of human activities in our days. To be more specific:

1) The amount of plastic consumption is growing annually (global plastic production in 2018 was determined to be 359 million metric tons per year [1]).
2) The packaging sector share amounts to 42% of the total plastic consumption.
3) The packaging sector presents the lowest main product lifetime (6 months).
4) The major countries that generate plastic waste are also the same countries that have inadequate plastic waste management.
5) The major countries that generate plastic waste are also those with minor to medium income.

Moreover, we are at a time where consumption habits associated with packaging and plastics are in the spotlight for their effect on the environment. In this context, the world of packaging is undergoing a real revolution. For example, in Europe, the new legislation related to packaging and packaging waste (Directive 852/2018) includes as a main focus a Circular Economy vision, which is establishing the following waste hierarchy (2008/9/EC): Prevention, Preparing for Reuse, Recycling, Other than Recovery (such as energy), and Disposal.

To be precise, the following definitions have been established:

Prevention means measures taken before a substance, material, or product has become waste, in order to reduce:

(a) The quantity of waste, including through the re-use of products or the extension of the life span of products.
(b) The adverse impacts of the generated waste on the environment and human health.
(c) The content of harmful substances in materials and products.

Preparing for re-use means checking, cleaning, or repairing recovery operations, by which products or components of products that have become waste are prepared, so that they can be re-used without any other pre-processing.

Recovery means a procedure according to which plastic waste can undergo re-use by mechanical recycling, or, generally, using plastic waste as an input material to create valuable products as new outputs.

Re-use means any operation by which products or components that are not waste are used again for the same purpose for which they were conceived.

Recycling means any recovery operation by which waste materials are reprocessed into products, materials, or substances, whether for the original purpose or for other purposes. It includes the reprocessing of organic material, but does not include energy recovery and reprocessing into materials that are to be used as fuels or for backfilling operations.

Disposal means any operation which is not recovery even where the operation has as a secondary consequence the reclamation of substances or energy.

Among other goals, the new plastics strategy establishes that in the year 2030, 100% of the containers must be reusable or recyclable (including compostable), while Directive 852/2018 sets ambitious recycling objectives (50% of packages) in 2025 and 55% in 2030.

In this frame, how can polyhydroxyalkanoates (PHA) be considered suitable materials for cost-effective solutions for low-income countries? How can PHA meet Europe's ambitions in terms of the Circular Economy? How can PHA be re-used? And how can we be sure that new innovative packaging based on PHA is complying with the legal requirements in terms of food contact, for example?

Indeed, there is a rigorous and technical way that serves as a reference framework to demonstrate compliance with the requirements that are presented to us. It is represented by the Harmonized Standards. The Harmonized Standards were published between 2000 and 2005 with the purpose of helping companies to comply with the essential requirements of Directive 94/62/EC on Packaging and Packaging Waste of the European Parliament and of the Council of December 20, 1994, and to contribute to European waste policies at that time. It considered that the management of packaging waste should focus its priority on the prevention of its production and assumes the following key principles: re-use, recycling, and other ways of recovery of packaging waste.

The European Standards that grant a presumption of conformity with the corresponding requirements of Directive 94/62/EC of Packaging and Packaging Waste that were developed and published are the following:

- EN 13427: 2004 Packaging – Requirements for the use of European Standards in the field of packaging and packaging waste.
- EN 13428: 2004 Packaging – Requirements specific to manufacturing and composition – Prevention by source reduction.
- EN 13429: 2004 Packaging – Reuse.
- EN 13430: 2004 Packaging – Requirements for packaging recoverable by material recycling.
- EN 13431: 2004 Packaging – Requirements for packaging recoverable in the form of energy recovery, including specification of minimum inferior calorific value.
- EN 13432: 2000 Packaging – Requirements for packaging recoverable through composting and biodegradation – Test scheme and evaluation criteria for the final acceptance of packaging.

The European Parliament and Council Directive 2018/852 still maintains the definition of packaging of the EU Directive 94/62/EC in which the packaging "shall mean all products made of any materials of any nature to be used for the containment, protection, handling, delivery, and presentation of goods, from raw materials to processed goods, from the producer to the user or the consumer".

The present chapter will focus on those technical aspects that are in good agreement with the transition towards a Circular Economy strategy. The market share of PHA will be reviewed, along with the barrier properties of PHB, the biodegradability of plastic items based on PHA, mechanical and thermal properties of PHA-based materials, and some applications based on PHA done within a Circular Economy vision.

12.1.2 Position of PHA in Food Packaging Applications: Present Status

The emerging pollution issues regard also the spreading of plastic material in the environment, and their resistance to biodegradation allows their accumulation in all environmental areas. The exploitation of biodegradable plastic and even more of bio-based plastics is well studied and well recognized. Among the biodegradable plastics, it is possible to find cellulose- and starch-based plastic and PHA as bio-based plastics. PHA are a family of natural thermoplastic and elastomeric bio-based polyesters synthesized as carbon storage material by a plethora of bacterial and archaeal strains. The monomeric units might be short-, medium-, or long-chain length carbons (*scl*-, *mcl*-, and *lcl*-PHA), and polymers might be homo- or copolymers if they are composed of more than one monomer type [2, 3].

One of the most important compartments for plastic products is the food packaging. The common requirements to be fulfilled are [4]: a proper wrapping and shielding from dust, moisture, dehydration, microbial and chemical contamination, UV radiation, preservation of sensory food characteristics, maintenance of sensory and stability of food under extreme storage conditions and resistance to biodegradation during the storage period, and enhanced biodegradation once the packaging is discarded.

Mechanical, thermal, chemical proprieties, permeability, flavor maintenance, migration, and biodegradability are the key characteristics for evaluating the possible exploitation of PHA in food packaging [5].

Thermal and mechanical properties of the polymer influence the production processes and the end product application. The crystallinity influences the stiffness and brittleness of the material. As thermal properties, glass transition temperature (T_g) and melting temperature (T_m) are the main observed and considered physical features. From the mechanical point of view, Young's modulus (polymer stiffness), tensile strength (total amount of force required to pull a material before it breaks), and elongation at break (stretchability of the material before it breaks) are the main parameters [6–8].

Permeability to water vapor, oxygen, and CO_2 is important for humidity equilibrium and protection from fungi like molds, food oxidation, and bacteriostatic properties, respectively [5]. Biodegradability of PHA was demonstrated in several bacterial strains having genes encoding for PHA depolymerases [9]. This is certainly one of PHA's major benefits in comparison with traditional plastics, and allows safe disposal of the plastic material into the environment.

Thermodegradation, migration of polymers or additives into the food ("leaching" of the packaging material), interaction, and production costs are the main concerns for PHA exploitation for food packaging. Even though polymers such as poly(3-hydroxybutyrate) (PHB, also known as P(3HB)) and its copolymer with 3-hydroxyvalerate (PHBV) showed properties comparable to fossil fuel-based plastic materials, which makes PHA interesting in general, it is still important to study them deeply in order to improve their quality and to use them in the food packaging field [5].

Israni *et al.* [5] reported recently (2019) an excellent comprehensive review regarding PHA in packaging applications. The author highlighted new aspects of packaging such as PHA as a paper and cardboard coating material and PHA multilayer films, and provided examples of demonstrations of PHA in food packaging applications. Arrieta *et al.* [10] studied PHA, covering key aspects on the development of biodegradable materials for packaging applications. The authors used PHB as a sustainable ingredient to improve processability, modulate mechanical properties, and improve barrier properties of poly(lactic acid) (PLA). They found that a binary composition based on 75% of PLA and 25% of PHB in a blend presents optimal miscibility in the melt state, as promoted by transesterification reactions. Another key aspect highlighted by these authors is related to the composition and the propensity to biodegradation in different environmental compartments; in this frame, PHB can be more suitable to withstanding home compost conditions than PLA. Koller [11] reported a comprehensive review of the use of PHA in food packaging applications. The author highlighted the biological aspects that make PHA properly defined as bioplastic with a complete green life cycle, and emphasized the reason why the

PHA present higher costs compared to other bio-based polymers (raw material, process design, method for recovery of PHA from biomass). Moreover, this author cited interesting PHA copolymers used for packaging applications; such as, for example, poly(3HB-*co*-16%-4HB), which has shown a melting temperature of 152°C, possesses a glass transition temperature of −8°C, a tensile modulus of 26 MPa, and an elongation at break of 444%. Placket *et al.* [12] and Bugnicourt *et al.* [13] have done a comprehensive review of PHA by focusing on synthesis, characteristics, and processing, and have presented challenges and prospects of PHA-based materials in the comprehensively reviewed literature.

Bucci *et al.* [14], Pérez Amaro *et al.* [15], and Greene [16] published scientific reports in which the feasibility of the use of PHA as a biodegradable polymer suitable for producing injected parts and films by using injection-molding, film-extrusion, and blow-molding have been demonstrated. The selection of the bio-based polymers and the additives, as well as the processing conditions and parameters of the evaluation of the final items were in good agreement with the valid standards for food packaging applications. Fabra *et al.* [17] studied the aging performance of PHA and relative composites by simulating multilayer film structures. The authors found that the humidity is one of the major factors that provokes an increment in the crystallinity of PHA, but meanwhile registers a lesser influence on the storage time on that property.

Weber *et al.* [18] analyzed the main characteristics of bio-based and biodegradable polymers including PHA with respect to the needs of niche market applications. They indicate that among the manifold requirements for food packaging applications, the barrier properties related to gas and water vapor permeability are the most relevant characteristics in which PHA can give a relatively consistent contribution. Nowadays, petrochemical-based plastic can satisfy in an excellent manner the following requirements: gas and water vapor permeability, mechanical properties, sealing capability, thermoforming properties, resistance to warm water, grease, acids, UV light, machinability, transparency, antifogging capacity, printability, availability in the market, cost, and disposal (for example, recycling or energy recovery). PHB possesses a high transmission of CO_2 with respect to O_2; this particular characteristic can be exploited in packaging of fruits and vegetables with high respiration rates, cheese, and short-term storage of non-carbonated beverages. In particular, the stabilization of high respiration rate fruits and vegetables is a complex duty, since either O_2 and CO_2 are inter-correlated. The liberation of CO_2 has been estimated to be 20–40 mg of CO_2/(kg•h) at 5°C for fast respiration, 40–60 mg for very fast, and over 60 mg of CO_2/(kg•h) at 5°C for extremely fast respiration [19]. In this context, PHA can be suitable for use in packaging applications for broccoli, mushrooms, spinach, sweetcorn, strawberry, avocado, cauliflower, and artichoke. Factors limiting the shelf-life of beverages include microbial growth, migration/scalping, oxidation of flavor components, nutrients, and pigments, non-enzymatic browning, and, in the case of carbonated beverages, loss of carbonation. For packaging of acidic beverages, the material must be resistant to acids. PHB has a much lower oxygen transmission rate than PLA and has high water resistance. Coating of PLA with PHB is expected to give a useful bio-based packaging material for beverages. Packaging materials based on 100% PHB are also expected to be useful for beverages. Paperboard coated with PLA, PHB, or modified starch in order to improve the moisture and oxygen barrier properties of the paperboard could also be of potential use in packaging beverages [20].

Rydz *et al.* [21] and Pérez Amaro *et al.* [22] pointed out two negative aspects of PHA. The first one is related to odor problems, while the second one is related to its high production cost. It is well established that polymers in food packaging applications must have high purity. During microbial production of PHA, the obtained material contains small amounts of impurities such as proteins and lipids derived from the cell walls. This small amount of impurities may cause a significant odor problem when using PHA as package materials. Nevertheless, a combination of chemical methods and high-pressure extraction can be used to remove the remaining impurities from the PHA. Unfortunately, the main reason why PHA materials are not widely used is their high cost. Several authors attribute the high cost of PHA to the cost of the raw materials necessary to feed the microorganisms; however, from an industrial point of view, another important fact emerges that is correlated with the application field for PHA. As biocompatible polymers, PHA also find applications in the pharmaceutical field, in which the high costs of the polymeric materials are absorbed and assimilated by the market. So, it seems obvious that the producer will prefer those applications in which the margin of economic return is higher. In this setting, the application of PHA in packaging also depends on the decision of the producer to gain a share of the market in mass consumption applications such as packaging.

To stimulate the use of PHA in packaging applications it has been reported that PHA exhibit printability, a flavor and odor barrier, heat sealability, grease and oil resistance, temperature stability, and are easy to dye; all of these characteristics improve its applications in the food industry. In particular, medium-chain-length (*mcl*) PHA can be applied as biodegradable cheese coverings, lawn and leaf bags, disposable diapers, fast-food service wares, single-use medical devices, biodegradable bottles, containers, sheets, and fibers. PHA's low values of water vapor permeability coefficients are similar to those of poly(ethylene terephthalate) (PET). This factor has a significant impact during the selection of packages for food when desiccation or water inflow needs to be avoided [21].

Rabnawaz *et al.* [23] agree that the expedient barrier properties, along with the biodegradable nature of PHA, have facilitated the development of many commercial products based on this family of polymers, including packaging applications. The authors have done a concise resume of PHA properties, highlighting those properties that depend upon the chemical compositions and stereochemical aspects. The authors reported that the T_g values of PHA depend on nature and composition in the case of copolymers. Indeed, PHB typically has a T_g of 1°C and a T_m of ~170°C, while PHBV at 60 mol-% of 3HV has a T_g of –7°C and a T_m of ~143°C. Similarly, the degree of crystallization encountered among PHA varies with their chemical nature and usually ranges between 40–70%. As is the case with their thermal properties, PHA offer a wide range of mechanical properties bound to their structural diversity. Overall, PHA offer mechanical properties similar to PP. Consequently, PHA are available with mechanical properties ranging from stiff and brittle to soft and elastic. For example, PHB is a very stiff and brittle polymer (with a Young's Modulus of 3.5 GPa) in comparison with PHBV (20 mol-% 3HV, with a Young's Modulus of 0.8 GPa). Similarly, the elongation at break increases from 5% to 50% when PHB is replaced by PHBV (20 mol-% 3HV). This greater flexibility of PHBV makes it more suitable for packaging applications than the relatively brittle PHB. PHA do not readily absorb UV light at wavelengths above 200 nm and thus they offer poor UV blocking properties.

From a commercial and market point of view, Danimer Scientific – USA recently launched and claimed a PHA that is totally biodegradable in all the environmental compartments (soil, freshwater, and marine water). Moreover, their innovative PHA-based material is compostable in home and industrial facilities. The producer claims that the material is food contact approved, and it can be processed to obtain a barrier film suitable to protect potato chips. The company envisages home composting as one of the final disposal strategies of the relevant packaging.

The packaging materials include fillers, plasticizers, and stabilizers. Fillers tend to maintain barrier and mechanical properties, while filling the polymer with inexpensive materials reduces the cost. Plasticizers are applied to strengthen flexibility, ductility, and toughness of polymers while at the same time reducing hardness and stiffness. Also, stabilizers are incorporated into the matrix to inhibit deterioration of mechanical properties due to UV light and oxygenation [24]. There are still many concerns about food packaging materials and their possible interactions with food, especially when food-spoiled plastics are re-used. Moreover, there are some issues about environmental problems due to the slow degradation of polymers and the importance of consumption of renewable sources. Therefore, there is a growing interest world-wide in replacing conventional mass plastics with biodegradable plastics, particularly in packaging applications. The use of biodegradable plastics and resources is seen as one of the many strategies to minimize the environmental impact of petroleum-based plastics. Plastic recycling is not often economically viable as there are problems of contamination of the food packaging. However, recycled materials can be used for food packaging only if they meet specific legislative standards. Such limitation leads to a cost- and time-consuming process. Despite a great amount of different recycling programs, in some places there is still no cohesive recycling program [25].

Food packaging is a combination of art, science, and technology for enclosing a product for achieving safe transportation and distribution of the products in wholesome conditions to the users at the lowest price [26]. Bio-based plastics and biopolymers have been widely studied in recent years, and a lot of research has been done on biopolymers for food packaging applications. Replacing the oil-based packaging materials with bio-based films and containers might give not only a competitive advantage due to a more sustainable and greener image, but also some improved technical properties. A sustainable economy utilizes mainly biomass-derived raw materials for high-volume applications, such as packaging. Special attention should be paid to barrier properties, which are extremely important, especially for bio-based food packaging materials. Moisture resistance with hydrophilic polymers is, in many cases, inadequate, and thus excessive water vapor transmission through packaging lowers the quality of food, resulting in shorter shelf-lives, increased costs, and eventually more waste. PLA, PHA, cellophane, etc., are typically and widely used bio-based plastics. As with all bio-based plastics, the barrier properties of, e.g., PLA films can be improved by adding fillers such as nanoclays [27].

PHB acts as a better light barrier in the visible and ultraviolet light regions than PLA. PHA properties may be improved by modifying the surface or combining PHA with other polymers, inorganic materials, and enzymes.

PHA were also developed as packaging films mainly for use as shopping bags, containers and paper coatings, disposable items such as razors, utensils, diapers, feminine hygiene products, cosmetic containers, and cups, as well as medical surgical garments, upholstery, carpets, packaging, compostable bags, and lids or tubs for thermoformed articles by P&G, Biomers, Metabolix, and several other companies.

12.1.3 Polyhydroxyalkanoates Production

The European Bioplastics Organization states that a plastic material is defined as a "bio-plastic" if it is either bio-based, biodegradable, or features both properties. The need for replacement of petroleum-based plastic with bio-based polymers is imperative because producing conventional plastics consumes 65% more energy, is unsustainable (due to environmental problems), and emits 30–80% more greenhouse gases than bio-based plastics [28, 29]. Biopolymers can be used as a solution to the problems posed by plastics, as they easily degrade in the environment and they have similar properties to conventional polymers like poly(ethylene) (PE), poly(propylene) (PP), poly(ethylene terephthalate) (PET), etc. The environmental impact of petroleum-based plastic usage is a very critical concern, and too significant to repair. The only problem with synthetic polymers is their resistance to degradation in the environment [30]. Thus, bio-based polymers, like PHA, in the form of packaging materials are key innovations that can help in reducing the environmental impact of conventional mass plastic production [31]. A shift toward biodegradable food packaging is one very important option for the environment.

Bio-based plastics production increased from 1.6 to 2.0 million tons during the period 2013–2015. Most biodegradable plastics are made up of PLA (10.9%), biodegradable polyesters (10.8%), biodegradable starch blends (9.4%), and PHA (3.6%). The majority of biodegradable bio-based plastics are used for flexible packaging and non-degradable bio-based plastics are used for rigid packaging. The future of bio-based plastics is focused on the market for compostable, semi-durable, and durable bio-plastics used in consumer and industrial applications [28]. Based on the industry report of the Nova-Institute (Germany) from 2015, it can be noted that the production of bio-based polymers for rigid packaging (including food service ware) will increase from 13% in the year 2014 to 40% in 2020 [32, 33]. PHA producers are optimistic and still see potential in these biomaterials, claiming that PHA represent a new generation of biopolymers and their market needs time to develop. It is estimated that demand for PHA will grow tenfold by 2020 [32, 33].

PHA are gaining increasing attention in the biodegradable polymer market due to their promising properties such as high biodegradability in different environments, not just in composting plants, and processing versatility. Indeed, among biopolymers, these polyesters represent a potential sustainable replacement for fossil fuel-based thermoplastics. Biodegradable polymers can be used for modified and improved atmospheric storage (MAP) of food, fruit, and vegetables instead of conventional polymers.

As mentioned before, the packaging industry consumes most petroleum-based polymers. This industrial branch has a considerable interest in biodegradability since packaging is only needed for short times, but in big quantities, which contributes to the accumulation of waste. It should be understood that not all bio-based polymers are biodegradable, but some important ones are, e.g., PHA, PLA, and starch blends. This characteristic is also interesting for agriculture and horticulture applications (e.g., mulch films).

In summary, production of PHA reflects a commercial interest in new "green" materials as replacements for synthetic non-degradable polymers in a wide range of applications. These applications include products for use in packaging, agriculture, medicine, biology, and fast food. The biocompatible nature of PHA will also continue to stimulate research and development to increase PHA's applications in even more sectors.

PHA constitute a family of 100% bio-based, biodegradable, natural polymers that have attracted major interest in recent years. PHA are linear polyesters produced in nature by bacterial fermentation of sugars, lipids, acids, alcohols, or gases like CH_4, CO, or CO_2. As they are produced by many bacteria and archaea as intracellular carbon and energy storage granules, "PHA" is not just one type of polymer, but a polymer family [34]. Depending upon the carbon substrates and the metabolism of the microorganisms, different monomers and thus (co)polymers could be obtained. More than 150 different monomers can be combined within this family to give materials with extremely different properties. PHA are mainly produced from renewable resources by fermentation; they are compatible with the world-wide increased interest in environmentally friendly materials and sustainable development. PHA are biodegradable (i.e., potentially suitable for short-term packaging uses), thermo-processable, making them attractive as biomaterials for different applications, and biodegradable even in cold seawater. PHA are natural polymers and represent an environmentally favorable route to the replacement of conventional fossil fuel-based thermoplastics such as poly(styrene) (PS) and polyolefins (PE and PP). PHA also exhibit biocompatibility when in contact with living tissues (e.g., for biomedical application such as tissue engineering) and have found an increasing application not just for the most assessed medical applications, but also considered for applications including packaging, agriculture, molded goods, paper coatings, non-woven fabrics, leisure, fast-food, adhesives, films, and performance additives.

PHA are generally classified into short-chain-length (*scl*) PHA and *mcl*-PHA by the different number of carbons in their repeating units. For instance, *scl*-PHA contain 4 or 5 carbons in their repeating units, whereas *mcl*-PHA contain 6 or more carbons in the repeating units. The term *mcl* was coined because the number of carbons in the monomers roughly corresponds to those of medium-chain-length carboxylic acids. PHA nomenclature and classification may still evolve as new structures continue to be discovered.

The main biopolymer of the PHA family is the homopolyester PHB, but different copolyesters also exist, such as poly(3-hydroxybutyrate-*co*-3-hydroxyvalerate) (PHBV or P(3HB-*co*-3HV)), poly(3-hydroxybutyrate-*co*-3-hydroxyhexanoate)

(P(3HB-*co*-3HHx)), poly(hydroxybutyrate-*co*-hydroxyoctanoate) (PHBO), or poly(3-hydroxybutyrate-*co*-3-hydroxyoctadecanoate) (PHBOD) [34].

The main companies producing PHA for use in the packaging industry are shown in Table 12.1.

The use of biodegradable polymers such as PHA in the production of food packaging is related to the demand for sustainable management of the environmental impact of plastics. These materials are already being used in disposable packaging with a short lifespan and for single-use, as well as in the production of service ware items and disposable non-woven fabrics, among others [37].

12.1.4 Barrier Properties of PHA Packaging

Nowadays, the main application of plastic products is in the packaging sector, which covers about 40% of Europe's plastic converter demand [38]. An ideal packaging

TABLE 12.1
Current Main PHA-Producing Companies for Use in the Packaging Industry

Company	Country	Trade name	Type of PHA	Production [tons/yr]
Metabolix	USA	Mirel	Several types of PHA	–
Kaneka Corporation	Japan	Nodax	Several types of PHA	–
Biotechnology Co.	Germany	Biomer®	P(3HB)	50 (2003) [35]
PHB Industrial S/A	Brazil	Biocycle	P(3HB)	60
Mitsubishi Gas Chemical Company Inc	Japan	Biogreen®	P(3HB)	30–60 000 (2010) [35]
Metabolix, (BASF, ADM GmbH)	USA, Germany	Biogreen®	P(3HB), P(3HO)	50 000 (2008) [35]
Jiangsu Nantian Group	China	Biogreen®	P(3HB)	–
Metabolix, (BASF, ADM GmbH)	USA, Germany	Biopol®	P(3HB-*co*-3HV)	1 100 (2003) [35]
PHB Industrial S/A	Brazil	Biopol®	P(3HB-*co*-3HV)	10 000 (2006) [35]
Tianan Biologic Material	China	ENMAT®	P(3HB-*co*-3HV)	1 000
Procter & Gamble, USA (Kaneka)	USA, Germany	Nodax	P(3HB-*co*-3HHx)	20–50 000 (2010) [35]
Jiangsu Nantian Group	China	Nodax™	P(3HB-*co*-3HHx)	2 000
Lianyi Biotech	China	Nodax™	P(3HB-*co*-3HHx)	2 000 [35]
Telles	USA	Mirel™	PHB	50 000 [36]
Meredian	USA	Meredian	PHB	272 000 [36]
Kaneka Corporation[2]	Japan	Kaneka	P(3HB-*co*-3HHx)	1 000 [36]
Biomer Inc	Germany	Biomer	PHBV+PHB	50 [36]
Tianjian Gree Bio-Science Co/DSM	China	Green Bio	P(3HB-*co*-4HB)	10 000 [36]
Tianan Biologic, Ningbo	China	ENMAT®	P(3HB-*co*-3HV) + Ecoflex blend	10 000 [36]

material should preserve and maintain the characteristics of the product contained. According to the type of product to be contained, e.g., food, beverages, medicine, or other kinds, the conditions in which the perishing or the alteration occur are different, and therefore several precautions are considered during packaging processing, like vacuum conditions or inert atmosphere conditions. Despite the manner in which the product is packaged, a poor barrier effect is one of the common ways in which products perish, causing the shelf-life to be shortened. The negative consequences of a weak barrier effect should be considered not only against gases in the environment outside, but also to avoid leakage of gases contained inside the package. For example, in the food industry, some packages are required to have a strong barrier against oxygen or water vapor, while in the beverages industry, a barrier against CO_2 is necessary to maintain carbonated drinks. Based on these premises, barrier properties are mandatory for polymers employed in several fields of packaging.

PHA are thermoplastic biopolyesters produced by a variety of bacteria and archaea as storage materials in response to particular environmental stresses. Within the PHA family, the most widely used material is the homopolymer PHB and its copolymers with 3-hydroxyvalerate (PHBV), but according to the combined microorganism–carbon source, other monomeric units are also available, like 4-hydroxybutyrate (4HB) and 3-hydroxyhexanoate (3HHx). Nowadays, PHA have become commercially available through many sources and are already being used in small disposable products and in packaging materials [39]. Actually, the most widely used so-called "bio-based plastic" in the packaging industry is polylactic acid (PLA). However, PHA, and particularly PHB, also represent good candidates, especially when associated with nanofillers [40].

Due to the promising properties of some types of PHA in the packaging sector, their barrier properties are the objects of current research. Many of these studies are focused on making PHA-based composites with biodegradable fillers, in order to extend the properties of PHA while maintaining their major feature: high biodegradability in different environmental compartments.

In a study conducted by Tanase *et al.* [41], composite materials were realized starting from a commercial PHB and different percentage of commercially available cellulose fibers. The aim of the research was to improve PHB's physical-mechanical behavior. The polymer employed was PHB type 319 E, manufactured by Biomer (Krailling, Germany), while the cellulose fibers (CFs) were type EFC 1000 (Rettenmeier & Söhne AG, Germany). The composites were prepared by melting in a BRABENDER Plastograph and contained a plasticizer (Bis(2-(2-butoxyethoxy) ethyl) adipate (DBEEA)) with a ratio between PHB and plasticizer of 9:1. Neat PHB was used as a reference, and tests were conducted on sheets and films obtained from a heated press. They measured the water vapor transmission rate (WVTR) at 23°C and 85% relative humidity and they expressed the results as permeability in g/(m^2·day). Neat PHB possessed the lowest permeability, with a value of about 6 g/(m^2·day). Plasticized PHB showed a permeability of about 7 g/(m^2·day), while composites exhibited higher permeability as a function of CFs' content. The authors concluded that CFs are not effective at improving the barrier properties of PHB, and they also correlated the lowest PHB permeability with the highest crystalline content. Plasticizer and CFs reduced the crystalline content, which resulted in more chain mobility and higher permeability.

Similarly, Dhar *et al.* [42] prepared composites based on PHB and cellulose, but they used laboratory-synthesized cellulose nanocrystals (CNCs) obtained from pristine cellulose bamboo pulp. The polymer used, PHB, was retrieved from Sigma Aldrich, India. Composites were not obtained by melt processing, but by mixing a solution of PHB and a dispersion of CNCs in the same solvent with different CNCs' content. Films were obtained from solution casting. The authors measured the oxygen transmission rate (OTR), the steady state rate at which oxygen gas permeates through a film at specified conditions of temperature and relative humidity, at different temperatures and 0% of relative humidity. Furthermore, they extrapolated permeability data, diffusion, and solubility coefficients. The authors observed a drastic reduction in OTR values upon incorporation of 1% CNCs: from 300 to 162 (cm^3/(m^2·day)). Reduction in values of both solubility and diffusion coefficients is also observed, which implies increased resistance to permeation of oxygen molecules through the CNCs network. They explained OTR reduction due to the CNCs percolation network within the polymer matrix. According to the authors, a possible intercalation of the PHB matrix into the CNCs could be another explanation for OTR reduction. However, at higher CNCs loadings (>5 wt-%), such intercalation effect and percolation network turned out to be negligible, because they observed the formation of voids due to a difference in polarity between the hydrophilic CNCs and the hydrophobic PHB polymer. Voids could possibly lead to a significantly higher solubility coefficient, and consequently to an increased permeability coefficient. Dhar *et al.* also studied the oxygen permeation over a range of temperatures. They observed that the activation energy of permeation increases with the increase in CNCs loading.

Follain *et al.* [43] published a work where the transport properties of PHA films prepared from commercially available PHA were investigated for three inert gas molecules. Authors studied two types of commercially available PHBs from Biomer, Germany (P209 and P226) and a PHBV copolymer (3 mol-% of 3-hydroxyvalerate units) from Tianan Biologic Material, China (Y1000P). Furthermore, they also investigated the influence of the film preparation method – a solvent-casting and hot compression-molding process – on the permeation of gases. Nitrogen (N_2), oxygen (O_2), and carbon dioxide (CO_2), differing in their van der Waals molar volume, were selected because they are known not to interact (or to interact only very weakly) with organic polymers. Authors revealed that the impact of the film-forming process on the thermal features displayed by the PHA films is negligible, but additional free volume and spaces between polymer chains in the cast films favored the transfer of gas molecules. However, although the cast films (for oxygen: $P_{Y1000P} = 0.35$ Barrer, $P_{P209} = 1.95$ Barrer, and $P_{P226} = 0.49$ Barrer) were more permeable than the compressed films with a similar parameter ranking (for oxygen: $P_{Y1000P} = 0.10$ Barrer, $P_{P209} = 1.39$ Barrer, and $P_{P226} = 0.50$ Barrer), their permeability was lower than that of nonpolar polymer films such as LDPE, and by well-studied polyester films, such as PLA films (both of them possess $P > 1$ Barrer for N_2, CO_2, and O_2). For the PHA, the P, D, and S coefficients were reduced when the degree of crystallinity of the polymers increased.

In their paper, Keskin *et al.* [40] reviewed the potential of PHA composites as substitutes of petroleum-based polymers for packaging applications. The authors reported several studies in which PHA are added to organic and inorganic fillers to increase barrier properties. For example, CNCs, keratin, and lignin are reported to decrease gas permeability with filler content due to the increasing tortuosity in the PHA matrices. However, at a high load of filler, agglomeration can occur and a reduction of tortuosity increases the permeability. Furthermore, zinc oxide, boron nitride, montmorillonite, and clays are reported by Keskin *et al.* to decrease the permeability of PHA. It's of interest that the study reported the influence of plasticizer on the barrier effect. Propylene glycol, glycerol, triethyl citrate, castor oil, epoxidized soybean oil, and poly(ethylene glycol) were the plasticizers studied. The results showed that plasticizers increased the crystallinity of PHBV and affected mechanical properties adversely, but, however, some plasticizers improved the barrier properties of PHBV. Plasticizers which contain ketone and ether groups were found as suitable plasticizers for PHBV, and they are capable of interacting with PHBV. Plasticizers with low molecular mass have no effect on PHBV. The most effective plasticizer found for oxygen and water vapor permeability was glycerol at 5 wt-%. Authors concluded that PHA are suitable nominees for the replacement of synthetic plastics in the future. The large-scale use of these polymers composites is limited, due to their cost. Currently, many efforts have been made to develop the properties of the PHA with different additives by lowering the production costs world-wide. In particular, PHA nanocomposites, which are produced with nanosized fillers, have reached the level at which they can compete with the properties of petroleum-based plastics, making them usable in the packaging industry.

Akin *et al.* [44] reported a study in which effects of temperature and organo-modified clay addition on the water vapor permeability (WVP) properties of both PHB and PHBV polymer matrices were evaluated. Particularly, PHB and PHBV (5 wt-% of 3HV content) were supplied by Sigma Aldrich, while natural modified montmorillonite with a quaternary ammonium salt (Cloisite 10A, Southern Clay Product, USA) was used as filler. The PHB or PHBV composites were prepared by a solution intercalation method, then films were obtained by solution-casting. The WVP of nanocomposites was determined accordingly at 37.8°C and 90% relative humidity (RH). Moreover, temperature dependence of WVP of PHB and PHBV nanocomposite films were evaluated at 10°, 20°, 30°, and 40°C. Akin *et al.* reported a critical limit of clay content to barrier properties due to the changes in nanostructure after certain nanoclay loadings. Lowest permeability values were obtained at 2 and 1 wt-% clay content for PHB and PHBV. Depending on the amount of clay loaded and its dispersion level, the WVP of pure PHB and PHBV decreased by 41% and 25%, respectively. The authors observed that the increase in the tortuosity is the main factor that affects the water vapor barrier performance of nanocomposites. Besides the tortuosity effect, the addition of filler in the polymer matrix resulted in the decrease of the free volume of polymer that affects the solubility of the water vapor and hence the barrier performance of nanocomposites. Akin *et al.* observed that the temperature dependence on gas transport through a polymer generally follows an Arrhenius-type plot, where there is a linear relationship between a transport parameter logarithm and the reciprocal of temperature. The authors observed an exponential increase in permeability with increasing temperature, and they explained this behavior as resulting from the dependence of WVP on both temperature and partial pressure of water vapor across the sample.

Sanchez-Garcia *et al.* [45] reported a study in which the barrier performance of PHB, PHBV, and composites was evaluated and compared to PET. Melt-processable semi-crystalline thermoplastic polymers, PHB (powder form, density of 1.25 g/cm^3) and PHBV (pellet form, 12 mol-% of 3HV), were obtained from Goodfellow Cambridge Ltd, UK. Composites with a Nanoter 2000 grade based on modified montmorillonite (NanoBioMatters, Spain) were prepared by melt-blending, then films were obtained by hot compression. Permeability of oxygen, D-limonene, and water was measured.

The authors observed that the permeability was seen to generally decrease with increasing MMT content, as would have been expected. They concluded that PHB and its nanobiocomposites show better water, aroma (limonene), and oxygen barrier than pure PET. Nevertheless, the PET nanocomposites' oxygen permeability is the lowest of all materials considered.

In their paper, Cretois *et al.* [46] focused the attention on the copolyester poly(3-hydroxybutyrate-*co*-4-hydroxybutyrate) (P3HB4HB). Its physical properties are strongly dependent upon the fraction of 4HB units in the copolyester. The presence of 4HB units usually improves the processability and the elongation at break. However, the 4HB units inhibit the crystallization so that the decrease of the crystallinity degree reduces the barrier properties of P3HB4HB compared to PHB and PHBV. P3HB4HB (4HB of 13 mol-%, $\overline{Mw}$ of 3.1×10^5 g/mol, supplied by Tinajin Green Bioscience Company, China) and montmorillonite (Cloisites 30B, Southern Clay Products, USA) were mixed by an extrusion process, then composite films were obtained by compression-molding. Cretois *et al.* measured permeability, diffusion, and solubility coefficient values for nitrogen, oxygen, and carbon dioxide. They reported that the barrier effect is indeed dependent upon the penetrant used and cannot be linked only to the tortuosity effect imposed by montmorillonite. However, the addition of nanofiller in general leads to a decrease in permeability.

Vandewijngaarden *et al.* [47] investigated the barrier properties of P(3HB-*co*-3HHx) and its composites. P(3HB-*co*-3HHx) is one type of PHA-based copolymer that shows great potential, because of its low brittleness and wider processing window, as compared to PHB and PHBV, which are the two most investigated types of PHA. The authors reported that P(3HB-*co*-3HHx) has relatively low WVT (1.42 g mm/(m^2 day)) and a moderate oxygen permeability (8.3 cm^3 mm/(m^2 day atm)). However, for actual use as a high barrier food packaging material, the gas permeability must be drastically reduced. Vandewijngaarden *et al.* improved P(3HB-*co*-3HHx)'s gas barrier properties by the incorporation of organo-modified montmorillonite clay. They retrieved P(3HB-*co*-3HHx) powder and a commercial P(3HB-*co*-3HHx) granulate (trade name Aonilex X151A, both with a 3HHx content of 10.5 mol-%), from Kaneka Corporation, and melt-mixed them with Nanocor I.34 TCN montmorillonite. Finally, composite films were obtained by compression-molding. Vandewijngaarden *et al.* measured the permeability of O_2, CO_2 dioxide, and water vapor. The authors observed that for the P(3HB-*co*-3HHx) sample set, the permeability of nanocomposites for all three gases shows a decreasing trend upon increasing the MMT content.

The sample set produced from Aonilex pellets was also subjected to the same type of behavior. Vandewijngaarden *et al.* concluded that the clay platelets create a tortuous path which gas molecules must follow instead of passing straight through, thus acting as a physical barrier for gas permeation. Overall, according to the authors, this kind of composite has a certain potential for innovative packaging applications.

12.2 BIODEGRADABILITY OF POLYHYDROXYALKANOATES

12.2.1 Biodegradability and Biodegradable Polymers: An Overview

With advances in technology and the increase in the global population, plastic items have found wide applications in every aspect of life and industries. However, most conventional plastics such as PE, PET, PS, and PVC, are not biodegradable (except for polyolefins re-engineered with addition of pro-oxidant/pro-degradant catalysts). Their increasing accumulation in the environment has been a threat to the planet [48]. In fact, terrestrial and aquatic are the environmental compartments that are most cluttered with plastic waste which, due to atmospheric mechanical and physical agents, can fragment and give rise to the formation of so-called microplastics. The marine environment is particularly affected by this problem. Evidence is increasing that plastic items interacting with environmental contaminants can transport these products to the highest levels of the food chain (including food for pets and humans).

To overcome all these problems, some steps have been undertaken. One of the most common strategies involved the production of plastics with a high degree of biodegradability. Biodegradable plastics are seen by many as a promising solution to this problem because they are environmentally friendly. They can be derived from renewable feedstocks, thereby reducing greenhouse gas emissions. For instance, PHA and PLA can be produced by biotech processes using agricultural products and selected microorganism strains for PHA and chemical processes for PLA. Biodegradable plastics offer a lot of advantages such as increased soil fertility, low accumulation of bulky plastic materials in the environment (which invariably will minimize injuries to wild animals), and reductions in the cost of waste management.

Biodegradability is the ability of organic substances and materials to be degraded (oxidized) into simpler compounds through microbial respiration of environmentally ubiquitous microorganisms. If this biological process is completed, there is a total conversion of the original organic substances into simple inorganic molecules such as water and carbon dioxide (aerobic process) or water and methane (anaerobic process), and eventually, cellular biomass.

In general, (bio)degradation processes evolved through successive stages, including both abiotic and biotic factors, which must be considered. Abiotic degradation precedes microbial assimilation, and must absolutely be considered when explaining the degradation mechanisms. The mechanism of the polymer biodegradation takes place in several stages and the process could be stopped at each stage. The degradation begins from abiotic attacks of an oxidative or hydrolytic nature. In these steps, external conditions such as weather, aging, sunshine, soil burying, and moisture could accelerate the degradation process. Polymers consequently undergo thermal, chemical, mechanical, and photo degradation, becoming synergistic factors useful for the acceleration of the biodegradation process. Once the polymer is fragmented, the microbial activity starts on the surface of the material and in the bulk of the material. Microorganisms can act as mechanical, chemical, and enzymatic promoters. Depending on the polymer's constitution, on the environmental conditions (temperature, humidity, weather, pollutants), and on the microorganism involved (bacteria, protozoa, algae, fungi, lichenaceae), the materials could be damaged in a different way. On the surface of the polymers, a biofilm of simple molecules, sources of carbon and nitrogen, is created, which could become nutrients for some microorganisms. This phase is called biodeterioration and depolymerization and is followed by biofragmentation. In abiotic degradation, this is different: in this step, microorganisms secrete enzymes and free radicals for cleaving the polymers' backbone to release monomer or oligomeric components that can be assimilated by microbial organism strains. The final stage is the assimilation. In this step, a direct interaction between the polymer fragments and the microbial cells is evident. In the last phase of the biodegradation process, the microorganism consortia metabolize fragmented macromolecules as energy sources and release CO_2 and water (aerobic condition) or CH_4 (anaerobic condition), and eventually biomass [49].

Fundamental aspects that determine biodegradability are linked, first of all, to the chemical composition of the polymeric material. The structure and chemical composition of the polymeric material regulates the chemical and physical properties of the material and its interaction with the physical environment. In turn, this aspect affects the biodegradability of the material according to particular mechanisms of both biotic and abiotic nature.

In fact, the biodegradability of a manufactured article is not governed exclusively by the chemical nature of the polymer, but the conditions and the environmental compartment in which the biodegradative processes take place, are also both crucial.

The environment is a determining factor in governing the speed and the degree of biodegradation of the matrices of bio-based polymers. The environmental compartments in which biodegradation processes can occur are attributable to two categories: aerobic (presence of oxygen), and anaerobic (lack of oxygen). The biodegradability of polymeric materials and related plastic items in general depends upon [50]:

- Presence of microbial strains susceptible to colonizing polymeric materials and related fragmented products.
- Temperature.
- Availability of oxygen.
- Moisture content.
- Chemical environment constitution (e.g., pH-value, presence of salts, etc.).
- Molecular weight and its distribution.
- Chemical cross-linking.
- Chemical structure of the polymer backbone.
- Morphology (difference between amorphous or semi-crystalline material).

In particular, the first three factors are of crucial importance in biodegradation processes and govern the relevant times and results.

Among all the kinds of biodegradable polymers, PHA are one of the most promising categories of biodegradable polymers. In fact, over 90 different types of PHA consisting of various monomeric components have been reported (and indeed, the number is increasing). Some PHA (mostly *scl*-PHA) behave similarly to conventional thermoplastics such as PE and PP, while others (*mcl*-PHA) are elastomeric [51]. The large number of possible combinations of different repeating units depending on the synthetic route, the microbial strains exploited, and the type of feed make the PHA production, and hence the application almost boundless.

12.2.2 Biodegradability of PHA

One of the unique properties of a lot of PHA materials is their ubiquitous biodegradability in natural or anthropic environments (fresh water, brackish and salt water, soil, sludge, and compost). This aspect is one of the commercially attractive features which distinguishes the PHA from fossil-based plastics. The main advantage of PHA over other types of biodegradable plastics are that they do not require special environmental conditions and can degrade in either aerobic or anaerobic environments through thermal degradation or enzymatic hydrolysis.

Biodegradation of PHA is performed by microorganisms, which inhabit a specific natural environment. But in the absence of biological agents (bacteria, algae, and fungi) PHA are barely subject to mass loss under normal conditions, and they are degraded in biological media to form products harmless to the environment [52].

Considering that PHA are high-molecular-weight solid polymers that cannot be transported through the cell wall, microorganisms, such as bacteria and fungi, excrete extracellular PHA-degrading enzymes (PHA depolymerases) that hydrolyze the solid PHA into water-soluble monomeric components and oligomers. These low-molecular-weight degradation products are then transported into the cell and subsequently metabolized as carbon and energy sources. Analyses of the structural genes of the extracellular PHA depolymerase from *Rhodospirillum pickettii* T1, which is the most studied PHA depolymerase, have shown that they have a multi-functional domain structure, i.e., a catalytic domain, a substrate-binding domain, and a linker region connecting the two domains. The substrate-binding domain is primarily responsible for the adsorption of the PHA depolymerase to the surface of water-insoluble polyester materials. In recent years, it has also been proposed that the substrate-binding domain is of great importance in the degradation process, disturbing molecular chain packing at the surface of PHB crystals, which induces the initial step of the enzymatic degradation. Rates of enzymatic degradation of PHB by PHA depolymerase are strongly dependent on the crystalline state of PHB, as well as by the concentration, properties, and reaction conditions of PHA depolymerase. The rate of enzymatic hydrolysis increased to a maximum value with the concentration of PHA depolymerase, followed by a gradual decrease [53]. End products of PHA degradation in aerobic environments are CO_2 and water, while CH_4 and water are produced under anaerobic conditions, as formerly stated.

12.2.3 PHA's Biodegradation in Different Environmental Compartments

Studies have been reported on biodegradation of pure PHA in aqueous environments, activated sludge, compost, and even soil [54].

One of the most important biodegradation processes is realized under composting conditions. In this case, PHA packaging (in many cases blended with natural filler and other bio-based and biodegradable polymer components) showed the most competitive performance in terms of biodegradation percentage and time compared with cellulose [55].

Wang *et al.* [56] studied the influence of chemical structure on the biodegradability of PHA films. In particular, the PHB, PHBV (40% mol 3HV), PHBV (20% mol 3HV), PHBV (3 mol-% 3HV), and P(3HB-*co*-4HB) (10 mol-% 4HB) were investigated under controlled composting conditions according to ISO 14855-1 [57]. In this study, it was found that PHA with different chemical structures can be biodegraded under controlled composting conditions. A preliminary disintegration test was carried out to determine the fragmentation of the different films. The result of this test was to observe that the size of all the PHA films with different chemical structures became smaller and smaller with the increase in composting time. After 10 days of composting, all the PHA films had lost their integrated appearance, with PHBV-40 and P(3HB-*co*-4HB) apparently degrading faster than the others. After 20 days, only 50% of the film could be found for all the samples. The degraded materials are either digested by the microorganisms or broken into such small fragments that no film residuals can be found in the compost. More material loss and a greater appearance change can be taken to reflect a faster degradation rate. Comparisons among the biodegradation curve observed that PHBV-40, PHBV-20, PHBV-3, and P(3HB-*co*-4HB) had a better propensity to biodegradation with respect to the reference (cellulose), while PHB was almost similar. At the end of this study, it is possible to determine the order of biodegradation degree, that is: PHBV-40 > P3HB4HB > PHBV-20 > PHBV-3 > PHB. The reason for this difference in the biodegradability propensity of PHA is that the amorphous part degrades faster than the crystalline part, and in the studied polymers, crystallinity of the copolymer decreases with the increase of 3HV and 4HB molar fractions. In another study published by Arrieta *et al.* [58], the biodegradability of PLA/PHB blend films plasticized with poly(ethylene glycol) and acetyl-tri-n-butyl citrate is evaluated. An important consideration is obtained from this test: despite the fact that PHB acts as a nucleating agent in PLA/PHB blends slowing down the PLA disintegration, the presence of plasticizers speeded it up. The high biodegradability of PHA (homopolymers, copolymers, or in blend with other biodegradable polymers) under industrial composting conditions make this polymer very desirable as material to produce packaging that can be disposed of in the organic fraction chain of urban waste. Moreover, one of the main criticisms that have been leveled at biodegradable plastics is that some materials will biodegrade in industrial composting plants, but less reliably in home composts or aquatic environmental conditions. Other materials have even been shown to merely fragment rather than fully biodegrade, which further exacerbates the negative environmental impacts of plastic items.

Considering environmental conditions where the concentration of microorganisms is lower than compost or municipal sludge, biodegradation of bio-based plastic items are more complicated.

For PHA, good biodegradability was observed also in soil and seawater. Studies carried out by Yousif *et al.* [59] showed that PHB films degrade completely after 3 weeks when buried in soil. In the same work it is reported that the biodegradation time can be increased with the addition of TiO_2 nanoparticles, or by treatment of PHB under UV light. This result is promising in order to understand the persistence of PHA packaging loaded with inorganic nanoparticles in the environment.

Similar results are reported by Narancic *et al.* [60]. In this work, neat polymers PLA, PHB, PHO, poly(butylene succinate), (PBS), thermoplastic starch, polycaprolactone (PCL), and blends thereof are tested for biodegradation across seven managed and unmanaged environments. PLA, which is one of the world's best-selling biodegradable plastics, showed very good biodegradability in industrial compost, but it is not home compostable and extremely recalcitrant to biodegradation under natural environmental condition (such as fresh and marine waters). On the other hand, PLA blended with PCL becomes home compostable. Another important result shown in this study is that the majority of the tested bio-based plastics and their blends degrade by thermophilic anaerobic digestion with high biogas output, but degradation times are 3–6 times longer than the retention times in commercial plants. PHA and their blends showed good biodegradation in soil and water. In fact, PHB degraded completely in all tested aquatic environments: marine pelagic (30°C), freshwater aerobic (21°C), and aquatic mesophilic anaerobic (35°C). Only 6% of PHO degraded in anaerobic aquatic conditions and up to 51% in freshwater in the testing time frame of 56 days. PHB/PHO (85/15 weight % ratio) blend showed a slower degradation rate in the first 10 days compared to PHB alone, but the blend reached the same level of biodegradation as PHB in marine conditions after 56 days. Unfortunately, the majority of polymers and their blends tested in this study (except PHB) failed to achieve ISO and ASTM biodegradation standards, and some failed to show any biodegradation.

Taking into account the literature that reports the mass loss of PHA in the natural marine environment, Dilkes-Hoffman *et al.* [61] studied the rate of biodegradation of PHA in the marine environment and applied this to the lifetime estimation of PHA products, such as food packaging and single-use objects. This provides the clarification required as to what marine biodegradation of PHA means in practice and allows the risks and benefits of using PHA to be transparently discussed. In order to determine average rates of biodegradation of PHA, a simple approach to biodegradation has been adopted. A simplified biodegradation process has been conceptualized with three key steps: biofilm formation, depolymerization, and mineralization. As a result, it was determined that the mean rate of biodegradation of PHA in the marine environment is 0.04–0.09 mg/(day•cm) ($p = 0.05$) and that, for example, a PHA water bottle could be expected to take between 1.5 and 3.5 years to completely biodegrade.

Based on the recent study published by researchers around the world, it is possible to conclude that PHA, and in particular PHB, are very important in plastic waste management. These polymers showed very high versatility in terms of biodegradability in different environmental compartments (natural or anthropic). However, biodegradable plastic blends (even if with PHA) need careful post-consumer management, and further design to allow more rapid biodegradation in multiple environments is needed, as their release into the environment can cause plastic pollution.

12.3 FUTURE PERSPECTIVES ON PHA IN FOOD PACKAGING: WITHIN THE CIRCULAR ECONOMY'S EXPECTATIONS

The principal function of packaging is protection and preservation from external contamination of the selected content. Packaging protects food from environmental influences such as heat, light, the presence or absence of moisture, oxygen, pressure, enzymes, odors, microorganisms, insects, dirt and dust particles, gaseous emissions, etc. [62].

In a study published by Geyer *et al.* [63] the plastic production data were combined with product lifetime distributions for eight different industrial use sectors, or product categories, to model how long plastics are in use before they reach the end of their useful lifetimes and are discarded. From this analysis, it appears that the packaging is the sector where plastics are largely consumed for the shortest use life. In addition, the consumption of plastic produced yearly for packaging is about 40% of total plastic converted [64] (for the European Union). Polyolefins are the most required polymers in the packaging sector. The reported data highlight the importance to consider the end-of-life of packaging materials, following a sustainable disposal for plastic waste, and alternative production which is not based on fossil fuel feedstocks as a supply. Nowadays packaging produced with PHA is a minority part of the all bio-based polymer in material productions. In fact, the non-biodegradable bio-based plastics are about 64% of production while the biodegradable ones are 36% (PHA are only 3% of the total bio-based polymers) [65].

The packaging, and in particular the food packaging, is becoming more and more technological, advanced, and "smart". A lot of different polymer blends, additives, nanofillers, and inorganic fibers make innovative packaging perform better, but in many cases, is an important barrier to recycling or to disposing of waste in a sustainable way. In fact, the most common packaging materials were planned to be useful, but were not conceived to have a low environmental impact at the end of their service life. Actually, the life of packaging is almost linear; after use, the plastic waste is collected in a landfill or incinerated (or released into the environment). Indeed, only a low amount of plastic products are re-introduced into the production chain by re-use or recycling at the end of their service life. The linear model follows the path of production, use, and disposal. This behavior is not sustainable considering the population growth, the consumption of the natural resources (in particular coal and oil), and the increasing of the amount of plastic in the environment (mainly in seas, oceans, and lakes). The circular economy represents an alternative model characterized by greater sustainability compared to the traditional linear economy. In a circular economy, on the contrary, the resources are used for the longest possible period of time, and the value is exploited to the maximum during use, and then products and materials are recovered and regenerated at the end of their service life. The unique characteristics of plastics put them among the protagonists of the path towards a more sustainable and resource-efficient future. Lightweight, versatile, and durable plastics can help save key resources such as energy and water in strategic sectors, including packaging, construction, automotive, and renewable energy, to give just a few examples. Furthermore, the use of plastics in packaging can contribute to the reduction of food waste. However, to improve the circularity of plastics, it is essential to ensure that an increasing amount of plastic-based waste is recovered and does not end up in landfills or in the environment [66]. Another solution is to substitute the commodity plastics, such as those based on polyolefins, with those based on biodegradable polymeric materials to produce biodegradable plastic products. The European Commission (May 2018) has published a proposal for a Directive of the European Council on the reduction of the impact of certain plastic products on the environment. The goal is to prevent and reduce marine litter from single-use plastics and plastic fishing gear. Some products are being banned from the market, e.g., plastic cutlery, straws, and plates, whereas the use of plastic cups and food containers needs to be reduced by the Member States. Among these actions, there are other rules that Member States and their producers must fulfill (European Commission 2018 [67]). One example of a national ordinance, which is a driving factor for increasing the use of bio-based polymers, is the French ordinance Décret no. 2016-1170 (2016). It requires all disposable tableware (e.g., cups and plates) to be home-compostable and produced from 50% renewable materials by January 2020. Furthermore, the required renewable content will be raised up to 60% by January 2025 [68].

Renewable plastics in the circular economy are aiming at replacing conventional recalcitrant plastics used in the current linear economy. The inputs for the linear path of recalcitrant plastics are fossil carbon feedstocks that required millions of years to accumulate, and recycle loops are small; the outputs either accumulate in landfills or into the environment. By contrast, renewable plastics fit within the circular economy vision with large recycle loops and feedstocks that are renewable over a short time (<10 years). In order to meet current plastic demands, a roughly 500-fold increase in renewable plastics production is required. In addition, production costs need to be decreased by a factor of 2–4 to achieve cost parity with recalcitrant fossil fuel feedstock-derived plastics. Renewable plastic will need to match the in-use functionality of current plastics while also safeguarding the environment and enabling disassembly at end-of-life. To that end, an 'ecocyclable' framework has been proposed in which new materials are compared to reference polymers in terms of degradability, susceptibility to bioaccumulation, and toxicity [67].

Considering PHA, critical bioprocess challenges in their production are feedstock and production costs, the need to tailor polymer structure for specific applications, maximizing productivity (g PHA/(L·h)), and optimizing downstream polymer extraction and purification. Costs can be decreased through economies of scale and use of waste-derived feedstocks. Actually, the most common production for PHA (and in particular, PHB) beginning from sugar (from sugar cane or beet) or starch (from corn, potato, wheat). PHB production requires a lot of land and water to grow different crops used to derive starting raw material: from 0.3 to 2 ha/t PHB of land and from 2000 to 13000 m^3 of water per ton of PHB [69]. Moreover, the conversion of sugar or starch into PHB is obtained through up to seven steps and the conversion is not quantitative.

In order to evaluate the most sustainable production processes, life cycle assessments (LCA) can be used. This evaluation instrument provides solid, comprehensive, and quantifiable information about the ecological performance of products, and thus answer the questions about any superiority of bio-polymers regarding their environmental impacts. In a classical meta-study published by Patel *et al.* [70], a total of 20 LCA studies about various bio-based polymers and natural fibers are compared, and among them five are about PHA. The studies show mixed results for total energy requirement (cradle-to-gate), depending on raw materials (starch, sugar beet) and the production (including PHA grown in genetically modified corn). This can vary from 10% better than fossil-based high-density PE, to dramatically worse (by a factor of almost 8). Similar results can be obtained for CO_2 emissions from the life cycle.

In 2010, Tabone *et al.* [71] compared 12 polymers of fossil and biological origin, among them PHA from corn stalks and corn grain. The study is of great interest insofar as it uses a wide range of problem-oriented ecological indicators like acidification, carcinogens, eutrophication, eco-toxicity, Global Warming Potential (GWP), and ozone and fossil fuel depletion. While it states the superiority of most bio-based materials (and PHA from corn stalks in particular) in GWP, it shows that they still have other impacts where they perform equally or even worse than many fossil-based materials. PHA from corn grain even tops the impact list in acidification by far, and both PHA from corn stalks and grain have a larger ozone depletion potential than fossil-based polymers and are on a par with or worse than PE, poly(carbonates) and PP in eco-toxicity. Many of these impacts are, however, caused by the agricultural production of the raw materials [72].

To avoid competition with cultivated lands for crops (and associated water, energy, and fertilizer requirements), to enhance the profitability of the system, and to facilitate its implementation in the plastic market, many operating alternatives have been proposed at lab-scale. One of the most promising alternatives is the use of industrial by-products and/or waste streams, i.e., agriculture feedstock, waste plant oils, or wastewater, as sources of carbon for PHA production. PHA production using waste streams can be considered an environmentally friendly management method, in line with the circular economy political vision [73]. In fact, this production consumes waste material to produce attractive polymers that can be re-introduced in the economic-productive circuits. However, pilot- and industrial-scale implementation require new technological advances to facilitate the employment of waste streams as valuable raw material.

Among all kinds of PHA, the PHBV copolymers are considered the most promising substitutes for oil-based synthetic polymers. Among their advantages, they can be biologically synthesized using various feedstocks such as agro-food and urban by-products, residues, and wastes, either liquid or solid. The following European-funded projects [74–78] demonstrate the feasibility, using food industry by-products as feedstock (olive wastewater or cheese whey) and mixed natural microbial cultures, of lab-scale production of a PHBV with a 3HV fraction (in the range of 10–25%) higher than the current commercial grade (3%). This higher 3HV fraction induces some polymer structural changes that can be advantageous to its processing and conversion into packaging. Higher 3HV contents could be achieved by using Volatile Fatty Acids (VFAs) precursors with a high propionic acid content. The incorporation of low-cost lignocellulosic fillers stemming from lignocellulosic solid residues into PHBV permitted tailoring functional properties, especially water vapor and oxygen permeability, while decreasing the overall cost of the final bio-composite packaging material and maintaining its biodegradability. Incorporation of lignocellulosic fillers tends to decrease the ultimate tensile properties because of a lack of adhesion between the hydrophobic matrix and hydrophilic fibers. Globally, the mechanical properties are governed by those of the PHBV matrix which is, for the commercial grade with low 3HV content, too brittle to be used for flexible packaging application. To go further in the industrial deployment of PHA-based material, PHA conversion must be scaled-up based on the use of an optimized eco-efficient mixed microbial culture-based process. This allows decreasing investments and operating costs of PHA conversion with respect to pure culture and are made easier by using non-costly by-products such as feedstock. This type of process will enable the bioconversion of agro-food residues (no competition with food usage) into value-added material that is a better alternative use for bio-waste rather than only energy or compost. Municipal and industrial bio-waste could also be converted into PHA [79]. Many studies exist that proposed different organic waste, such as sugar industry wastewater [80], corn cob, straw, and banana peel [81]. In general, many different systems exist that have been proposed to convert organic and food waste into PHA, as there are numerous waste streams generated by food production and municipal waste collection. In food waste, each source has its own complexities and requires different pre-treatments, bacterial strains, culturing conditions, and downstream processing. Often, the organics associated with food wastes are complex compounds that cannot be directly used by PHA-producing organisms. In these cases, a pre-treatment or processing method is necessary to convert the complex molecules found in waste into PHA precursors. Precursors include simple sugars like glucose or lactose and fatty acids like acetic or propionic acids. Many of the simpler food wastes are hydrolyzed to convert the food waste into suitable precursor molecules and then fed directly to a pure culture of an appropriate microbial strain. Whey, starch, oils, lignocellulosic materials, and legume and sugar wastes each have methods proposed to produce PHA [82]. All these processes are at laboratory scale. Valentino *et al.* [83] in their study proposed PHA production by mixed microbial culture using pre-treated organic fraction of municipal solid waste and excess thickened secondary sludge, applying the feast–famine approach in the traditional three-step process scheme, developed at pilot scale. As a result, the overall process yield of PHA production from selected organic waste has been estimated to be 6.5 wt-% with respect to the total volatile solids on untreated waste streams. This solution allows for the use of one main technology for the conversion of urban bio-wastes into PHA, while also minimizing any residual or consequent waste to be disposed of. This work demonstrated how the innovation in urban waste management could contribute to better waste collection and processing for the production of higher value bio-based products.

Sustainable food packaging in the circular economy vision may only be achieved by combining such efforts, considering any conflicting goals, and involving all stakeholders, including food and packaging manufacturers, recyclers, decision-makers, civil society, and consumers. Innovative processes to produce bio-polymers by sourcing alternative sources of food and feed products must be studied and applied. As soon as the technological barriers related to the process scale-up are overcome, the production of PHA can become environmentally and economically advantageous. Considering the potential of PHA (in terms of performance, environmental impact, and market), they are the main candidates for packaging production that does not increase the quantity of plastic in the environment and that respects all the dictates of the circular economy and sustainability concepts.

From this perspective, PHA have recently been produced by microbial bioconversion of oxidatively fragmented plastics based on mass full carbon backbone polymeric materials such as PS [84], PP [85], and PE [86]. The attained results are opening a new challenging scenery in the management of plastic items, based on mass fossil fuel feedstock polymeric materials, once they reached the end of their relevant service life.

12.4 CONCLUSIONS AND OUTLOOK

PHA are good candidates to be employed in packaging. Many studies about barrier properties of PHA are focused on PHB and PHBV, and some literature on P(3HB-*co*-4HB) or P(3HB-*co*-3HHx) is known as well. Although PHA exhibit good and comparable barrier properties with respect to traditional polyolefins, for certain applications, permeability values need to be lowered. The most studied approach is aimed at reducing permeability by obtaining composite materials with PHA, and also obtaining better mechanical performances and lowering the polymer price, which, among others, is a heavy drawback for PHA applications.

Conventional plastics for food packaging are neither renewable nor biodegradable, resulting in serious environmental problems. Plastic recycling is often not economical and practical due to contamination of the food packaging. The move towards biodegradable food packaging is the only positive way, even if polymer characteristics such as permeability to gases, humidity, and odors remain a concern. The cost is also the most important limiting factor for PHA applications in the food industry.

A further significant problem for the application of PHA in food packaging is also represented by their limited flexibility. Although blending them may reduce brittleness, the incorporation of nanofillers into PHA may result in higher Young's modulus and increased toughness. Further comprehensive investigation on the improvement of PHA barrier properties is represented by the incorporation of suited nanofillers. Also, decreased water permeability and antimicrobial effects of nanoparticles, such as MMTs, reduce the biodegradation rate. Solutions for this limitation cannot be fixed without continuing research and development over the incoming years. Further investigations are required regarding the potential migration of degradation products produced during either processing or biodegradation.

Moreover, incorporation of sufficient dispersed nanofillers into packaging materials suggests a new strategy for preserving and extending the antimicrobial shelf-life of foods sensitive to low temperatures. Nanocomposite packaging is forecast to make up a significant element of the food packaging market in the near future.

REFERENCES

1. Statista – Global plastic production from 1950 to 2018 (in million metric tons). https://www.statista.com/statistics/282732/global-production-of-plastics-since-1950/, accessed 09/06/2020.
2. Keshavarz T, Roy I. Polyhydroxyalkanoates: bioplastics with a green agenda. *Curr Opin Microbiol* 2010; 13: 321–326.
3. Tan GYA, Chen CL, Li L, *et al.* Start a research on biopolymer polyhydroxyalkanoate (PHA): a review. *Polymer* 2014; 6: 706–754.

4. Prasad P, Kochhar A. Active packaging in food industry: a review. *IOSR J Environ Sci Toxicol Food Technol* 2014; 8: 01–07.
5. Israni N, Shivakumar S. Polyhydroxyalkanoates in packaging. In: Kalia V (ed.), *Biotechnological Applications of Polyhydroxyalkanoates*, 1 edn. Springer, Singapore, 2019; 363–388.
6. Chen GQ. Plastics completely synthesized by bacteria: polyhydroxyalkanoates. In: Guo G, Chen Q (eds.), *Plastics from Bacteria: Natural Functions and Applications*, 2010; 17–37. Springer, Berlin.
7. Chanprateep S. Current trends in biodegradable polyhydroxyalkanoates. *J Biosci Bioeng* 2010; 110: 621–632.
8. Rai R, Keshavarz T, Roether J, *et al.* Medium chain length polyhydroxyalkanoates, promising new biomedical materials for the future. *Mater Sci Eng R* 2011; 72: 29–47.
9. Khanna S, Srivastava A. Recent advances in microbial polyhydroxyalkanoates. *Process Biochem* 2005; 40: 607–619.
10. Arrieta, MP, Samper MD, Aldas M, *et al.* On the use of PLA-PHB blends for sustainable food packaging applications. *Materials* 2017; 10(9): 1008.
11. Koller M. Poly(hydroxyalkanoates) for food packaging: application and attempts towards implementation. *Appl Food Biotechnol* 2014; 1(1): 3–15.
12. Plackett D, Siro I. Polyhydroxyalkanoates (PHAs) for food packaging. In: Lagarón JM (ed.), *Multifunctional and Nanoreinforced Polymers for Food Packaging*. Woodhead Publishing, Cambridge, UK, 2011: 498–526.
13. Bugnicourt E, Cinelli P, Lazzeri A, *et al.* Polyhydroxyalkanoate (PHA): Review of synthesis, characteristics, processing and potential applications in packaging. *Express Polym Lett* 2014; 8(11): 791–808.
14. Bucci DZ, Tavares LBB, Sell I. PHB packaging for the storage of food products. *Polym Test* 2005; 24(5): 564–571.
15. Pérez Amaro L, Chen H, Barghini A, *et al.* High performance compostable biocomposites based on bacterial polyesters suitable for injection molding and blow extrusion. *Chem Biochem Eng Q* 2015; 29(2): 261–274.
16. Greene J. PHA biodegradable blow-molded bottles: compounding performance. *Plast Eng* 2013; 69(1): 16.
17. Fabra MJ, Sanchez G, Lopez-Rubio A, *et al.* Microbiological and ageing performance of polyhydroxyalkanoates-based multilayer structures of interest in food packaging. *LWT-Food Sci Technol* 2014; 29(2): 760–767.
18. Weber CJ, Haugaard V, Festersen R, *et al.* Production and applications of biobased packaging materials for the food industry. *Food Addit Contam* 2002; 19(S1): 172–177.
19. Saltveit ME. Is it possible to find an optimal controlled atmosphere? *Postharvest Biol Tech* 2003; 27(1): 3–13.
20. Weber CJ. Biobased packaging materials for the food industry. Status and perspectives. *Conference Proceeding, The Food Biopack Conference*. The Royal Veterinary and Agricultural University, Denmark, 2000.
21. Rydz J, Musiol M, Zawidlak-Wegrzynska B, *et al.* Present and future of biodegradable polymers for food packaging applications. In: Mihai Grumezescu AM, Holban AM (eds.), *Biopolymers for Food Design*, 1 edn. Academic Press, MA, 2018: 431–467.
22. Pérez Amaro L, Abdelwahab MA, Morelli A, *et al.* Bacterial polyesters: the issue of their market acceptance and potential Solutions. In: Koller M (ed.), *Recent Advances in Biotechnology*, Vol. 2: 3–74, Bentham Science, Sharjah, United Arab Emirates, 2016.
23. Rabnawaz M, Wyman I, Auras R, *et al.* A roadmap towards green packaging: the current status and future outlook for polyesters in the packaging industry. *Green Chem* 2017; 19: 4737–4753.
24. Hourston DJ. *Degradation of Plastics and Polymers. Shreir's Corrosion*, 3rd ed. Elsevier, Amsterdam, 2010.

25. Khosravi-Darani K, Bucci DZ. Application of poly(hydroxyalkanoate) in food packaging: improvements by nanotechnology. *Chem Biochem Eng Q* 2015; 29(2): 275–285.
26. Robertson GL. *Food Packaging: Principles and Practice*, 3rd ed. CRC Press, Boca Raton, FL, 2012.
27. Vartiainen J, Vaha-Nissi M, Harli A. Biopolymer films and coatings in packaging applications - a review of recent developments. *Mater Sci Appl* 2014; 5: 708–718.
28. Ahvenainen R. *Novel Food Packaging Techniques*. Wood head Publishing, Cambridge, 2003.
29. Yadav A, Mangaraj S, Ranjeet S, *et al.* Biopolymers as packaging material in food and allied industry. *Int J Chem Stud* 2018; 6(2): 2411–2418.
30. Webb HK, Arnott J, Crawford RJ, Ivanova EP. Plastic degradation and its environmental implications with special reference to poly(ethylene terephthalate). *Polymers* 2013; 5: 1–18.
31. Halley P. Biodegradable packaging for the food industry. *Packag Bottling Int* 2002; 4(4): 56–57.
32. Aeschelmann F, Carus M. Bio-based building blocks and polymers in the world – capacities, production and applications: status quo and trends toward 2020. *Industrial Biotechnology* 2015; 11(3): 154–159.
33. Brown H, Williams J. Packaged product quality and shelf life. In: Coles R, McDowell D, Kirwan MJ (eds.), *Food Packaging Technology*. Blackwell Publishing Ltd./CRC Press LLC, Boca Raton, FL, 2003; 65–94.
34. Thakur VK, Thakur MK. *Handbook of Sustainable Polymers: Processing and Applications*. CRC Press. Boca Raton, FL, 2015.
35. Jacquel L, Lo CW, Wei YH, *et al.* Isolation and purification of bacterial poly(3-hydroxyalkanoates). *Biochemical Eng J* 2008; 39(1): 15–27.
36. Alavi S, Thomas S, Sandeep KP, *et al.* (eds.) *Polymers for Packaging Applications*, 1 edn. CRC Press, Taylor & Francis Group, Boca Raton, FL, 2014.
37. Kaneka Weins Japan Bioindustry award 2019 for seawater biodegradable bioplastics. *Bioplastics News*, 08/08/2019. https://window-to-japan.eu/2019/08/06/jbas-bioindustry-award-2019-goes-to-kaneka-and-riken-researchers-for-the-development-of-a-phbh-polymer-biodegradable-on-soil-and-in-seawater/
38. *Plastics – The Facts 2019. An Analysis of European Plastics Production, Demand and Waste Data*. https://www.plasticseurope.org/application/files/9715/7129/9584/FINAL_web_version_Plastics_the_facts2019_14102019.pdf [accessed 9/06/2020].
39. Pardo-Ibanez P, Lopez-Rubio A, Martinez-Sanz M, *et al.* Keratin–polyhydroxyalkanoate melt-compounded composites with improved barrier properties of interest in food packaging applications. *J Appl Polym Sci* 2014; 131: 39947.
40. Keskin G, Kizil G, Bechelany M, *et al.* Potential of polyhydroxyalkanoate (PHA) polymers family as substitutes of petroleum based polymers for packaging applications and solutions brought by their composites for form barrier materials. *Pure Appl Chem* 2017; 89: 1841–1848.
41. Tanase EE, Popa ME, Rapa M, *et al.* PHB/cellulose fibers based materials: physical, mechanical and barrier properties. *Agric Sci Procedia* 2015; 6: 608–615.
42. Dhar P, Bhardwaj U, Kumar A, *et al.* Poly(3-hydroxybutyrate)/cellulose nanocrystal films for food packaging applications: barrier and migration studies. *Pol Eng Sci* 2015; 55: 2389–2395.
43. Follain N, Chappey C, Dargent E, *et al.* Structure and barrier properties of biodegradable polyhydroxyalkanoate films. *J Phys Chem* 2014; 118: 6165–6177.
44. Akin O, Tihminlioglu F. Effects of organo-modified clay addition and temperature on the water vapour barrier properties of polyhydroxy butyrate homo and copolymer nanocomposite films for packaging applications. *J Polym Environ* 2018; 26(3): 1121–1132.

45. Sanchez-Garcia MD, Gimenez E, Lagaron JM. Novel PET nanocomposites of interest in food packaging applications and comparative barrier performance with biopolyester nanocomposites. *J Plast Film Sheeting* 2007; 23: 133–148.
46. Cretois R, Follain N, Dargent E, *et al.* Poly(3-hydroxybutyrate-*co*-4-hydroxybutyrate) based nanocomposites: influence of the microstructure on the barrier properties. *Phys Chem Chem Phys* 2015; 17: 11313–11323.
47. Vandewijngaarden J, Wauters R, Murariu M, *et al.* Poly(3-hydroxybutyrate-co-3-hy droxyhexanoate)/organomodified montmorillonite nanocomposites for potential food packaging applications. *J Polym Environ* 2016; 24(2): 104–118.
48. Tokiwa Y, Calabia BP, Ugwu CU, *et al.* Biodegradability of plastics. *Int J Mol Sci* 2009; 10: 3722–3742.
49. Siracusa V. Microbial degradation of synthetic biopolymers waste. *Polymers* 2019; 11(6): 1066–1084.
50. Kale G, Kijchavengkul T, Auras R, *et al.* Compostability of bioplastic packaging materials: an overview. *Macromol Biosci* 2007; 7: 255–277.
51. Ghanbarzadeh B, Almasi H. *Biodegradable polymers, biodegradation – Life of Science*, IntechOpen, London, UK, 2013.
52. Shabina M, Afzal M, Hameed S. Bacterial polyhydroxyalkanoates-eco-friendly next generation plastic: production, biocompatibility, biodegradation, physical properties and applications. *Green Chem Lett Rev* 2015; 8: 56–77.
53. Numata K, Abe H, Iwata T. Biodegradability of poly(hydroxyalkanoate) materials. *Materials* 2009; 2: 1104–1126.
54. Arcos-Hernandez MV, Laycock B, Pratt S, *et al.* Biodegradation in a soil environment of activated sludge derived polyhydroxyalkanoate (PHBV). *Polym Degrad Stabil* 2012; 97: 2301–2312.
55. Mohee R, Unmar GD, Mudhoo A, *et al.* Determining biodegradability of plastic materials under controlled and natural composting environments. *Waste Manag* 2007; 27: 1486–1493.
56. Weng YX, Wang XL, Wang YZ, *et al.* Biodegradation behavior of PHAs with different chemical structures under controlled composting conditions. *Polym Test* 2011; 30: 372–380.
57. ISO 14855-1:2005 Determination of the ultimate aerobic biodegradability of plastic materials under controlled composting conditions – method by analysis of evolved carbon dioxide – part 1: general method.
58. Arrieta MP, López J, Rayónc E, *et al.* Disintegrability under composting conditions of plasticized PLA–PHB blends. *Polym Degrad Stabil* 2014; 108: 307–318.
59. Yousif E, Altaee N, El-Hiti GA, *et al.* Biodegradation of different formulations of polyhydroxybutyrate films in soil. *Springerplus* 2016; 5(1): 762–774.
60. Narancic T, Verstichel S, Reddy Chaganti S, *et al.* Biodegradable plastic blends create new possibilities for end-of-life management of plastics but they are not a panacea for plastic pollution. *Environ Sci Technol* 2018; 52: 10441–10452.
61. Dilkes-Hoffman LS, Lant PA, Laycock B, *et al.* The rate of biodegradation of PHA bioplastics in the marine environment: a meta-study. *Mar Pollut Bull* 2019; 142: 15–24.
62. Kour H, Ahmad N, Wani T, *et al.* Advances in food packaging – a review. *Stewart Posthar Rev* 2013; 4: 1–7.
63. Geyer R, Jambeck JR, Lavender Law K. Production, use and fate of all plastics ever made. *Sci Adv* 2017; 3(7): 1–5.
64. Plastics – the facts 2018 – Plastics Europe. https://www.plasticseurope.org/it/resources/publications/619-plastics-facts-2018 [accessed 9/06/2020].
65. Improving markets for recycled plastics trends, prospects and policy responses. OECD Publishing, Paris, 2018. https://www.oecd.org/environment/improving-markets-for-recycled-plastics-9789264301016-en.htm

66. https://www.plasticseurope.org/, cited 2019.
67. European Commission. *Bio-Based Products*. European Commission, 2017. http://ec.europa.eu/growth/sectors/biotechnology/bio-based-products_en.
68. Helanto K, Matikainen L, Talja R, *et al.* Bio-based polymers for sustainable packaging and biobarriers: a critical review. *BioRes* 2019; 14(2): 4902–4951.
69. Biopolymers, facts and statistics 2016 – *IfBB*. https://www.ifbb-hannover.de/files/IfBB/downloads/faltblaetter_broschueren/Biopolymers-Facts-Statistics-2018.pdf [accessed 09/06/2020].
70. Patel M, Bastioli C, Marini M, *et al.* Life-cycle assessment of bio-based polymers and natural fiber composites. *Biopoly Gen Aspects Spec Appl* 2005; 10: 1–45.
71. Tabone MD, Cregg JJ, Beckman EJ, *et al.* Sustainability metrics: life cycle assessment and green design in polymers. *Environ Sci Technol* 2010; 44: 8264–8269.
72. Narodoslawsky M, Shazad K, Kollmann R, *et al.* LCA of PHA production - identifying the ecological potential of bio-plastic. *Chem Biochem Eng Q* 2015; 29(2): 299–305.
73. Rodriguez-Perez S, Serrano A, Pantion AA, *et al.* Challenges of scaling-up PHA production from waste streams. A review. *J Environ Manage* 2018; 205: 215–230.
74. EU/FP5 (2001–2004) Growth Program on *Diary Industry Waste as Source for Sustainable Polymeric Material Production – WHEYPOL*. https://cordis.europa.eu/project/id/G5RD-CT-2001-00591
75. EU/FP5 (2001–2004) LIFE Program on *Recyclable and Biodegradable Eco-Efficient Packaging Solutions for the Food Industry – ECOPAC*. https://cordis.europa.eu/project/id/QLK1-CT-2001-01823
76. EU/FP6 (2006–2008) Craft POLYMER COOP-CT-2006-032967 *Production of Poly(Hydroxyalkanoate)s from Olive Mills Wastewater*. https://cordis.europa.eu/project/id/32967/it
77. EU/FP7 (2010–2012) KBBE-2009-3-5-02 Collaborative Medium-scale focused research project on *Biotechnological Conversion of Carbon Containing Wastes for Eco-Efficient Production of High Added Value Products – ANIMPOL*. https://cordis.europa.eu/project/id/245084
78. EcoBioCAP FP7 (2011–2015) *ECOefficient BIOdegradable Composite Advanced Packaging (2011–2015)*. https://cordis.europa.eu/project/id/265669/it
79. Guillard V, Gaucel S, Fornaciari C, *et al.* The next generation of sustainable food packaging to preserve our environment in a circular economy context. *Front Nutr* 2018; 5: 121–134.
80. Singh G, Kumari A, Mittal A, *et al.* Cost effective production of poly-β- hydroxybutyrate by *Bacillus subtilis* NG05 using sugar industry waste water. *J Polym Environ* 2012; 21: 1–12.
81. Getachew A, Woldesenbet F. Production of biodegradable plastic by polyhydroxybutyrate (PHB) accumulating bacteria using low cost agricultural waste material. *BMC Res Notes* 2016; 9(1): 509–518.
82. Nielsen C, Rahman A, Rehman AU, *et al.* Food waste conversion to microbial polyhydroxyalkanoates. *Microb Biotechnol* 2017; 10(6): 1338–1352.
83. Valentino F, Moretto G, Lorini L, *et al.* Pilot-scale polyhydroxyalkanoate production from combined treatment of organic fraction of municipal solid waste and sewage sludge. *Ind Eng Chem Res* 2019; 58(27): 12149–12158.
84. Johnston B, Radecka I, Hill D, *et al.* The microbial production of polyhydroxyalkanoates from waste polystyrene fragments attained using oxidative degradation. *Polymers* 2018; 10: 957–976.
85. Johnston B, Radecka I, Chiellini E, *et al.* Mass spectrometry reveals molecular structure of polyhydroxyalkanoates attained by bioconversion of oxidized polypropylene waste fragments. *Polymers* 2019; 11: 1580–1599.
86. Ekere AI, Johnston B, Zieba M, *et al.* Environment cleaning mission-bioconversion of oxidatively fragmented polyethylene plastic waste to value-added copolyesters. *Chim Oggi* 2019; 37(6): 36–39.

Part IV

Degradation of PHA and Fate of Spent PHA Items

13 Aerobic and Anaerobic Degradation Pathways of PHA

Jorge A. Ferreira and Dan Åkesson

CONTENTS

13.1 INTRODUCTION

Environmental concerns related to the exploitation of fossil fuels, together with the slow biodegradation of derived plastic materials and shortage of resources, have motivated the production of bioplastics. A variety of bioplastics has already reached the market (see Figure 13.1), although their contribution to the sector (≈1%) is still far from that of fossil fuel-based plastics (*cf.* European Bioplastics), considering a global production of 348 million tons in 2017 (*cf.* Plastics Europe). However, the market for bioplastics is expanding and, hence, a concomitant continuous increase of bioplastic-based waste is expected. This, in turn, calls for comprehensive studies on their biodegradability in the environment and on efficient waste management routes.

The label "bioplastics" on the products that have reached the market sparks an intuitive association with biodegradability. However, as it is presented in Figure 13.1,

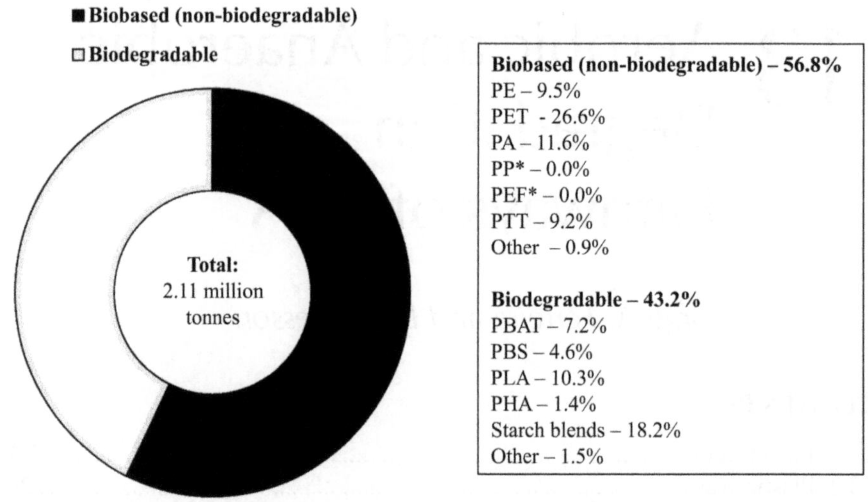

FIGURE 13.1 Global production, by material type, of bioplastics in 2018 (adapted, *c.f.* European Plastics). Key: PE – poly(ethylene); PET – poly(ethylene terephthalate); PA – polyamide; PP – poly(propylene); PEF – poly(ethylene furanoate); PTT – poly(trimethylene terephthalate)e; PBAT – poly(butylene adipate terephthalate); PBS – poly(butylene succinate); PLA – poly(lactic acid). * Bio-based PP and PEF were reported to be under development and predicted to be available at commercial scale in 2023.

less than 50% of the bioplastics produced are actually biodegradable; hence, the wide definition of bioplastics including both bio-based and biodegradable plastics [1]. Consequently, there is a need for competent waste collecting systems, to avoid non-biodegradable plastics reaching the environment, and intensified research efforts towards diversification of the application for biodegradable bioplastics.

Bio-based plastics include those originating from renewable resources such as starch-, cellulose-, protein-based substrates, etc. or from the metabolism of microorganisms, growing in renewable substrates, such as poly(lactic acid) (PLA) and polyhydroxyalkanoates (PHA) [1]. PHA are polyesters accumulated as granules inside microbial cells, mostly in bacteria [2], the extraction and purification of which need to follow to obtain the polymer for further processing [3]. The family of PHA is diverse, as a result of the huge number of different PHA-producing microorganisms; over 300 species of mainly bacteria-producing PHA [2] and over 155 different PHA monomer subunits (of variable carbon chain lengths) have been reported [4]. This opens up the range of potential applications, hence the interest surrounding PHA. Polyhydroxybutyrate (PHB), a polymer composed of short-length monomers of four carbon atoms, has been the most studied PHA; PHB-based products, in both pure form or as blends with other types of PHA monomers, are available in the market [2]. In Table 13.1, the overall molecular structure of the PHA monomeric unit, representative examples of the PHA family, and marketable products are provided.

Presently, research on PHA is focused, to a great extent, on improving the profitability of the commercial processes towards an increasing development rate. Strategies include the substitution of refined and food-grade substrates by waste

TABLE 13.1
An Overview of the Molecular Structure of the Monomeric Unit in PHA, of Different PHA Polymers Based on the Substitution of the Functional R-Group and on the Number of Carbon Atoms in the Monomeric Unit, and of the Composition of Different PHA-Based Products Available in the Market (adapted from [2, 7])

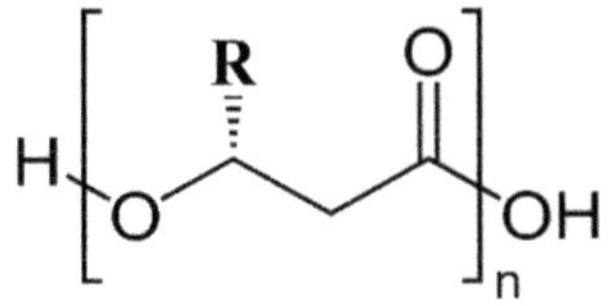

PHA polymers based on R-group and carbon no. in the monomeric unit

R-group	Number of carbons in the monomeric unit	Name of the PHA polymer
Methyl	C4	Poly(3-hydroxybutyrate)
Ethyl	C5	Poly(3-hydroxyvalerate)
Propyl	C6	Poly(3-hydroxyhexanoate)
Butyl	C7	Poly(3-hydroxyheptanoate)
Pentyl	C8	Poly(3-hydroxyoctanoate)
Hexyl	C9	Poly(3-hydroxynonanoate)
Heptyl	C10	Poly(3-hydroxydecanoate)
Octyl	C11	Poly(3-hydroxyundecanoate)
Nonyl	C12	Poly(3-hydroxydodecanoate)
Decyl	C13	Poly(3-hydroxytridecanoate)
Undecyl	C14	Poly(3-hydroxytetradecanoate)
Dodecyl	C15	Poly(3-hydroxypentadecanoate)
Tridecyl	C16	Poly(3-hydroxyhexadecanoate)

Industrial PHA

Product	Composition
Biogreen™	Poly(3-hydroxybutyrate)
Biopol™	poly(3-hydroxybutyrate-*co*-3-hydroxyvalerate) P(3HB-*co*-3HV) or PHBV
Nodax™	3HB and smaller amounts of 3-hydroxyhexanoate, 3-hydroxyoctanoate, and 3-hydroxydecanoate
Degra pol™	polyhydroxybutyrate-diol (hard segment) and α,ω-dihydroxy-poly(ε-caprolactone-block-diethyleneglycol-block-ε-caprolactone) (soft segment)

materials, research on more cost-effective extraction and purification methods, the establishment of clear links between application and degree of purification needed [3], and the use of non-sterile mixed bacterial cultures [5]. However, emphasis needs also to be given to the environmental aspects of PHA when their products reach their end-of-life and are disposed of. This should follow, in general terms, two lines of understanding, namely, the impact of PHA if discarded in or reaching the environment, or if included in common waste management systems, mostly composting and biogas-producing plants. The latter is expected to play a major role in the future management of plastic waste considering a society moving towards a circular economy, where wastes need to be converted into energy and chemicals [6].

This chapter aims to provide a comprehensive description of insights available in the literature concerning the biodegradation of PHA under aerobic and anaerobic conditions, including biodegradation in soil and aquatic systems, but also through composting, and anaerobic digestion towards biogas production. Emphasis is also given to the microorganisms able to degrade PHA and the related necessary enzymatic machinery, as well as factors that can influence the polymer biodegradability rate. In accordance with the dominance of research and development using PHB, the overview provided herein will be mostly based on this polymer. Finally, the chapter proposes future directions within the area based on identified research gaps.

13.2 PLASTIC WASTE

13.2.1 Accumulation in the Environment

The global production of plastics reached 348 million tons in 2017 (*cf.* Plastics Europe) and it is expected to keep increasing in line with population growth. In addition to environmental concerns related to the exploration of fossil fuels for the production of plastics, the increasing amount of plastic waste, as a result of increased production, exacerbates the negative impact of the plastic sector. Despite the presence of waste disposal systems, the efficiency of which varies worldwide [8], plastic materials can reach both soil and aquatic systems and their accumulation, as a result of slow biodegradation, can cause acute problems. Plastic accumulation in soil systems can lead to infertility, not allowing crop growth, and diseases and even death in animals as a result of plastic consumption mistaken for food. In sewage systems, plastic accumulation can lead to blockages, creating suitable shelter conditions for potential disease-spreading flies, insects, and parasites. In aquatic systems, plastic accumulation can also lead to death of wildlife through ingestion or entanglement [7]. This reality calls for efficient strategies to avoid plastic accumulation in the environment that can include improvements in waste disposal systems and the motivation for the alternative production of biodegradable bioplastics such as PHA. Although biodegradable, PHA will still have a negative impact on the environment, considering that a certain period of time is needed for total degradation. Accordingly, at first glance and in the absence of fast biodegradable polymers, it is argued that the production of PHA should meet the specifications for ease management of the originated waste using other systems such as composting and anaerobic digestion, contributing to the protection of the environment. This is in full agreement with Agenda 2030's goals for sustainable development.

13.2.2 Disposal

In general, collected plastic waste can end in landfills, in incinerators for energy recovery, and in recycling or biological degradation routes (see Figure 13.2). Landfilling is the cheapest disposal strategy but environmentally unsustainable due

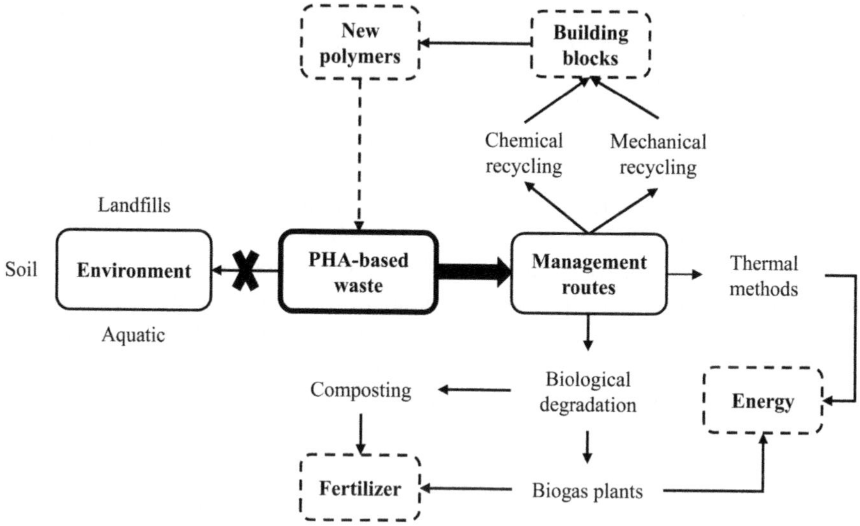

FIGURE 13.2 Overview of potential end-fates of PHA-based waste. The scheme highlights the need to minimize the accumulation of plastic waste in the environment and motivation towards other routes such as those including thermal, mechanical, chemical, and biological methods. The dashed line towards the PHA-based waste box is due to the possibility of the new polymers produced being either PHA or another type of polymer.

to the production of greenhouse gases, lack of space, and probability of soil and water contamination by the leachate produced [9]. Nonetheless, landfills continue to be one of the main disposal systems; 40% of global plastic waste ended up in landfills in 2016 (*c.f.* Statistica). Waste thermal treatment through incineration is a mature and efficient technology for waste reduction through combustion and conversion to heat, flue gas, and ash [10]. However, it also has related environmental and health concerns [11]. Alternative waste thermal treatment strategies, including gasification, pyrolysis, or their combination, have been investigated but their present development level seems to be limited to laboratory and demonstration scales [8].

Recycling strategies are considered, in this discussion, as those aiming to recover the building blocks of plastics. These can then be used for the production of new polymers, saving resources. Recycling strategies generally include mechanical and chemical methods. In particular, mechanical methods through extrusion and chemical methods through pyrolysis have been the most investigated recycling routes for PHA [3]. The use of enzymes, isolated from PHA-degrading microorganisms, have also been investigated for chemical recycling, but there is an absence of investigations in the last decade [3]. The type of PHA has been found to influence the output of mechanical recycling. Research studies have shown that the recycling of PHBV [12] and a blend of PHBV/PLA [13] could be repeated up to five and six times, respectively, whereas the physical properties of PHB were significantly reduced after two cycles [14]. The main end-products from chemical recycling, that are also dependent on the type of PHA, have included crotonic acid, 2-pentenoic acid, 2-decenoic acid, and PHB-derived oligomers and monomers [3].

Biological degradation can include home and industrial composting, as well as biogas plants operated under aerobic and anaerobic conditions, respectively. These are two established technologies with a huge impact on the reduction of the organic portion of the municipal solid waste (e.g., food waste) [6]. Composting results in carbon dioxide, water, and a remaining soil-like material (humus) known as compost, while anaerobic digestion converts waste mainly into methane and also into a solid-like material known as digestate. Both the compost and the digestate, depending on the composition, can be used as fertilizers [6], while biogas (a mixture of mainly methane and carbon dioxide) can be used as a transportation fuel or for cooking, heat, and electricity [15]. Thus, biogas-producing processes can potentially have a higher contribution to energy security goals in comparison to that of composting. Due to their present industrial establishment, it becomes clear that the relevance of composting and anaerobic digestion for degradation of plastic waste will depend on the production of materials that can be degraded within the same timeframe (that is, retention time) as for the organic fraction of the municipal solid waste.

13.3 BIOPLASTICS

The group of bioplastics includes both bio-based and biodegradable bioplastics (see Figure 13.1). Around 2.11 million tonnes of bioplastics were produced in 2018, an amount that is expected to increase to 2.62 million tons in 2023 (*c.f.* European Plastics). Bio-based plastics are produced from renewable resources such as polysaccharides (e.g., starch, cellulose, pectin, and chitin), proteins (e.g., wheat gluten, wool, casein), and lipids (e.g., plant oils), originated as a product of microbial metabolism (e.g., PHA), or chemically synthesized from bio-derived products (e.g., PLA, produced from lactic acid, a by-product of microbial metabolism) [1]. It is hypothesized that polymers such as PLA and PHA will play an important role in the future of the sector of bioplastics, considering their microbial origin. As is presented in Figure 13.1, starch blends represent the highest fraction of produced bioplastics. However, as for the sector of biofuels, e.g., bioethanol, where starch-based fuel cannot hold the responsibility of fulfilling fossil fuel-free energy security goals [16], alternatives to starch-based bioplastics should be made a reality, considering population growth forecasts. However, the contribution of microbial-based bioplastics will only be fully exploited if non-food-grade nutrients are used for their growth, as is the case in most industrially established processes. Therefore, the replacement of food-grade starch and other nutrients by wastes materials (e.g., food industry-derived waste and sidestreams, lignocellulosic waste, etc.), as it is being extensively investigated for the production of PHA [3], is encouraged. While considering alternative non-food-grade substrates for direct production of bioplastics, cellulose and chitin have potential. The former is the most common polymer on earth, while the latter can be obtained from the cell walls of various fungal strains which have the ability to grow in non-food-grade substrates [17].

A fraction of bioplastics, although bio-based, are, however, not biodegradable including the so-called "drop-ins" (e.g., bio-PE, bio-PP, bio-PET), while biodegradable plastics, being biodegradable to some extent, can also be obtained from petrochemical sources, e.g., poly(ε-caprolactone) (PCL) [1]. As is presented in Figure 13.1, bio-based PE, PET, and PA made up to 48% (1 million tonnes) of the global production of bioplastics.

Standards in order to classify plastics as biodegradable are presently available considering aerobic (composting) and anaerobic conditions (anaerobic digestion towards biogas production). For instance, according to the ISO 17088:2012 standard [18] or as defined by the American Society for Testing and Materials (ASTM) [19], a plastic is considered biodegradable if significant modification of its chemical structure, that is, degradation, occurs leading to production of carbon dioxide, water, inorganic compounds, and microbial biomass without visible or toxic residues, under composting conditions [18]. Therefore, a compostable plastic is biodegradable, whereas a biodegradable plastic is not always compostable [20, 21]. According to the CEN standard, EN 13432:2000, a polymeric material is biodegradable if 90% of it is converted to carbon dioxide in a period of six months of composting, while 50% conversion to biogas, based on the theoretical maximum, is needed under anaerobic conditions [22].

13.4 BIODEGRADATION OF PHA

Microorganisms, through competent enzymatic depolymerase systems, play the most determinant role in the full degradation of polymers. However, their action can also occur in combination with natural degradation via abiotic factors including temperature, water, and sunlight [23]. Reasonably, PHA-accumulating microorganisms are the first strong candidates for the degradation of PHA-based materials since they naturally have the enzymatic machinery for both biosynthesis and degradation of PHA granules [7].

Sunlight, particularly UV light, contributes to the initiation of natural degradation by providing the needed activation energy for the addition of oxygen atoms to the polymer, that is, photo-degradation with thermo-oxidative degradation taking place [24]. This pre-step leads to the breakdown of the polymers into shorter oligomers, dimers, or monomers that ease passage through microbial cell walls into the intracellular space. Here, their use as carbon and energy sources for biochemical processes within the cells takes place, leading to the full degradation by intracellular depolymerases (see Figure 13.3) [23]. Alternatively, extracellular depolymerases are responsible for the role played by the abiotic factors.

Biodegradation of polymers, carried out by bacteria, fungi, or algae, can be considered a three-step process: (1) microbial growth on or inside the surface of the polymer leading to modification of mechanical, chemical, and physical properties, known as bio-deterioration; (2) polymer depolymerization into oligomers, dimers, and monomers by microbial enzymatic pathways; and (3) assimilation or mineralization of polymer fragmentation products that serve as carbon, energy, and nutrient sources to microorganisms, ultimately leading to the conversion of carbon in the polymer into CO_2 and H_2O, under aerobic conditions, and additional CH_4/H_2S, under anaerobic conditions [23, 25]. The hydrolysis of the polymer by enzymes is a two-step process, namely the binding of the enzyme to the polymer followed by catalytic cleavage. Reasonably, microbial adherence and colonization is the rate limiting step [23].

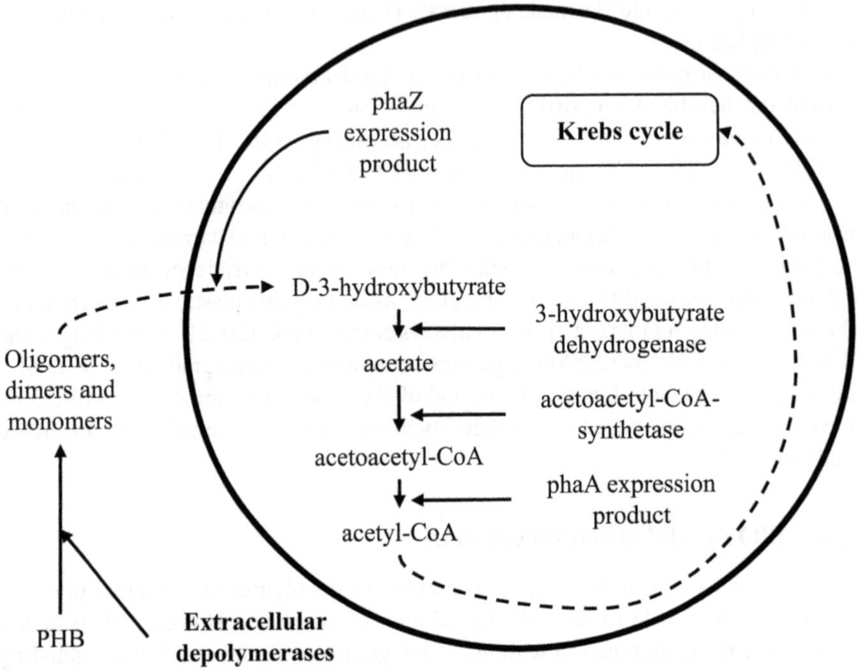

FIGURE 13.3 Proposed enzymatic degradation steps of PHB, catalyzed by both extra- and intracellular enzymes, according to the current knowledge available in the literature.

13.4.1 Factors Influencing Biodegradability

The biodegradability rate of bioplastics is a result of the combined effect of polymer characteristics and environmental conditions. The recalcitrance level of polymers, that is, the difficulty level to be degraded, is a result of their chemical and physical properties, including surface characteristics (hydrophobicity or hydrophilicity, and area), first-order structures (molecular weight, distribution, chemical structure), and higher-order structures (crystallinity, crystal structure, modulus of elasticity, glass transition temperature, melting temperature) [26]. Environmental parameters include the location where the bioplastics are present that, in turn, will influence the type and number of potential degrading microorganisms. For instance, the amount of bacteria in a sewage works is higher than that in the sea, therefore, longer biodegradation times will be needed in the latter [27]. Moreover, physical parameters, including temperature, pH, moisture, and concentration of oxygen content, will influence microbial growth, expression, and consequent concentration and activity of enzymes, and hence, the biodegradation of polymers [21, 28].

Generally, polymers of a shorter chain, more amorphous, and with a simpler chemical formula will be more easily degraded [29], while an increase in the melting temperature of the polymer decreases the enzymatic degradability [7]. Here, the type of monomeric units composing the PHA polymer will have a great influence since the enzymatic activity is determined by the interaction of enzymes and the side groups of the PHA-composing monomers (see Table 13.1) [29]. It has been reported that the microbial assimilation of polyesters containing side chains is lower than that of those with absent side chains [7]. PHB, in pure form, is characteristically stiff and brittle and difficult to process for further applications. Therefore, it is commonly blended with other PHA monomers, an example of which is poly(3-hydroxybutyrate-*co*-3-hydroxyvalerate) (PHBV) [30]. A further example includes PHA copolymers produced under the trade name Nodax (see Table 13.1); PHA copolymers comprising medium-chain-length PHA monomers have been reported to be of lower crystallinity than PHB homopolymers or copolymers (PHBV), leading to faster degradation [31]. Therefore, comprehensive investigations need to be carried out to clearly describe the effects of different PHA-based materials on the biodegradability rate.

13.4.2 PHA-Degrading Microorganisms

Several genera of bacteria and fungi have been identified with the capability of degrading bioplastics, being soil and compost valuable sources [32–34], in accordance with their higher microbial diversity, in comparison to, e.g., aquatic environments. The first genera of microorganisms reported to be able to degrade PHA belonged to *Bacillus*, *Pseudomonas*, and *Streptomyces* [35]; today, a wide range of microorganisms are known to be able to degrade PHA (see Table 13.2), with a variable range of polymer specificity, where co-culture for enhanced degradation should not be neglected [29]. In particular, PHB depolymerases represent a large family with predominance in *Alcaligenes*, *Comamonas*, and *Pseudomonas* spp. [36]. The level of investigation leading to characterization of PHB depolymerases is much higher for those produced by bacteria than those produced by fungi [37].

Both intracellular and extracellular depolymerases have been found in PHA-degrading microorganisms [7], and several genes have been identified playing a role in PHA depolymerization, including *phaA*, *phaZ*, etc. [38]. The action mode of intracellular depolymerases is suggested to be dissimilar from that of extracellular depolymerases. The former act on amorphous native intracellular granules leading to their fast depolymerization as a result of their content of proteins and lipids, whereas extracellular depolymerases attack crystalline, denatured PHA, leading to their conversion to water-soluble products [7, 39]. Intracellular depolymerase is suggested to be an exotype hydrolase acting at the carbonyl terminus of the polymer chain [39]. Despite their important role in the depolymerization of PHB, little is known about intracellular depolymerases, related to the complexity of their action [39]. Extracellular depolymerases are responsible for the breakdown of the polymers into shorter molecules that can then be transported into the cells for further degradation by intracellular depolymerases [23]. Extracellular depolymerases are divided into two classes according to their substrate specificity, namely, short-chain-length depolymerases, degrading only PHB and related copolyesters, and medium-chain-length depolymerases, acting primarily on aliphatic and aromatic PHA composed of monomeric units with 6 to 14 carbon atoms [7]. Although most PHA-degrading microorganisms produce only one type of PHA depolymerase, short- or medium-chain-length [7], exceptions exist, such as *Pseudomonas lemoignei*, where different types of extracellular depolymerases of different biochemistry were found [40].

TABLE 13.2
Taxonomic Overview of PHA-Degrading Microorganisms (adapted from [7, 29])

Bacteria	Fungi
Actinomadura sp.	*Acremonium recifei*
Alcaligenes faecalis	*Aspergillus niger*
Bacillus megaterium	*Aspergillus penicilloides*
Bacillus sp.	*Aspergillus* sp.
Bacillus subtilis	*Aspergillus ustus M-224*
Burkholderia cepacia	*Aspergillus ustus T-221*
Caenibacterium thermophilum	*Candida albicans*
Clostridium acetobutylicum	*Candida guilliermondii*
Clostridium botulinum	*Cogronella* sp.
Cupriavidus sp.	*Debaryomyces hansenii*
Entrobacter sp.	*Emericellopsis minima W2*
Gracilibacillus sp.	*Fusarium oxysporum*
Leptothrix sp.	*Fusarium solani*
Microbispora sp.	*Paecilomyces lilacinus*
Micrococsus sp.	*Penicillium oxalicum DSYD05-1*
Mycobacterium sp.	*Penicillium* sp.
Nocardiopsis sp.	*Pseudozyma antarctica JCM*
Pseudomonas aerogusina	*Pseudozyma japonica*
Pseudomonas fluorescens	*Trichoderma pseudokoningii*
Pseudomonas lemoignei	
Pseudomonas putida	
Pseudomonas sp. *AKS2*	
Rhodospirillum rubrum	
Saccharomonospora genus	
Schlegelella thermodepolymerans	
Streptomyces ascomycinicus	
Streptomyces bangladeshensis	
Streptomyces exfoliates K10	
Streptomyces sp.	
Streptomyces venezuelae	
Thermoactinomyces	
Variovorax sp.	

13.4.3 PHB Depolymerases and Action Mechanism

PHB depolymerases have a molecular weight ranging from 40 to 90 KD, consisting mostly of only one polypeptide chain and having alkaline isoelectric points with maximum activity within 7.5–9.5. Many are inhibited by reducing agents, most do not establish bonds with ion exchangers, while all are composed of two functionally separate domains, namely, a binding domain and a catalytic domain [7]. Considering the functional arrangement, PHB depolymerases are similar to other enzymes acting on water-insoluble complex polysaccharides such as cellulase, chitinase, and xylanase [7]. PHB depolymerases are characterized by containing three well-conserved amino acids, namely, serine, aspartate, and histidine. Furthermore, serine makes part of the so-called lipase-box penta-peptide Gly-Xaa1-Ser-Xaa2-Gly, common for several serine hydrolases [41]; Xaa1 and Xaa2 stand for naturally occurring amino acids. The oxygen atom present in the serine side chain is responsible for the nucleophilic attack of the ester bond [38]. In the case of extracellular PHB depolymerases, the substrate-binding domain has specificity for water-insoluble PHB polymers [41] and plays a crucial role on enzyme adsorption to PHB [42]. Studies have shown that PHB depolymerases can bind to both enzymatically degradable and non-biodegradable polyesters [43], with the substrate-binding domain possessing broader specificity than that of the catalytic domain [41].

The enzymatic degradation of the PHB polymer includes two stages, namely, adsorption, where binding of the enzyme to the substrate using the binding domain occurs, and hydrolysis by the catalytic domain [43]. After depolymerization by extracellular enzymes, and passage of the PHB oligomers, dimers, and monomers into the cells, further depolymerization to full monomeric units can take place by intracellular enzymes. The arising D-3-hydroxybutyrate monomers are then oxidized to acetoacetate by the action of 3-hydroxybutyrate dehydrogenase, which, in turn, is esterified to acetoacetyl-CoA by the acetoacetyl-CoA synthetase. The latter is further hydrolyzed into acetyl-CoA by the phaA gene expression protein and acetyl-CoA then enters the Krebs cycle (see Figure 13.3) [38].

13.5 BIODEGRADATION OF PHA UNDER AEROBIC AND ANAEROBIC CONDITIONS

The biodegradation of PHA has been studied in different environments, including soil and aquatic samples, and composting and anaerobic digestion conditions. However, the number of studies available in the literature is dissimilar among the groups of conditions studied (see Table 13.3). Furthermore, biodegradation in landfills has not overall been studied at great extent [29]. The following sections are dedicated to an overview of research findings on biodegradation of PHA in soil, aquatic, composting, and anaerobic digestion environments; a summary of the main achievements is provided in Table 13.3.

13.5.1 Soil

In general, soil environments should have higher potential for biodegradation of PHA than that in air or aquatic systems, in view of the microbial biodiversity [29]. Indeed, during screening initiatives from soil, the fraction of PHB-degrading microorganisms varied between 5% and 86% of the overall amount of microorganisms growing on solid agar medium [35]. However, with the exception of a few studies, such a claim cannot be made strongly, in view of the results presented in Table 13.3. The biodegradation studies carried out in soil samples point towards the effect of the PHA composition and location. Significant biodegradation differences (from 47% to up to 98% of weight loss) were observed while testing the same PHB and PHBV

TABLE 13.3
Overview of the Main Output from Investigations Carried Out on Biodegradation of PHA under Different Conditions

PHA polymer	Environment	Method	Biodegradability	Ref.
PHB films	Microbial culture from soil	Weight loss	ca. 18% after 18 days	[44]
PHBV-8 mol-% films	Microbial culture from soil	Weight loss	ca. 41% after 18 days	
PHBV-12 mol-% films	Soil/compost (90/10%)	Produced CO_2	86% after 9 months*	[45]
PHBV-43 mol-% films	Soil/compost (90/10%)	Produced CO_2	84% after 9 months*	
PHBV-47 mol-% films	Soil/compost (90/10%)	Produced CO_2	72% after 9 months*	
PHBV-52 mol-% films	Soil/compost (90/10%)	Produced CO_2	87% after 9 months*	
PHBV-64 mol-% films	Soil/compost (90/10%)	Produced CO_2	74% after 9 months*	
PHBV-72 mol-% films	Soil/compost (90/10%)	Produced CO_2	77% after 9 months*	
PHB films	Soil in Hoa Lac	Weight loss	98% after 300 days	[46]
PHB pellets	Soil in Hoa Lac	Weight loss	55% after 300 days	
PHBV films	Soil in Hoa Lac	Weight loss	61% after 300 days	
PHBV pellets	Soil in Hoa Lac	Weight loss	35% after 300 days	
PHB films	Soil in Dam Bai	Weight loss	47% after 400 days	
PHB pellets	Soil in Dam Bai	Weight loss	28% after 400 days	
PHBV films	Soil in Dam Bai	Weight loss	14% after 400 days	
PHBV pellets	Soil in Dam Bai	Weight loss	8% after 400 days	
PHA films	Soil	Produced CO_2	48.5% after 280 days	[47]
PHA pellets	Soil	Weight loss	10% after 168 days	[48]
PHA/DDGS pellets	Soil	Weight loss	41% after 168 days	
PHB/CAB films	Soil	Weight loss	31.5% after 180 days	[49]
PHB films	Soil	Weight loss	64.3% after 180 days	
PHB films	Soil	Weight loss	62% after 42 days	[50]
PHB films UV treated	Soil	Weight loss	ca. 68% after 42 days	
PHB-TiO_2 films	Soil	Weight loss	ca. 51% after 42 days	
PHB-TiO_2 films UV treated	Soil	Weight loss	56% after 42 days	
PHB nanofiber	Soil	Weight loss	100% after 28 days	
PHB-TiO_2 nanofiber	Soil	Weight loss	100% after 28 days	
PHB films	Brackish water sediment	Weight loss	100% after 56 days	[51]
PHB and PHBV films	Seawater (static)	Weight loss	88–99% after 49 days	[52]
PHB and PHBV films	Seawater (dynamic)	Weight loss	30–73% after 90 days	
PHB films	Marine water	Weight loss	42% after 160 days	[53]
PHBV-11 mol-% films	Marine water	Weight loss	46% after 160 days	
PHB and PHBV-11 mol-% pellets	Marine water	Weight loss	<20% after 160 days	
PHB films	Seawater	BOD	80% after 14 days	[54]
PHB films	Compost	Produced CO_2	79.7% after 110 days	[55]
P(3HB-*co*-4HB) films	Compost	Produced CO_2	90.3% after 110 days	
PHBV-3 mol-% films	Compost	Produced CO_2	80.2% after 110 days	
PHBV-20 mol-% films	Compost	Produced CO_2	89.3% after 110 days	
PHBV-40 mol-% films	Compost	Produced CO_2	90.5% after 110 days	

(*Continued*)

TABLE 13.3 (CONTINUED)
Overview of the Main Output from Investigations Carried Out on Biodegradation of PHA under Different Conditions

PHA polymer	Environment	Method	Biodegradability	Ref.
PHA/Rice husk films	Compost	Weight loss	>90% after 60 days	[56]
PHB films	Compost	Produced CO_2	ca. 80% after 28 days	[57]
PHB/PBAT films	Compost	Produced CO_2	ca. 50% after 28 days	
PHB	Anaerobic digestion	Produced CO_2 and CH_4	87% after 16 days	[58]
PHBV-13 mol-%	Anaerobic digestion	Produced CO_2 and CH_4	96% after 16 days	
PHBV-20 mol-%	Anaerobic digestion	Produced CO_2 and CH_4	83% after 16 days	
PHBV-8 wt.-% films	Anaerobic digestion	Produced CO_2 and CH_4	85% after 20 days	[59]
PHBO powder	Anaerobic digestion	Produced CO_2 and CH_4	80% after 25 days	[60]
PHB powder	Anaerobic digestion	Produced CO_2 and CH_4	90% after 9 days	[61]
PHB powder	Anaerobic digestion	Produced CO_2 and CH_4	100% after 40 days	[62]

* Predicted by the model developed by the authors based on experimental results.

films in two different locations in Vietnam, possibly due to different pH values [46]. The biodegradation of the composite of PHB containing residue-based fibers, originated from the fermentation of potato peel waste, was found to be more efficient than that of neat PHB due to reduced crystallinity [63]. The addition of TiO_2 nanoparticles [50] and cellulose acetate butyrate (CAB) [49] to PHB reduced the biodegradability rate of the biocomposites. Inversely, the addition of distiller's dried grains with solubles (DDGS), a product from grain-based ethanol facilities, improved both the biodegradation and the mechanical properties of the biocomposite [48], where degradation of both crystalline and amorphous regions took place [48].

13.5.2 Aquatic Environment

Biodegradation of PHA such as that of PHB in aquatic systems takes place mostly in the interface water-sediments as a result of their tendency to sink [27]. Similarly to the studies in soil, the location can influence the biodegradation of bioplastics, where a pelagic environment was found to be more efficient than a eutrophic environment [64]. Additional influencing factors include water temperature, turbulence, and the microbial community present, and the shape of the polymer. The weight loss of PHB and PHBV was lower in a dynamic system than in a static system using seawater, where the addition of sediments had a beneficial effect [52]. The biodegradation of

PHA films was found to vary across different periods of the year as a result of the water temperature, and possibly the microbial community present [65]. The variation in strength retention of PHB also varied among different seawaters, supporting the influence of the existing microbial community on biodegradation [66]. Similarly to the studies in soil [46], PHA films were also found to degrade faster than PHA pellets as a result of a higher surface area that enabled microbial attachment and colonization [53].

13.5.3 Composting

Composting is an efficient aerobic technology for organic waste reduction, requiring lower investment, operation costs, and expertise in comparison to that of anaerobic digestion [67]. In addition to CO_2 and H_2O, the degradation process gives rise to a soil-like, nutrient-rich product, the compost, that can be used alternatively to chemical fertilizers due to its rich content in macro- and micro-nutrients [68]. During composting, a diversified community of microorganisms cooperates for the success of biodegradation, and therefore maintaining their health status is of utmost importance. The main factors influencing the activity of microorganisms and, hence, biodegradation during composting, include the type of raw materials, temperature, moisture content, oxygen supply, C/N ratio, and pH [69]. Failing to provide those factors at the right levels will impair biodegradation, leading to a compost of lower quality or even environmental pollution [70]. Reasonably well-controlled conditions are obtained at industrial scale in comparison to home composting, which can lead to biodegradation differences [29]. The conversion of organic matter to CO_2, H_2O, and compost is presented in Equation (13.1) below considering industrial composting conditions (high-oxygen environment ($\geq$6%)) [21]:

$$\text{Organic matter} + S + O_2 \rightarrow CO_2 + H_2O + NO_2 + SO_2 + \text{Heat} + \text{Compost} \qquad (13.1)$$

Composting in industrial conditions generally includes warm temperatures of around 60–70°C, moisture of ca. 60%, and pH of around 8.5 [71]. Some of the heat produced during composting, which originated from the release of the energy present in the organic matter, needs to be removed to keep the healthy state of the microbial community; this is carried out by continuous mixing.

Studies on composting of PHA have almost exclusively considered the effect of the polymer composition. The amount of rice husk added to PHA and acrylic acid-grafted PHA had a direct relationship with the biodegradation of the produced biocomposites. The latter were characterized by having higher tensile strength, but leading to slightly lower biodegradation due to their higher resistance to water absorption [56]. The blending of PLA and PHB with poly(butylene adipate-*co*-terephthalate) (PBAT) ended in lower biodegradation in comparison to that of neat PLA and PHB [57]. By varying the amount of 3-hydroxyvalerate added to 3-hydroxybutyrate, it was found that its addition reduced the biocomposites crystallinity and enhanced biodegradability [55].

A modified version of composting, called vermicomposting, has attracted research attention from the scientific community in the past few decades [72]. In vermicomposting, the organic matter is converted to a nutrient-rich final product, the vermicompost, by the combined action of earthworms and microorganisms [73]. In addition to the vermicompost – that, similarly to compost, can be used as fertilizer – vermicomposting also gives rise to earthworms, with potential medical and feed applications [74]. Presently, no research studies are available in the literature on biodegradation of PHA under vermicomposting conditions.

13.5.4 Anaerobic Digestion

Similarly to composting, anaerobic digestion is an industrially established and efficient technology for the reduction of organic waste. It leads to the production of biogas (a mixture of mostly CH_4 and CO_2), and a solid residue, the digestate, that can also be used as fertilizer [6], according to Equation (13.2) below:

$$\text{Organic matter} + H_2O + \text{Nutrients} \rightarrow \text{Digestate residue} + CO_2 + CH_4 + NH_3 + H_2S + \text{Less heat} \tag{13.2}$$

At industrial-scale biogas plants, the anaerobic digestion is normally carried out in mesophilic (37°C) or thermophilic (55 °C) conditions. In comparison to composting, less heat and microbial biomass are produced under anaerobic conditions, since the energy contained in the organic matter is mainly released in the form of methane. The process of anaerobic digestion has been extensively studied, where the production of biogas is the final result of a complex combination of metabolic activities involving various groups of microorganisms (Figure 13.4) [71].

It has been reported that the degradation of PHA is faster under anaerobic conditions than under aerobic conditions [75]. According to the research studies available in the literature, different types of PHA were found to degrade at similar rates. For instance, PHBV (8% 3HV) was 85% degraded within 20 days in anaerobic sludge [59], poly(3-hydroxybutyrate-*co*-3-hydroxyoctanate) (PHBO) reached 80% degradation after 25 days under anaerobic conditions [60], and between 83% and 96% of the carbon present in PHB and PHBV (13% and 20% 3HV) was converted to methane and carbon dioxide after 16 days in anaerobic sewage sludge [58]. Furthermore, PHA were reported to degrade faster than other biodegradable polyesters including PCL, PLA, and PBS. A degradation of 90% of PHB was obtained after 9 days, whereas 22% and 49% degradation of PLA and PCL, respectively, were obtained after 277 days, and PBS was rather recalcitrant to biodegradation [61]. The inclusion of a pretreatment step, using alkaline methods, prior to anaerobic digestion of PHB and PLA was, more recently, found to enhance biodegradation, and concomitantly, the production of methane [62].

13.5.5 Scrutinizing the Potential of Anaerobic Digestion

The biogas produced during anaerobic digestion processes is a versatile fuel considering that, in addition to its use in the transportation sector, it can also play roles in cooking, heat, and electricity [6]. Thus, considering that a potential fertilizer is also produced, biogas plants give an important contribution towards sustainable circular economies.

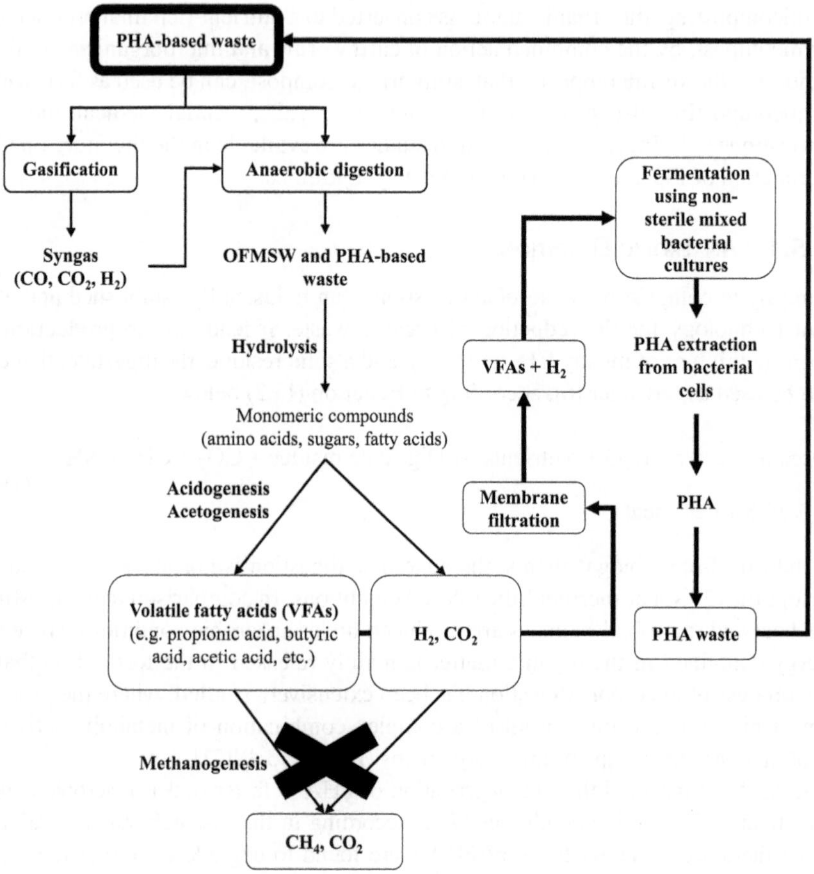

FIGURE 13.4 Overview of the process of anaerobic digestion leading to the production of mostly methane and CO_2, with integration of syngas produced from gasification, as well as the production of new PHA polymers through fermentation of volatile fatty acids (VFAs) and H_2. OFMSW stands for organic fraction of the municipal solid waste (adapted from [3]).

Reasonably, the potential of including PHA in biogas plants treating, e.g., the organic fraction of municipal solid waste will be enhanced if their degradation occurs at the same retention time (15–30 days) under mesophilic and thermophilic conditions. As discussed above, PHA can be degraded within this timeframe. Therefore, if an efficient sorting-at-source takes place, there is a high potential for mixing PHA-based end-of-life products with the normal stream of organic waste towards the biogas-producing established plants.

Biogas plants can also be considered for integration strategies (Figure 13.4). For instance, a product of gasification is syngas that, in turn, can be used for anaerobic digestion. Therefore, a joint development of both technologies can aid industrial establishment of gasification plants. Furthermore, despite being an established technology, the biogas production from organic waste still faces bottlenecks, such as high investment and maintenance costs, hampering wider application [70]. Therefore, the research community has been investigating ways to enhance the value of anaerobic digestion. One of those strategies has been the enhancement of the production of volatile fatty acids (VFAs) and hydrogen via inhibition of the methanogenesis stage (Figure 13.4), that is, carrying out an acidogenic fermentation [5, 76]. Volatile fatty acids have the potential to serve as a platform for the production of biofuels and biochemicals, including the production of new PHA [77]; the use of H_2 for production of PHA has also been reported [78]. Thus, membrane filtration can be applied for *in situ* separation of VFAs and H_2 [79] that, in turn, can be used for bacterial fermentation for the production of new PHA polymers (Figure 13.4). Indeed, it has been reported that the addition of VFAs to non-sterile mixed bacterial cultures, through medium acidification, motivated PHA-producing bacteria, hampering the habitual alternative production of polysaccharides and lipids [5]. Altogether, an in-line development of gasification, anaerobic digestion, and bacterial fermentation for PHA production can lead to important contributions for the establishment of closed-loop processes for PHA-based waste.

13.6 CONCLUSIONS AND OUTLOOK

The contribution of bioplastics to the global plastics market is expected to continue increasing in the future, as a result of higher application versatility and environmental concerns. However, the positive environmental contribution of bioplastics will be fully attained when solely biodegradable materials are produced. Here, substantial research and development are needed, considering the present >50% share of non-biodegradable materials within the group of bioplastics.

PHA have a high contribution potential in view of their wider family leading to materials with a range of properties and, hence, a range of possible applications. Furthermore, PHA were found to be biodegradable in all environments considered, although different time periods were needed for substantial biodegradation.

Research shows that PHA can be biodegraded within the same timeframe used for anaerobic digestion of the organic fraction of the municipal solid waste into biogas. However, it should be highlighted that the polymers were used in powder form, calling for comprehensive studies on the impact of the polymer shape on anaerobic biodegradation. This is of utmost importance, since if a grinding pre-step is needed, it will increase energy demands and concomitantly impact negatively on the process' profitability. Research intensification on anaerobic biodegradation of PHA is considered to be of high importance for the future realization of circular economies in view of the present establishment of biogas plants for reduction of waste and the related production of energy and fertilizer.

The biodegradation studies of PHA in soil and aquatic environments need a wider geographical reach in order to establish clearer relationships between polymer composition and environmental and microbial conditions. A great part of the studies available in the literature point out long periods of time for substantial or almost full biodegradation of PHA, a situation shared, to some extent, by biodegradation studies under composting conditions. Close collaboration among academic research teams and industries producing PHA-based materials, in this regard, can motivate further optimization of polymer construction, attaining the gradually shorter biodegradation periods needed, and hence lowering the environmental impact.

The studies reported have, overall, used neat PHA, mostly PHB homopolymer and its copolymers. The addition of other compounds, e.g., fillers, to improve the mechanical properties of the final material, is common practice during polymer processing. Therefore, future investigations should also address this aspect.

Finally, independently of the route followed by PHA-based end-of-life materials, there is a probability of microplastics accumulating in the environment, either during biodegradation in soil and water, or as a result of fertilizer originating from composting and anaerobic digestion being administered to land. Thus, the research also needs to address this aspect in the near future, since the problem of microplastics accumulation has already been perceived.

REFERENCES

1. Bátori V, Åkesson D, Zamani A, et al. Anaerobic degradation of bioplastics: A review. Waste Manage 2018; 80: 406–413.
2. Anjum A, Zuber M, Zia KM, et al. Microbial production of polyhydroxyalkanoates (PHAs) and its copolymers: A review of recent advancements. In J Biol Macromol 2016; 89: 161–174.
3. Vu DH, Åkesson D, Taherzadeh MJ, et al. Recycling strategies for polyhydroxyalkanoate-based waste materials: An overview. Bioresource Technol 2019; 298: 122393.
4. Agnew DE, Pfleger BF. Synthetic biology strategies for synthesizing polyhydroxyalkanoates from unrelated carbon sources. Chem Eng Sci 2013; 103: 58–67.
5. Ivanov V, Stabnikov V, Ahmed Z, et al. Production and applications of crude polyhydroxyalkanoate-containing bioplastic from the organic fraction of municipal solid waste. Int J Environ Sci Technol 2015; 12(2): 725–738.
6. Ferreira JA, Agnihotri S, Taherzadeh MJ. Waste biorefinery. In: Taherzadeh MJ, Bolton K, Wong J, Pandey A, Eds., Sustainable Resource Recovery and Zero Waste Approaches. St. Louis, MO, Elsevier:2019; pp. 35–52.
7. Roohi, Zaheer MR, Kuddus M. PHB (poly-β-hydroxybutyrate) and its enzymatic degradation. Polym Advan Technol 2018; 29(1): 30–40.
8. Rajendran K, Lin R, Wall DM, et al. Influential aspects in waste management practices. In: Taherzadeh MJ, Bolton K, Wong J, Pandey A, Eds., Sustainable Resource Recovery and Zero Waste Approaches. St. Louis, MO, Elsevier:2019; pp. 65–78.
9. Awasthi MK, Zhao J, Soundari PG, et al. Sustainable management of solid waste. In: Taherzadeh MJ, Bolton K, Wong J, Pandey A, Eds., Sustainable Resource Recovery and Zero Waste Approaches. St. Louis, MO, Elsevier:2019; pp. 79–99.
10. Pan S-Y, Du MA, Huang IT, et al. Strategies on implementation of waste-to-energy (WTE) supply chain for circular economy system: A review. J Clean Prod 2015; 108: 409–421.
11. Hopewell J, Dvorak R, Kosior E. Plastics recycling: Challenges and opportunities. Philos Trans R Soc Lond B Biol Sci 2009; 364(1526): 2115–2126.
12. Zaverl M, Seydibeyoğlu MÖ, Misra M, et al. Studies on recyclability of polyhydroxybutyrate-co-valerate bioplastic: Multiple melt processing and performance evaluations. J Appl Polym Sci 2012; 125(S2): E324–E331.
13. Zembouai I, Bruzaud S, Kaci M, et al. Mechanical recycling of poly(3-hydroxyb utyrate-co-3-hydroxyvalerate)/polylactide based blends. J Polym Environ 2014; 22(4): 449–459.
14. Rivas LF, Casarin SA, Nepomuceno NC, et al. Reprocessability of PHB in extrusion: ATR-FTIR, tensile tests and thermal studies. Polímeros 2017; 27: 122–128.

15. Kabir MM, Forgács G, Taherzadeh MJ. Biogas from wastes: Processes and applications. In: Taherzadeh MJ, Richards T, Eds., Resource Recovery to Approach Zero Municipal Waste. Boca Raton, FL, CRC Press, Taylor & Francis Group:2016; pp.107–140.
16. Ferreira JA, Brancoli P, Agnihotri S, et al. A review of integration strategies of lignocelluloses and other wastes in 1st generation bioethanol processes. Process Biochem 2018; 75: 173–186.
17. Ferreira JA, Mahboubi A, Lennartsson PR, et al. Waste biorefineries using filamentous ascomycetes fungi: Present status and future prospects. Bioresource Technol 2016; 215: 334–345.
18. International Organization for Standardization. Specification for Compostable Plastics, ISO 17088:2012.
19. American Society for Testing and Materials. Compostable plastics, ASTM D6400 2004.
20. Kale G, Auras R, Singh SP, et al. Biodegradability of polylactide bottles in real and simulated composting conditions. Polym Test 2007; 26(8): 1049–1061.
21. Kale G, Kijchavengkul T, Auras R, et al. Compostability of bioplastic packaging materials: An overview. Macromol Biosci 2007; 7(3): 255–277.
22. European Committee for Standardization. Packaging – Requirements for Packaging Recoverable Through Composting and Biodegradation – Test Scheme and Evaluation Criteria for the Final Acceptance of Packaging, EN13432:2000.
23. Shah AA, Hasan F, Hameed A, et al. Biological degradation of plastics: A comprehensive review. Biotechnol Adv 2008; 26(3): 246–265.
24. Andrady AL. Microplastics in the marine environment. Mar Pollut Bull 2011; 62(8): 1596–1605.
25. Lucas N, Bienaime C, Belloy C, et al. Polymer biodegradation: Mechanisms and estimation techniques – A review. Chemosphere 2008; 73(4): 429–442.
26. Tokiwa Y, Calabia BP, Ugwu CU, et al. Biodegradability of plastics. Int J Mol Sci 2009; 10(9): 3722–3742.
27. Sharma M, Dhingra HK. Poly-β-hydroxybutyrate: A biodegradable polyester, biosynthesis and biodegradation. Br Microbiol Res J 2016; 14(3): 1–11.
28. Massardier-Nageotte V, Pestre C, Cruard-Pradet T, et al. Aerobic and anaerobic biodegradability of polymer films and physico-chemical characterization. Polym Degrad Stabil 2006; 91(3): 620–627.
29. Emadian SM, Onay TT, Demirel B. Biodegradation of bioplastics in natural environments. Waste Manage 2017; 59: 526–536.
30. Pagliano G, Ventorino V, Panico A, et al. Integrated systems for biopolymers and bioenergy production from organic waste and by-products: A review of microbial processes. Biotechnol Biofuels 2017; 10(1): 113.
31. Philip S, Keshavarz T, Roy I. Polyhydroxyalkanoates: Biodegradable polymers with a range of applications. J Chem Technol Biotechnol 2007; 82(3): 233–247.
32. Lee KM, Gimore DF, Huss MJ. Fungal degradation of the bioplastic PHB (poly-3-hydroxy-butyric acid). J Polym Environ 2005; 13(3): 213–219.
33. Kumaravel S, Hema R, Lakshmi R. Production of polyhydroxybutyrate (bioplastic) and its biodegradation by *Pseudomonas lemoignei* and *Aspergillus niger*. J Chem 2010; 7(S1): S536–S542.
34. Accinelli C, Saccà ML, Mencarelli M, et al. Deterioration of bioplastic carrier bags in the environment and assessment of a new recycling alternative. Chemosphere 2012; 89(2): 136–143.
35. Tokiwa Y, Ando T, Suzuki T, et al. Biodegradation of synthetic polymers containing ester bonds. In: Glass JE, Swift G, Eds., Agricultural and Synthetic Polymers. Washington, DC, American Chemical Society:1990; pp. 136–148.

36. Merrick JM, Doudoroff M. Depolymerization of poly-β-hydroxybutyrate by an intracellular enzyme system. J Bacteriol 1964; 88(1): 60–71.
37. Gilkes NR, Henrissat B, Kilburn DG, et al. Domains in microbial beta-1, 4-glycanases: Sequence conservation, function, and enzyme families. Microbiol Rev 1991; 55(2): 303–315.
38. Trainer MA, Charles TC. The role of PHB metabolism in the symbiosis of rhizobia with legumes. Appl Microbiol Biotechnol 2006; 71(4): 377–386.
39. Doi Y, Kawaguchi Y, Koyama N, et al. Synthesis and degradation of polyhydroxyalkanoates in *Alcaligenes eutrophus*. FEMS Microbiol Rev 1992; 9(2–4): 103–108.
40. Schöber U, Thiel C, Jendrossek D. Poly(3-hydroxyvalerate) depolymerase of Pseudomonas lemoignei. Appl Environ Microb 2000; 66(4): 1385–1392.
41. Shinomiya M, Iwata T, Doi Y. The adsorption of substrate-binding domain of PHB depolymerases to the surface of poly(3-hydroxybutyric acid). In J Biol Macromol 1998; 22(2): 129–135.
42. Doi Y, Mukai K, Kasuya K, et al. Biodegradation of biosynthetic and chemosynthetic polyhydroxyalkanoates. In: Doi Y, Fukuda K, Eds., Studies in Polymer Science. Amsterdam, Elsevier:1994; pp. 39–51.
43. Hadad D, Geresh S, Sivan A. Biodegradation of polyethylene by the thermophilic bacterium *Brevibacillus borstelensis*. J Appl Microbiol 2005; 98(5): 1093–1100.
44. Woolnough CA, Charlton T, Yee LH, et al. Surface changes in polyhydroxyalkanoate films during biodegradation and biofouling. Polym Int 2008; 57(9): 1042–1051.
45. Arcos-Hernandez MV, Laycock B, Pratt S, et al. Biodegradation in a soil environment of activated sludge derived polyhydroxyalkanoate (PHBV). Polym Degrad Stabil 2012; 97(11): 2301–2312.
46. Boyandin AN, Prudnikova SV, Karpov VA, et al. Microbial degradation of polyhydroxyalkanoates in tropical soils. Int Biodeterior Biodegr 2013; 83: 77–84.
47. Gómez EF, Michel FC. Biodegradability of conventional and bio-based plastics and natural fiber composites during composting, anaerobic digestion and long-term soil incubation. Polym Degrad Stabil 2013; 98(12): 2583–2591.
48. Madbouly S, Schrader J, Srinivasan G, et al. Biodegradation behavior of bacterial-based polyhydroxyalkanoate (PHA) and DDGS composites. Green Chem 2014; 16: 1911–1920.
49. Jain R, Tiwari A. Biosynthesis of planet friendly bioplastics using renewable carbon source. J Environ Health Sci Eng 2015; 13(1): 11.
50. Altaee N, El-Hiti G, Fahdil A, et al. Biodegradation of different formulations of polyhydroxybutyrate films in soil. SpringerPlus 2016; 5: 762.
51. Sridewi N, Bhubalan K, Sudesh K. Degradation of commercially important polyhydroxyalkanoates in tropical mangrove ecosystem. Polym Degrad Stabil 2006; 91(12): 2931–2940.
52. Thellen C, Coyne M, Froio D, et al. A processing, characterization and marine biodegradation study of melt-extruded polyhydroxyalkanoate (PHA) films. J Polym Environ 2008; 16(1): 1–11.
53. Volova TG, Boyandin AN, Vasiliev AD, et al. Biodegradation of polyhydroxyalkanoates (PHAs) in tropical coastal waters and identification of PHA-degrading bacteria. Polym Degrad Stabil 2010; 95(12): 2350–2359.
54. Tachibana K, Urano Y, Numata K. Biodegradability of nylon 4 film in a marine environment. Polym Degrad Stabil 2013; 98(9): 1847–1851.
55. Weng Y-X, Wang X-L, Wang Y-Z. Biodegradation behavior of PHAs with different chemical structures under controlled composting conditions. Polym Test 2011; 30(4): 372–380.
56. Wu C-S. Preparation and characterization of polyhydroxyalkanoate bioplastic-based green renewable composites from rice husk. J Polym Environ 2014; 22(3): 384–392.

57. Tabasi RY, Ajji A. Selective degradation of biodegradable blends in simulated laboratory composting. Polym Degrad Stabil 2015; 120: 435–442.
58. Budwill K, Fedorak PM, Page WJ. Methanogenic degradation of poly(3-hydroxyalkanoates). Appl Environ Microb 1992; 58(4): 1398–1401.
59. Shin PK, Kim MH, Kim JM. Biodegradability of degradable plastics exposed to anaerobic digested sludge and simulated landfill conditions. J Environ Polym Degrad 1997; 5(1): 33–39.
60. Federle TW, Barlaz MA, Pettigrew CA, et al. Anaerobic biodegradation of aliphatic polyesters: Poly(3-hydroxybutyrate-*co*-3-hydroxyoctanoate) and poly(ε-caprolactone). Biomacromolecules 2002; 3(4): 813–822.
61. Yagi H, Ninomiya F, Funabashi M, et al. Mesophilic anaerobic biodegradation test and analysis of eubacteria and archaea involved in anaerobic biodegradation of four specified biodegradable polyesters. Polym Degrad Stabil 2014; 110: 278–283.
62. Benn N, Zitomer D. Pretreatment and anaerobic co-digestion of selected PHB and PLA bioplastics. Front Environ Sci 2018; 5(93).
63. Wei L, Liang S, McDonald AG. Thermophysical properties and biodegradation behavior of green composites made from polyhydroxybutyrate and potato peel waste fermentation residue. Ind Crop Prod 2015; 69: 91–103.
64. Tosin M, Weber M, Siotto M, et al. Laboratory test methods to determine the degradation of plastics in marine environmental conditions. Front Microbiol 2012; 3(225).
65. Volova TG, Gladyshev MI, Trusova MY, et al. Degradation of polyhydroxyalkanoates in eutrophic reservoir. Polym Degrad Stabil 2007; 92(4): 580–586.
66. Sekiguchi T, Saika A, Nomura K, et al. Biodegradation of aliphatic polyesters soaked in deep seawaters and isolation of poly(ε-caprolactone)-degrading bacteria. Polym Degrad Stabil 2011; 96(7): 1397–1403.
67. Onwosi CO, Igbokwe VC, Odimba JN, et al. Composting technology in waste stabilization: On the methods, challenges and future prospects. J Environ Manage 2017; 190: 140–157.
68. Wong JWC, Karthikeyan OP, Selvam A. Biological nutrient transformation during composting of pig manure and paper waste. Environ Technol 2017; 38(6): 754–761.
69. Bernal MP, Alburquerque JA, Moral R. Composting of animal manures and chemical criteria for compost maturity assessment. A review. Bioresource Technol 2009; 100(22): 5444–5453.
70. Wang Q, Awasthi MK, Zhang Z, et al. Sustainable composting and its environmental implications. In: Taherzadeh MJ, Bolton K, Wong J, Pandey A, Eds., Sustainable Resource Recovery and Zero Waste Approaches. St. Louis, MO, Elsevier:2019; pp. 115–132.
71. Mohee R, Unmar GD, Mudhoo A, et al. Biodegradability of biodegradable/degradable plastic materials under aerobic and anaerobic conditions. Waste Manage 2008; 28(9): 1624–1629.
72. Hussain N, Abbasi SA. Efficacy of the vermicomposts of different organic wastes as "clean" fertilizers: State-of-the-art. Sustainability 2018; 10(4): 1205.
73. Pramanik P, Chung YR. Changes in fungal population of fly ash and vinasse mixture during vermicomposting by *Eudrilus eugeniae* and *Eisenia fetida*: Documentation of cellulase isozymes in vermicompost. Waste Manage 2011; 31(6): 1169–1175.
74. Sharma K, Garg VK. Vermicomposting of waste: A zero-waste approach for waste management. In: Taherzadeh MJ, Bolton K, Wong J, Pandey A, Eds., Sustainable Resource Recovery and Zero Waste Approaches. St. Louis, MO, Elsevier:2019; pp. 133–164.
75. Siracusa V, Rocculi P, Romani S, et al. Biodegradable polymers for food packaging: A review. Trends Food Sci Technol 2008; 19(12): 634–643.
76. Wainaina S, Lukitawesa L, Awasthi MK, et al. Bioengineering of anaerobic digestion for volatile fatty acids, hydrogen or methane production: A critical review. Bioengineered 2019; 10(1): 437–458.

77. Cerrone F, Choudhari SK, Davis R, et al. Medium chain length polyhydroxyalkanoate (mcl-PHA) production from volatile fatty acids derived from the anaerobic digestion of grass. Appl Microbiol Biotechnol 2014; 98(2): 611–620.
78. Volova TG, Kiselev EG, Shishatskaya EI, et al. Cell growth and accumulation of polyhydroxyalkanoates from CO2 and H2 of a hydrogen-oxidizing bacterium, *Cupriavidus eutrophus* B-10646. Bioresource Technol 2013; 146: 215–222.
79. Wainaina S, Parchami M, Mahboubi A, et al. Food waste-derived volatile fatty acids platform using an immersed membrane bioreactor. Bioresource Technol 2019; 274: 329–334.

14 Factors Controlling Lifetimes of Polyhydroxyalkanoates and their Composites in the Natural Environment

Bronwyn Laycock, Steven Pratt, Alan Werker, and Paul A. Lant

CONTENTS

14.1 INTRODUCTION

Polyhydroxyalkanoates (PHA) are a valuable family of biopolymers due to their ability to degrade in a very wide range of environments under ambient conditions, while still being processible as drop-in replacements for petroleum-derived polymers and having mechanical and physical properties that can be tailored for a wide range of applications. However, in using these biopolymers for commercial applications, and for understanding their lifetimes and degradation, it is important to understand the mechanisms and kinetics of polymer mechanical failure, as well as complete breakdown via hydrolytic degradation. Yet in most of the published studies on polyhydroxyalkanoate biodegradation, the primary focus has been on the time taken for complete mineralization, as indicated by the evolution of carbon dioxide or even just mass loss, rather than the time taken for the embrittlement of the polymers (loss of mechanical integrity), and hence the end of their practical utility. In this chapter we will be exploring PHA lifetimes from both perspectives.

While much of the underlying background and theory relating to biopolymer degradation has been covered in detail in our recent review paper [1], it has recently emerged that some of the critical drivers for degradation and loss of mechanical integrity were not given the focus that they deserve. This is particularly the case for PHA composites and blends, where the underlying drivers for polymer failure following biodegradation are both mechanical and chemical, as opposed to being driven just by the enzymatically catalyzed hydrolytic scission processes typically modeled.

This chapter will walk through the definitions of lifetimes, followed by the drivers of PHA degradation, starting with the very basics of uncatalyzed hydrolysis, followed by catalyzed hydrolysis, enzymatic processes, biofilm formation, and fungal degradation through to the increasingly complex drivers encountered in natural environments, initially for neat PHA and then for PHA-based blends and composites. This is then followed by a brief overview of methods for controlling PHA lifetimes and accelerated lifetime estimation. It should be noted that the scope of this review does not include higher temperature thermal degradation processes.

14.2 OVERVIEW OF BIOPOLYMER DEGRADATION

Biopolymer degradation in the environment occurs via a complex combination of mechanisms, and it is important to understand these in order to model and predict lifetimes for both mechanical integrity and ultimate mineralization. This degradation can be classified as photo-oxidative, thermo-oxidative, ozone-induced, mechanochemical, hydrolytic, catalytic and/or biodegradation, depending on the mechanism, with definitions of these terms as in Laycock *et al.* [1]. For a neat PHA polymer (i.e., a material made from the polymer alone, not a composite or blend) the key process is the reaction of water with esters within the polymer main chain, whether uncatalyzed or catalyzed.

In this chapter, we define biodegradation as the breakdown of materials through biological activity, such as through the action of microorganisms such as fungi, bacteria, archaea, and algae, primarily via enzymatically driven hydrolysis resulting in chain scission and loss of molecular weight. This is driven under both aerobic and anaerobic conditions by PHA degraders present in most natural environments [2, 3]. Under aerobic conditions, the polymer should ultimately be converted into biomass, CO_2, and water, while under anaerobic conditions the products should ultimately be biomass, CO_2, methane, and water [4].

It should be noted that for composites and blends, the biodegradation process is driven by this hydrolytic chain scission operating in combination, frequently synergistically, with many of the other degradation processes, particularly mechanochemical degradation.

14.2.1 Definitions of Lifetimes

If the rate of change of a property (such as mechanical properties or degree of CO_2 evolution) is known in a given environment, then the lifetime should, in principle, be predictable. For mechanical properties, this could be, for example, the point at which fracture energy has reached 50% of the initial value [5], although in practice this may be beyond the point at which the mechanical integrity of the polymer is still serviceable. Other approaches are determining if the ultimate elongation has fallen to 5% of the initial, or measuring the actual fracture toughness, since this is a direct measure of the extent of loss of mechanical integrity. For PHA, the critical molecular weight for chain entanglement, below which the polymer no longer has any strength and thus is no longer of use as a structural material in an application, is ~13 000 g/mol [6].

For CO_2 evolution, the lifetimes of interest are typically those that match with the relevant international standards [7]. For example, the new European Standard specification for biodegradable plastic mulch film for use in soils, EN 17033 (2018), specifies that these films must achieve at least 90% conversion to CO_2 under aerobic conditions in natural agricultural or forest topsoil at 20°C to 28°C within a maximum period of 2 years, without any negative environmental effects or ecotoxicity. Sometimes indirect measures of lifetime, such as mass loss or changes in dimensions, are used. These are, however, only proxies for biodegradation, since polymer fragments and oligomers may be formed that are not yet completely converted to CO_2.

14.2.2 Surface versus Bulk Erosion

While there are a great many simple, and also increasingly complex, models of the hydrolytic biodegradation of PHA, particularly in the area of controlled (chemical) release, at its heart this process is a competition between the rate of the hydrolysis reaction of the main chain ester groups and the rate of water diffusion into the bulk of the polymer. If the diffusion rate is much slower than the rate of reaction, then the polymer experiences degradation via surface erosion. If the opposite is true, and the water diffuses into the bulk more rapidly than the hydrolysis reaction, then a bulk erosion process dominates. There are four key variables that underlie this process and the relationship between them is critical to the mechanism of polymer degradation. These are:

- The rate of water diffusion into the polymer (D),
- The pseudo-first-order rate of hydrolysis (λ'),
- The thickness of the specimen (L), and
- The critical thickness (L_{crit}).

When $\lambda' > D$ and $L > L_{crit}$, polymer is degraded from the surface and the core polymeric material remains intact (retaining average molecular weight (M_W) and mechanical properties). As the polymer erodes and becomes thinner, its load-bearing capacity decreases, until it reaches L_{crit}. At this point the mechanism of biodegradation shifts to bulk erosion ($\lambda' < D$; $L < L_{crit}$) (Figure 14.1). The time to failure becomes dominated by the rate of end group (acid) catalyzed auto-acceleration of hydrolysis, until M_n reaches a critical value (M_e at around 8–13 000 g/mol for PHA). From this point, the polymer depolymerizes into water-soluble oligomers and monomers.

But there are only limited studies exploring the rate of water diffusion into PHA [8–16]. Some of the experimentally determined coefficients of diffusion (D_0), for a range of PHBV samples, measured at close to room temperature and zero water concentration, are summarized in Table 14.1.

It should be noted that the diffusion constant does change over time for PHA following immersion in water, due to the plasticization effect of water – with diffusivity

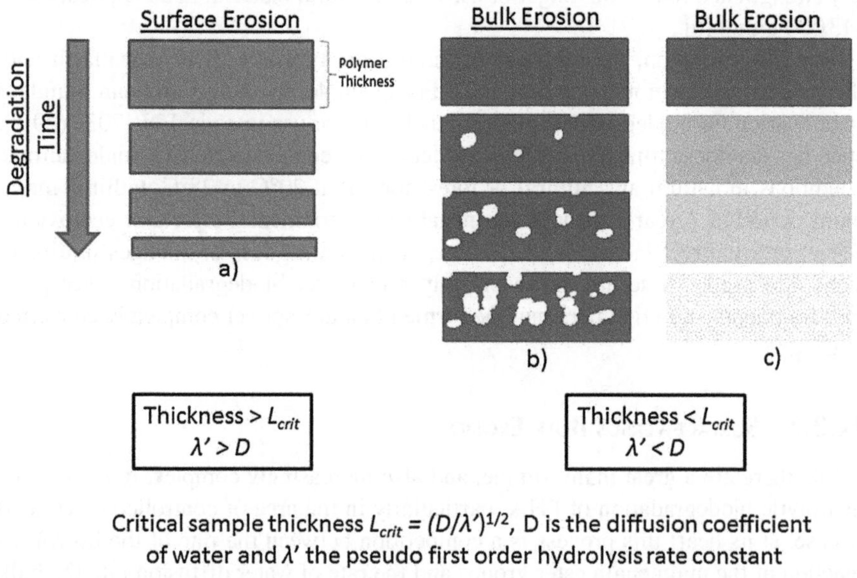

FIGURE 14.1 Schematic illustration of three types of erosion phenomena: (a) surface erosion with a growing hydrolysis front (e.g., enzymes), (b) bulk erosion with autocatalysis due to retained degradation products (e.g., PLA), (c) bulk erosion without autocatalysis (e.g., PLA-*co*-PCL) where water diffusion and catalyst is faster than the reaction rate. (Reproduced with permission [1]. Copyright 2017, Elsevier).

TABLE 14.1
Diffusion Coefficients at Zero Water Concentration (D0) for PHBV Samples Measured at Approximately Ambient Temperatures

Temperature (°C)	3HV content [mol-%]	Processing method	$D_0 \times 10^{12}$ $m^2\ s^{-1}$	Ref.
25	3	Compression-molded	0.46	[15]
25	3	Solvent-cast	1.31	[15]
25	3	Compression-molded	0.33	[17]
20	12	Compression-molded	0.44	[13]
37	12	Compression-molded	1.41	[13]
20	8	Injection-molded	1.9	[16]
30	8	Injection-molded	3.6	[16]
40	0	Solvent-cast	1.73	[10]
40	14	Solvent-cast	1.85	[10]
30	0	Solvent-cast	0.92	[10]

roughly following an exponential law related to the local water concentration (also called water activity, C), as free volume increases following water permeation (Equation (14.1)):

$$D = D_0 e^{\gamma C} \tag{14.1}$$

where D is the diffusion coefficient at a given water concentration, D_0 is the diffusion coefficient at zero water concentration, and γC is the plasticization factor for that water. This is noted as being typical behavior for hydrophobic polymers [15].

However, more complex models comprising several underlying and competing mechanisms for water uptake and diffusion in hydrophobic polymers such as PHA have also been developed [10]. As noted by Follain *et al.* [15], the Park model of water sorption by polymers is a useful one when a sigmoidal water uptake is observed (Equation (14.2)):

$$C = \frac{A_L b_L}{1 + b_L a_w} a_w + K_H a_w + K_H^n K_a n a_w^n \tag{14.2}$$

where A_L and b_L are the Langmuir concentration in specific sites and the affinity between water and specific sites, respectively; K_H is the Henry-type solubility coefficient; and n and K_a are the mean number of water molecules in aggregates and the equilibrium constant of water clustering, respectively [18].

The water sorption behavior was found to be virtually independent of the 3HV content of PHBV (from 0 to 24 mol-% 3HV) in a dynamic uptake study at 30°C [10]. However, there is a clear dependency of water permeability on the polymer crystallinity, with diffusion being through the amorphous phase and thus decreasing with an increase in overall crystallinity, due to the tortuous pathway imposed by the crystalline components [12, 15].

The processing methodology also plays a significant role, related to two competing factors. Firstly, solvent-cast films have higher free volume than compression-molded or melt-processed films, and thus higher water diffusion coefficients overall. However, the plasticization factor is lower for these solvent-cast films, hypothesized to be due to the fact that they had an initially higher free volume which enabled ready water vapor condensation initially, making the plasticization effect of the additional water at the highest water concentration less significant [12, 15]. There is significant variation in measured water permeabilities, with P values for melt-pressed PHBV (3 mol-% 3HV) estimated to be 149±6 Barrer [17], and by another group as 486 Barrer (2.7×10^{-12} g m^{-1} s^{-1} Pa^{-1}) at 50% relative humidity for the same polymer [8] – likely due to differences in sample preparation and experimental conditions (such as a difference in relative humidity gradient).

It should be noted that the partial pressure of water in the external environment does not greatly affect the water diffusion kinetics in a polymer [11].

14.2.3 Uncatalyzed Hydrolytic Chain Scissioning of PHA

The rate of abiotic hydrolysis of PHA in the absence of any promoter at neutral pH is very slow when in air or water. For example, Doi *et al.* [6] found that PHB films in buffer solution at 37°C showed no mass loss until after 180 days, when the critical molecular weight (13 000 g/mol) was reached. It was not until after about 80 days that molecular weight started to decrease [6], with this induction period being attributed to the time required for water to permeate the polymer matrix, although the possibility of microbial contamination over this period cannot be excluded in this, or other non-sterile tests. Bonartsev *et al.* [19] similarly saw a long induction period of >100 days at 37°C for PHB and PHBV, except for very low molecular weight PHB (170 kDa). Likewise, Freier *et al.* observed a relatively slow hydrolysis of solvent-cast PHB films, with a half-life in molecular weight loss of about 1 year at 37°C [20]. It was noted that while the addition of PHA-derived oligomers did not have an effect, the presence of hydrophobic plasticizers slowed this degradation, not surprisingly. Related to this, Marois *et al.* [21] conducted long-term hydrolysis studies of the medium-chain-length (*mcl*)-PHA, polyhydroxyoctanoate, and found that after 24 months in buffer, there was less than 1 wt-% mass loss, although the M_w and M_n had both fallen to 30% of the original, so internal (bulk) chain scissioning was occurring. There are a couple of recent reports that contradict these outcomes, with around 55% mass loss over 7 weeks for 50–60 micron solvent-cast PHBV (1 mol-% 3HV) films in one [22] and 6% mass loss over 8 weeks for solvent-cast PHB (thickness unknown) in the other [23]. However, solvent-cast films have much more potential for forming voids and porosity, and the fact that these films took up around 16% by weight water in 30 days in one study is indicative of a high level of porosity for such a hydrophobic material – thus providing a pathway for internal water and bulk degradation [22]. Further, the hydrolytic studies were conducted at 37°C under non-sterile conditions in buffer, with one of the teams noting that biofilms may have been forming during their study; the high porosity could therefore have also facilitated enzyme transport into the material [23].

Temperature, surface porosity, specific surface area, and polymer crystallinity can also have a significant effect on this rate [24–26]. Mechanistically, the hydrolysis process involves random chain scission in both the crystalline and amorphous regions of the polymer, with an initial increase in degree of crystallization being due to recrystallization of the fragments in the hydrolyzed amorphous regions [6], although there is evidence that over much longer times (>200 days) the mechanism may shift to non-random (chain end) scission, if the pH is lowered following ester group cleavage to form acid end groups and/or free crotonic acid [27].

Deroiné *et al.* [16] used the time–temperature superposition principle to predict lifetimes of PHBV (8 mol-% 3HV) plaques in water, undertaking accelerated aging tests on injection-molded 4 mm thick tensile bars in distilled water with bactericide added to prevent biologically driven degradation. Tests were conducted at 2°, 30°, 40°, and 50°C, well above the T_g of these samples, over 12 months, and changes in molecular weight and mechanical properties were assessed (Table 14.2, Figure 14.2). For all samples there was typical Fickian diffusion behavior initially, followed by a saturation plateau, then an increase in water uptake associated with polymer chain scission (more free volume) and consequent carboxyl group formation (increasing the affinity for water) – obviously accelerated by temperature. The control samples aged in air showed the effects of typical secondary crystallization on mechanical properties (slightly decreased elongation, increased Young's modulus, and stress at break). It is also noted that water had a plasticizing effect on the samples, with a distinct reduction in strain at break for samples on drying, so these results are for dry samples. There was a molecular weight threshold of 110000 g/mol below which the mechanical properties were no longer maintained.

Overall, though, a more rapid degradation of the material properties of PHA requires acid, base, or enzyme catalysis.

14.2.4 Acid or Base Catalyzed Hydrolytic Chain Scissioning of PHA

There have been a number of studies investigating the effect of acids or bases on the rate of PHA hydrolysis. Yu *et al.* [28], for example, compared the percentage

TABLE 14.2
Evolution of PHBV Mechanical Properties before and after Aging in Distilled Water at Different Temperatures (Reproduced with permission [16]. Copyright 2014, Elsevier)

	T [°C]	Aging time [days]	E [MPa]		σ_{uts} [MPa]		E_{uts} [%]	
Unaged			4390 ± 102		35 ± 1.2		1.4 ± 0.1	
Aged in air		360	5005 ± 104	+14%	37 ± 0.6	+6%	1.1 ± 0.04	−22%
Aged in water	25	360	5065 ± 58	+15%	36.5 ± 0.9	+4%	1.1 ± 0.1	−22%
	30	360	4612 ± 269	+5%	35.7 ± 1.6	+2%	1.0 ± 0.1	−28%
	40	360	4946 ± 236	+13%	26.5 ± 1	−25%	0.62 ± 0.05	−55%
	50	360	5686 ± 181	+30%	2.5 ± 0.5	−92%	0.26 ± 0.04	−82%

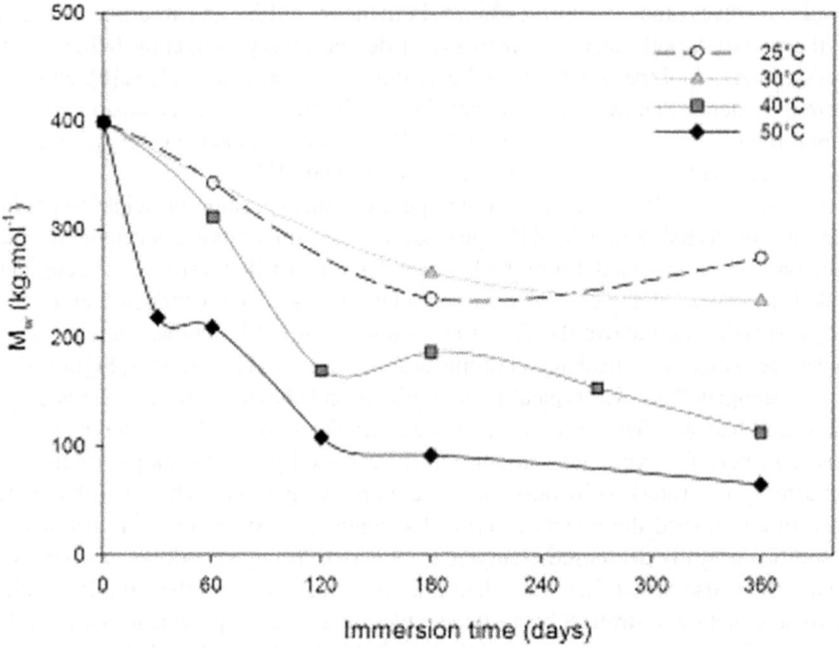

FIGURE 14.2 Evolution of the molecular weight, Mw, of PHBV specimens aged in distilled water at 25°, 30°, 40°, and 50°C after 360 days. (Reproduced with permission [16]. Copyright 2014, Elsevier).

mass loss and product distribution for the acid and base catalyzed hydrolysis for native (predominantly amorphous) granules, precipitated (semi-crystalline) extracts and cast films of PHB. They found that there was very limited degradation in 0.1–4 N sulfuric acid over a 14 h period, even at 80°C for the most reactive, amorphous granules, while samples were completely decomposed by concentrated acid. By contrast, the hydroxyl anions present in 0.1–4 N NaOH solutions efficiently lowered the barrier to hydrolysis, with rates ranging from 1.96% to 84.5% mass loss h^{-1} at 30° to 70°C, following a quasi 0th order kinetic with linear mass loss over time; the activation energy being lowered to that comparable to the activation energy of enzymatic hydrolysis (see below). A second process, the formation of crotonic acid via a simultaneous cyclic process involving dehydration of the 3-hydroxyl group to form double bonds adjacent to the ester groups with subsequent hydrolysis to crotonic acid, was less thermodynamically favored. The rate of hydrolysis of the amorphous PHB was found to be around 80–100 times faster than that of either semi-crystalline form [29]. It is also surface area dependent, with a centrifugally spun "wool" of PHA having very enhanced rates in comparison with films [30].

However, acid or base catalysis is not the common pathway for degradation of PHA in the natural environment or in use. Instead, biologically driven degradation is the primary mechanism, and this is principally an enzymatic process.

14.3 BIOLOGICALLY DRIVEN BIOPOLYMER DEGRADATION

The ability of PHA to biodegrade comes from their susceptibility to naturally occurring hydrolases, which are ubiquitous enzymes in our environment. While there have been many proposed and often conflicting descriptions of the stages of biodegradation [1], they can be effectively and simplistically modeled as comprising several overall steps, these being a) colonization of the polymer surface by microorganisms such as bacteria and fungi, including the lag time for film formation, b) enzyme catalyzed hydrolytic depolymerization, and c) the diffusion of small molecules out of the polymer followed by the uptake by microorganisms in a bioassimilation process used for either growth and reproduction or mineralization (Figure 14.3) [31]. The first two steps will be discussed in more detail for a bulk polymer PHA matrix, since it can be assumed that there is typically a relatively much faster enzymatic cleavage of oligomers and low M_n polymers in the final stage [32].

14.3.1 Microbial Colonization of Polymer Surfaces

In most natural environments, microorganisms tend to grow as biofilms rather than as freely dispersed planktonic cells [33]. A biofilm is a complex community comprising a single microbial species or, more commonly, a mixed community of bacteria

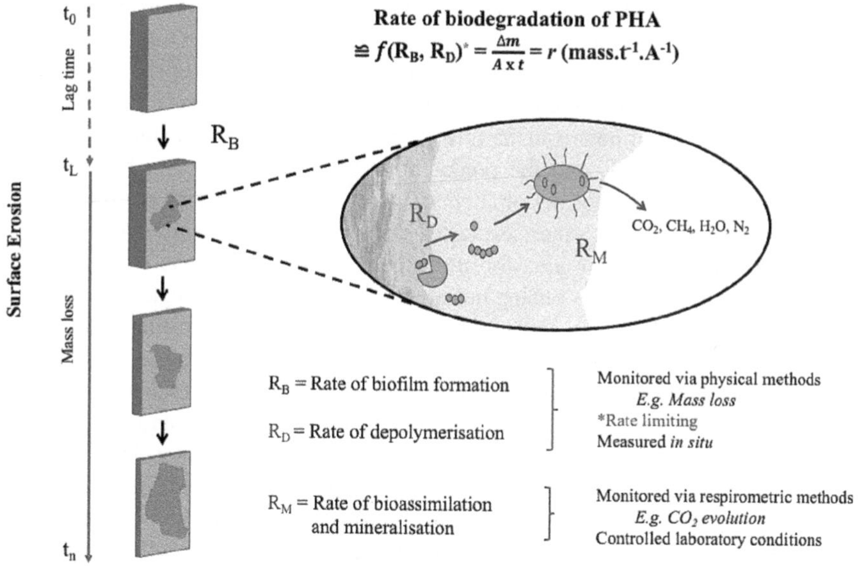

FIGURE 14.3 Rate of biodegradation of PHA. RB = Rate of biofilm formation; RD = Rate of depolymerisation, RM = Rate of bioassimilation and mineralisation; t0, tL, tn = initial time point, lag time, and final time point respectively; A = surface area, m = mass. It should be noted that whilst the steps of biodegradation are shown as occurring sequentially for means of communication, all processes are taking place concurrently; one process does not occur in totality before the next commences. (Reproduced with permission [33]. Copyright 2019, Elsevier).

and fungi, which are embedded within a self-produced hydrophobic exopolymeric substance matrix (EPS) and attached to one another or to a solid surface [34]. The first step in PHA biodegradation is the colonization of the polymer surface by both fungi and unicellular organisms to form a biofilm, with many of these organisms having the ability to secrete extracellular depolymerases.

The formation of a biofilm on a biopolymer surface is complex and still not fully understood [35]. However, it can be roughly characterized as having a number of stages: an initial coating of inorganic and organic matter (the "conditioning" film), followed by initial reversible microbial attachment, which depends on the surface properties of both the biopolymer substrate and the microbial cells, followed by increase in mass of the biofilm, which depends on irreversible cell attachment and the specific growth rate of the attached cells [36]. This irreversible attachment occurs through the formation of pili, adhesion proteins, and EPS and typically results in changes to the substrate surface. Finally, there is a "maturation phase", at which point the mass of the biofilm remains relatively constant, which is characterized by a range of competitive or synergistic interactions between cells, with either further recruitment or loss of species. All of these stages are affected by both the microbial strains present and the biopolymer material properties. With respect to bacterial cells, the underlying properties driving their attachment behavior are their relative hydrophobicity as well as the presence of charged species on the bacterial cell surface, such as carboxyl, phosphate, and amino groups [36]. It is believed that the more hydrophobic the cell surface, the more readily this attachment can take place.

With regard to the polymer properties, there is some dispute in the literature as to the primary controlling factors for bacterial adhesion [35]. However, most authors consider surface roughness and texture, porosity, fibrousness, surface hydrophobicity/wettability/polarity, and the preconditioning of the surface to be important for adhesion and biofilm formation, depending on their importance for the specific microorganism [33, 35]. An increase in polymer surface roughness leads to both an increase in exposed surface area for attachment as well as pockets where bacteria can shelter from shear forces during initial attachment [37]. With regards to polymer surface chemistry, there is considerable variability in reported results, although in general, hydrophobic surfaces have been regarded as resulting in more rapid biofilm formation, sometimes by orders of magnitude [35]. However, a recent study using carefully modified surfaces based on a range of ligand attachments to a gold substrate have shown that this effect may again rely on a complex interplay of factors, with adhesion of the bacterial pathogens *Staphylococcus aureus* and *Escherichia coli* O157:H7 being greatest on hydrophilic substrates characterized with positive surface charge [33]. This was followed by hydrophobic surface substrates with negative surface charge characteristics, and was smallest on hydrophilic surface substrates with negative surface charge characteristics. It was also found that the kinetics of adhesion shifted from a power law relationship for the hydrophilic surface substrates with positive surfaces to an exponential relationship for the other types of surfaces. This shift was interpreted as being due to initial elastic energy effects due to adhesion-induced cell deformation followed by the DLVO effect (the standard model integrating van der Waals attraction and electrostatic repulsion due to double-layer charge effects). While such an effect may vary with bacteria and substrate types, this does point to the need to understand the effects of surface chemistry changes on biofilm formation. Reinforcing this conclusion, the properties of the "conditioning" film on the polymer surface are dependent on the inorganic and organic matter present in the environment that can be sorbed, and can significantly affect the relative hydrophobicity of the surface and thus also affect the processes of microbe adhesion and biofilm development [36].

In terms of the kinetics of biofilm formation on PHA, in a recent study comparing biofilm formation on a range of polymers in seawater, Dussud *et al.* [34] showed that the reversible attachment phase took 7 days for PHBV films in the marine environment, with 1.3×10^5 cells cm^{-2}. After 22 days, a "maturation" phase was reached with cell counts of 16.3×10^5 cells cm^{-2}. However, as noted above, this rate is likely to be highly variable, depending as it does on many material and environmental factors, and, consistent with this, significant variations in the lag time for polymer degradation were noted in a recent metastudy of PHA degradation in the marine environment [31].

One further consideration with respect to bacterial biofilms and their effect on the rate of biodegradation is the size of the microorganism. The typical size is of the order of 0.1 to 10 micron (depending on species) [38] – so much larger than most surface or bulk cracks that may be present in polymers, at least before biodegradation. It should also be noted that it has been observed that there is a threshold of concentration of bacteria in the environment beyond which the biodegradation kinetics are no longer affected, linked to saturation of the surface by enzymes (see Section 14.3.3) [39].

14.3.2 Fungal Processes

A specific subclass of PHA biodegradation processes is that of fungal-promoted biodegradation – either alone or as part of a biofilm. For polymeric materials subjected to outdoor weathering, particularly in warm, humid locations, fungal growths are a common feature, whether the polymer is degradable or nondegradable. A fibrous network and/or isolated star-shaped features are commonly observed, and these are usually attributable to the presence of fungi, including mildew, and are surface phenomena, not cracks. Pitting is also commonly associated with these features, attributed to extracellular compounds such as acids or enzymes that are secreted by fungi [40, 41].

Sang *et al.* [42, 43] have modeled the biodegradation of polyhydroxyalkanoate (PHBV) solvent-cast films by fungi in soil and have shown that there are two growth phases for the fungal hyphae – an initial exponential phase, and a linear phase. During this linear phase, some of the growing microcolonies contact each other at the growing hyphae tips, and as a result, some tips stop extending and branching. It is these non-growing hyphae that release the depolymerases that degrade the PHA. Once the whole film surface is occupied, the fungi continue to grow by making a denser and thicker mat, with some depolymerases then being released into the aqueous medium in the pores and voids of this mat and with some hyphae becoming totally inactive due to senility.

Of particular note is that, in contrast to bacterial degradation that is only characterized by pitting and eventual erosion into the bulk of the polymer from these holes, fungal degradation of biopolymers is characterized by surface erosion directly adjacent to the hyphae, since the entire length of the hyphae can excrete depolymerases, thus forming a network of eroded surface channels. In addition, when comparing degradation rates in soil with bacteria, actinomycetes, and fungi, one study found that the fungi spread much more rapidly across the surface of the biopolymer, and as a result caused much more rapid biodegradation [44]. This is not an uncommon finding – other studies have shown that in a range of soils, the communities isolated from the polymer surface were rapidly enriched in specific fungi (often of the phylum Ascomyceta), with fewer fungal species present than in the bulk soil, indicating some selectivity [45–47].

Overall, this process of fungal biodegradation can be an important factor for PHA biodegradation, particularly in the early stages, but also over the long term [48]. For example, a range of bacterial and fungal taxa capable of degrading PHB has recently been isolated from PHA/PLA mulch films following extended soil burial for one year [49]. Overall, there is increasing evidence of a high level of complex interactions within fungal/bacterial biofilms [50], with little work to date on the evolution of such mixed-species biofilms on PHA.

It has also been noted that some microbes such as fungi and bacteria are able to produce a biosurfactant [51] that may affect the biodegradation processes, particularly through altering the surface hydrophobicity of polymer materials and thus influencing biofilm formation, as well as through affecting stress crack promotion and growth – discussed in more detail later.

14.3.3 Enzyme-Promoted Degradation of PHA

As noted by Jendrossek, "the ability to degrade short-chain-length (*scl*) PHA is widely distributed among bacteria and filamentous fungi and a large number of depolymerases have been purified and characterized" [52]. PHA are water-insoluble, but PHA depolymerases are water-soluble, meaning that these enzyme-promoted degradation reactions are heterogeneous.

The approximate diameter of a PHB depolymerase molecule is 4–5 nm [53, 54]. As such, this molecule is much too large to diffuse into a bulk polymer matrix in its native state. For this reason, PHA degrade primarily via an apparent surface-degrading mechanism, at least until they reach their critical thicknesses or form microcracks – although there will also be a much slower abiotic hydrolysis process happening in parallel in the bulk. For this reason, the processes of enzymatic degradation are critical in understanding PHA degradation.

PHA depolymerases are carboxylesterases that catalyze the hydrolysis of the ester group in the main chain. There are two overall types of PHA depolymerases – intracellular (responsible for conversion of native PHA granules in the cell back into accessible, water-soluble, lower-molecular-weight carbon molecules when needed by the organism) and extracellular (responsible for degradation of denatured, i.e., semi-crystalline, extracellular PHA granules in the environment). These PHA depolymerases are very diverse in sequence and substrate specificity, but typically share a common α/β-hydrolase fold and a catalytic triad (serine-histidine-aspartic acid), which is also found in other α/β-hydrolases [55]. Most of the extracellular depolymerases also have multi-functional domains, including this catalytic domain, as well as a substrate-binding domain and a linker domain that connects the other two domains [56].

The process of enzyme-catalyzed hydrolysis of PHA has been very well studied, both for single crystals as well as bulk polymers [56, 57]. As outlined in a detailed review by Numata *et al.* [56], the depolymerase firstly binds to the chain folding surface of the PHA polymer crystal, deforming it in the process and resulting in increased disorder of the chains and disruption of the molecular chain packing. The enzyme then primarily hydrolyzes polymer chains at the edges and ends of the lamellar stacks rather than on the chain-folding surface, attacking first the amorphous regions, before subsequently hydrolyzing the polymer chains in the crystalline state. This produces grooves and thinning of the tip of the lamellar crystals, fibrillar structures that ultimately degrade into fragmented low molecular weight products (see Figure 14.4).

The rate of enzyme-catalyzed hydrolysis is strongly affected by the level of crystallinity (with the rate increasing as the crystallinity decreases and being much faster for chains in the amorphous region than in the crystalline region), although this rate is unaffected by the size of the spherulites [58]. The rate of hydrolysis also decreases with an increase in the long period (lamellar thickness) of the lamellar crystals, possibly due to thinner lamellae being more easily disordered at the edge following enzyme-binding, leading to more ready hydrolysis [59]. Likewise, higher molecular weight PHB single crystals degraded more slowly than lower molecular weight ones,

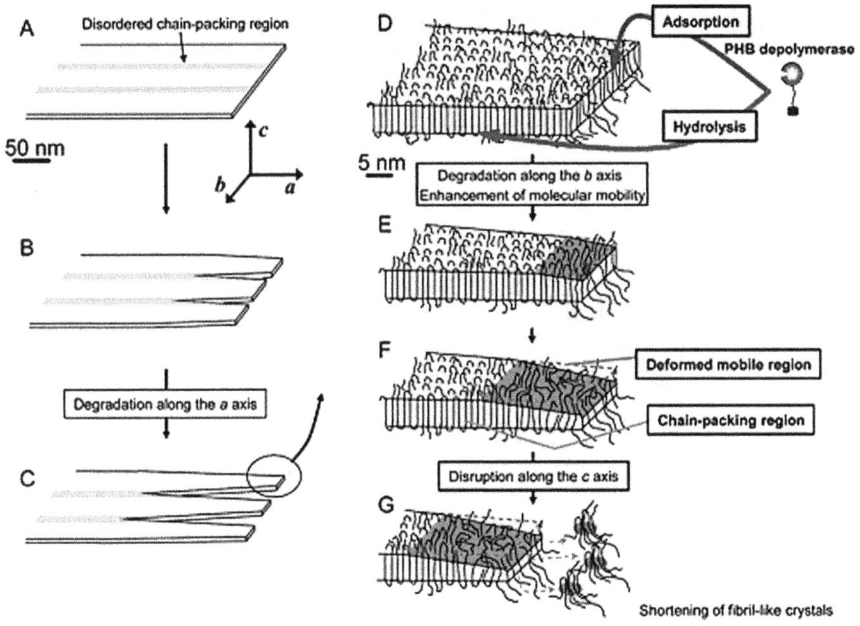

FIGURE 14.4 Schematic model of the enzymatic degradation process of PHA lamellar crystals by the PHA depolymerase from *R. pickettii*. (Reproduced with permission [56]. Copyright 2008, NRC Canada).

with the erosion rate at the point of the grooves along the a-axis increasing with a decrease in the molecular weight, while the rates of erosion at the tip of fibril-like crystals along the a- and b-axes had relatively similar values, independent of molecular weight [59]. This was attributed to an increase in the relative number of chain ends at lower molecular weight, again leading to an increase in disorder.

In addition, the concentration, properties, and reaction conditions of the PHA depolymerase in the environment play a strong role in the kinetics, with the rate increasing with the increase in enzyme concentration up to a point, likely determined by the concentration at which there is no space available at the surface for more enzymes following adsorption [53, 60]. Scandola *et al.* [61] provided a simple two-step kinetic model of the degradation of compression-molded films and suspended powders of PHB using isolated PHB-depolymerase A from *Pseudomonas lemoignei* (Equation (14.3)), which predicts that the hydrolysis rate per unit substrate surface area will reach a plateau at high enzyme concentrations:

$$A/V = \left(1/Kk_2\right)\left(1/C_e\right) + 1/k_2 \tag{14.3}$$

where A is the polymer substrate surface area, V is the rate of mass loss, K is the adsorption equilibrium constant, C_e is the initial enzyme concentration, and k_2 is the hydrolysis rate constant. The k_2 values for films (1.48 μg cm^{-2} min^{-1}) and powder suspensions (1.42 μg cm^{-2} min^{-1}) based on experimental data (see Figure 14.5) were similar [61].

The copolymer composition of the PHA has a very strong effect on the degradation rate, which is in part related to overall crystallinity, but also to molecular

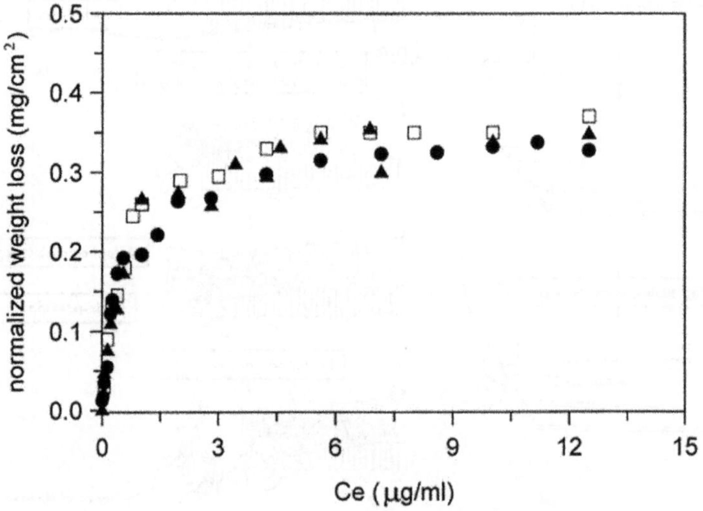

FIGURE 14.5 Weight loss per unit surface area of PHB films (area: ▲ = 1.92 cm^2; □ = 2.80 cm^2; ● = 4.00 cm^2), as a function of enzyme concentration (Ce). (Reproduced with permission [62]. Copyright 1998, American Chemical Society).

structure (molecular weight and distribution, degree of branching) and higher order structures (glass transition temperature, melting temperature, elastic modulus) as well as surface properties such as roughness and specific surface area. For example, in one study, PHBV (8 mol-% 3HV content) films degraded five times faster than PHB films [62]. Similarly, Abe *et al.* [59] showed that the rate of erosion increased markedly with an increase in comonomer content for a range of PHA copolymers, up to 10–20 mol-%, then decreased after that – and the rate of erosion of PHBV (6 mol-% 3HV) was several times faster than for PHB films at the same level of crystallinity. They hypothesized that increased polymer chain mobility due to the presence of side groups (as reflected in lower melt temperatures) may be a factor. In another study [63], the rate of degradation of films was found to be PHBV (18 mol-% 3HV content) > PHB > poly(3-hydroxybutyrate-*co*-3-hydroxyhexanoate) P(3HB-*co*-3HHx) (19 mol-% 3HHx) > P(HB-*co*-2HA) (15 mol-% HA, where HA refers to mixed *mcl*-PHA), and it was proposed that an increase in length of the side chain has a slowing effect on enzymatically driven chain hydrolysis, with these hydrophobic side chains interfering with enzyme-binding at the surface at these higher copolymer contents. By contrast, in another a study on PHA film degradation by lipase in phosphate buffer saline, it was found that P(3HB-*co*-HHx) (12 mol-% 3HHx) was degraded faster compared with PHB, P(3HB-*co-3*HHx) (5 mol-% HHx) and P(33HB-*co*-3HHx) (20 mol-% 3HHx), which all degraded at about the same rate. In this case, it was hypothesized that the 12 mol-% 3HHx sample combined low crystallinity and a comparatively rougher surface, leading to accelerated degradation [64]. Cao *et al.* noted that poly(3-hydroxybutyric-*co*-3-hydroxypropionic acid) (PHBP) samples were only degradable by depolymerase (and river water) if PHB-type crystallites were present, and these had much higher degradation rates than for samples with PHP type crystallites – and that the amorphous copolyester was almost unable to be degraded [65].

In studies on drawn PHB and PHBV fibers, X-ray patterns showed reflections from both the usual α-form (an orthorhombic crystal system with a 2_1 helix conformation space) and the β-form (with a planar zigzag conformation) simultaneously. The enzymatic erosion rate of the β-form was found to be much faster in comparison with the α-form [62, 66, 67].

The processing method also has an effect on the rate of enzymatic hydrolysis [68], with solvent-cast films degrading more slowly than those prepared by compression-molding in one study, possibly because of residual solvent effects, or differences in morphology or degree of degradation due to thermal processing. The authors claimed that the materials were amorphous but this is unlikely under the preparation conditions used, so that differences in crystallinity or in local stress cracking and voids associated with the interspherulitic regions may also be responsible.

There is also evidence of repression of PHA esterase activity in the presence of a catabolite, such as the more readily utilized glucose [69].

Unlike enzyme catalyzed processes in solution, that are homogeneous and follow classic Michaelis–Menten kinetics, the enzymatic degradation of a solid polymer in an aqueous environment is a heterogeneous process. SEM imaging of the surfaces of PHA materials following bacterial degradation consistently shows hemispherical pits and pores associated with enzyme activity, with the amorphous region of the polymer spherulites being eroded first, exposing the crystalline lamellae within the spherulites [70]. It can be modeled as comprising two consecutive steps, viz. the initial adsorption of the enzyme to the surface to then form an enzyme-substrate complex followed by either progression to the actual catalysis of PHB degradation/hydrolysis, or desorption of the enzyme from the surface without catalysis [71]. Using an intracellular PHB depolymerase that was modified to be incapable of catalyzing the ester hydrolysis reaction but still retained a binding function, Polyák *et al.* [71] were able to analyze the kinetics of depolymerase adsorption and desorption to and from the surface of PHB particles of around 10 micron in diameter, and found the rate constants for adsorption and desorption to be 0.052 min^{-1} and 0.025 min^{-1}, respectively. This shows that adsorption is thermodynamically favored and twice as fast as the reverse process, so that under steady state conditions, most enzymes present will be bound to the polymer surface. Moreover, these values are orders of magnitude slower than the estimated 10^1–10^6 min^{-1} rate of polymer chain scission (i.e., the number of substrate molecules that can be converted to product by a single enzyme molecule per minute) for the catalyzed hydrolysis. Thus, the enzyme adsorption step is the rate limiting stage of the depolymerization process. In terms of the area occupied by these extracellular PHA depolymerases: the enzyme molecule from *Bacillus megaterium* (ATCC 11561) was determined to occupy 13.1 nm^2 of a PHB film surface – a similar area to that found for the extracellular depolymerase of *Alcaligenes faecalis* (17 ± 8 nm^2) [53].

14.4 PHA BIODEGRADATION IN NATURAL ENVIRONMENTS

In the natural environment, the factors controlling biodegradation and polymer lifetimes are much more complex than for "simple" controlled enzyme studies in the laboratory, where multiple chemical, physical, and biological processes can take place in parallel or sequentially, such as stress cracking, creep, swelling, leaching, oxidation, reduction, chewing, abrasion, dehydration, dissolution, abiotic hydrolysis, cyclic loading, local tension and stress, osmotic effects, ozonolysis, and biodegradation. Thus, there is a very wide range of processes and environmental agents that a biopolymer will be exposed to – and not all of these will be covered in this review.

In general, the methods used to determine the extent of biodegradation include mass loss, visual and microscopic observation, changes in tensile properties and molecular weight, microbial community analysis (in the medium and/or on the polymer surface, and including growth as well as composition), and indicators of microbial activity such as oxygen consumption and carbon dioxide evolution. Tests can be conducted in the laboratory using simulation of field conditions or in the field.

Field testing of PHA degradation is challenging and expensive. There is massive variation within the literature across studies and across different sites in terms of temperature, type and concentration of microbial community in the environment, oxygen levels, pH, sediment loads, UV exposure, mechanical forces on samples, and PHA sample variation (morphology, composition, processing method, crystallinity, glass transition temperature, mechanical properties (particularly modulus of elasticity), surface properties (such as hydrophobicity, specific surface area, and roughness), etc.). Furthermore, there is inconsistency in terms of data reported, with few studies including sufficient data to determine the rate of degradation relative to exposed surface area. However, the following sections will summarize key relevant outcomes to date with respect to PHA biodegradation in the natural environment and cover the main drivers for environmental degradation across a range of environments studied, thereby providing an overview of the underlying drivers.

14.4.1 Degradation of PHA in Freshwater Environments

There have been a number of studies of PHA degradation in freshwater environments, ranging from laboratory-based studies in water inoculated with PHA-degrading microbes through to direct riverine and lake immersion/degradation studies [72–77]. In one laboratory-based study using Japanese river water as an inoculum, the biodegradation of a very wide range of PHA copolymer film samples (thickness unknown) were compared, based on weight loss and biochemical oxygen demand (BOD). Only poly(3-hydroxypropionate) (PHP) and PHBV (80 mol-% 3HV) were not 100% degraded (by mass) after 28 days [77]. Voinova *et al.* [78] explored the biodegradation of 100 micron PHBV (8 mol-% 3HV) discs in two freshwater lakes in Russia, placing these discs in both the aerobic littoral zone (i.e., the zone between the high water mark and the areas below the shoreline that are permanently submerged) and in the anaerobic silt. These lakes, 6 km apart, had the same pH, temperature, and dissolved oxygen, but different inorganic phosphorus levels and thus different microbial biomass abundances. The half-lives varied considerably, with the fastest (21.6 days) being in the aerobic zone of the phosphorus abundant lake. Biodegradation was significantly slower in the anaerobic silt zone of this lake. The lack of phosphorus in the other lake, which restricted phytoplankton growth, meant that biopolymer degradation was limited, with a half-life of 194.5 days even in the aerobic zone. Similarly, Mergaert *et al.* found that the biodegradation of 2 mm thick injection-molded dog bone samples of PHB and PHBV (10 and 20 mol-% 3HV) was slow in freshwater ponds, with less than 7% mass loss for all samples after 6 months and 34%, 77%, and 100% mass loss, respectively, after 358 days. In another study, shampoo bottles made from Biopol (a commercialized PHBV, mol-% 3HV unknown) were placed at various depths in Lake Lugano, Switzerland, for 254 days. On this basis, the estimated life-span was a maximum of 5–10 years, with 8–10% weight loss after 250 days. The rate of mass loss decreased progressively, from around 17.2 mg/day at 20 m depth to 10.6 mg/day at 85 m depth – probably due to the local microbial environment plus dissolved oxygen and temperature effects.

Overall, however, there was insufficient data available of the quality needed to undertake a higher order analysis of the rate of biodegradation in the freshwater environment in order to normalize the rate relative to the surface area across a number of studies.

The effect of biodegradation on mechanical properties in these environments has not yet been characterized, to our knowledge.

14.4.2 Degradation of PHA in Marine Environments

The marine biodegradation of PHA is an area of considerable current interest, due to the large amounts of our plastic debris polluting the oceans (with an estimated hundred million tons of plastic waste to enter the oceans between the years 2010 to 2025 [79, 80]). The marine environment is a particularly challenging one for biodegradation, being low in microbial community density, and with low oxygen concentrations, UV exposure near the surface, and temperatures once such debris becomes submerged below the surface. In addition, as with other studies cited herein, it is very challenging to compare across studies, given the very large amount of diversity in terms of polymer form/composition/processing/morphology as well as experimental methodologies and differences between sites. As such, we have recently collated the best available information on biodegradation of PHA in the marine environment, where the minimum data (dimensions of initial sample and mass loss over a given time) were given (Figure 14.6) [31]. On the basis of the collated analysis, the mean rate of biodegradation of PHA in the marine environment was found to range between 0.04–0.09 mg·day^{-1}·cm^{-2} initial exposed surface area ($p = 0.05$) – and based on that, a water bottle made from PHA could be expected to take between 1.5 and 3.5 years to completely biodegrade. However, while a summary was also given of the known factors influencing the degradation rate, no single environmental or polymer-related parameter emerged as the key factor – and there is more work to be done in understanding this in more detail. The reader is referred to this paper for more detailed analysis, including references for those key studies [31].

One detailed study explored the changes in mechanical properties in 4 mm thick PHBV (8 mol-% 3HV) dog bone samples following exposure to seawater, comparing a laboratory-based study, using continuously renewed and filtered natural seawater pumped into the laboratory at different temperatures of 4°, 25°, and 40°C, with aging in the sea [11]. The most significant change over 360 days was for the elongation at

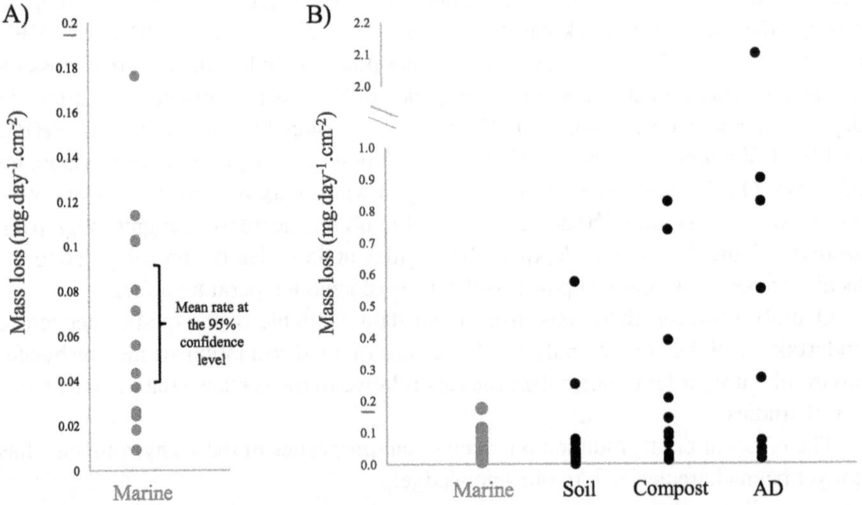

FIGURE 14.6 Normalised biodegradation rate of PHA in different environments: (A) Marine data in detail including the 95% confidence interval of the mean; (B) Degradation rate in different environments (marine data is the same as in A). Note the break in the y-axis for B. Reproduced with permission [31]. Copyright 2019, Elsevier.

break, dropping by 58% at 40°C, with M_w dropping to 32% of original in parallel. At lower temperatures there was less change in properties, with the effects decreasing, as expected, with temperature. Enzymatic degradation at the surface was shown to be occurring in parallel with ester hydrolysis in bulk, making the determination of an activation energy through an Arrhenius relationship invalid. It should be noted that this filtration of seawater in itself may be problematic, with the possibility that some of the native microbial community was removed in the process.

14.4.3 Degradation of PHA in Soil Environments

Compared with the aqueous environments discussed so far, soils are very complex and highly variable environments comprising mineral particles, organic matter, water, air, and a vast number and variety of living organisms [81]. Factors that play into biopolymer degradation in soil can include soil moisture content, porosity, soil temperature, soil pH, O_2 availability, presence of suitable microbes, presence of contaminants and their concentrations, sand/clay ratio, cation exchange capacity, pH, organic matter content, NPK loads, $CaCO_3$ content, availability of nutrients, presence of other electron acceptors, redox potential, etc. [81, 82], with nutrient management, particularly of nitrogen, being critical. Specifically, for samples containing 1 g of C in 300 g of soil, the addition of 0.1 g of N significantly improved the reproducibility of the test, almost regardless of soil type, although sandy soils did need additional nutrients [81]. In support of this conclusion, Hoshino *et al.* [83] also showed that the biodegradation of biodegradable polymers in soil, including PHBV, is mainly affected by the soil temperature and the total available nitrogen, which is likely linked to the needs of the soil microbial community. However, there are few systematic studies on the effects of polymer- and soil-specific factors on soil microbial function and polymer degradation rates [83, 84].

The effect of soil microbial community composition on biodegradation is likely significant, with Song *et al.* showing that the PHBV-biodegrading capacity of a soil was related to the number of PHBV degraders in that soil [85]. In addition, a thick mycelial growth is often observed around the surface of degrading plastic items [86]. Sang *et al.* analyzed the succession of microbial consortia around a PHB film and showed that there was a rapid increase in fungi and actinomycetes both in soil and on the film surface but not of PHBV-degrading bacteria, resulting in a dominance of fungi in the population [44]. In this study, the number of PHBV-degrading microbes present in garden soil and paddy field soil was estimated to be 4.30×10^5 and 5.06×10^5 degraders per gram of dry soil [87].

Also, fungi can direct their hyphae to grow across air gaps between soil particles, meaning that direct contact with a polymer is not necessary, unlike for bacteria [44, 88], so fungi may outcompete bacteria in colonizing and degrading biopolymers in soils.

In terms of reported rates of biodegradation of PHA in soil – once again, these are enzyme-driven, surface erosion processes, with rates being dependent, as above, on a combination of copolymer composition, crystallinity, microstructure, and surface morphology, along with soil and environmental parameters as noted, particularly microbial population and distribution, with enzymatic depolymerization commonly considered to be the rate-limiting step [44, 89]. A typical biodegradation profile (in

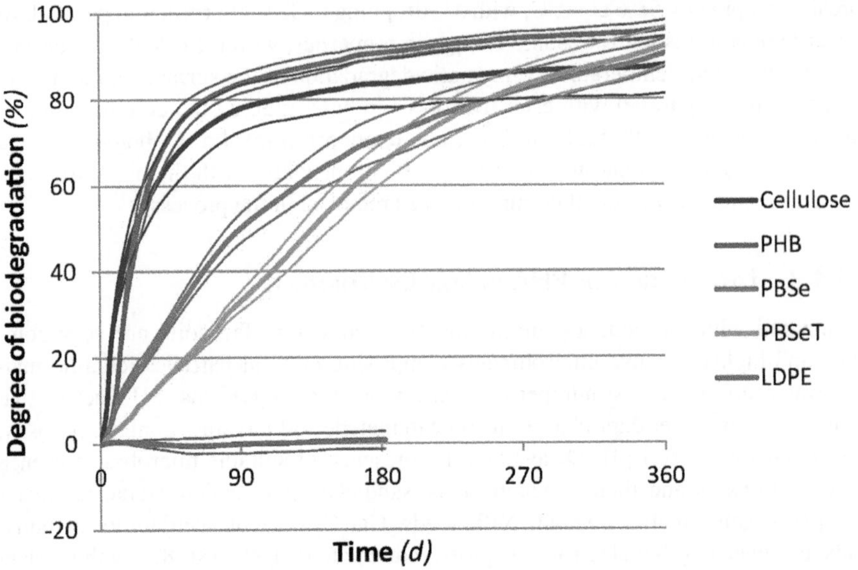

FIGURE 14.7 Evolution of the average degree of biodegradation of polymer films degraded in soil as a function of time. PHB = polyhydroxybutyrate; PBSe = polybutylene sebacate; PBSeT = polybutylene terephthalate co-sebacate; LDPE = low density poly(ethylene). Thin lines represent standard deviations. Colour version available in original. (Reproduced with permission [7]. Copyright 2020, Elsevier).

this case for 80-micron PHB film) is given in Figure 14.7, with PHB degrading rapidly initially and in a similar fashion to cellulose, based on CO_2 evolution, before degrading completely over around 360 days.

Under extremely unfavorable environmental/soil conditions, however, degradation rates can be reduced to nearly zero. Also, because of the many different degradation protocols, types of processing and polymer morphologies adopted for the different biodegradation studies in soils, there is a wide reported range of biodegradation rates (see Figure 14.6 for a non-comprehensive snapshot of degradation rates reported in soils relative to reported surface areas, ranging from <0.1 to ~0.6 mg. $day^{-1}.cm^{-2}$ initial surface area). As expected, the reported studies show preferential degradation in the amorphous, interspherulitic regions as well as the crystal nucleii of polymers at the surface, although bulk mechanical and thermal properties are typically not affected, and there is preferential loss of the comonomer unit for PHB-based copolymers [90, 91] although this is not always the case [89]. Also, it should be noted that the processing step has a large influence on polymer properties and for realistic lifetime assessment, it is preferable to assess biodegradability in soils on the final, processed products [92].

Only a few studies have attempted to link polyester biodegradation in soils to soil esterase activities [93, 94]. For example, nitrogen-limited microorganisms may not produce esterases given that their secretion and subsequent hydrolysis of polyesters does not increase N-availability. Secondly, the general availability of carbon substrates may regulate esterase production in colonizing microorganisms. Quantifying esterase expression and secretion in soils during PHA degradation is a challenge that has not yet, to our knowledge, been addressed.

As expected, biodegradation in soil is surface area dependent. Modelli *et al.* [95] showed biodegradation of PHA powder (≥100 μm) followed first-order kinetics, leading to 90% biodegradation in 65 days, while film (0.2–0.3 mm) showed a constant slow CO_2 evolution leading to 26% biodegradation in 210 days – about 90 times slower than the initial rate for the PHB powder.

One other area of investigation that merits further consideration is that of analyzing the residual PHA content in soils following fragmentation/biodegradation [96]. To our knowledge, this has not yet been done for PHA, although there are test methods being developed for other degradable polymers, which are improvements on the existing flawed gravimetric methods. These include the use of Soxhlet and accelerated solvent extraction, to recover residual polymer from soils [97, 98], where the properties of the extracted polyesters could be quantified.

With respect to mechanical properties, there are few studies that report changes in mechanical properties of bulk PHA over time following exposure to soil. The elongation at break is again usually the factor that is most significantly affected, and it is important to recognize that this measure is strongly influenced by surface flaws, which can be readily induced by a number of different environmental processes, apart from enzymatic degradation, such as environmental stress cracking, water cycling, mechanical abrasion, animal attack, and root processes, just to name a few. Given that, Mergaert *et al.* [99] saw a 50% reduction in elongation over 150 days at 15°C and 70% and 81% reduction over 200 days at 28° and 40°C, respectively, for PHB tensile test samples in hardwood forest soils. Other mechanical properties were not reported.

An additional point to note is that the effects of PHA degradation on soils and their associated microbial communities may be significant, with some reports showing that the diversity and population of the soil microbial community were correlated with the degradation of PHA [100]. The stimulatory effect on microbial activity may be more than can be accounted for by direct carbon input from these polymers, so the impacts on nutrient biogeochemistry in soils need to be understood [101].

14.4.4 Degradation of PHA in Aerobic Compost Environments

Compost delivers accelerated degradation of biopolymers, given that it is a microbially rich aerobic environment operating under high humidity and temperature conditions [32]. As such, PHA lifetimes in these environments are much shorter than in normal soil under ambient conditions, although the same drivers apply. A recent review of aerobic biodegradation of biopolymers in simulated composting conditions (focusing on evolved CO_2 studies) provides a useful summary of factors controlling biodegradation – with strict control needed over moisture content, organic matter, C/N ratio, and pH for reproducibility, along with sufficient nutrients for microbe function [102, 103]. There have been a number of studies of PHA biodegradation in controlled composting environments at elevated humidities, degrees of aeration/ventilation, and temperatures, typically around 55–65°C [27, 32, 103–114], showing again that these are primarily microbially driven, enzymatic, surface erosion processes, with the same relationship to composition and crystallinity as previously evidenced – with some exceptions, likely related to local microbial substrate preferences. However, in general, the order of rate of degradation is P(3HB-*co*-4HB) (40 mol-% 4HB) ≈PHBV (40 mol-% 3HV) > PHBV (30 mol-% 3HV) > PHBV (20 mol-% 3HV) > PHBV (3 mol-% 3HV) > PHB [106, 115]. Hydrolytic degradation is also taking place in parallel, but even at 60°C is relatively slow [27]. Just one report to our knowledge has considered the effect of mass relative to surface area (i.e., product form) in aerobic biodegradation in compost [105] – using melt-pressed PHB plaques of different thicknesses. It was found that for an initial mass:surface area of <67, there was a lag time of ~12 weeks, with complete degradation in less than 4 months. The mechanical properties were retained for 18 weeks, before onset to failure. It was noted that there was a positive correlation between rate of mass loss relative to exposed surface area versus initial mass:surface area, which was attributed to the fact that the larger and thicker samples had more polymer available internally for pores and channels to form, leading to much larger relative increases in exposed surface area following degradation – and hence a faster rate [116]. At the same time, other studies show a shorter (7 day) lag time and associated shorter total time to degradation [117]. It was also noted in another study [106, 115] that there was a difference between laboratory and pilot-scale results based on ISO 14855 standard, with the 3% 3HV copolymer being 100% converted to CO_2 after 12 weeks in the pilot study and only 81% in the laboratory reactor. This difference points to the need to be cautious in extrapolating laboratory-based estimations of lifetimes to the larger scale. Given that, we have summarized the range of degradation rates based on surface area across a number of studies (Figure 14.6), showing that PHA degradation rates in compost ranged from <0.1 to ~0.8 $mg.day^{-1}.cm^{-2}$ initial surface area.

One study to our knowledge has explored the change in mechanical properties of PHA in compost, looking at the degradation of 0.5 mm PHBV melt-pressed films. Surface erosion via enzymatic processes was the dominant mechanism, with little evidence of bulk hydrolysis, and Table 14.3 shows the changes in tensile properties

TABLE 14.3
Tensile Properties of PHBV in a Composting Medium over Time. Numbers inside the Parentheses Are Coefficient of Variation (%) (Reproduced with permission [109]. Copyright 2003, Elsevier)

Composting time (days)	σ_{uts} (MPa)	*E* (MPa)	ε_{uts} (%)
Control	**31.3 (0.5)**	**1315 (5.9)**	**5.2 (4.3)**
10	28.3 (2.5)	1272 (4.0)	4.4 (4.1)
20	18.2 (21.5)	1198 (6.6)	2.2 (25.0)
30	16.8 (14.8)	1260 (3.6)	1.9 (17.8)
40	8.4 (7.7)	1276 (6.7)	1.1 (38.2)
50	0.9 (59.8)	1242 (7.4)	1.0 (61.0)

over time, with little change in the Young's modulus but significant decreases in elongation and ultimate tensile strength [109].

14.4.5 Degradation of PHA in Anaerobic Environments

The anaerobic biodegradation of PHA and other biopolymers in landfills and anaerobic digestion systems is well studied [118–127] and was recently reviewed [128], with a focus on anaerobic digestion (AD) – a process that could couple well with one of the possible areas of application of PHA, which is in food packaging with end-of-life management of contaminated packaging through AD. In terms of lifetimes, this equates to around 77% conversion to CO_2 in 85 days for a P(3HB-*co*-3HHx) 1 mm thick sheet [118], or between 83–96% conversion to CO_2 for PHB and PHBV (13 and 20 mol-% 3HV) films over 16 days [121], or 85% conversion to CO_2 for PHBV (8 mol-% 3HV) films over 25 days [119]. PHB powders degraded even faster, reaching 90% degradation in 9 days [129]. Likewise, Morse *et al.* showed that for two 0.3 mm melt-pressed P(3HB-*co*-3HHx) films, there was 80% mass loss within 7 days for a 10 mol-% 3HHx sample, and 28% mass loss for a 3.8 mol-% HHx sample [122]. Noda also showed 80% conversion for poly(3-hydroxybutyrate-*co*-3-hydroxyoctanoate) (PHO) films after 25 days [120]. While non-exhaustive, this list does show consistency of outcome, which is also consistent with older studies, and the available relevant data has been collated (Figure 14.6) to show that relative to the other environments (apart from compost), this form of degradation is rapid – as a result of high concentrations of inoculum and nutrients – with degradation rates ranging from <0.1 to as high as 2.1 $mg.day^{-1}.cm^{-2}$ initial surface area.

Ryan *et al.* have also recently published a kinetic model of anaerobic digestion of PHBV [130]. The model and physical microscopy showed that microbes were attached to the PHA surface after 10 days, with complete conversion to methane within 42 days.

It is noted that for anaerobic digestion it is not relevant to consider changes in mechanical properties over time.

14.5 BIODEGRADATION OF PHA-BASED BLENDS

The blending of two or more polymers is an easy strategy for modification of material properties of these polymers, readily achieved in existing processing equipment such as melt blenders or extruders. Blends can deliver significant changes in physical properties, including melting temperature, glass transition temperature, crystallinity, and mechanical properties (modulus, elongation at break, flexibility, etc.) compared to the initial polymer components – although usually these properties fall as intermediate between the extremes of the individual components. Two reviews have provided an oversight of PHA-based blends and their biodegradability [131, 132], showing that the biodegradation of both miscible and immiscible blends of PHA with other polymers has been studied quite extensively over the years. For immiscible blends, the biodegradation performance is effectively governed by independent biodegradation of the separate phases, with phase distribution and substrate accessibility to enzymes playing a critical role [131, 132]. For example, the slow abiotic hydrolysis of highly crystalline PHBV has also been shown to be accelerated by the blending of this material with low molecular weight PLA (750 g/mol), which was attributed to increased water penetration following dissolution and migration of these components from the bulk [133]. Similarly, the addition of hydrophilic PEG has been shown to increase the degradation rate of PHA blends [23, 134]. In addition, there is often poor interfacial adhesion between immiscible polymer phases, which can lead to local stress cracking and other mechanical loads at the interspherulitic junctions, thus opening up cracks and pathways for water and enzyme ingress, in turn accelerating the bulk degradation processes. However, when the blends are miscible, then it is the properties of the amorphous phase in these blends and the mobility of the chains within this phase that are the controlling factors in influencing the rate.

Biodegradability of blends is also strongly influenced by the processing methodology, with solvent-cast blends degrading more rapidly than melt-processed. The addition of plasticizers, including oxidized cottonseed oil and licowax, has also been shown to significantly increase biodegradation rates in PHA blends, possibly by decreasing the crystallinity or by increasing the mobility of the polymer chains and thus increasing water diffusivity [135]. It is clear that blending can make a very significant difference to the rate of biodegradation and loss of mechanical properties on aging, compared to the bulk polymer components on their own.

The biodegradation of PHB-PHA blends has been explored, given that PHB is usually incompatible with other PHA or copolymers, except for PHBV when at less than 15 mol-% 3HV content. It was found that when PHB-PHBV blends were hydrolytically degraded in sterile buffer, the rate appeared to be dependent primarily on the crystallinity of the combined materials [136], although there would be some question over the effect of the interfaces between phases for phase-separated materials, and whether or not these act as zones of weakness and/or pathways for more rapid water passage. Incompatible PHB-polyhydroxyoctanoate (PHO)-PHB blends were also degraded by surface erosion via the amorphous domains [137]. When atactic PHB and normal isotactic PHB were blended, there was a significant increase in the rate of enzymatic degradation – five times higher for the 75% atactic/25% isotactic PHB compared to the isotactic PHB alone. However, on its own, the atactic PHB barely degraded. The authors attributed the difference to a decrease in crystallinity [138].

The incompatible PHA elastomer (rubber) blends are also increasingly under investigation. If a co-continuous morphology is formed through a balance of composition and processing, then the addition of the elastomer has no effect or even results in an increase in biodegradability of the PHA phase, again attributed to an increase in the exposed surface area internally [139, 140]. However, if the PHA phase is encapsulated in the rubber, again biodegradation is severely inhibited [141].

The biodegradation of poly(propylene) (PP) and poly(ethylene) (PE) blends with PHA have been explored a number of times [131, 132]. These materials form incompatible blends, and the main degradation observed is in the PHA phase. In some cases, such as with an 80/20 PP/PHA mix [142], the PHA phase is so encapsulated that biodegradation cannot occur.

Other interesting blended materials include: PHA-lignin, where the lignin inhibits the degradation, possibly through inhibition of biofilm formation [143];

PHBV – poly(propylene carbonate) (PPC) blends, where the PPC phase degrades through random hydrolytic chain scission given the absence of natural degraders in soil, and hence the blends behave in a similar fashion to PHA-PLA blends [144]; and PHB-poly(ethylene oxide) (PEO) blends, where enzymatic degradation of the blends was accelerated relative to the native PHB due to dissolution of the PEO and pore formation, enabling enzyme access to the bulk of the polymer [145].

In addition, the blending in of compounds with an acidic or basic side chain, such as carboxylic acids, has been shown to promote the hydrolysis of PHA, presumably by promoting water flux as well as accelerating the ester cleavage [133].

Altogether, there is a wide range of PHA-polymer blends which have been explored, many of which behaved in a similar fashion to those reported above and which will therefore not be covered herein. However, to our knowledge, there are few studies on the changes in mechanical properties of blends on biodegradation. In one such study, looking at 51-micron-thick extruded starch-PHBV (12 mol-% 3HV) blends degrading in the marine environment, the more hydrophilic, incompatible starch component (which also had 10–50 times more microbial degraders present in the environment) degraded much more rapidly than the PHBV, so that its biodegradation dominated, including in the loss of tensile properties [146]. A predictive model based on mass loss found that the starch and the PHA phases in the blends had half-lives of 19 and 158 days, respectively.

Dharmalingam *et al.* [147] also followed the changes in tensile strength and molecular weight for spunbond and melt-blown P(3HB-*co*-4HB) (17 mol-% 4HB)/PLA-based nonwoven mulches on exposure to soil. For spunbond blends, with larger fiber diameters and higher degree of orientation, there was very little change in tensile properties or molecular weight over 30 weeks. However, the melt-blown samples showed 78–79% loss in tensile properties in just 10 weeks, with reduction in molecular weight, but to a lesser extent – indicative of surface erosion. No change was observed in sterilized soil. Likewise Iannace *et al.* [148] looked at the effect of solvent-cast-blend films of PLA and PHBV (Biopol) (thickness unknown) in non-sterile PBS at 37°C over 180 days and found that the presence of PLA accelerated the degradation of mechanical properties relative to bulk PHBV, and hypothesized that this was in part due to partial miscibility of (acidic) degradation byproducts. The presence of PHBV slowed the loss of mechanical properties for PLA, presumably through acting as a barrier.

14.6 BIODEGRADATION OF PHA-BASED COMPOSITES

So far, this chapter has been reviewing the biodegradation of bulk polymer and polymer-blend matrices. However, in many cases the biodegradable product is a composite, comprising a polymer matrix with solid filler/s – whether water-soluble or not. The additives used in such biocomposites are often more hydrophilic than the bulk polymer matrix. Whether this is the case or not, the introduction of solids into these systems introduces complex dynamics around water percolation and diffusion, moisture effects on interface chemistries, the different degrees of swelling/relative water uptake in the two phases – and the mechanical stresses associated with this, plus crack propagation at local stress-points due to solid additives, and enzyme penetration through cracks and pores, to name just a few of the many factors at play [149, 150]. Given biodegradation is ultimately driven by enzyme activity, and cracks can facilitate enzyme transport into the bulk, crack formation is discussed in some detail in below.

Microcracking occurs as a result of impact and internal stresses and is a major cause of material failure. Cracks, structural defects, and delamination that occur deep in the structure of polymer composites are extremely difficult to detect. Microcracks may evolve into macrocracks, moisture swelling, and de-bonding. They also enable biodegradation if the crack size allows for the passage of enzymes or even microorganisms.

Macrocracks can form due to hygrothermal aging, with moisture transport in composites occurring via diffusion through the polymer matrix (through voids between the polymer chains (*cf.* Section 14.2.2)), through capillary processes at microgaps formed during processing at the filler–polymer interface, and through microcracks formed during processing, with the major mechanism (and most rapid by orders of magnitude) being along the filler–matrix interface [150].

Environmental Stress Cracking (ESC) is a particular challenge for biocomposites, and follows exposure to corrosive environmental chemicals while under tensile stress, causing crazing, and ultimately resulting in brittle failure [150]. This process has recently been reviewed by Robeson [151], with the mechanisms being discussed in detail. As mentioned in Section 14.3.2, surfactants can have this effect on biopolymers, as can solvents and many other organic chemicals, particularly caustic ones (NaOH) [152].

According to Awaja *et al.* [150]:

> For a given fibre reinforced composite where the fiber is gripped by the polymer matrix, a matrix crack is halted by the fiber. Upon increasing the load, the crack starts to pass around the fibre without breaking the interfacial bond. Interfacial shearing and lateral contraction of the fibre result in de-bonding and a further increment of crack extension. After considerable de-bonding the fibres break at some weak points within the matrix and further crack extension occurs. The total failure of the composite happens when the broken fibre end is pulled out against the frictional grip of the matrix.

Fiber orientation in the matrix also plays a significant role in modulating crack propagation, as well as mechanical properties and tortuosity of water (and enzyme, if the crack size permits) penetration through the bulk, and thus can strongly influence the degradation rate [153, 154].

PHA-based composites have been exposed to a wide variety of environments, including the usual soil, marine, and natural waters, but also to more unique situations such as industrial composting facilities [155] and accelerated weathering in QUV chambers [156, 157]. As a general rule, the addition of hydrophilic additives results in accelerated biodegradation, although the degree to which this occurs can be variable [158]. And likewise, the addition of more impermeable components, such as cork [155], slowed down the degradation rate, possibly due to both inhibiting enzyme attack as well as water permeation.

For PHA-nanoclay composites, for example, the clay particles are partially exfoliated into layered sheets with intercalation of the polymer between these clay sheets, particularly for modified nanoclays, resulting in improved dispersion leading to better mechanical and barrier properties. The presence of these relatively hydrophilic nanoclays has been found to significantly increase the total water solubility in a PHA matrix in buffer [12], while slowing the overall rate of diffusion due to increased tortuosity. In addition, the local pockets of water (water clusters) can slow diffusion by acting as a trap/reservoir for permeating species such as water molecules and breakdown products, and there may also be restricted chain mobility inhibiting water diffusion. A similar effect was observed for nanocellulose particles [156].

Once again, the processing method can make a large difference to the biodegradation rates, with Thomas *et al.* [159] showing that PHA-filler composites (made from clay, peat, or birch wood flour) that were cold pressed at 36 Bar or screw-granulated at ambient temperature from a solvent-based paste degraded more rapidly than melt-mixed samples from the literature. Likewise, Zhu *et al.* [22] saw rapid biodegradation for solvent-cast nanocelluose/PHBV composites. For PHBV-organoclay nanocomposites, this barrier effect resulted in a lower rate of degradation for solvent-cast composite films in both water and compost compared to the control [104].

In terms of changes in mechanical properties over time for PHA-based composites, there are limited studies to date. Mazur *et al.* [158] showed that for injection-molded dumbbells (4 mm thick) of PHBV and basalt fiber composites subjected to aging in distilled water at 40°C, there was a slight decrease in mechanical properties over 2 weeks, more so for the composites than for the bulk matrix. Ventura *et al.* [160] saw a similar result for flax composites in ambient and high temperature (65°C), with stiffness decreasing over 100 h, again more so for the composites. Seggiani *et al.* saw a similar result in the marine biodegradation of seagrass/PHBV/$CaCO_3$/acetyl-tri-n-butyl citrate composites produced in the form of dog bones for tensile tests using melt mixing/extrusion. Over 13 months of incubation in marine sediments the bulk sample maintained its tensile properties, while the composites lost all mechanical properties after 10 months [161]. The samples with 20 wt-% seagrass showed well-developed biofilm formation (comprising bacteria and fungi) after only 3 months. Joyyi *et al.* [162] likewise found that for compression-molded composites of P(3HB-*co*-3HHx) with kenaf fibers, after 60 days in room temperature water, the flexural strength for control sample (no filler), composite with untreated kenaf fibers and composite with treated kenaf fibers fell to 42%, 19%, and 15% of the original, respectively, and the modulus fell less, to 69%, 42%, and 43% of original, respectively. Slightly greater changes were seen following 6 weeks of soil burial. Finally, in some of our recent work, the authors have explored the biodegradation in soil and under natural weathering of extruded PHBV (1 mol-% 3HV)/pine wood composites [163–165]. The tensile modulus and tensile strength of the bulk PHBV plaques increased slightly initially (perhaps due to secondary crystallization), then remained relatively constant over 12 months, while the tensile strain at break decreased by around 50%, which is unsurprising given its sensitivity to surface flaws and local cracks. By contrast, the 50% wood sample lost all mechanical integrity over 12 months and was significantly degraded after 6 months [165]. The 20 wt-% wood sample lost 17% of its tensile strength over 12 months [164].

However, the mass loss results did not reflect this loss in mechanical properties, with the 50 wt-% wood sample showing only a 12.7% weight loss over the 12-month

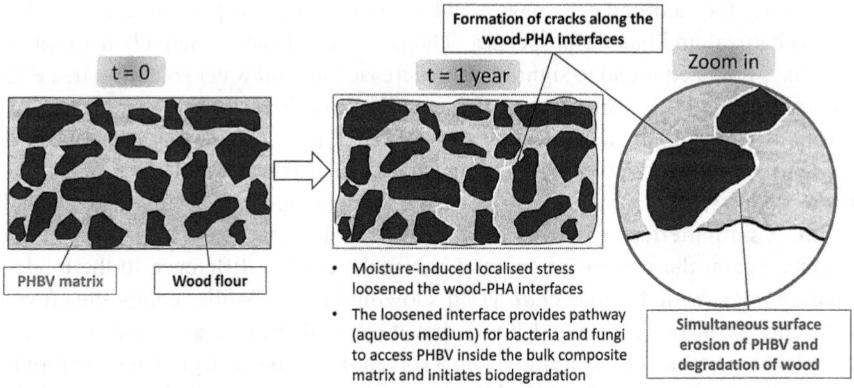

FIGURE 14.8 Schematic diagram showing the biodegradation mechanism of PHA wood plastic composites. (Reproduced with permission [165]. Copyright 2019, Elsevier).

degradation period, despite completely failing. In explaining this outcome, we have proposed a synthesis of the concepts discussed to date, as shown in Figure 14.8, whereby crack formation and propagation due to the range of processes discussed above, particularly filler swelling and water sorption, allows water ingress and causes localized stress that loosens the interface and provides a pathway for enzymes and, ultimately, microorganisms to penetrate into the bulk, loosening the fiber–matrix interface and accelerating the degradation in mechanical properties. This interpretation is in line with the model suggested by Azwa *et al.* [149], which also suggested the degradation/dissolution of water soluble components from the fillers as an additional mechanism for loosening the matrix–filler adhesion.

14.7 UV DEGRADATION

The sensitivity of polymers to photodegradation is related to their ability to absorb the harmful (high energy) part of the tropospheric solar radiation, including the ultraviolet (UV)-B terrestrial radiation (~295–315 nm) and UV-A radiation (~315–400 nm).

In predicting the lifetimes of PHA exposed to such UV radiation, the effect of exposure is frequently simulated using accelerated weathering devices (typically under abiotic environmental conditions). These devices combine UV light irradiation with controlled humidity, temperature, and "rainfall" exposure, to assess the combined effects of photodegradation, photooxidation, and abiotic hydrolysis. The ASTM standard G155–13, for example, specifies the protocols for using a xenon-arc lamp to simulate the natural sunlight spectrum. These tests assume that an Arrhenius relationship can be derived to estimate polymer lifetime under natural weathering based on the total irradiation exposure leading to failure at different temperatures in the device.

Such accelerated aging studies are limited for biopolymers, given that their biodegradation is primarily driven by catalyzed hydrolytic chain scission, as outlined above. However, there have been some reports of PHA and composite photodegradation under such accelerated weathering conditions [156, 157, 166–169]. In this way it has been shown that, unlike PLA and PCL, there are many competing reactions for PHA – including free radical initiation, Norrish Type I and Type II reactions, and crosslinking, as well as the underlying abiotic hydrolysis. Wei and McDonald, for example, found that solvent-cast PHBV (33 mol-% 3HV) films degraded following exposure to a QSun xenon-arc weatherometer primarily via β-chain scission reactions following free radical formation, leading to loss of molecular weight [170]. The exposed surfaces showed increased crystallinity and thus crazing due to crosslinking, while the 3HV units were preferentially cleaved. In addition, the ultimate strength and elongation at break decreased while the Young's modulus increased, again ascribed to increases in crystallinity. Sadi *et al.* [166] found similar results for injection-molded tensile and impact bars of PHB exposed to 12 weeks in a QUV chamber. Surface cracking and whitening due to surface crosslinking/crystallization were observed, along with an associated delay in biodegradation time following soil burial. Again, scission reactions from free radical initiation dominated in the bulk, with Norrish I and crosslinking possibly occurring in parallel, but little evidence of Norrish II reactions. Strain at break and tensile strength had decreased to a third of the original after 12 weeks, but the samples had not yet reached embrittlement. In a recent study, it was also found that the addition of an orange-red colored extract from annatto seeds could slightly increase the rate of photodegradation of PHA films under UVA exposure [171].

Overall, though, photodegradation of PHA through terrestrial UV irradiation under ambient, not accelerated, conditions is a slow process leading to limited impact on polymer properties in the absence of hydrolytic processes. This was observed in a recent 6-month aging study in sub-tropical Brisbane, where PHBV and wood-PHBV plaques were exposed to sunlight on a rack under ambient outdoor conditions for 6 months. The bulk PHBV extruded plaque sample showed some bleaching, but was otherwise unchanged, while a fungal attack on the back of the wood containing composites (including PLA- and PE-based controls) seemed to be the dominant mechanism controlling the (limited) degradation observed [163].

While not relevant to normal outdoor aging, short-wave irradiation (i.e., at wavelengths not found in terrestrial sunlight) can affect biodegradable polymers such as PHA differently than the longer wavelength irradiation, through Norrish Type I/II reactions and/or crosslinking reactions and oxidative processes [172], with the C=O bond in the main chain of polyesters having weak absorption bands at 280 nm (due to n-π^* excitation) and ~190 nm (due to n-σ^* excitation) [173]. One interesting aspect of such short-wave irradiation of PHA was recently revealed in a study by Heitmann *et al.* [174], wherein free $HO^\bullet$ radicals generated at the surface of PHA films following 254 nm UV light exposure from a mercury vapor lamp were able to oxidize methylene blue dye (as a proxy for pollutants). This may have implications for film biodegradation in practice, if exposure of polymers to such UV lamps affects the biofilm formation and/or durability.

14.8 MECHANICAL DEGRADATION

PHA products can be subjected to the same types of mechanical stresses as today's common commodity polymers during processing, storage, and use, including tensile, compressive, shearing, or bending stresses, in the form of bending, stretching, oscillations (vibrations), compression, constant strain/load, grinding, or hard extrusion [175]. Physical forces such as heating/cooling, freezing/thawing, or wetting/drying, as well as air and/or water turbulence, can also cause mechanical damage, such as the cracking of polymeric materials [176]. Environmental Stress Cracking (ESC) has already been discussed for composites (Section 14.6), but is an issue for neat polymers as well. A recent review by Li *et al.* [177] summarizes the effects of external stress on biodegradable orthopedic materials, effects which can be generalized to biodegradable polymers such as PHA. At the base level, many of these forces result in polymer chain cleavage with molecular weight loss. Such forces can act in combination with enzymatic hydrolysis to accelerate the rate of polymer biodegradation and polymer failure, particularly when associated with crack propagation (see Section 14.6).

The influence of static tensile loading on polymers can be described using the Zhurkov equation (Equation (14.4)), as follows:

$$K_f = K_0 e^{[-(E_a - \phi\sigma_x)/RT]} \tag{14.4}$$

where K_0 is the Arrhenius frequency factor, K_f is the rate of bond rupture events, E_a is the activation energy, σ_x is the tensile stress, and ϕ is a coefficient linked to the activation volume. This equation can be added into other models of biodegradation so as to include static stress effects in lifetime predictions, thus factoring in a lowered activation energy for chain scission due to applied stress.

Soares *et al.* [178, 179] used this approach to develop a model to account for the accelerated breakdown of PLA stents under uniaxial extension, that explicitly took into account both surface and bulk erosion. On that basis, failure from biodegradation was found to be most likely to occur at stress points such as stent rings and junction points, and characteristics such as stress relaxation and creep were also taken into account. Hayman *et al.* [180] also studied the effect of static and dynamic load on the degradation of PLLA stent fibers *in vitro* over 15 months. There was more significant increase in the rate of loss of mechanical properties under dynamic load, although both had an effect. As discussed in other sections, the methods of processing can also have a major impact on the rate of failure under mechanical stresses. Melt-processing (injection-molding, extrusion, compression-molding), for example, is a higher temperature process, performed in the melt at high shear, at temperatures close to the degradation temperature for PHB in particular. As such, loss of molecular weight through thermal degradation and mechanoscission is commonly experienced, particularly if pretreatment following biosynthesis, such as with acids, is not conducted. There can also be some polymer chain orientation, particularly for injection-molding, and there is typically more orientation at the skin than in the bulk, leading to slower degradation in the skin [181].

For a detailed analysis of the fracture mechanics of polymer composites, the reader is referred to the review article on cracks and fractures in polymer structures by Awaja *et al.* [150].

14.9 STRATEGIES FOR MODIFICATION OF DEGRADATION RATE

Many approaches have been taken to PHA modification in order to control the rate of biodegradation. If the aim is to accelerate the rate, then, since PHA are relatively hydrophobic and resistant to water permeation, the usual strategy is to introduce more hydrophilic groups, either by polymer modification, blending, or composite production using more hydrophilic fillers. The introduction of hydrophobic groups will have the opposite effect, slowing degradation (as already touched on above).

The introduction of polar groups can be done through the introduction of reactive groups such as pendant double bonds during biopolymer synthesis (through the addition of appropriate hydroxyacid monomers, such as 3-hydroxyundecenoic acid) [133, 182–185]. This functional group can then be oxidized to form hydroxyl or carboxylic acid groups on the side chain – delivering solubility in polar solvents such as ethanol. The same side groups can be used to prepare cationic PHA, through epoxidation of the double bond followed by amination of the epoxy group with diethanolamine, delivering a tertiary amine group. In this case the polymer can become water-soluble at pH values below the pK_a, i.e., at <10. In all of these cases, the increase in hydrophilicity was linked to an increase in the rate of enzymatic biodegradation.

Grafting of side chains can be accomplished by using the "grafting onto", "grafting from", "transfer to", or "grafting through" strategies (Figure 14.9), with the first involving reaction of a polymer block with a reactive group onto the polymer side chains with reactive groups. The second approach involves synthesis of side chain polymer blocks from the main chain from active sites on that main chain, the third involves a chain transfer reaction that is used to attach actively growing polymer chains onto the main chain, and the last involves polymerization of a reactive end group of a macromonomer [186]. Such strategies have delivered, for example, hydrophilic poly(ethylene glycol) (PEG) side chains grafted onto PHA main chains, where the solubility of the graft copolymer in acetone/water mixtures increased with the length of the PEG chain [185]. The biopolymer biodegradation rate was correlated with the length of the PEG chain.

Modification of the surface chemistry is another strategy that has been attempted a number of times. For example, the addition of the anti-fouling agent 4,5-dichlo ro-2-n-octyl-4-isothiazolin-3-one (at less than 10 wt-%) restricted microbial colonization onto PHA films and delayed the onset of weight loss by 100 days, as well as reducing the rate of weight loss by up to 150% [187]. A recent review by Vigneswari *et al.* [188] provides a good overview of other approaches to surface modification of PHA, which are also summarized in Table 14.4. Although these strategies were originally intended for the management of interfaces, such as in composites, and in biomedical applications, they will also likely have a significant effect on control of biodegradation.

There is a range of other more complex approaches to modification of PHA as well, such as through synthesis of amphiphilic block copolymers based on PHA and PEG [185]. Such strategies have effectively delivered unique functionality and increased the hydrophilicity for PHA-based polymers. However, covering these approaches in detail is beyond the scope of this review.

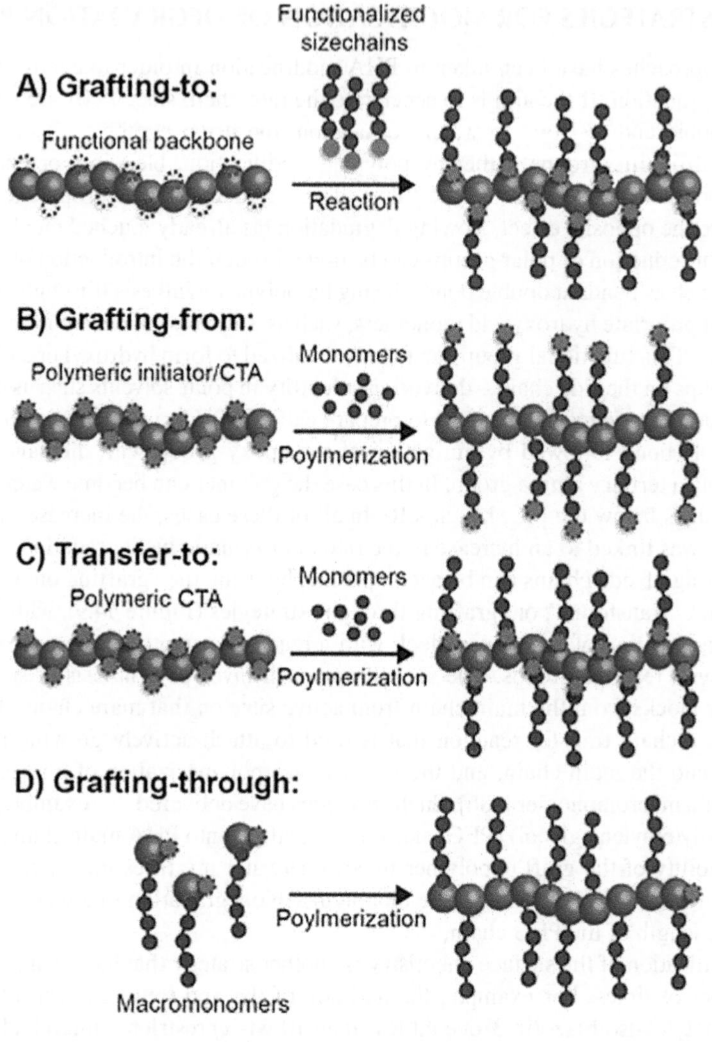

FIGURE 14.9 Common methods of synthesis for production of side chains on polymers: A) grafting to; B) grafting from; C) transfer to; and D) grafting through. (Reproduced with permission [186]. Copyright 2019, American Chemical Society).

14.10 ACCELERATED AGING FOR LIFETIME ESTIMATION

One of the major issues faced in using all polymers, both nondegradable and biodegradable, in practice is developing robust strategies for estimating polymer lifetimes in the environment using accelerated aging tests – both in terms of mechanical properties and ultimate fate/conversion to CO_2 and water. Most often, this is done by conducting the degradation processes under a range of temperatures and assuming

TABLE 14.4
Surface Modification Techniques for Functionalizing Biomaterials, Modified from [188]

Types of modifications	Methodology
Mechanical	• Creation of physical micro-rough surfaces, grooves, or pores by enzymatic degradation, hydrolysis, particulate leaching, or freeze-drying methods
Physicochemical	• Treatment with active gases, vapors, or ion treatments to form hydrophilic groups on the surface • Plasma treatment and grafting to form functional groups on the surface • Photo-induced grafting to produce charged surfaces
Biological	• Immobilizing biomolecules (peptides, heparin, collagen, chitosan, and gelatin) on the surface • Coating and entrapment of biomolecules on the surface

an Arrhenius relationship for extrapolation back to in-use temperature. For this acceleration to be appropriate, the polymer should undergo the same microstructural changes as under ambient conditions and the degradation should involve a single chemical process activated by temperature.

However, for biodegradable polymers, there is an added level of complexity in this assumption, in that microorganisms typically have very constrained operating windows with respect to viability and efficiency, particularly with respect to temperature, pH, dissolved oxygen content, and UV exposure. As such, only limited acceleration studies have been performed, typically in distilled water at neutral pH-value, and with temperature ranges that are constrained to 50°C and below. In addition, there is often a polymer-related reason for non-Arrhenius behavior, such as for PLA where the glass transition temperature is at around 56°C, and hence many properties, particularly water diffusivity, are significantly affected as this temperature is approached. Deroiné *et al.*, for example, evaluated the activation energy for PHBV hydrolysis in distilled water using just such a time-temperature method, as discussed already in Section 14.2.3.

14.11 CONCLUSIONS AND OUTLOOK

This chapter has summarized the key drivers underlying biopolymer degradation in the natural environment – with environmental factors such as microbial community composition and density, local nutrient availability, temperature, and pH being of critical importance, as well as polymer-related properties such as surface chemistry and roughness, crystallinity and exposed surface area (through sample dimensions/morphology, as well as polymer porosity and cracks), all coupled to mechanical forces. While there is a wealth of evidence supporting the ultimate biodegradability of PHA in these environments, there remain many gaps in our understanding of the mechanisms and processes driving PHA degradation kinetics. In summary, we need to better understand:

- The effects of sample dimensions on degradation across environments – given the complex interactions between surface and bulk erosion and diffusion effects for water and degradation products;
- The interplay between mechanical (stress/deformation) effects, crack propagation, and degradation, as these are likely to be significant controllers of lifetimes;
- The effects of additives such as plasticizers, as well as sorbed compounds at the surface on biofilm formation, and ultimately, the biodegradation rate;
- The effects of biopolymers on the local microbial and larger communities in the soil and aqueous environments and on the nutrient biogeochemistry in soils; and
- The mechanical as well as the ultimate biological lifetimes of products disposed to different environments with the context of application or lifecycles, particularly when processed in a form relevant for actual use (such as extrusion as opposed to solvent-casting).

In addition, it is critical that researchers conducting biodegradation studies provide sufficient details of sample preparation, composition, and dimensions for an adequate interpretation of results to be made.

ACKNOWLEDGMENTS

This work was supported by the Australian Research Council through Linkage Grants LP140100596 and 160101763, as well as Discovery Grant DP15010306.

REFERENCES

1. Laycock B, Nikolić M, Colwell JM, *et al.* Lifetime prediction of biodegradable polymers. *Prog Polym Sci* 2017; 71: 144–189.
2. Jendrossek D and Handrick R. Microbial degradation of polyhydroxyalkanoates. *Annu Rev Microbiol* 2002; 56: 403–432.
3. Shah AA, Hasan F, Hameed A, and Ahmed S. Biological degradation of plastics: A comprehensive review. *Biotechnol Adv* 2008; 26(3): 246–265.
4. Gu J-D. Microbiological deterioration and degradation of synthetic polymeric materials: Recent research advances. *Int Biodeterior Biodegrad* 2003; 52(2): 69–91.
5. Fayolle B, Colin X, Audouin L, and Verdu J. Mechanism of degradation induced embrittlement in polyethylene. *Polym Degrad Stabil* 2007; 92(2): 231–238.
6. Doi Y, Kanesawa Y, Kawaguchi Y, and Kunioka M. Hydrolytic degradation of microbial poly(hydroxyalkanoates). *Makromol Chem-Rapid* 1989; 10(5): 227–230.
7. Briassoulis D, Mistriotis A, Mortier N, and Tosin M. A horizontal test method for biodegradation in soil of bio-based and conventional plastics and lubricants. *J Clean Prod* 2020; 242: 113892.
8. Corre YM, Bruzaud S, Audic JL, and Grohens Y. Morphology and functional properties of commercial polyhydroxyalkanoates: A comprehensive and comparative study. *Polym Test* 2012; 31(2): 226–235.

9. Thellen C, Coyne M, Froio D, Auerbach M, Wirsen C, and Ratto JA. A processing, characterization and marine biodegradation study of melt-extruded polyhydroxyalkanoate (PHA) films. *J Polym Environ* 2008; 16(1): 1–11.
10. Miguel O and Iruin JJ. Water transport properties in poly(3-hydroxybutyrate) and poly(3-hydroxybutyrate-co-3-hydroxyvalerate) biopolymers. *J Appl Polym Sci* 1999; 73(4): 455–468.
11. Deroine M, Le Duigou A, Corre YM, *et al.* Seawater accelerated ageing of poly(3-hyd roxybutyrate-co-3-hydroxyvalerate). *Polym Degrad Stabil* 2014; 105: 237–247.
12. Follain N, Cretois R, Lebrun L, and Marais S. Water sorption behaviour of two series of PHA/montmorillonite films and determination of the mean water cluster size. *Phys Chem Chem Phys* 2016; 18(30): 20345–20356.
13. Tang CY, Chen DZ, Yue TM, Chan KC, Tsui CP, and Yu PHF. Water absorption and solubility of PHBHV/HA nanocomposites. *Compos Sci Technol* 2008; 68(7–8): 1927–1934.
14. Balac I, Tang CY, Tsui CP, *et al.* Nanoindentation of in situ polymers in hydroxyapatite/poly-L-lactide biocomposites. *Recent Dev Adv Mater Processes* 2006; 518: 501–506.
15. Follain N, Chappey C, Dargent E, Chivrac F, Cretois R, and Marais S. Structure and barrier properties of biodegradable polyhydroxyalkanoate films. *J Phys Chem C* 2014; 118(12): 6165–6177.
16. Deroine M, Le Duigou A, Corre YM, *et al.* Accelerated ageing and lifetime prediction of poly(3-hydroxybutyrate-co-3-hydroxyvalerate) in distilled water. *Polym Test* 2014; 39: 70–78.
17. Cretois R, Follain N, Dargent E, *et al.* Microstructure and barrier properties of PHBV/ organoclays bionanocomposites. *J Membr Sci* 2014; 467: 56–66.
18. Park GS. Transport principles-solution, diffusion and permeation in polymer membranes. In: Bungay, PM, Ed., *Synthetic Membranes: Science, Engineering and Applications*, Holland: Reidel, 1986. pp. 57–107.
19. Bonartsev AP, Boskhomodgiev AP, Iordanskii AL, *et al.* Hydrolytic degradation of poly(3-hydroxybutyrate), polylactide and their derivatives: Kinetics, crystallinity, and surface morphology. *Mol Cryst Liq Cryst* 2012; 556(1): 288–300.
20. Freier T, Kunze C, Nischan C, *et al. In vitro* and *in vivo* degradation studies for development of a biodegradable patch based on poly(3-hydroxybutyrate). *Biomaterials* 2002; 23(13): 2649–2657.
21. Marois Y, Zhang Z, Vert M, Deng XY, Lenz R, and Guidoin R. Mechanism and rate of degradation of polyhydroxyoctanoate films in aqueous media: A long-term *in vitro* study. *J Biomed Mater Res* 2000; 49(2): 216–224.
22. Zhu JY, Chen YX, Yu HY, *et al.* Comprehensive insight into degradation mechanism of green biopolyester nanocomposites using functionalized cellulose nanocrystals. *ACS Sustain Chem Eng* 2019; 7(18): 15537–15547.
23. Silva RdN, da Silva LRC, de Morais ACL, Alves TS, and Barbosa R. Study of the hydrolytic degradation of poly-3-hydroxybutyrate in the development of blends and polymeric nanocomposites. *J Thermoplast Compos Mater* 2019: 1–18. doi:10.1177/ 0892705719856044
24. Grassie N, Murray EJ, and Holmes PA. The thermal-degradation of poly(–(D)-beta-hydroxybutyric acid). 1. Identification and quantitative-analysis of products. *Polym Degrad Stabil* 1984; 6(1): 47–61.
25. Lauzier C, Revol JF, Debzi EM, and Marchessault RH. Hydrolytic degradation of isolated poly(beta-hydroxybutyrate) granules. *Polymer* 1994; 35(19): 4156–4162.
26. Iwata T, Doi Y, Kokubu F, and Teramachi S. Alkaline hydrolysis of solution-grown poly[(R)-3-hydroxybutyrate] single crystals. *Macromolecules* 1999; 32(25): 8325–8330.

27. Eldsater C, Karlsson S, and Albertsson AC. Effect of abiotic factors on the degradation of poly(3-hydroxybutyrate-co-3-hydroxyvalerate) in simulated and natural composting environments. *Polym Degrad Stabil* 1999; 64(2): 177–183.
28. Yu J, Plackett D, and Chen LXL. Kinetics and mechanism of the monomeric products from abiotic hydrolysis of poly[(R)-3-hydroxybutyrate] under acidic and alkaline conditions. *Polym Degrad Stabil* 2005; 89(2): 289–299.
29. Chen LXL and Yu J. Abiotic hydrolysis of poly [(R)-3-hydroxybutyrate] in acidic and alkaline media. *Macromol Symp* 2005; 224(1): 35–46.
30. Foster LJR and Tighe BJ. Centrifugally spun polyhydroxybutyrate fibers: Accelerated hydrolytic degradation studies. *Polym Degrad Stab* 2004; 87(1): 1–10.
31. Dilkes-Hoffman L, Lant PA, Laycock B, and Pratt S. The rate of biodegradation of PHA bioplastics in the marine environment: A meta-study. *Mar Pollut Bull* 2019; 142: 15–24.
32. Salomez M, George M, Fabre P, *et al.* A comparative study of degradation mechanisms of PHBV and PBSA under laboratory-scale composting conditions. *Polym Degrad Stabil* 2019; 167: 102–113.
33. Oh JK, Yegin Y, Yang F, *et al.* The influence of surface chemistry on the kinetics and thermodynamics of bacterial adhesion. *Sci Rep* 2018; 8(1): 17247.
34. Dussud C, Hudec C, George M, *et al.* Colonization of non-biodegradable and biodegradable plastics by marine microorganisms. *Front Microbiol* 2018; 9: 1571.
35. De-la-Pinta I, Cobos M, Ibarretxe J, *et al.* Effect of biomaterials hydrophobicity and roughness on biofilm development. *J Mater Sci Mater M* 2019; 30: 77.
36. Mauclaire L, Brombacher E, Bunger JD, and Zinn M. Factors controlling bacterial attachment and biofilm formation on medium-chain-length polyhydroxyalkanoates (*mcl*-PHAs). *Colloids Surf B* 2010; 76(1): 104–111.
37. Cheng YF, Feng GP, and Moraru CI. Micro- and nanotopography sensitive bacterial attachment mechanisms: A review. *Front Microbiol* 2019; 10: 191.
38. Narihiro T and Sekiguchi Y. Microbial communities in anaerobic digestion processes for waste and wastewater treatment: A microbiological update. *Curr Opin Biotechnol* 2007; 18(3): 273–278.
39. Deroine M, Cesar G, Le Duigou A, Davies P, and Bruzaud S. Natural degradation and biodegradation of poly(3-hydroxybutyrate-co-3-hydroxyvalerate) in liquid and solid marine environments. *J Polym Environ* 2015; 23(4): 493–505.
40. Pickett JE, Hall ML, de Heer J, Kuvshinnikova O, and Boven G. Microbial growth on outdoor-weathered plastics. *Polym Degrad Stabil* 2019; 163: 206–213.
41. Wachtendorf V, Schulz U, Geburtig A, and Stephan I. Mildew growth on automotive coatings influencing the results of outdoor weathering. *Mater Corros* 2012; 63(2): 140–147.
42. Sang BI, Hori K, Tanji Y, and Unno H. A kinetic analysis of the fungal degradation process of poly(3-hydroxybutyrate-co-3-hydroxyvalerate) in soil. *Biochem Eng J* 2001; 9(3): 175–184.
43. Sang BI, Hori K, and Unno H. A mathematical description for the fungal degradation process of biodegradable plastics. *Math Comput Simulat* 2004; 65(1–2): 147–155.
44. Sang BI, Hori K, Tanji Y, and Unno H. Fungal contribution to in situ biodegradation of poly(3-hydroxybutyrate-co-3-hydroxyvalerate) film in soil. *Appl Microbiol Biotechnol* 2002; 58(2): 241–247.
45. Koitabashi M, Noguchi MT, Sameshima-Yamashita Y, *et al.* Degradation of biodegradable plastic mulch films in soil environment by phylloplane fungi isolated from gramineous plants. *AMB Express* 2012; 2(1): 40.
46. Muroi F, Tachibana Y, Kobayashi Y, Sakurai T, and Kasuya K. Influences of poly(butylene adipate-co-terephthalate) on soil microbiota and plant growth. *Polym Degrad Stabil* 2016; 129: 338–346.

47. Kasuya K, Ishii N, Inoue Y, *et al.* Characterization of a mesophilic aliphatic-aromatic copolyester-degrading fungus. *Polym Degrad Stabil* 2009; 94(8): 1190–1196.
48. Kim DY and Rhee YH. Biodegradation of microbial and synthetic polyesters by fungi. *Appl Microbiol Biotechnol* 2003; 61(4): 300–308.
49. Jeszeova L, Puskarova A, Buckova M, *et al.* Microbial communities responsible for the degradation of poly(lactic acid)/poly(3-hydroxybutyrate) blend mulches in soil burial respirometric tests. *World J Microb Biot* 2018; 34(7): 101.
50. Montelongo-Jauregui D, Saville SP, and Lopez-Ribot JL. Contributions of *Candida albicans* dimorphism, adhesive interactions, and extracellular matrix to the formation of dual-species biofilms with Streptococcus gordonii. *mBio* 2019; 10(3): e01179–19.
51. Phukon P, Saikia JP, and Konwar BK. Bio-plastic (P-3HB-co-3HV) from *Bacillus circulans* (MTCC 8167) and its biodegradation. *Colloids Surf B* 2012; 92: 30–34.
52. Jendrossek D. Peculiarities of PHA granules preparation and PHA depolymerase activity determination. *Appl Microbiol Biotechnol* 2007; 74(6): 1186–1196.
53. Kasuya K, Inoue Y, and Doi Y. Adsorption kinetics of bacterial PHB depolymerase on the surface of polyhydroxyalkanoate films. *Int J Biol Macromol* 1996; 19(1): 35–40.
54. Erickson HP. Size and shape of protein molecules at the nanometer level determined by sedimentation, gel filtration, and electron microscopy. *Biol Proced Online* 2009; 11(1): 32–51.
55. Knoll M, Hamm TM, Wagner F, Martinez V, and Pleiss J. The PHA Depolymerase Engineering Database: A systematic analysis tool for the diverse family of polyhydroxyalkanoate (PHA) depolymerases. *BMC Bioinformatics* 2009; 10: 89.
56. Numata K, Abe H, and Doi Y. Enzymatic processes for biodegradation of poly(hydroxyalkanoate)s crystals. *Can J Chem* 2008; 86(6): 471–483.
57. Numata K, Abe H, and Iwata T. Biodegradability of poly(hydroxyalkanoate) materials. *Materials* 2009; 2(3): 1104–1126.
58. Kumagai Y, Kanesawa Y, and Doi Y. Enzymatic degradation of microbial poly(3-hydroxybutyrate) films. *Makromol Chem* 1992; 193(1): 53–57.
59. Abe H, Doi Y, Aoki H, and Akehata T. Solid-state structures and enzymatic degradabilities for melt-crystallized films of copolymers of (*R*)-3-hydroxybutyric acid with different hydroxyalkanoic acids. *Macromolecules* 1998; 31(6): 1791–1797.
60. Iwata T, Doi Y, Nakayama S, Sasatsuki H, and Teramachi S. Structure and enzymatic degradation of poly(3-hydroxybutyrate) copolymer single crystals with an extracellular PHB depolymerase from Alcaligenes faecalis T1. *Int J Biol Macromol* 1999; 25(1–3): 169–176.
61. Scandola M, Focarete ML, and Frisoni G. Simple kinetic model for the heterogeneous enzymatic hydrolysis of natural poly(3-hydroxybutyrate). *Macromolecules* 1998; 31(12): 3846–3851.
62. Tanaka T and Iwata T. Physical properties, structure analysis, and enzymatic degradation of poly[(*R*)-3-hydroxybutyrate-*co*-(*R*)-3-hydroxyvalerate] films and fibers. In: Khemani, K and Scholz, C, Eds., *Degradable Polymers and Materials: Principles and Practice* (2nd edition), Washington, DC: ACS Symposium Series, American Chemical Society, 2012. pp. 171–185.
63. Li ZG, Lin H, Ishii N, Chen GQ, and Inoue Y. Study of enzymatic degradation of microbial copolyesters consisting of 3-hydroxybutyrate and medium-chain-length 3-hydroxyalkanoates. *Polym Degrad Stabil* 2007; 92(9): 1708–1714.
64. Wang YW, Mo WK, Yao HL, Wu Q, Chen JC, and Chen GQ. Biodegradation studies of poly(3-hydroxybutyrate-co-3-hydroxyhexanoate). *Polym Degrad Stabil* 2004; 85(2): 815–821.
65. Cao A, Arai Y, Yoshie N, Kasuya KI, Doi Y, and Inoue Y. Solid structure and biodegradation of the compositionally fractionated poly(3-hydroxybutyric acid-co-3-hydroxypropionic acid)s. *Polymer* 1999; 40(24): 6821–6830.

66. Iwata T, Aoyagi Y, Tanaka T, *et al.* Microbeam X-ray diffraction and enzymatic degradation of poly[(*R*)-3-hydroxybutyrate] fibers with two kinds of molecular conformations. *Macromolecules* 2006; 39(17): 5789–5795.
67. Tanaka T, Yabe T, Teramachi S, and Iwata T. Mechanical properties and enzymatic degradation of poly[(*R*)-3-hydroxybutyrate] fibers stretched after isothermal crystallization near T_G. *Polym Degrad Stabil* 2007; 92(6): 1016–1024.
68. Polyak P, Dohovits E, Nagy GN, Vertessy BG, Voros G, and Pukanszky B. Enzymatic degradation of poly-[(R)-3-hydroxybutyrate]: Mechanism, kinetics, consequences. *Int J Biol Macromol* 2018; 112: 156–162.
69. Takaku H, Kimoto A, Kodaira S, Nashimoto M, and Takagi M. Isolation of a Gram-positive poly(3-hydroxybutyrate) (PHB)-degrading bacterium from compost, and cloning and characterization of a gene encoding PHB depolymerase of Bacillus megaterium N-18-25-9. *FEMS Microbiol Lett* 2006; 264(2): 152–159.
70. Molitoris HP, Moss ST, deKoning GJM, and Jendrossek D. Scanning electron microscopy of polyhydroxyalkanoate degradation by bacteria. *Appl Microbiol Biot* 1996; 46(5–6): 570–579.
71. Polyak P, Urban E, Nagy GN, Vertessy BG, and Pukanszky B. The role of enzyme adsorption in the enzymatic degradation of an aliphatic polyester. *Enzyme Microb Technol* 2019; 120: 110–116.
72. Kasuya K, Takagi K, Ishiwatari S, Yoshida Y, and Doi Y. Biodegradabilities of various aliphatic polyesters in natural waters. *Polym Degrad Stabil* 1998; 59(1–3): 327–332.
73. Doi Y, Kasuya K, Abe H, *et al.* Evaluation of biodegradabilities of biosynthetic and chemosynthetic polyesters in river water. *Polym Degrad Stabil* 1996; 51(3): 281–286.
74. Mergaert J, Wouters A, Anderson C, and Swings J. In-situ biodegradation of poly(3-hydroxybutyrate) and poly(3-hydroxybutyrate-co-3-hydroxyvalerate) in natural-waters. *Can J Microbiol* 1995; 41: 154–159.
75. Puechner P, Mueller WR, and Bardtke D. Assessing the biodegradation potential of polymers in screening and long-term test systems. *J Environ Polym Degr* 1995; 3(3): 133–143.
76. Brandl H and Puchner P. Biodegradation of plastic bottles made from 'biopol' in an aquatic ecosystem under *in situ* conditions. *Biodegradation* 1992; 2(4): 237–243.
77. Doi Y and Abe H. Structural effects on biodegradation of aliphatic polyesters. *Macromol Symp* 1997; 118(1): 725–731.
78. Voinova O, Gladyshev M, and Volova TG. Comparative study of PHA degradation in natural reservoirs having various types of ecosystems. *Macromol Symp* 2008; 269(1): 34–37.
79. Geyer R, Jambeck JR, and Law KL. Production, use, and fate of all plastics ever made. *Sci Adv* 2017; 3(7): e1700782.
80. Jambeck JR, Geyer R, Wilcox C, *et al.* Plastic waste inputs from land into the ocean. *Science* 2015; 347(6223): 768–771.
81. Briassoulis D and Mistriotis A. Key parameters in testing biodegradation of bio-based materials in soil. *Chemosphere* 2018; 207: 18–26.
82. Kyrikou I and Briassoulis D. Biodegradation of agricultural plastic films: A critical review. *J Polym Environ* 2007; 15(2): 125–150.
83. Hoshino A, Sawada H, Yokota M, Tsuji M, Fukuda K, and Kimura M. Influence of weather conditions and soil properties on degradation of biodegradable plastics in soil. *Soil Sci Plant Nutr* 2001; 47(1): 35–43.
84. Baba T, Tachibana Y, Suda S, and Kasuya K. Evaluation of environmental degradability based on the number of methylene units in poly(butylene n-alkylenedionate). *Polym Degrad Stabil* 2017; 138: 18–26.
85. Song CJ, Wang SF, Ono S, Zhang BH, Shimasaki C, and Inoue M. The biodegradation of poly(3-hydroxybutyrate-*co*-3-hydroxyvalerate) (PHB/V) and PHB/V-degrading microorganisms in soil. *Polym Advan Technol* 2003; 14(3–5): 184–188.

86. Tosin M, DegliInnocenti F, and Bastioli C. Effect of the composting substrate on biodegradation of solid materials under controlled composting conditions. *J Environ Polym Degr* 1996; 4(1): 55–63.
87. Song CJ, Uchida U, Ono S, Shimasaki CH, and Inoue M. Estimation of the number of polyhydroxyalkanoate (PHA)-degraders in soil and isolation of degraders eased on the method of most probable number (MPN) using PHA-film. *Biosci Biotech Bioch* 2001; 65(5): 1214–1217.
88. Kubowicz S and Booth AM. Biodegradability of plastics: Challenges and misconceptions. *Environ Sci Technol* 2017; 51(21): 12058–12060.
89. Arcos-Hernandez MV, Laycock B, Pratt S, *et al.* Biodegradation in a soil environment of activated sludge derived polyhydroxyalkanoate (PHBV). *Polym Degrad Stabil* 2012; 97(11): 2301–2312.
90. Baidurah S, Kubo Y, Kuno M, *et al.* Rapid and direct compositional analysis of poly(3-hydroxybutyrate-co-3-hydroxyvalerate) in whole bacterial cells by thermally assisted hydrolysis and methylation-gas chromatography. *Anal Sci* 2015; 31(2): 79–83.
91. Wen X and Lu XP. Microbial degradation of poly(3-hydroxybutyrate-co-4-hydroxy butyrate) in soil. *J Polym Environ* 2012; 20(2): 381–387.
92. Li G, Shankar S, Rhim JW, and Oh BY. Effects of preparation method on properties of poly(butylene adipate-co-terephthalate) films. *Food Sci Biotechnol* 2015; 24(5): 1679–1685.
93. Sakai Y, Isokawa M, Masuda T, Yoshioka H, Hayatsu M, and Hayano K. Usefulness of soil p-nitrophenyl acetate esterase activity as a tool to monitor biodegradation of polybutylene succinate (PBS) in cultivated soil. *Polym J* 2002; 34(10): 767–774.
94. Yamamoto-Tamura K, Hiradate S, Watanabe T, *et al.* Contribution of soil esterase to biodegradation of aliphatic polyester agricultural mulch film in cultivated soils. *AMB Express* 2015; 5: 10.
95. Modelli A, Calcagno B, and Scandola M. Kinetics of aerobic polymer degradation in soil by means of the ASTM D 5988-96 standard method. *J Environ Polym Degr* 1999; 7(2): 109–116.
96. Sander M. Biodegradation of polymeric mulch films in agricultural soils: Concepts, knowledge gaps, and future research directions. *Environ Sci Technol* 2019; 53(5): 2304–2315.
97. Siotto M, Zoia L, Tosin M, Degli Innocenti F, Orlandi M, and Mezzanotte V. Monitoring biodegradation of poly(butylene sebacate) by gel permeation chromatography, H-1-NMR and P-31-NMR techniques. *J Environ Manage* 2013; 116: 27–35.
98. Rychter P, Biczak R, Herman B, *et al.* Environmental degradation of polyester blends containing atactic poly(3-hydroxybutyrate). Biodegradation in soil and ecotoxicological impact. *Biomacromolecules* 2006; 7(11): 3125–3131.
99. Mergaert J, Webb A, Anderson C, Wouters A, and Swings J. Microbial-degradation of poly(3-hydroxybutyrate) and poly(3-hydroxybutyrate-Co-3-Hydroxyvalerate) in soils. *Appl Environ Microbiol* 1993; 59(10): 3233–3238.
100. Ong SY and Sudesh K. Effects of polyhydroxyalkanoate degradation on soil microbial community. *Polym Degrad Stabil* 2016; 131: 9–19.
101. Bandopadhyay S, Martin-Closas L, Pelacho AM, and DeBruyn JM. Biodegradable plastic mulch films: Impacts on soil microbial communities and ecosystem functions. *Front Microbiol* 2018; 9: 819.
102. Castro-Aguirre E, Auras R, Selke S, Rubino M, and Marsh T. Insights on the aerobic biodegradation of polymers by analysis of evolved carbon dioxide in simulated composting conditions. *Polym Degrad Stabil* 2017; 137: 251–271.
103. Grima S, Bellon-Maurel V, Feuilloley P, and Silvestre F. Aerobic biodegradation of polymers in solid-state conditions: A review of environmental and physicochemical parameter settings in laboratory simulations. *J Polym Environ* 2000; 8(4): 183–195.

104. Iggui K, Le Moigne N, Kaci M, Cambe S, Degorce-Dumas JR, and Bergeret A. A biodegradation study of poly(3-hydroxybutyrate-co-3-hydroxyvalerate)/organoclay nanocomposites in various environmental conditions. *Polym Degrad Stabil* 2015; 119: 77–86.
105. Gutierrez-Wing MT, Stevens BE, Theegala CS, Negulescu II, and Rusch KA. Aerobic biodegradation of polyhydroxybutyrate in compost. *Environ Eng Sci* 2011; 28(7): 477–488.
106. Weng YX, Wang XL, and Wang YZ. Biodegradation behavior of PHAs with different chemical structures under controlled composting conditions. *Polym Test* 2011; 30(4): 372–380.
107. Rutkowska M, Krasowska K, Heimowska A, *et al.* Environmental degradation of blends of atactic poly[(R,S)-3-hydroxybutyrate] with natural PHBV in Baltic Sea water and compost with activated sludge. *J Polym Environ* 2008; 16(3): 183–191.
108. Rosa DD, Calil MR, Guedes CDF, and Rodrigues TC. Biodegradability of thermally aged PHB, PHB-V, and PCL in soil compostage. *J Polym Environ* 2004; 12(4): 239–245.
109. Luo S and Netravali AN. A study of physical and mechanical properties of poly(hydroxybutyrate-co-hydroxyvalerate) during composting. *Polym Degrad Stabil* 2003; 80(1): 59–66.
110. Henze M, Dircks K, van Loosdrecht MCM, Mosbaek H, and Aspegren H. Storage and degradation of poly-beta-hydroxybutyrate in activated sludge under aerobic conditions. *Water Res* 2001; 35(9): 2277–2285.
111. Imam SH, Chen L, Gordon SH, Shogren RL, Weisleder D, and Greene RV. Biodegradation of injection molded starch-poly (3-hydroxybutyrate-*co*-3-hydroxyvalerate) blends in a natural compost environment. *J Environ Polym Degr* 1998; 6(2): 91–98.
112. Mergaert J, Anderson C, Wouters A, and Swings J. Microbial degradation of poly(3-hydroxybutyrate) and poly(3-hydroxybutyrate-co-3-hydroxyvalerate) in compost. *J Environ Polym Degr* 1994; 2(3): 177–183.
113. Ohtaki A and Nakasaki K. Report. Ultimate degradability of various kinds of biodegradable plastics under controlled composting conditions. *Waste Manag Res* 2000; 18(2): 184–189.
114. Yue CL, Gross RA, and McCarthy SP. Composting studies of poly (beta-hydroxybu tyrate-co-beta-hydroxyvalerate). *Polym Degrad Stabil* 1996; 51(2): 205–210.
115. Weng YX, Wang Y, Wang XL, and Wang YZ. Biodegradation behavior of PHBV films in a pilot-scale composting condition. *Polym Test* 2010; 29(5): 579–587.
116. Gutierrez-Wing MT, Stevens BE, Theegala CS, Negulescu II, and Rusch KA. Anaerobic biodegradation of polyhydroxybutyrate in municipal sewage sludge. *J Environ Eng ASCE* 2010; 136(7): 709–718.
117. Arrieta MP, Fortunati E, Dominici F, Rayon E, Lopez J, and Kenny JM. PLA-PHB/cellulose based films: Mechanical, barrier and disintegration properties. *Polym Degrad Stabil* 2014; 107: 139–149.
118. Wang SL, Lydon KA, White EM, *et al.* Biodegradation of poly(3-hydroxybutyrate-*co*-3-hydroxyhexanoate) plastic under anaerobic sludge and aerobic seawater conditions: Gas evolution and microbial diversity. *Environ Sci Technol* 2018; 52(10): 5700–5709.
119. Shin PK, Kim MH, and Kim JM. Biodegradability of degradable plastics exposed to anaerobic digested sludge and simulated landfill conditions. *J Environ Polym Degr* 1997; 5(1): 33–39.
120. Noda I, Lindsey SB, and Nodax Caraway D. (TM) class PHA copolymers: Their properties and applications. *Microbiol Monogr* 2010; 14: 237–255.
121. Budwill K, Fedorak PM, and Page WJ. Methanogenic degradation of poly(3-hydroxyalkanoates). *Appl Environ Microbiol* 1992; 58(4): 1398–1401.
122. Morse MC, Liao Q, Criddle CS, and Frank CW. Anaerobic biodegradation of the microbial copolymer poly(3-hydroxybutyrate-*co*-3-hydroxyhexanoate): Effects of

comonomer content, processing history, and semi-crystalline morphology. *Polymer* 2011; 52(2): 547–556.
123. Abou-Zeid DM, Muller RJ, and Deckwer WD. Degradation of natural and synthetic polyesters under anaerobic conditions. *J Biotechnol* 2001; 86(2): 113–126.
124. Ishigaki T, Sugano W, Nakanishi A, Tateda M, Ike M, and Fujita M. The degradability of biodegradable plastics in aerobic and anaerobic waste landfill model reactors. *Chemosphere* 2004; 54(3): 225–233.
125. Federle TW, Barlaz MA, Pettigrew CA, *et al.* Anaerobic biodegradation of aliphatic polyesters: Poly(3-hydroxybutyrate-*co*-3-hydroxyoctanoate) and poly(ε-caprolactone). *Biomacromolecules* 2002; 3(4): 813–822.
126. Lee SY and Choi JI. Production and degradation of polyhydroxyalkanoates in waste environment. *Waste Manag* 1999; 19(2): 133–139.
127. Reischwitz A, Stoppok E, and Buchholz K. Anaerobic degradation of poly-3-hydroxybutyrate and poly-3-hydroxybutyrate-co-3-hydroxyvalerate. *Biodegradation* 1997; 8(5): 313–319.
128. Batori V, Akesson D, Zamani A, Taherzadeh MJ, and Horvath IS. Anaerobic degradation of bioplastics: A review. *Waste Manag* 2018; 80: 406–413.
129. Yagi H, Ninomiya F, Funabashi M, and Kunioka M. Mesophilic anaerobic biodegradation test and analysis of eubacteria and archaea involved in anaerobic biodegradation of four specified biodegradable polyesters. *Polym Degrad Stabil* 2014; 110: 278–283.
130. Ryan CA, Billington SL, and Criddle CS. Assessment of models for anaerobic biodegradation of a model bioplastic: Poly(hydroxybutyrate-co-hydroxyvalerate). *Bioresour Technol* 2017; 227: 205–213.
131. Li ZB, Yang J, and Loh XJ. Polyhydroxyalkanoates: Opening doors for a sustainable future. *NPG Asia Mater* 2016; 8(4): e265.
132. Ha CS and Cho WJ. Miscibility, properties, and biodegradability of microbial polyester containing blends. *Prog Polym Sci* 2002; 27(4): 759–809.
133. Renard E, Walls M, Guerin P, and Langlois V. Hydrolytic degradation of blends of polyhydroxyalkanoates and functionalized polyhydroxyalkanoates. *Polym Degrad Stabil* 2004; 85(2): 779–787.
134. Wang SA, Ma PM, Wang RY, Wang SF, Zhang Y, and Zhang YX. Mechanical, thermal and degradation properties of poly(*d,l*-lactide)/poly(hydroxybutyrate-co-hydroxyvale rate)/poly(ethylene glycol) blend. *Polym Degrad Stabil* 2008; 93(7): 1364–1369.
135. Brunel DG, Pachekoski WM, Dalmolin C, and Agnelli JAM. Natural additives for poly (hydroxybutyrate-co-hydroxyvalerate) - PHBV: Effect on mechanical properties and biodegradation. *Mater Res* 2014; 17(5): 1145–1156.
136. Satoh H, Yoshie N, and Inoue Y. Hydrolytic degradation of blends of poly(3-hydroxybutyrate) with poly(3-hydroxybutyrate-co-3-Hydroxyvalerate). *Polymer* 1994; 35(2): 286–290.
137. Basnett P, Ching KY, Stolz M, *et al.* Novel poly(3-hydroxyoctanoate)/poly(3-hydroxyb utyrate) blends for medical applications. *React Funct Polym* 2013; 73(10): 1340–1348.
138. Abe H, Matsubara I, and Doi Y. Physical-properties and enzymatic degradability of polymer blends of bacterial poly[(R)-3-hydroxybutyrate] and poly[(R,S)-3-hydroxybutyrate] stereoisomers. *Macromolecules* 1995; 28(4): 844–853.
139. Bhatt R, Shah D, Patel KC, and Trivedi U. PHA-rubber blends: Synthesis, characterization and biodegradation. *Bioresour Technol* 2008; 99(11): 4615–4620.
140. Calvao PS, Chenal JM, Gauthier C, Demarquette NR, Bogner A, and Cavaille JY. Understanding the mechanical and biodegradation behaviour of poly(hydroxybutyrate)/rubber blends in relation to their morphology. *Polym Int* 2012; 61(3): 434–441.
141. Avella M, Calandrelli L, Immirzi B, *et al.* Novel synthesis blends between bacterial polyesters and acrylic rubber - A study on enzymatic biodegradation. *J Environ Polym Degr* 1995; 3(1): 49–60.

142. Goncalves SPC, Martins-Franchetti SM, and Chinaglia DL. Biodegradation of the films of PP, PHBV and its blend in soil. *J Polym Environ* 2009; 17(4): 280–285.
143. Mousavioun P, George GA, and Doherty WOS. Environmental degradation of lignin/poly(hydroxybutyrate) blends. *Polym Degrad Stab* 2012; 97(7): 1114–1122.
144. Tao J, Song CJ, Cao MF, *et al.* Thermal properties and degradability of poly(propylene carbonate)/poly(beta-hydroxybutyrate-co-beta-hydroxyvalerate) (PPC/PHBV) blends. *Polym Degrad Stabil* 2009; 94(4): 575–583.
145. Kumagai Y and Doi Y. Enzymatic degradation of poly(3-hydroxybutyrate)-based blends - poly(3-hydroxybutyrate) poly(ethylene oxide) blend. *Polym Degrad Stabil* 1992; 35(1): 87–93.
146. Imam SH, Gordon SH, Shogren RL, Tosteson TR, Govind NS, and Greene RV. Degradation of starch-poly(beta-hydroxybutyrate-co-beta-hydroxyvalerate) bioplastic in tropical coastal waters. *Appl Environ Microbiol* 1999; 65(2): 431–437.
147. Dharmalingam S, Hayes DG, Wadsworth LC, *et al.* Soil degradation of polylactic acid/polyhydroxyalkanoate-based nonwoven mulches. *J Polym Environ* 2015; 23(3): 302–315.
148. Iannace S, Ambrosio L, Huang SJ, and Nicolais L. Effect of degradation on the mechanical-properties of multiphase polymer blends - PHBV/PLLA. *J Macromol Sci Pure* 1995; A32(4): 881–888.
149. Azwa ZN, Yousif BF, Manalo AC, and Karunasena W. A review on the degradability of polymeric composites based on natural fibres. *Mater Des* 2013; 47: 424–442.
150. Awaja F, Zhang SN, Tripathi M, Nikiforov A, and Pugno N. Cracks, microcracks and fracture in polymer structures: Formation, detection, autonomic repair. *Prog Mater Sci* 2016; 83: 536–573.
151. Robeson LM. Environmental stress cracking: A review. *Polym Eng Sci* 2013; 53(3): 453–467.
152. Farias RF, Canedo EL, Wellen RMR, and Rabello MS. Environmental stress cracking of poly(3-hydroxibutyrate) under contact with sodium hydroxide. *Mater Res* 2015; 18(2): 258–266.
153. Rolland H, Saintier N, Wilson P, Merzeau J, and Robert G. *In situ* X-ray tomography investigation on damage mechanisms in short glass fibre reinforced thermoplastics: Effects of fibre orientation and relative humidity. *Compos B Eng* 2017; 109: 170–186.
154. Luo S and Netravali AN. Interfacial and mechanical properties of environment-friendly "green" composites made from pineapple fibers and poly(hydroxybutyrate-co-valerate) resin. *J Mater Sci* 1999; 34(15): 3709–3719.
155. Jurczyk S, Musiol M, Sobota M, *et al.* (Bio)degradable polymeric materials for sustainable future-part 2: Degradation studies of P(3HB–co–4HB)/cork composites in different environments. *Polymers* 2019; 11(3): 547.
156. Malmir S, Montero B, Rico M, Barral L, and Bouza R. Morphology, thermal and barrier properties of biodegradable films of poly(3-hydroxybutyrate-co-3-hydroxyvalerate) containing cellulose nanocrystals. *Compos A* 2017; 93: 41–48.
157. Yatigala NS, Bajwa DS, and Bajwa SG. Compatibilization improves performance of biodegradable biopolymer composites without affecting UV weathering characteristics. *J Polym Environ* 2018; 26(11): 4188–4200.
158. Mazur K and Kuciel S. Mechanical and hydrothermal aging behaviour of polyhydroxybutyrate-co-valerate (PHBV) composites reinforced by natural fibres. *Molecules* 2019; 24(19): 3538.
159. Thomas S, Shumilova AA, Kiselev EG, *et al.* Thermal, mechanical and biodegradation studies of biofiller based poly-3-hydroxybutyrate biocomposites. *Int J Biol Macromol* 2020; 155: 1373–1384. doi: 10.1016/j.ijbiomac.2019.11.112.
160. Ventura H, Claramunt J, Rodriguez-Perez MA, and Ardanuy M. Effects of hydrothermal aging on the water uptake and tensile properties of PHB/flax fabric biocomposites. *Polym Degrad Stabil* 2017; 142: 129–138.

161. Seggiani M, Cinelli P, Mallegni N, *et al.* New bio-composites based on polyhydroxyalkanoates and *Posidonia oceanica* fibres for applications in a marine environment. *Materials* 2017; 10(4): 326.
162. Joyyi L, Ahmad Thirmizir MZ, Salim MS, *et al.* Composite properties and biodegradation of biologically recovered P(3HB–co–3HHx) reinforced with short kenaf fibers. *Polym Degrad Stabil* 2017; 137: 100–108.
163. Chan CM, Pratt S, Halley P, *et al.* Mechanical and physical stability of polyhydroxyalkanoate (PHA)-based wood plastic composites (WPCs) under natural weathering. *Polym Test* 2019; 73: 214–221.
164. Chan CM, Vandi LJ, Pratt S, *et al.* Insights into the biodegradation of PHA/wood composites: Micro- and macroscopic changes. *Sustain Mater Techno* 2019; 21: e00099.
165. Chan CM, Vandi L-J, Pratt S, *et al.* Mechanical stability of polyhydroxyalkanoate (PHA)-based wood plastic composites (WPCs). *J Polym Environ* 2020; 28: 1571–1577.
166. Sadi RK, Fechine GJM, and Demarquette NR. Photodegradation of poly(3-hydroxybutyrate). *Polym Degrad Stabil* 2010; 95(12): 2318–2327.
167. Rosa DD, Calil MR, Guedes CDF, and Santos CEO. The effect of UV-B irradiation on the biodegradability of poly-beta-hydroxybutyrate (PHB) and poly-ε-caprolactone (PCL). *J Polym Environ* 2001; 9(3): 109–113.
168. Ikada E, Tanahashi A, and Morisaka N. Relationship between photo degradability of polyesters and their molecular-structures - Photo degradability of a polyester having phenylene groups in the chain skeleton. *Kobunshi Ronbunshu* 1995; 52(8): 472–477.
169. Michel AT and Billington SL. Characterization of poly-hydroxybutyrate films and hemp fiber reinforced composites exposed to accelerated weathering. *Polym Degrad Stabil* 2012; 97(6): 870–878.
170. Wei LQ and McDonald AG. Accelerated weathering studies on the bioplastic, poly(3-hydroxybutyrate-*co*-3-hydroxyvalerate). *Polym Degrad Stabil* 2016; 126: 93–100.
171. Pagnan CS, Mottin AC, Orefice RL, Ayres E, and Camara JJD. Annatto-colored poly(3-hydroxybutyrate): A comprehensive study on photodegradation. *J Polym Environ* 2018; 26(3): 1169–1178.
172. Nakamura H, Nakamura T, Noguchi T, and Imagawa K. Photodegradation of PEEK sheets under tensile stress. *Polym Degrad Stabil* 2006; 91(4): 740–746.
173. Sakai W and Tsutsumi N. Photodegradation and radiation degradation. In: R. Auras L-TL, S. E. M. Selke, and H. Tsuji, Eds., *Poly(Lactic Acid): Synthesis, Structures, Properties, Processing, and Applications*, John Wiley, Hoboken, NJ, 2010. pp. 413–421.
174. Heitmann AP, Rocha IC, de Souza PP, Oliveira LCA, and Patricio PSdO. Photoactivation of a biodegradable polymer (PHB): Generation of radicals for pollutants activation. *Catalysis Today* 2020; 344: 171–175. doi: 10.1016/j.cattod.2018.12.024.
175. Badia JD, Gil-Castell O, and Ribes-Greus A. Long-term properties and end-of-life of polymers from renewable resources. *Polym Degrad Stabil* 2017; 137: 35–57.
176. Kockott D. Natural and artificial weathering of polymers. *Polym Degrad Stabil* 1989; 25(2–4): 181–208.
177. Li X, Chu CL, and Chu PK. Effects of external stress on biodegradable orthopedic materials: A review. *Bioact Mater* 2016; 1(1): 77–84.
178. Soares JS, Moore JE, and Rajagopal KR. Constitutive framework for biodegradable polymers with applications to biodegradable stents. *ASAIO J* 2008; 54(3): 295–301.
179. Soares JS, Moore JE, and Rajagopal KR. Modeling of deformation-accelerated breakdown of polylactic acid biodegradable stents. *J Med Devices* 2010; 4(4): 041007-1.
180. Hayman D, Bergerson C, Miller S, Moreno M, and Moore JE. The effect of static and dynamic loading on degradation of PLLA stent fibers. *J Biomech Eng T ASME* 2014; 136(8): 081006-1.

181. Li Q, *Biodegradation Study of Polymeric Material with Various Levels of Molecular Orientation Induced via Vibration-Assisted Injection Molding*, Master of Science in Mechanical Engineering and Mechanics, Lehigh University, Bethlehem, Pennsylvania, 2011.
182. Renard E, Langlois V, and Guerin P. Chemical modifications of bacterial polyesters: From stability to controlled degradation of resulting polymers. *Corros Eng Sci Techn* 2007; 42(4): 300–311.
183. Renard E, Poux A, Timbart L, Langlois V, and Guerin P. Preparation of a novel artificial bacterial polyester modified with pendant hydroxyl groups. *Biomacromolecules* 2005; 6(2): 891–896.
184. Renard E, Tanguy PY, Samain E, and Guerin P. Synthesis of novel graft polyhydroxyalkanoates. *Macromol Symp* 2003; 197(1): 11–18.
185. Guerin P, Renard E, and Langlois V. Degradation of natural and artificial poly[(R)-3-hydroxyalkanoate]s: From biodegradation to hydrolysis. *Microbiol Monogr* 2010; 14: 283–321.
186. Foster JC, Varlas S, Couturaud B, Coe Z, and O'Reilly RK. Getting into shape: Reflections on a new generation of cylindrical nanostructures' self-assembly using polymer building blocks. *J Am Chem Soc* 2019; 141(7): 2742–2753.
187. Woolnough CA, Yee LH, Charlton TS, and Foster LJR. A tuneable switch for controlling environmental degradation of bioplastics: Addition of isothiazolinone to polyhydroxyalkanoates. *PLOS ONE* 2013; 8(10): p. e75817.
188. Vigneswari S, Chai JM, Shantini K, Bhubalan K, and Amirul AA. Designing novel interfaces via surface functionalization of short-chain-length polyhydroxyalkanoates. *Adv Polym Tech* 2019; 2019: 3831251.

Index

Milton Keynes UK
Ingram Content Group UK Ltd.
UKHW021825071024
449327UK00021B/1436

9 780367 541156